Handbook of Irrigation Hydrology and Management

Ever-increasing population growth has caused a proportional increased demand for water, and existing water sources are depleting day by day. Moreover, with the impact of climate change, the rates of rainfall in many regions have experienced a higher degree of variability. In many cities, government utilities have been struggling to maintain sufficient water for the residents and other users. The *Handbook of Irrigation Hydrology and Management: Irrigation Case Studies* examines and analyzes irrigated ecosystems in which water storage, applications, or drainage volumes are artificially controlled in the landscape and the spatial domain of processes varies from micrometers to tens of kilometers, while the temporal domain spans from seconds to centuries. The continuum science of irrigation hydrology includes the surface, subsurface (unsaturated and groundwater systems), atmospheric, and plant subsystems. Further, the book includes practical case studies from around the world, including locations such as Africa, Australia, China, India, the Middle East, the United States, and more.

Features:

- Offers water-saving strategies to increase the judicious use of scarce water resources
- Presents strategies to maximize agricultural yield per unit of water used for different regions
- Compares irrigation methods to offset changing weather patterns and impacts of climate change

Handbook of Irrigation Hydrology and Management
Irrigation Case Studies

Edited by
Saeid Eslamian and Faezeh Eslamian

CRC Press
Taylor & Francis Group
Boca Raton London New York

CRC Press is an imprint of the
Taylor & Francis Group, an **informa** business

Designed cover image: Shutterstock

First edition published 2023
by CRC Press
6000 Broken Sound Parkway NW, Suite 300, Boca Raton, FL 33487-2742

and by CRC Press
4 Park Square, Milton Park, Abingdon, Oxon, OX14 4RN

CRC Press is an imprint of Taylor & Francis Group, LLC

Library of Congress Cataloging-in-Publication Data
Names: Eslamian, Saeid, editor. | Eslamian, Faezeh A., editor.
Title: Handbook of irrigation hydrology and management :
irrigation management and optimization / Edited by Saeid Eslamian and Faezeh Eslamian.
Description: Boca Raton, FL : CRC Press, 2023. | Includes bibliographical references and index. | Also available online. | Description based on print version record and CIP data provided by publisher; resource not viewed.
Identifiers: LCCN 2022050740 (print) | LCCN 2022050741 (ebook) | ISBN 9780429290152 (ebook) |
ISBN 9780367258306 (hardback) | ISBN 9781032457468 (paperback) | ISBN 9780367258191(v. 1 ;hardback) |
ISBN 9781032457451(v. 1 ;paperback) | ISBN 9781032457468(v. 2;paperback) | ISBN 9780429290152(v. 2 ;ebook) |
ISBN 9781032406077(v. 3 ;hardback)| ISBN 9781032429106(v. 3 ;paperback) | ISBN 9781003353928(v. 3 ;ebook)
Subjects: LCSH: Irrigation. | Irrigation engineering.
Classification: LCC TC805 (ebook) | LCC TC805 .H36 2023 (print) |
DDC 627/.52 23/eng/20221–dc16
LC record available at https://lccn.loc.gov/2022050740

ISBN: 978-1-032-40607-7 (hbk)
ISBN: 978-1-032-42910-6 (pbk)
ISBN: 978-1-003-35392-8 (ebk)

DOI: 10.1201/9781003353928

Typeset in Minion Pro
by codeMantra

To

Henri Goblot (1896–1988),

A French engineer who explored the genesis of Qanat for the first time. He published a book entitled Qanats; a Technique for Obtaining Water.

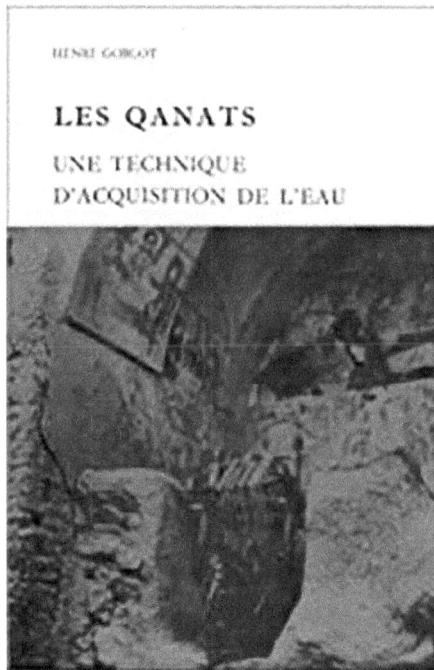

The ancient Iranians made use of the water that the miners wished to get rid of it, and founded a basic system named Qanat or Kariz to supply the required water to their farm lands, dating back to 600–800 BC (Goblot, 1979).

Contents

Preface

Water is known as the most important input required for plant growth for agricultural production. Irrigation can be defined as the completion of soil water storage in the root zone of the plant by methods other than natural rainfall. Irrigation seems to have the roots in human history from the beginning. This makes it possible to reduce uncertainties in agricultural practices, particularly climatic ones.

Archaeological research has found evidence of irrigation where natural rainfall was not sufficient to support farming. The ancient Egyptians, for example, practiced basin irrigation by using the Nile flood to flood land surrounded by dykes. Another example is the aqueduct, which was built in ancient Persia around 800 BC and is one of the oldest known methods of irrigation still in use.

Irrigation hydrology is constrained to analysis of irrigated ecosystems in which water storage, applications, or drainage volumes are artificially controlled in the landscape and the spatial domain of processes varies from micrometers to tens of kilometers, while the temporal domain spans from seconds to centuries. The continuum science of irrigation hydrology includes the surface, subsurface (unsaturated and groundwater systems), atmospheric, and plant subsystems.

Irrigation management involves many different decisions: selection of economically viable cropping patterns, land allocation by crop, water resource allocation by crop, irrigation scheduling, deficit management irrigation, etc. Plants need adequate amounts of water, and its distribution throughout the growth cycle has a huge influence on the final yield of the crop. This means that the management of soil water content is crucial to obtain an optimal allocation of water resources, provided that the other production factors are adequate.

Our book has several merits on the previous published books as follows:

- A comprehensive book on majority methods of Irrigation
- A new focus on Irrigation Hydrology, Landscape, Scales and Social Context
- A robust tool for computational analysis of Irrigation Methods
- A updated book including global warming, adaptation, resilience, and sustainability associated with Irrigation
- Deficit and Over-Irrigation Merits and Disadvantages
- Inclusion of Smart and Precision Irrigation
- Offering Solutions for Water Scarcity and Soil Salinity in Irrigation
- Satellite Measurements for Irrigation Management
- Selected case studies from different climates across world.

Volume 3 of this handbook, entitled *Irrigation Case Studies*, includes the following topics:

- African Case Studies
- America and Europe Irrigation Developments
- Australian Experiences
- Chinese Irrigation History

- Middle East Irrigation and Deficit Irrigation
- Water Scarcity and Irrigation in India

The information contained in this *Handbook of Irrigation Hydrology and Management* can be beneficial to students at the following levels of the study: Undergraduate, Postgraduate, Research Students, and Short Courses Programs, and could also be a useful resource for courses such as Surface Hydrology, Water Resources, Climatology, Agrometeorology, Irrigation Principals, Surface Irrigation, Irrigation System Design, Drip Irrigation, Sprinkler Irrigation, Water, Soils and Plants Relationships, Evapotranspiration and Water Requirement, Drainage Engineering, Irrigation and Drainage Networks, Irrigation Hydraulic Structures, and Special Problems in Irrigation.

Scholars and students of Applied Geography, Geosciences, Environmental Engineering, Environmental Health, Natural Resources, Agricultural Engineering, Irrigation Engineering and the related courses, as well as professionals, will find this handbook of great value.

Three-volume *Handbook of Irrigation Hydrology and Management* could be recommended not only for universities and colleges, but also for research centers, governmental departments, policy makers, engineering consultants, federal emergency management agencies and the related bodies.

Saeid Eslamian, Faezeh Eslamian
Isfahan University of Technology, McGill University

Editors

Saeid Eslamian is a full professor of environmental hydrology and water resources engineering in the Department of Water Engineering at Isfahan University of Technology, where he has been serving since 1995. His research focuses mainly on statistical and environmental hydrology in a changing climate. In recent years, he has worked on modeling natural hazards, including floods, severe storms, wind, drought, pollution, water reuses, sustainable development and resiliency, etc. Formerly, he was a visiting professor at Princeton University, New Jersey, and the University of ETH Zurich, Switzerland. On the research side, he started a research partnership in 2014 with McGill University, Canada. He has contributed to more than 600 publications in journals, books, and technical reports. He is the founder and chief editor of both the *International Journal of Hydrology Science and Technology (IJHST)* and the *Journal of Flood Engineering (JFE)*. Eslamian is now associate editor of four important publications: *Journal of Hydrology* (Elsevier), *Eco-Hydrology and Hydrobiology* (Elsevier), *Journal of Water Reuse and Desalination* (IWA), and *Journal of the Saudi Society of Agricultural Sciences* (Elsevier). Professor Eslamian is the author of approximately 35 books and 350 chapter books.

Dr. Eslamian's professional experience includes membership on editorial boards, and he is a reviewer of approximately 150 Web of Science (ISI) journals, including the *ASCE Journal of Hydrologic Engineering*, *ASCE Journal of Water Resources Planning and Management*, *ASCE Journal of Irrigation and Drainage Engineering*, *Advances in Water Resources*, *Groundwater*, *Hydrological Processes*, *Hydrological Sciences Journal*, *Global Planetary Changes*, *Water Resources Management*, *Water Science and Technology*, *Eco-Hydrology*, *Journal of American Water Resources Association*, *American Water Works Association Journal*, etc. UNESCO has also nominated him for a special issue of the *Eco-Hydrology and Hydrobiology Journal* in 2015.

Professor Eslamian was selected as an outstanding reviewer for the *Journal of Hydrologic Engineering* in 2009 and received the EWRI/ASCE Visiting International Fellowship in Rhode Island (2010). He was also awarded outstanding prizes from the Iranian Hydraulics Association in 2005 and Iranian Petroleum and Oil Industry in 2011. Professor Eslamian has been chosen as a distinguished researcher of Isfahan University of Technology (IUT) and Isfahan Province in 2012 and 2014, respectively. In 2016, he was a candidate for national distinguished researcher in Iran.

He has also been the referee of many international organizations and universities. Some examples include the U.S. Civilian Research and Development Foundation (USCRDF), the Swiss Network for International Studies, the Majesty Research Trust Fund of Sultan Qaboos University of Oman, the Royal Jordanian Geography Center College, and the Research Department of Swinburne University of Technology of Australia. He is also a member of the following associations: American Society of Civil Engineers (ASCE), International Association of Hydrologic Science (IAHS), World Conservation Union (IUCN), GC Network for Drylands Research and Development (NDRD), International Association for

Urban Climate (IAUC), International Society for Agricultural Meteorology (ISAM), Association of Water and Environment Modeling (AWEM), International Hydrological Association (STAHS), and UK Drought National Center (UKDNC).

Professor Eslamian finished Hakimsanaei High School in Isfahan in 1979. After the Islamic Revolution, he was admitted to IUT for a BS in water engineering and graduated in 1986. After graduation, he was offered a scholarship for a master's degree program at Tarbiat Modares University, Tehran. He finished his studies in hydrology and water resources engineering in 1989. In 1991, he was awarded a scholarship for a PhD in civil engineering at the University of New South Wales, Australia. His supervisor was Professor David H. Pilgrim, who encouraged him to work on "Regional Flood Frequency Analysis Using a New Region of Influence Approach." He earned a PhD in 1995 and returned to his home country and IUT. In 2001, he was promoted to associate professor and in 2014 to full professor. For the past 26 years, he has been nominated for different positions at IUT, including university president consultant, faculty deputy of education, and head of department. Eslamian is now director for center of excellence in Risk Management and Natural Hazards (RiMaNaH).

Professor Eslamian has made three scientific visits to the United States, Switzerland, and Canada in 2006, 2008, and 2015, respectively. In the first, he was offered the position of visiting professor by Princeton University and worked jointly with Professor Eric F. Wood at the School of Engineering and Applied Sciences for one year. The outcome was a contribution in hydrological and agricultural drought interaction knowledge by developing multivariate L-moments between soil moisture and low flows for northeastern U.S. streams.

Recently, Professor Eslamian has published the 14 handbooks by Taylor & Francis (CRC Press): three-volume *Handbook of Engineering Hydrology* (2014), *Urban Water Reuse Handbook* (2016), *Underground Aqueducts Handbook* (2017), three-volume *Handbook of Drought and Water Scarcity* (2017), *Constructed Wetlands: Hydraulic Design* (2019), *Handbook of Irrigation System Selection for Semi-Arid Regions* (2020), *Urban and Industrial Water Conservation Methods* (2020), and three-volume *Flood Handbook* (2022).

An Evaluation of Groundwater Storage Potentials in a Semiarid Climate and *Advances in Hydrogeochemistry Research* by Nova Science Publishers are also his book publications in 2019 and 2020, respectively. Two-volume *Handbook of Water Harvesting and Conservation* (Wiley) and *Handbook of Disaster Risk Reduction and Resilience* (New Frameworks for Building Resilience to Disasters) are early 2021 book publications of Professor Eslamian. *Handbook of Disaster Risk Reduction and Resilience* (Disaster Risk Management Strategies) and two-volume *Earth Systems Protection and Sustainability* are early 2022 handbooks of Professor Eslamian.

Professor Eslamian has been appointed as World Top 2-Percent Researcher by Standford University, USA, in 2019 and 2020. He has also been a Grant Assessor/Report Referee/Award Jury/Invited Researcher for international organizations such as United States Civilian Research and Development Foundation (2006), Intergovernmental Panel on Climate Change (2012), World Bank Policy and Human Resources Development Fund (2021), and Stockholm International Peace Research Institute (2022), respectively.

Faezeh Eslamian is a PhD holder of bioresource engineering from McGill University. Her research focuses on the development of a novel lime-based product to mitigate phosphorus loss from agricultural fields. Faezeh completed her bachelor's and master's degrees in civil and environmental engineering from Isfahan University of Technology, Iran, where she evaluated natural and low-cost absorbents for the removal of pollutants such as textile dyes and heavy metals. Furthermore, she has conducted research on the worldwide water quality standards and wastewater reuse guidelines. Faezeh is an experienced multidisciplinary researcher with research interests in soil and water quality, environmental remediation, water reuse, and drought management.

Contributors

Md. Maniruzzaman Bin A. Aziz
University of Technology
  Malaysia (UTM)
Iskandar Puteri, Malaysia

Mahmoud Abdellaoui
ENET'COM
Sfax, Tunisia

Cassio Hamilton Abreu-Junior
Center of Nuclear Energy in Agriculture
University of São Paulo
São Paulo, Brazil

Enyew Adgo
Bahir Dar University
Bahir Dar, Ethiopia

Seifu Admasu Tilahun
Bahir Dar University
Bahir Dar, Ethiopia

Eric Samuel Adu-Dankwa
Planning, Monitoring, Evaluation and
  Coordination
Ghana Irrigation Development
  Authority
Accra, Ghana

Ivan Aidarov
Department of Agricultural
  Science
Russian Academy of Sciences
Moscow, Russia

Abdulrasoul M. Al-Omran
King Saud University
Riyadh, Saudi Arabia

Abdul-Rauf Malimanga Alhassan
Department of Water, Sanitation and Hygiene
University of Environment and Sustainable
  Development
Somanya, Ghana

Arafat Alkhasha
King Saud University
Riyadh, Saudi Arabia

Patricia Amankwaa-Yeboah
Agricultural Engineering and
  Transport Division
CSIR-Crops Research Institute
Kumasi, Ghana

William Amponsah
Department of Agricultural and Biosystems
  Engineering
Kwame Nkrumah University of Science and
  Technology
Kumasi, Ghana

Chang Ao
State Key Laboratory of Water Resources and
  Hydropower Engineering Science
Wuhan University
Wuhan, China

Desale Kidane Asmamaw
Ghent University
Ghent, Belgium
and
Bahir Dar University
Bahir Dar, Ethiopia

Flávia Rosana Barros da Silva
Center of Nuclear Energy in Agriculture
University of São Paulo
São Paulo, Brazil

Bhasker Vijaykumar Bhatt
Arvindbhai Patel Institute of Environmental
  Design
Vallabh Vidyanagar, India

Ageel I. Bushara
Zuhair Fayez Partnership Consultants
Riyadh, Saudi Arabia

Jeff Camkin
The University of Western Australia
Perth, Australia
and
Institute for Study and Development Worldwide
Sydney, Australia

Lihua Chen
Guangxi University
Nanning, China

Rubens Duarte Coelho
Luiz de Queiroz College of Agriculture
São Paulo, Brazil

Neil A. Coles
University of Leeds
Leeds, United Kingdom
and
The University of Western Australia
Perth, Australia
and
Verdant Earth
Cambridge, United Kingdom

W. M. Cornelis
Ghent University
Ghent, Belgium

R. Deepa
Florida Agricultural and Mechanical University
Tallahassee, Florida

Mekete Desse
Bahir Dar University
Bahir Dar, Ethiopia

Ian Charles Dodd
Lancaster Environment Centre
Lancaster University
Lancaster, England

Ahmed Hayaty Elshaikh
University of Khartoum
Khartoum, Sudan

Saeid Eslamian
Department of Water Science and Technology
College of Agriculture
Isfahan University of Technology
Isfahan, Iran

Younis A. Gismalla
Ministry of Irrigation and Water Resources
Khartoum, Sudan

Achintya Kumar Sen Gupta
Institute for Hygiene and Environmental
  Sanitation
New Delhi, India
and
World Health Organization
Geneva, Switzerland

Rares Hălbac-Cotoară-Zamfir
Politehnica University of Timişoara
Timişoara, Romania

Monzur A. Imteaz
Swinburne University of Technology
Melbourne, Australia

Rajendra Kumar Isaac
Sam Higginbottom Institute of Agriculture,
  Technology and Sciences
Deemed University
Allahabad, India

P. Janssens
Ghent University
Ghent, Belgium
and
Soil Service of Belgium
Leuven, Belgium

Prerna Jasuja
Department of Architecture and Planning
Malaviya National Institute of
    Technology
Jaipur, India

Bright Mayinl Laboan
Soil Research Centre
CSIR-Water Research Institute
Accra, Ghana

Guoqing Lei
State Key Laboratory of Water Resources and
    Hydropower Engineering Science
Wuhan University
Wuhan, China

Jonathan Vasquez Lizcano
Corporación Colombiana de Investigación
    Agropecuaria
Municipio de Zona Bananera,
    Columbia

Ibrahim Louki
King Saud University
Riyadh, Saudi Arabia

A. R. Malimanga
CSIR-Crops Research
    Institute
Kumasi, Ghana

Y. Mohamed
Ministry of Irrigation and Water
    Resources
Khartoum, Sudan

Never Mujere
University of Zimbabwe
Harare, Zimbabwe

Ruaa A. Nasreldeen
Ministry of Irrigation and
    Water Resources
Khartoum, Sudan

Yuri Nikolskii
Postgrado en
    Hidrociencias
Colegio de Postgraduados
Texcoco, Mexico

Nkem J. Nwosu
University of
    Ibadan
Ibadan, Nigeria

J. Nyssen
Ghent University
Ghent, Belgium

Gilbert Osei
Sanitation and Environmental
    Management Division
CSIR-Institute of Industrial
    Research
Accra, Ghana

Saurau O. Oshunsanya
University of Ibadan
Ibadan, Nigeria

AbdelNassir Osman
Ministry of Irrigation and Water Resources
Khartoum, Sudan

Shovan K. Saha
Institution for Hygiene and Environmental
    Sanitation
New Delhi, India

Luca Salvati
University of Macerata
Macerata, Italy

Abdallah Shanableh
University of Sharjah
Sharjah, United Arab Emirates

Neerajkumar D. Sharma
GIDC Degree Engineering College
Navsari, India

Timóteo Herculino da Silva Barros
Center of Nuclear Energy in Agriculture
University of São Paulo
São Paulo, Brazil

S. Sundaram
New York University Abu Dhabi
Abu Dhabi, United Arab Emirates

Rina Surana
Department of Architecture and Planning
Malaviya National Institute of Technology Jaipur
Jaipur, India

Muhammad Touseef
Guangxi University
Nanning, China

K. Walraevens
Ghent University
Ghent, Belgium

Stephen Yeboah
Cereals Division
CSIR-Crops Research Institute
Kumasi, Ghana

Wenzhi Zeng
State Key Laboratory of Water Resources and
    Hydropower Engineering Science
Wuhan University
Wuhan, China

Australian Experiences

1

Irrigation Developments in Australia: Historical Development of Irrigation

Neil A. Coles
University of Leeds
The University of
Western Australia.

Jeff Camkin
The University of
Western Australia

1.1 Introduction

Australia has always been subject to climate-driven extremes in weather, where floods and droughts are part of the rhythm of water redistribution in the landscape (Eslamian, 1995). Australia's biodiversity and river systems have adapted to this reality. For at least 60,000 years, Aboriginal peoples have lived in Australia (Clarkson et al., 2017), during which time water has always played a critical role – not only to promote subsistence in an often arid and unforgiving landscape, but also for its importance to Aboriginal culture and identity. By developing a comprehensive knowledge of its resources and needs, through their continuous relationship with the land, Aboriginal and Torres Strait Islander peoples have helped to create the Australian landscape (Rose and Australian Heritage Commission, 1996).

Since the first introduction of European agricultural techniques that accompanied the first settlers in 1778, however, management of water and the Australian landscape has largely been dictated by the need for increased food production, water for irrigators, rural communities and urban expansion, initially with a limited understanding of the fragility of the landscape. During the 1850–1890s, European settlers initially farmed small areas, followed in later years by larger broadacre irrigation schemes, mainly in the Murrumbidgee and Murray River valley regions of Victoria and NSW, with small developments in southern Tasmania.

DOI: 10.1201/9781003353928-2

The land area developed for irrigation in Australia continued to expand over the next 150 years, largely in the Murray–Darling Basin (MDB; Figure 1.1). More recently, other pressures such as climate change, loss of biodiversity, dryland salinity and soil degradation, and changing community expectations for both the environment and recognition of inter-cultural needs, have driven extensive reforms to how water is managed and used in Australia, particularly for irrigated agriculture. In this chapter, the history of the development and expansion of irrigated agriculture in Australia is discussed, and in the following chapter the productive capacity, governance and water reforms associated with these schemes are covered.

1.2 Water Resources in Australia

Since colonisation and the resultant disruption in the natural hydrological cycle through clearing, redirection of runoff, damming of river systems, expansion of agriculture (including irrigation), and increased urban drawdown of river systems and water resources, managing water in Australia has become increasing challenging (Khan and Hanjra, 2008). Water availability in Australia, while always subject to climate variability, has in recent decades become increasingly unreliable as global heating has altered rainfall distribution patterns, delivery frequency and quantity (Kirby, 2011, Coles et al. 2021). These changes compounded our initial limited understanding of water availability, and our often very optimistic irrigation allocations based on wetter years, that were never realised even in average rainfall years, never mind low rainfall years. Increasingly, to retain some semblance of water security, manage costs and investments associated with irrigation, and provide some certainty around supply reliability, Australian governments, farmers and environmentalists have championed the need to rethink the allocation of this scarce resource, although often from different perspectives (Coles, 2017).

1.3 Sources of Water

Australians were, until recently, amongst the highest per capita water users in the world, with the agricultural sector (largely irrigation), accounting for about 70% of total extracted water use (ABS, 2020 Ritchie, 2017). This is an unusual dichotomy considering that Australia is recognised as the world's driest permanently inhabited continent, receiving on average around 465.2 mm of rainfall per year based on the 1961–1990 period (Bureau of Meteorology, 2020). This rainfall is mainly redistributed though evapotranspiration (88%), runoff (11%) and groundwater recharge (1%), although this varies spatially, temporally and seasonally (Batten et al., 2003). Water for irrigation in Australia is nominally derived from two main sources: surface waters (i.e., rivers, lakes and wetlands) in the main drainage river basins (Figure 1.1) or groundwater aquifers (e.g., Great Artesian Basin [GAB]; Figure 1.2).

1.3.1 Surface Waters

Apart from the MDB, which contains the three longest inland rivers in Australia (Figure 1.3), most of the water resources for irrigation in Australia are extracted from coastal river systems, in conjunction with the construction of dams and large reservoirs. Examples include Lake Argyle (Figure 1.4), on the Ord River in the Kimberley Region of Western Australia (WA), and Lake Dalrymple, on the Burdekin River in North Queensland (Figure 1.1), or groundwater aquifers.

1.3.2 Groundwaters

The main source of groundwater in Australia is the GAB, which underlies extensive areas (~1,700,000 km^2) of eastern and central Australia (Figure 1.2). Other groundwater sources play an important role in supplying smaller but regionally important horticulture developments. For example, the Carnarvon

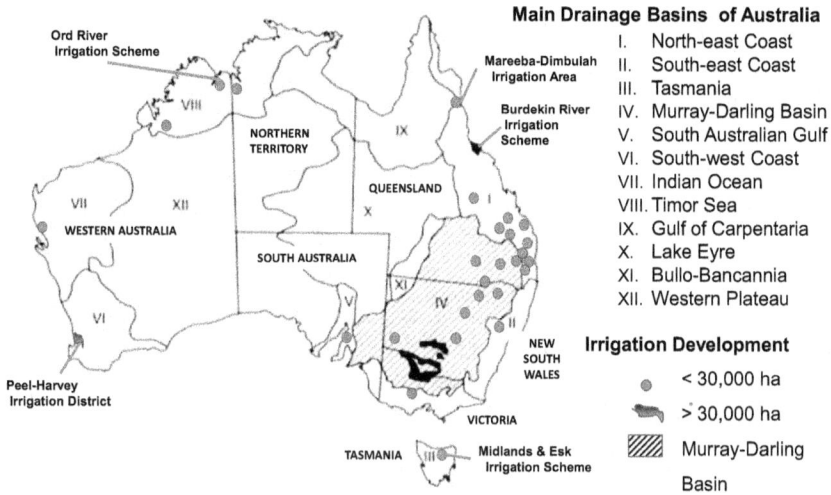

FIGURE 1.1 The main drainage basins of Australia. The larger irrigation developments in each state are also shown (adapted from Batten et al., 2003).

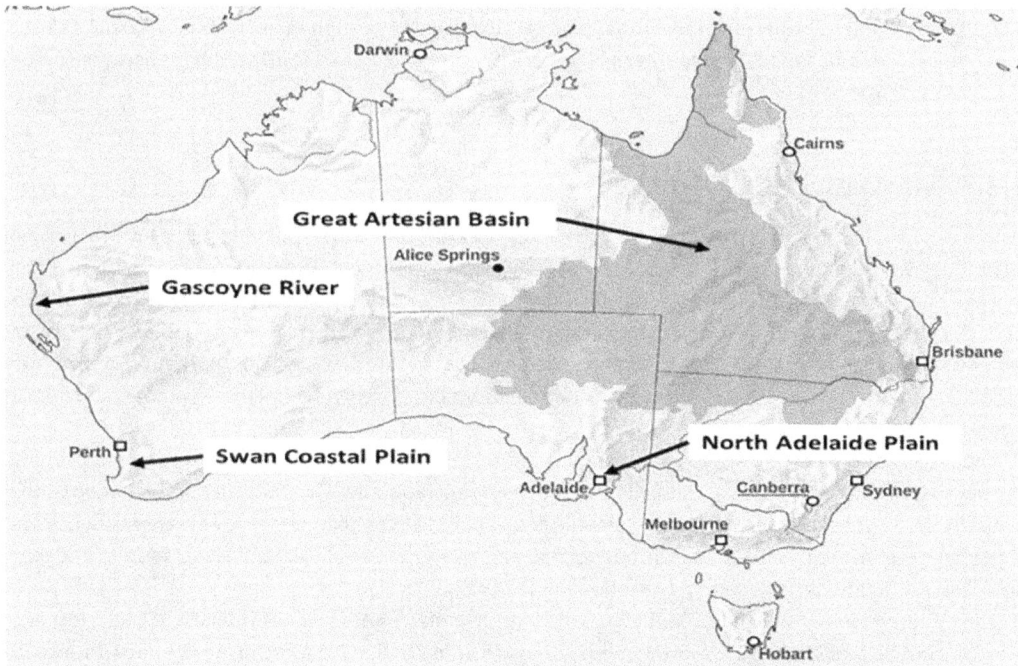

FIGURE 1.2 The Great Artesian Basin (GAB) and other groundwater irrigation regions in Australia. The GAB is by far the most extensive, underlying four states, and covering an area of approximately 1,700,000 km² (adapted from https://commons. wikimedia.org/wiki/File:Great_Artesian_Basin.png).

Irrigation Scheme (Gascoyne, WA), Swan Coastal Plain (southwestern WA) and the Northern Adelaide Plain, South Australia (SA) (Figure 1.2).

In Queensland, the extensive groundwater resources of the Burdekin delta have been utilised by the sugar industry since the late 1800s, and since the 1970s, surface water was being used to artificially recharge the aquifer. Construction of the Burdekin Falls Dam in 1986 (Figure 1.5) alleviated issues of over-extraction and the need to artificially recharge the surficial aquifers, in the process of making the delta region the most important area for irrigated sugarcane production in Australia (Qureshi et al., 2001).

In the Australian context, there are three main sources of groundwater for irrigation. These are as follows.

1. *Near-surface, unconsolidated sediments*, located in the principal river and lake systems, or as coastal dunes, deltas and narrow shoreline deposits. Good-quality groundwater resources are available in the inland drainage systems of NSW and Victoria within the MDB, for example the alluvium deposits in the Namoi, Lachlan, Murrumbidgee and Murray River valleys (Figure 1.6).
2. *Sedimentary basins* can be comprised of one or more major aquifer (Figure 1.7). Broad areas of Australia are underlain by aquifers; however, the water quality is often of limited value for irrigation due to high salt concentrations, particularly sodium. The GAB, which underlies large areas (~23%) of eastern Australia, is an example, with low water quality, meaning that most of the water resources are only suitable for domestic and livestock purposes (Figure 1.7). Australia's groundwater is, however, often critically important for local mining operations and the only reliable source of water in some remote areas.
3. *Fractured rock systems* yield variable water quantities and qualities and are mainly located in the uplands of southeast Australia, in Tasmania, parts of SA, central Australia and WA. These aquifers are generally low yielding. Water quality is expected to be high in north and eastern Australia, but low in much of SA, the southern Northern Territory (NT) and southwestern Australia (Batten et al., 2003).

1.3.3 Alternative Water Sources

Alternative sources of water have been evaluated and harnessed, where possible, to ensure security of agricultural production. Recycled water has become more prominent as a water source, with artificial storage and recovery or managed aquifer recharge being deployed extensively in SA (Salisbury Plains) and WA (Gnangara Aquifer)[1] (Kretschmer, 2017; Department of Water, 2010). Although recycled wastewater has been used for many years in regional Australia to irrigate playing fields, parks and community gardens, recent initiatives have utilised wastewater more effectively and in larger volumes to support large-scale irrigated production.

The integration of recycled water can be managed through conjunctive use strategies which promote the best use of water resources at the scale of river basins and aquifers (Groundwater Governance Global Framework for Action, 2015, Alley, 2016). Conjunctive management can improve climate resilience and water supply security by enhancing the best spatial and temporal use of different water sources and storage (Ross, 2018) and is discussed in the following sections.

The Willunga Basin scheme in McLaren Vale, South-eastern SA, is one of the largest and most successful examples of recycled wastewater for vineyard irrigation (Department of Primary Industries and Regions, 2019). The scheme services more than 220 irrigators across more than 2,000 ha through a 120 km network that delivers treated wastewater from the Christies Beach Wastewater treatment plant (The Lead, 2019). Expansion of the scheme in 2019–2020 increased the recycled water supply to around 6,000 ML/year, which is supplemented with groundwater to provide an additional 6,000 ML/year to irrigate the region's 7,000 ha of vineyards. The region is Australia's fifth largest wine producer by value, worth approximately $58 million in 2019 (Sustainable Australia Winegrowing, 2019).

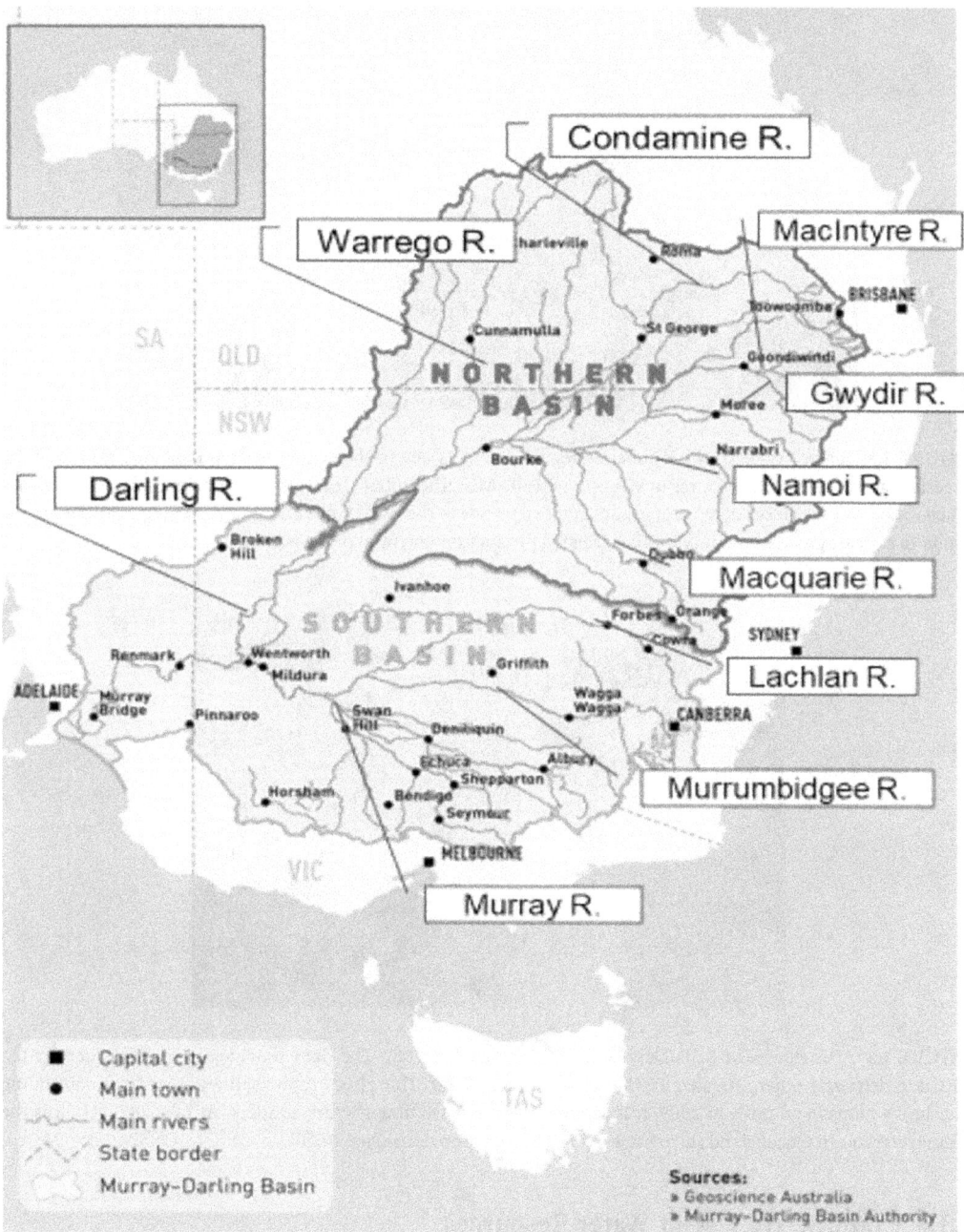

FIGURE 1.3 Division of the Murray–Darling Basin into the northern and southern basins. The more productive and intensive irrigation occurs in the southern Basin on the Murray, Darling, Lachlan and Murrumbidgee Rivers. The northern basin rivers centre on the Upper Darling, Warrego, Barwon and Balonne Rivers, which experience drier conditions and ephemeral flows, with additional contributions form upland rivers of the Great Dividing Range via the Condamine, MacIntyre, Gwydir, Namoi and Macquarie Rivers (Murray–Darling Basin Authority, 2019).

FIGURE 1.4   Lake Argyle in the Kimberley region of WA is one of the largest reservoirs in Australia, and the largest single surface water resource in WA (~10,600,000 ML). The initial Ord River development scheme was established in the 1960s with subsequent upgrades and extensions in the 1970s. Soils and crops research continues in the region to provide greater certainty for the irrigated production potential of the region.

FIGURE 1.5   The Burdekin Falls Dam was commissioned in 1987. The dam provides irrigation water for the Burdekin delta and other irrigated areas in North Queensland. This photo shows spillway overflows continuing after heavy monsoonal rains in 2009. It is also used to replenish downstream aquifers. At 1,860,000 ML, it is the largest reservoir in Queensland (http://www.scienceimage.csiro.au/image/10816).

1.3.4 Conjunctive Use of Water Resources

The conjunctive or co-ordinated use of water aims to optimise the use of multiple water resources to satisfy the various water demand/supply needs (Ross, 2018). This includes surface water, groundwater sources and alternative sources (e.g., recycled or wastewater) in combination to harmoniously manage water resources, storages or delivery pathways and avoid undesirable physical, environmental and economic effects of extraction (Ticehurst and Curtis, 2019). In its passive form, surface water is used during wet years, with groundwater and alternative sources substituted for surface water during drier years (Water Education Foundation, 2021). Conjunctive water management provides an opportunity to manage water resources to address competing demands such as water scarcity, water security and conflicts around access increases (Ticehurst and Curtis, 2019). Managed conjunctive use can also involve more

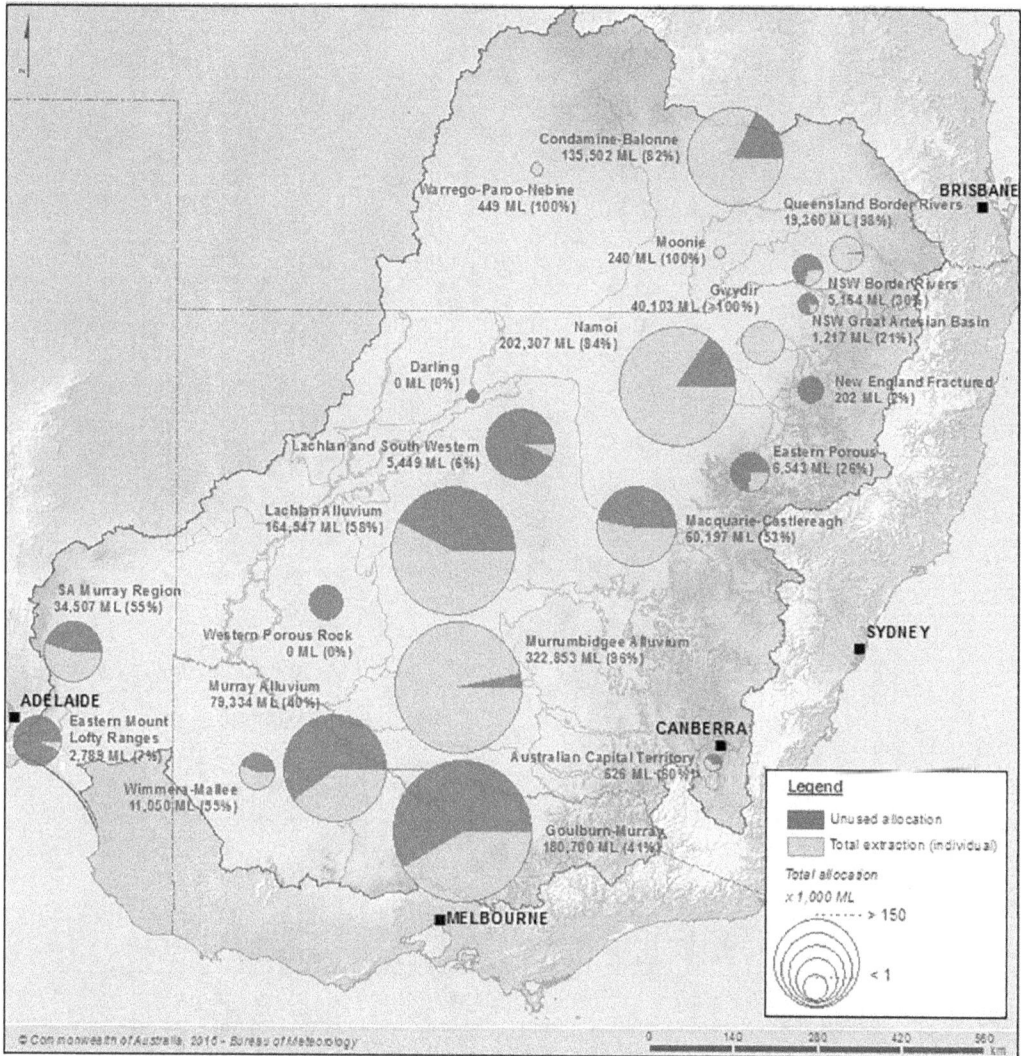

FIGURE 1.6 Strategic near-surface managed groundwater allocations within the MDB main irrigation areas. Note the GAB makes only limited contributions in this region (Commonwealth of Australia 2018- Bureau of Meteorology).

interactive approaches, where surface water (or recycled water) is directly injected into aquifers (e.g., Salisbury Plain, SA; Gnangara Aquifer, Perth, WA) to be stored as part of groundwater banking of water for later use as part of an integrated water resource management scheme (Dillon and Ashard, 2016). Alternatively, groundwater water is stored in a pond or basin and then allowed to percolate naturally into surficial aquifers.

Satisfying competing values and demands has become increasingly important for irrigation areas in Australia, in particular in the MDB where there have been progressive moves towards restoring the balance between consumptive use and environmental flows since reforms were introduced in 1994 (Ticehurst and Curtis, 2019). However, while there have been steps taking in the right direction to introduce better, more integrated groundwater and surface water accounting and planning, there are limited systematic efforts to plan and manage surface water and groundwater storage and use at a regional scale (Ross, 2018).

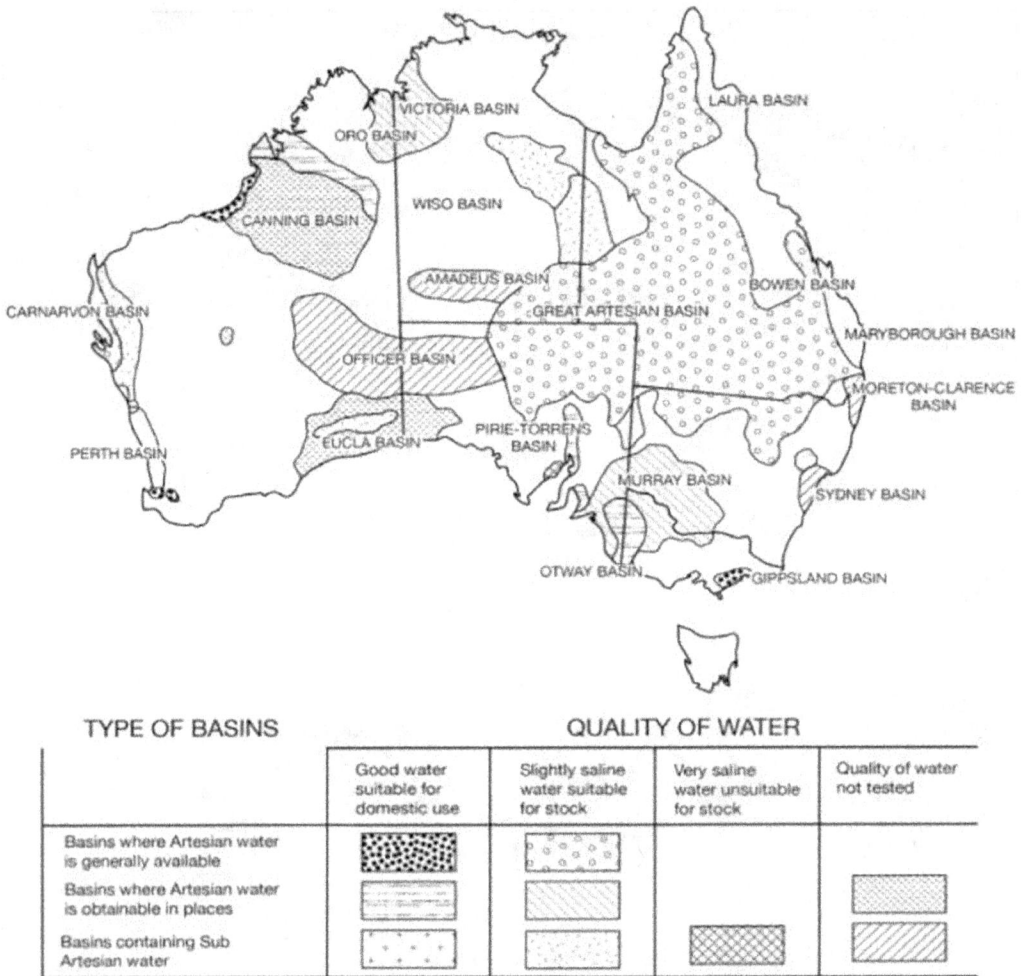

TYPE OF BASINS

QUALITY OF WATER

	Good water suitable for domestic use	Slightly saline water suitable for stock	Very saline water unsuitable for stock	Quality of water not tested
Basins where Artesian water is generally available				
Basins where Artesian water is obtainable in places				
Basins containing Sub Artesian water				

FIGURE 1.7 Sedimentary basins of Australia showing designated type and expected water quality (Anon, 1983).

1.4 Water Security

After 150 years of expansion, by the beginning of the 21st century, increasing limitations were placed on water allocations, owing to shifts in the amount and quality of water available and existing licenced abstraction in streams, wetlands, rivers and aquifers. To address issues of water scarcity and over-allocation, The National Plan for Water Security was announced in 2007. Under the Plan, $6 billion was provided over 10 years to modernise irrigation infrastructure both on- and off-farm to conserve water and increase water use efficiency. This was to be achieved primarily through piping or lining irrigation distribution channels, adopting more efficient watering methods on the farm, installing more effective water meters to improve measurement and revising river release operations and storage management (Coles, 2019).

A further $3 billion was provided to address over-allocation in the MDB. This funding provided assistance to irrigation districts to reconfigure irrigation systems, retire non-viable areas, help non-viable or inefficient irrigators to relocate or exit the industry, and, where necessary, purchase entitlements on the water market (Coles, 2019). Funds were also made available for establishing new governance arrangements in the MDB, upgrading the water information system, extending the bore-capping and piping programme to reduce water wastage in the GAB, establishing a Northern Australia Land and Water Taskforce and undertaking a Northern Australia Land and Water Futures Assessment.

1.4.1 Recent Impacts of Global Heating

Droughts, high temperatures and evaporative losses are now occurring in Australia in greater frequency. The most recent widespread drought, from late 2016 to 2019, included some of the hottest years on record (Bureau of Meteorology, 2019). Not only was 2019 the warmest year, it was also the driest on record, with the nationally averaged rainfall of 277.6 mm in 2019 being 40% below average (Bureau of Meteorology, 2020).

Rainfall for 2018–2019 was below to very much below average over most of Australia, greatly affecting key irrigated agricultural areas, particularly the MDB, with large areas of the basin states (Qld, NSW, VIC, SA and the ACT) drought-declared (Figure 1.8). This resulted in reduced water allocations and severe declines in production within the MDB.

Other regions of Australia, including pastoral SA, the central and southern Northern Territory, and south-eastern WA, received the lowest annual totals on record (Bureau of Meteorology, 2020). Given this variability in temperature, and increasing uncertainty in reliably forecasting rain, any large-scale irrigation project in Australia is even more dependent on access to a secure water supply at a reasonable cost to ensure viability. This security has become increasingly difficult to deliver, without formalised water allocation frameworks, government and industry support, and with community backing that also address environmental considerations.

1.4.2 Importance of Water Security

The major sources of water for agricultural production are given in Figure 1.9. As highlighted by the two financial years (FY)[2] shown in the figure (2017–2018 and 2018–2019 FY), the amount of water that

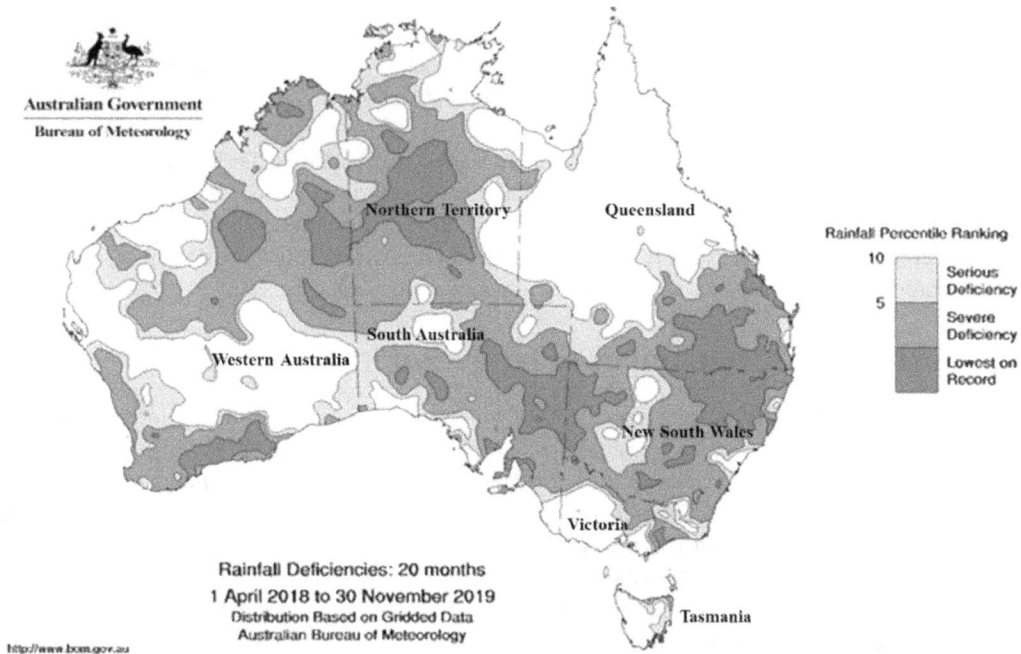

FIGURE 1.8 Rainfall deficiencies for the period from April 2018 to November 2019 showing severe or lowest on record conditions experienced across the Australian continent affecting all states (Bureau of Meteorology, 2019).

Sources of Water for Agricultural Production

Source	2018-19 (ML)	2017-18 (ML)
Reticulated mains	49125	54163
Recycled	114997	157680
Groundwater	2280404	2156385
Rivers, creeks, lakes	2013122	3005302
On-farm dams or tanks	759565	1165687
Irrigation	2743857	3949492

■ 2018-19 (ML)   ■ 2017-18 (ML)

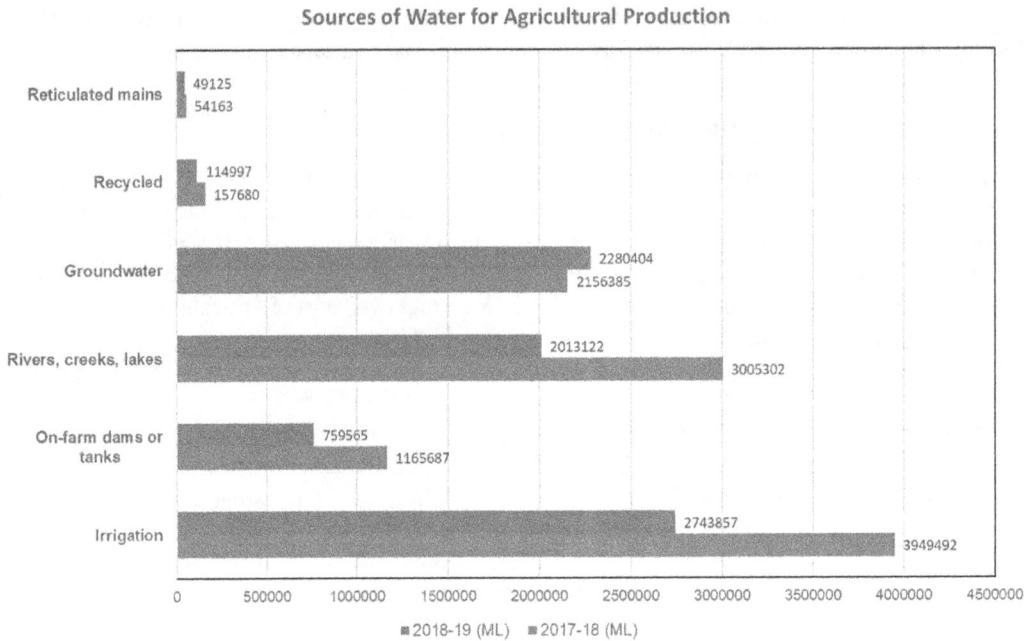

FIGURE 1.9 Comparison of water sources used on Australian Farms for the 2017–2018 and 2018–2019 financial years. A reduction in irrigation uses and surface waters is shown in 2019 (ABS, 2020).

is available for irrigation can vary markedly in any given year depending on the rainfall (which also impacts on irrigation needs), abstraction rates, water trading, environmental flow requirements and other demands placed on the river networks and water supply systems. Water allocated for irrigation, and agriculture in general, significantly declined in 2019. Although more water was abstracted from groundwater sources, this did not compensate for the reduction in water available owing to reduced flows in the MDB caused by the extensive drought. This was reflected in a substantial decline in agricultural productivity.

This demonstrates the need to manage, monitor and assess water available for irrigation, in line with climatic fluctuations, given there are other external higher priority demands for this resource (including environmental, community and cultural requirements). In the following sections, how irrigation has developed in Australia, how it expanded and how in recent years it is moving towards a more sustainable footing through the inclusion of cultural and environmental flows are discussed.

1.5 Historical Development of Irrigation in Australia

There is limited knowledge[3] currently available of indigenous irrigation systems prior to European settlement. However, it is being increasingly recognised that through the 'Dreaming', which forms the foundation of indigenous or First Nation Australians' ancient relationship to land and water, key aspects of traditional knowledge about locating and using water are passed on through oral histories, markers and artworks (Tan and Jackson, 2013). For many decades, in assessing the issues around water management in Australia, there has been an abject failure to include indigenous cultural values of water, compounded by a lack of knowledge of patterns of past water use, pre-contact, through to needs and requirements of present-day indigenous communities (Jackson et al., 2010). The failure to effectively incorporate this extensive history and knowledge about Australia's natural environment has been detrimental to efforts to manage and use Australia's land and water resources in a sustainable way.

In the last decade, indigenous customary systems of resource governance and management are being recognised for their value in managing the terrestrial environment, including water (Langton, 2002; Hill and Williams, 2009). Initiatives to provide new institutional configurations and include indigenous-led methodologies to define and apply cultural flows, and to ensure the flow regimes are adequate to deliver the socio-cultural, economic and health benefits have been undertaken (MLDRIN et al., 2020, 2014). However, these changes are slow to permeate the western standards and practices that have been applied in the past, even though it is widely recognised that change is inevitable if greater climate resilience is to be effectively introduced in the management of water in Australia.

1.5.1 Irrigation Schemes in the Late 19th Century

Modern irrigation has been practised in Australia since the mid-late 19th century with its early beginnings centred on the MDB. That region has grown to become the epicentre of Australian irrigated agriculture (Figure 1.10), with irrigation for crops and pasture consuming about two-thirds of the water extracted (Batten et al., 2003; Creighton et al., 2004). The first small irrigation schemes were probably established independently during the mid-1800s on rivers in the regions of early colonial settlements in Tasmania, Victoria and NSW that preceded the formation of the Commonwealth of Australian States in 1901. For example, the first recorded irrigation scheme in Tasmania was a small-scale development in the Derwent and Clyde River valleys, north of Hobart, in the early 1840s (Mason-Cox, 1994). Later activities included the building of a dam across the Tooms River, forming small lake or water storage,

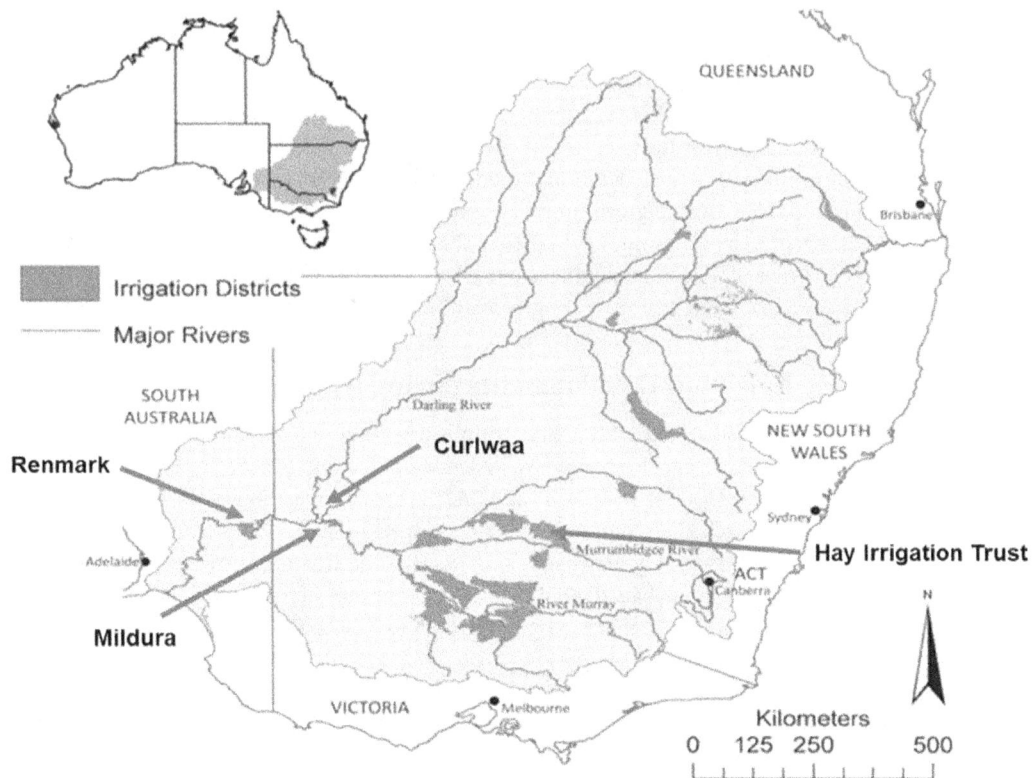

FIGURE 1.10 The Murray–Darling Basin covers 1,059,000 km² (or about 14% of Australia) but delivers only 6.1% of Australia's mean annual runoff or useable flow (Goss, 2003). Initial irrigation districts were established in three Basin states between 1889 and 1892, near Renmark (Riverland, SA), Mildura (Chaffey Brothers, Victoria) and Curlwaa and Hay (NSW) (indicated by arrows) (adapted from Skinner and Langford, 2013).

which promoted a broader acceptance of the potential benefits of irrigation to the Tasmanian colony (CTHS, 2006).

1.5.2 The Basin States

In Victoria, a small independent scheme was established on the Yarra River, near Melbourne, in 1859. These early developments in Victoria were followed some 25 years later by more uniform governance to manage the expansion of irrigation through the *Irrigation Act of Victoria (1886)*,[4] which vested in the Crown the rights to the use of and flow of water in rivers and streams, and furthermore, the control of all water in a watercourse, lake, lagoon, swamp or marsh in Victoria (Department of Sustainability and Environment, 2006). This enabled the formation of irrigation trusts, which were deemed necessary to administer the various works, allocation and distribution systems, and provide a mechanism for irrigators to apply for annual permits for water to be used for irrigation and other purposes (Department of Water Resources Victoria, 1989).

NSW followed suit, with limited small developments on the Murrumbidgee and at Curlwaa, near the junction of the Murray and Darling Rivers in the late 1880s, with similar schemes introduced on the Murray River in Mildura (VIC) and Renmark (SA) (Figure 1.10). Further investigation of irrigation potential in NSW was supported by the appointment of Sir William Lyne to a royal commission in 1884, which was to run for 4 years, although most of its recommendations were largely ignored (Langford-Smith and Rutherford, 1966).

The royal commission did, however, recommend further expansion, and around this time, more small developments were initiated in NSW with the establishment of the Wentworth Irrigation Trust (~400 acres c1902) by an Act of Parliament in 1890, and later the Hay Irrigation Trust on the Murrumbidgee River (~900 acres c1903) in 1892 (Figure 1.10). Neither of these schemes was particularly successful (Langford-Smith and Rutherford, 1966). Most of the early development in MDB occurred on the Victorian side of the border due to easier access, with only limited developments in NSW due to conflicts of interest, the impacts of the federation drought from 1885 to 1902 (Foley, 1957) and objections around water ownership from SA (Langford-Smith and Rutherford, 1966). In contrast to the slow progress of government schemes, by the turn of the century the private YANCO development in NSW had constructed nearly 60 miles of irrigation channels to transfer water from the Murrumbidgee to irrigate more than 2,000 acres of crops and pastures (Langford-Smith and Rutherford, 1966).

1.5.3 Large-Scale Irrigation Developments during the 20th Century

Despite recurrent problems associated with a lack of understanding of the growing conditions, crop performance, hydrology and hydrogeology, effective systems for land and water allocation, and financing, in the late 1890s and early 1900s in NSW, VIC and SA, the area of irrigation in the MDB expanded steadily from 1912. Irrigation schemes continued to be developed from the 1920s through to the mid-1950s, supported by state and federal government construction and financial programmes. Tables 1.1 and 1.2 provide a list (although not exhaustive) of the major dams and weirs constructed in each state, and Tables 1.3a and 1.3b list some of the major irrigation schemes in each state.

1.5.3.1 Irrigation Schemes in the MDB

Large-scale irrigation in the MDB initially developed on the three large southern valleys of the Murray, Murrumbidgee and Goulburn-Broken Rivers of the Southern Basin (Kirby, 2011), with further developments in the Northern Basin rivers, along the Lower Darling, Namoi, Macquarie, Lachlan and Gwydir (Murray–Darling Basin Authority, 2019) (Figure 1.11). Further expansion occurred in the Riverland region of SA, in western Victoria centred on the Murray River, and in the Goulburn Valley in central Victoria. These schemes, which are listed in Tables 1.3a and 1.3b, were supported by the construction of dams, weirs and other containment structures (Figure 1.12), on the most of the 'reliable' major river

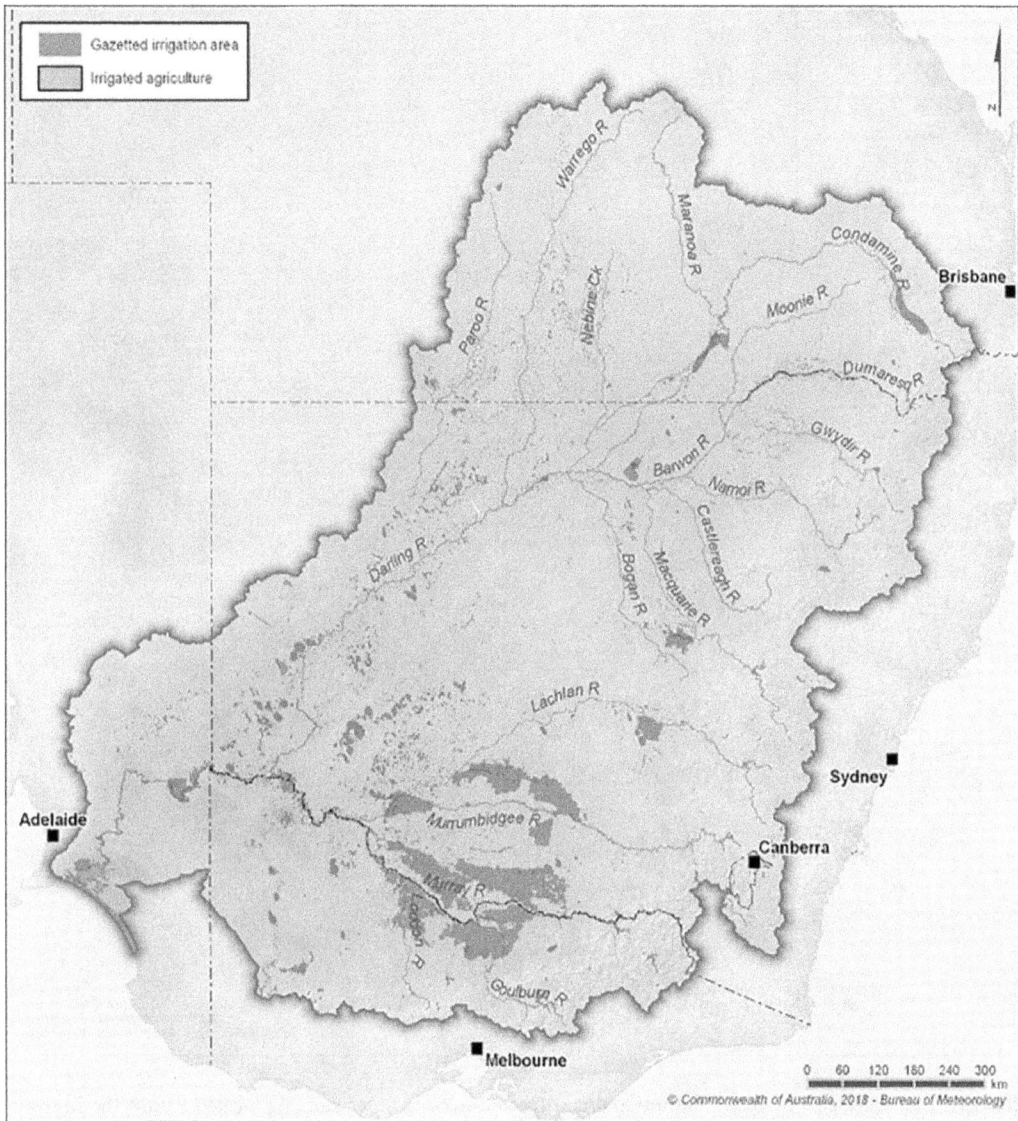

FIGURE 1.11 Major gazetted irrigation areas, irrigated agricultural districts and river systems in the Murray–Darling Basin (Commonwealth of Australia, 2018 – BoM).

systems of the MBD, with this expansion aided by the introduction of new water legislation (see the later sections). Apart from the construction of the Snowy Mountain Hydro-electric Scheme (1949–1972) – which contributed flows into the Murray and Murrumbidgee Rivers irrigation areas[5] – irrigation expansion in the 1960s–1980s mainly occurred outside the MDB. Nevertheless, the MDB continues to support most of the major irrigation areas in Australia (Figure 1.11).

1.5.3.2 Irrigation Schemes in Tasmania

After the initial developments along the Derwent and Clyde Rivers in southern Tasmania in the later part of the 19th century, it wasn't until the middle of the 20th century that attention turned toward the river systems of the midlands and the north of the state, and towards the development of larger integrated irrigation schemes. Three main large-scale irrigation schemes were commissioned between 1974

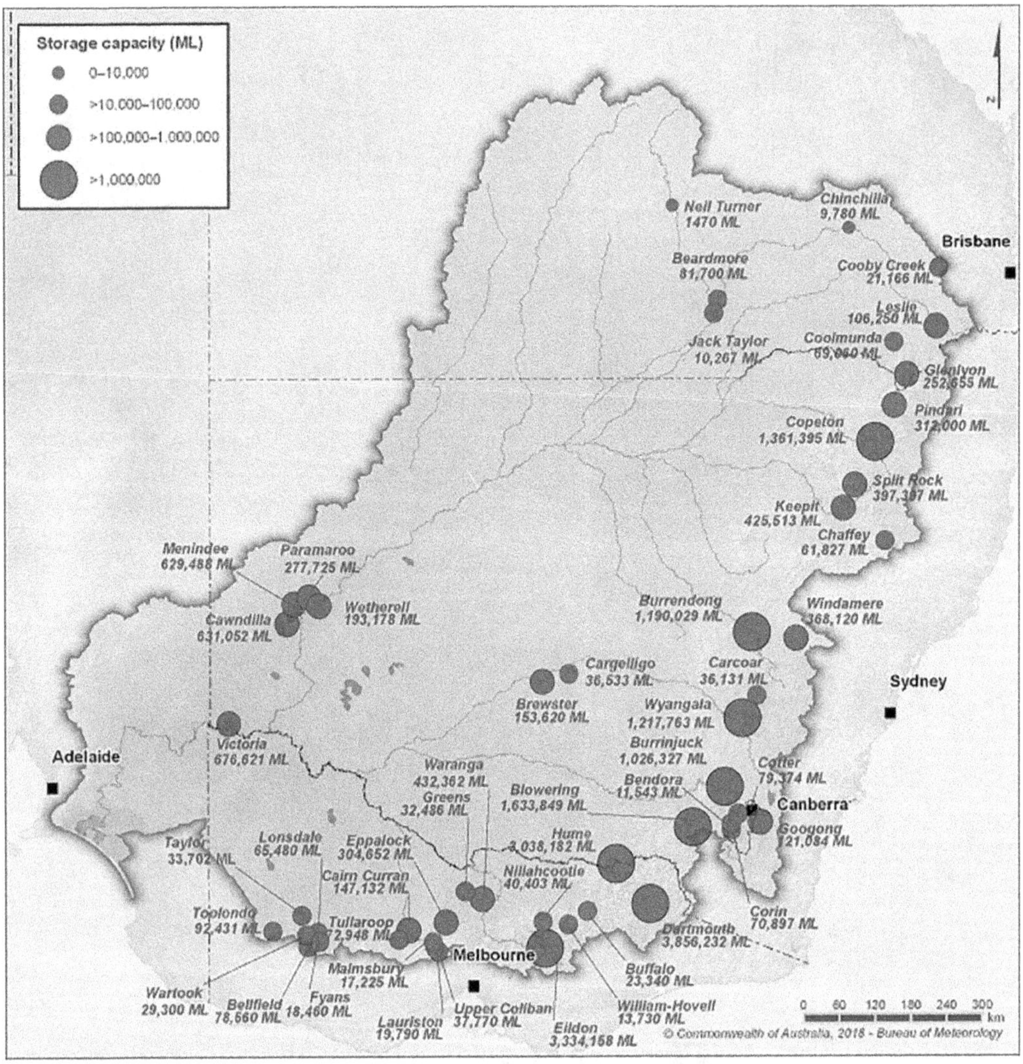

FIGURE 1.12 Strategic managed natural and constructed surface water storages located within the Murray–Darling Basin on the main river systems. More details are listed in Tables 1.1 and 1.2 for some of the major storages (Bureau of Meteorology, 2018).

and 1987 in Tasmania: the Cressy-Longford Irrigation Scheme (1974), The South-East Districts Scheme (1986) and The Winnaleah Irrigation Scheme (1987), which was later expanded by the Winnaleah (Augmentation) Irrigation Scheme in 2012 (Table 1.2).

1.5.3.3 Irrigation Schemes in Queensland

In Queensland, by contrast, large-scale irrigation developments were not supported until the mid-1900s, apart from a small version of the Dawson River experiment first mooted in 1920 and commissioned in 1926 (Sunwater, 2020). Major expansion was heralded with the Mareeba-Dimbulah Irrigation District established in 1959 in North Queensland, the southern Darling Downs in 1961 (further improved in 1986), and subsequent developments on the Nogoa and Kolan Rivers in Central Queensland in the 1970s (Tables 1.3a and 1.3b). The larger Burdekin River Scheme, which was supported by the creation of Lake Dalrymple by the Burdekin Falls Dam (Figure 1.5) with a capacity of 285,000 ML in 1987, greatly increased sugarcane production, amongst other crops (Batten et al., 2003). The most recent development to be approved is the lower Fitzroy River in central Queensland, with the construction of the Rookwood Weir with a capacity of 78,000 ML due for completion in 2021 (Tables 1.1, 1.3a and 1.3b).

1.5.3.4 Irrigation Schemes in Western Australia

In WA, the Harvey Irrigation Scheme was established in 1916, with expansion supported by the government in the late 1930s to provide work for the unemployed during the great depression. This scheme was augmented by the construction of the Wellington Dam in 1933 as the population in southwestern WA expanded and the need for local produce increased. Extensive drainage and irrigation channels were constructed in what is now known as the Peel-Harvey Irrigation District.

TABLE 1.1 A List of the Major Water Resources and Dams Constructed to Support Irrigation Activities in New South Wales and Queensland

State[a]	Dam	River System	Capacity (×100 ML)	Region	Year Completed
New South Wales	Burrinjuck	Murrumbidgee River	1,026	MBD	1927 (1956)
	Hume	Murray River	3,038	MBD	1936 (1961)
	Wyangala	Lachlan River	1,218	MBD	1936 (1971)
	Lake Victoria	Murray River	680	MBD	1928
	Lake Brewster	Lachlan River	150	MBD	1952
	Eucumbene	Eucumbene	4,807	MBD	1958
	Glenbawn	Hunter River	750	MBD	1958
	Keepit	Namoi River	425	MBD	1960
	Menindee Lakes	Darling River	1,794	MBD	1960
	Tantangara	Murrumbidgee River	254	MBD	1960
	Burrendong	Macquarie River	1,188	MBD	1967
	Jindabyne	Snowy River	688	MBD	1967
	Blowering	Tumut River	1,628	MBD	1968
	Talbingo	Tumut River	921	MBD	1971
	Copeton	Gwydir River	1,364	MBD	1976
	Glennies Creek	Hunter Valley	284	MBD	1983
	Windamere	Cudgegong River	368	MBD	1984
	Split Rock Dam	Manilla River	397	MBD	1987 (2012)
Queensland	Tinaroo Falls	Barron River	439	Tablelands, North Qld	1958
	Wuruma	Nogo River	165	Central Qld	1968
	Beardmore	Balonne River	81	MDB	1972
	Fairbairn	Nogoa River	1,301	Central Qld	1972
	Fred Haigh	Kolan River	562	Central Qld	1975
	Glenlyon	Pike Creek	254	MDB	1976
	Boondooma	Boyne River	204	South East	1983
	Leslie	Sandy Creek	106	MDB	1985
	Burdekin	Burdekin River	1,860	North Qld	1987
	Bjelke-Petersen Dam	Barker Creek	135	South East	1988
	Peter Faust Dam	Proserpine Rive	491	North Qld	1990
	Paradise Dam	Burnett River	300	Central Qld	2005
	Rookwood Weir	Lower Fitzroy River	78	Central Qld	2021

Year in parentheses denotes the year a later expansion of the existing scheme was completed. Only those irrigation dams exceeding 50,000 ML capacity are listed.

[a] Derived from multiple sources. https://www.sunwater.com.au/; https://en.wikipedia.org/wiki/List_of_dams_and_reservoirs_in_Australia.

TABLE 1.2 Major Water Resources and Dams Constructed to Support Irrigation Activities in Victoria, Western Australia, Tasmania and South Australia (Major Dams Listed)

State[a,b]	Dam	River System	Capacity (ML) 000's	Region	Year Completed
Victoria	Goulburn Weir	Goulburn River	??	MDB	1891
	Yarrawonga	Murray River	117	MDB	1939
	Rocklands Dam	Glenelg River	261	Wimmera Mallee	1953
	Cairn Curran	Lodden River	149	MDB	1958
	Eppalock	Campaspe River	304	MDB	1964
	Dartmouth	Mitta Mitta River	3,865	MDB	1979
	Thomson	Thomson River	1,123	West Gippsland	1984
	Waranga	Rushworth River	432	MDB	1915, 1926
	Glenmaggie	Macalister River	190	West Gippsland	1927 (1958)
	Eildon	Upper Goulburn River	3,390	MDB	1928 (1955)
	Toolondo	Horsham	107	Wimmera Mallee	1952 (1960)
Western Australia	Wellington	Collie River	185	SW WA	1933 (1960, 2012)
	Logue Brook Dam	Logue Brook	??	SW - WA	1963
	Ord (Lake Kununurra)	Ord River	90	Kimberley	1963
	Ord (Lake Argyle)	Ord River	10,760	Kimberley	1971 (1990)
Tasmania	Great Lake	Brumby Creek	10	Midlands	1974
	Craigbourne Dam	Coal River	??	SW Tasmania	1986
	Cascade Dam	Cascade River	6.9	NE Tasmania	1987
	Meander Dam	Meander River	43	North Tasmania	2007
	Frome Dam	Frome River	12?	NE Tasmania	2012
South Australia	Willunga Storage Dam	Recycled Wastewater Storage	6	South East SA	2020

Year in parentheses denotes the year a later expansion of the existing scheme was completed. Only those irrigation dams exceeding 50,000 ML capacity are listed. Tasmania is exception with the largest dams listed and Hydro Schemes not included.

[a] Amount of water available for irrigation may change over time depending on water availability, community needs, water quality or technical viability. ? denotes estimated volume, and ?? denotes volume not known.

[b] Derived from multiple sources. https://en.wikipedia.org/wiki/List_of_dams_and_reservoirs_in_Australia; https: https://web.archive.org/web/20061123113332/http://www.watercorporation.com.au/D/dams_all.cfm

As population continued to grow, attention turned again to the development of northern WA, with the Camballin Irrigation Scheme on the Fitzroy River in the West Kimberley and initial pilot schemes at Kununurra on the Ord River in the East Kimberley during the 1950s. In the early 1970s, construction of the Ord River Dam created Lake Argyle to support the Ord Irrigation Scheme (Figure 1.6), with further expansion of the Ord River scheme through phases II (2012) and III (2020) (Figure 1.13).

While Ord Stages I and II were wholly developed within WA, Ord Stage III is linked to developments across the WA-NT border at the Keep River Plains in the NT. At 67,000 ha of both irrigated and dryland farming, this represents the NT's first large-scale irrigation development. In WA's southwest, irrigation expanded through 1950–2000 and became more heavily dependent on groundwater resources, including in the peri-urban areas surrounding the capital city of Perth on the Swan Coastal Plain and in the Peel-Harvey area to the south of the City. Water was abstracted by independent self-supply irrigators who accounted for nearly 60% of all water allocations in these areas, of which ~90% was sourced from groundwater (McCrea and Rivers, 2003).

In the early 2000s, most of the state-owned surface-supplied irrigation schemes were transferred to private hands, such as grower-owned irrigation cooperatives, as an incentive to improve water resource management, capital and operating efficiencies (Roberts and Henneveld, 2003). This was consistent with national water and competition policy reforms. By 2005, about 90% of the surface water utilised in

TABLE 1.3A Historical Irrigation Development (1859–1962) in Australia by Region and State with Only Initial State Irrigation Developments and Later Large-Scale Irrigation Schemes Listed[a]

Year	State	Irrigation Scheme	Region
1859	VIC	Yarra River Scheme, near Melbourne Small-scale irrigation scheme	Victoria
1889	VIC	Mildura Irrigation Trust Still in operation	MDB
1889	SA	Renmark Irrigation District Established by the Chaffey Brothers from California	MDB
1889	NSW	North Yanco Estate established on the banks of the Murrumbidgee River by Sir Samuel McCaughey	MDB
1890	NSW	Curlwaa Irrigation Scheme Established at the junction of the Murray and Darling rivers	MDB
1912	NSW	Murrumbidgee Irrigation Areas Commissioned Base around resumed lands in Mirrool and including the YANCO property, ~300,000 acres	MDB
1916	WA	Development of the Harvey Irrigation Scheme Extended with the building of the Wellington Dam in 1933 The region now covers 112,000 ha. Covering a network of channels and pipes to the Harvey, Waroona and Collie River districts. Water is drawn from the from Waroona, Drakesbrook, Logue Brook, Harvey and Wellington dams. https://www.harveywater.com.au/	SW Coastal Western Australia
1923–1926 Upgraded 1940	Qld	Dawson River Irrigation Scheme (latterly the Dawson River Valley Scheme). Initially limited success due to distance to markets, disease, isolation and high costs, but eventually through consolidation of blocks and hard work the district became one of the state's leading irrigated cotton, farming and grazing production areas	SE Central Queensland
1935–1946	NSW	Irrigation districts formed on the Murrumbidgee in NSW. Tabbita (6,300 ac), Benerembah (112, 000 ac) and Wah (580,000 ac)	Southern NSW
1949–1972	NSW	Snowy Mountain Hydro-Scheme Comprised both irrigation (2100 GL) diversion water and hydro-electricity generation. The largest scheme in Australia and on completion in 1972 provides flows into the Murray and Murrumbidgee Irrigation Schemes	Southern NSW
1950s	Qld	Lower Burdekin River irrigation district established Soldier settlement scheme on ~7,500 ha	North Qld
1959	Qld	Mareeba-Dimbulah Irrigation Area created Supplies about 225,000 ML per annum	North Qld
1961–1965	Qld	Darling Downs Construction of the Leslie Dam increase water security for irrigation. Expanded in 1986 to double strong capacity	MDB-Southern Queensland
1960–1962	WA	Ord River Pilot irrigation farms	Kimberley, WA

[a] Derived from multiple sources. https://www.sunwater.com.au/. https://en.wikipedia.org/wiki/List_of_dams_and_reservoirs_in_Australia; https://web.archive.org/web/20061123113332/http://www.watercorporation.com.au/D/dams_all.cfm

TABLE 1.3B Historical Irrigation Development (1972–2020) in Australia by Region and State with Only Initial State Irrigation Developments and Later Large-Scale Irrigation Schemes Listed[a, b]

Year	State	Irrigation Scheme	Region
1972–1980	WA	Ord River Irrigation Scheme Ord scheme created Lake Argyle the largest dam reservoir in Australia covering an area of 741 km^2	Kimberley, WA
1974	TAS	The Ivanhoe Plains and Packsaddle Irrigation Districts	Midlands, TAS
1987	Qld	Burdekin River Scheme commissioned Delivers about 280,000 ML annually to 103,000 ha (c2007). Lake Reservoir with a capacity of 1,860,000 ML	North Qld
1986	TAS	The South-East Districts Scheme established	SE TAS
1987	TAS	Winnaleah Scheme commissioned Since 2012, the Winnaleah (Augmentation) Irrigation Scheme, with the addition of the Frome Dam	NE TAS
2008	TAS	Greater Meander Valley Irrigation Scheme	Midlands TAS
2012	WA	Ord Stage II Further expansion of the Ord scheme Goomig lands (Weaber Plain)	Kimberley, WA
2020	WA	Ord Stage III Ord East Kimberley Development Plan Further expansion of the Ord Scheme cockatoo sands development, ~6,000 ha	Kimberley, WA
2020	QID	Rookwood Weir-Fitzroy River Construction Approved. Capacity 76,000 ML	Central Qld
2020	NT	*Keep Plains Agricultural Development:* Considered as part of rebranded Ord III covering about 67,500 ha. Adjoining WA's Ord River agricultural developments on the NT side of the border *Wildman Agricultural Precinct, 26,000 ha:* Located 135 km east of Darwin, north of the Arnhem Highway along the Point Stuart Road, the large-scale agricultural precinct. *Larrimah Agricultural Precinct, 5,712 ha:* Located 180 km south of Katherine, the area suitable for mangoes, citrus and melons, plus opportunities for dryland and irrigated crops and the development of intensified beef operations	NT

[a] This is not a comprehensive list of all irrigation schemes or trusts created in Australia with the water resources managed by some 800+ irrigation authorities, water boards, local governments and private companies.

[b] Derived from multiple sources. *https://www.sunwater.com.au/.*https://en.wikipedia.org/wiki/List_of_dams_and_reservoirs_in_Australia; https://web.archive.org/web/20061123113332/ http://www.watercorporation.com.au/D/dams_all.cfm

irrigation in WA was supplied by four main irrigation schemes – the Ord River, Carnarvon (Gascoyne), Preston Valley and Harvey Irrigation Area (Figure 1.13). Note that the Gascoyne Irrigation Scheme is mainly sourced from alluvial sands in the Gascoyne River (Figure 1.14), with supplementary water extracted during large river flows resulting from sporadic rainfall events.

1.6 Irrigation for the 21st Century

By the turn of the millennium, there was approximately 2.4 M ha of land under irrigation in Australia. In all Australian states, the options for irrigation developments in the 21st century are limited, with a greater reliance placed on improving existing irrigation developments through water conservation,

FIGURE 1.13 Irrigation zones in Western Australia showing the three main areas of the Ord River, Gascoyne and Southwest (Harvey Water) Irrigation Areas, along with a smaller development around the Preston River Valley. Regions highlighted in blue are areas of potential irrigation development within the state (Water and Rivers Commission, 2000).

FIGURE 1.14 Intermittent flows on the Gascoyne River. Water is sourced from the river during infrequent large rainfall events but is usually extracted from the alluvial sands of the dry riverbed which are recharged during rainfall events during the remainder of the year.

Photo source R. Thomas. https://en.wikipedia.org/wiki/Gascoyne_River

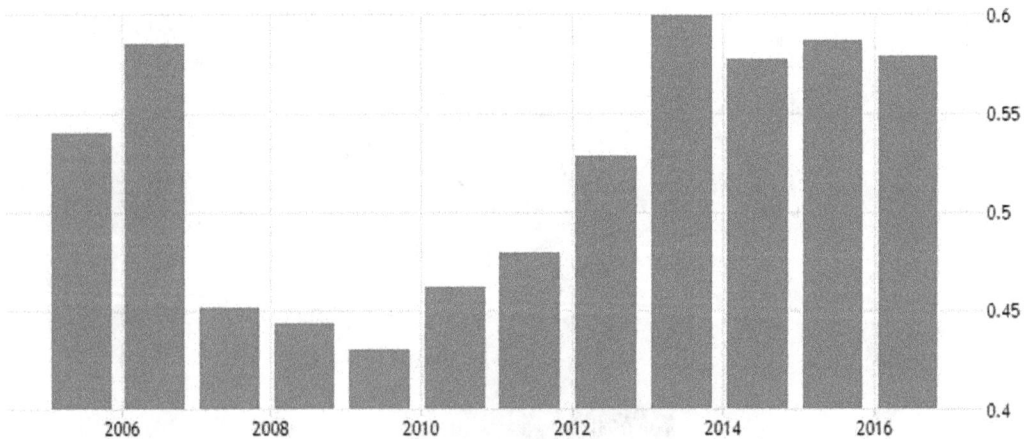

FIGURE 1.15 Agricultural irrigated land (% of total agricultural land) in Australia was reported at 0.578% in 2016, according to the World Bank data from officially recognised sources (World Bank, 2021).

better irrigation techniques, closer monitoring of water applications, improved business management and water use efficiency. At both a policy and farm operational level, there has been increased emphasis on the use of financial instruments, such as trading in water entitlements. The introduction of new water conservation techniques, irrigation technologies and improved monitoring systems enabled the irrigated area in Australia to expand by 30% between 1985 and 2000 onto previously unirrigated dryland crop and pasture lands (Creighton et al., 2004).

Since 2000, the area under irrigation has fluctuated between 2.0 and 2.5 million ha (~0.45% and 0.6%), influenced by seasonal water availability (Figure 1.15), and more recently the introduction of water management plans, water trading and water allocation mechanisms, particularly in the MDB.

1.6.1 The Murray–Darling Basin

Even with the extensive developments in other regions of Australia, the MDB continues to be the most important irrigation region; although it covers only 14% of the Australian landmass, the MDB contains 71% (1.47M ha) of all irrigated lands in Australia, including the major irrigation areas and districts in southwestern NSW and northern Victoria (MDBC, 2000). The basin contains 20 major rivers, including the three longest inland rivers in Australia: the Darling (2,740 km), the Murray (2,530 km) and the Murrumbidgee (1,690 km) (Figure 1.3). The MDB is a region with rivers and groundwater systems (Figure 1.6) that are notably hydrologically interdependent, but which are governed different authorities within each of the Basin states in accordance with the Australian Constitution. This applied 'artificial' political borders and various ownership entitlements across the region that were not always complimentary or beneficial to the water resources they impacted (Coles, 2017).

As described in the previous sections, not long after Federation in 1901, a complex system of diversions, locks, dams and weirs was built to encourage irrigation and regional rural development along the Murray and Murrumbidgee Rivers (Kingsford, 2000). More recently, it was realised that this extensive concentration on the MDB comes at a significant cost, and it has been increasingly recognised that in future, less water will be available for use because of the need to return some water to restore environmental values; because of lower river flows as a result of climate change, bushfires and changing land use (Kirby, 2011), and to address changing social and cultural values.

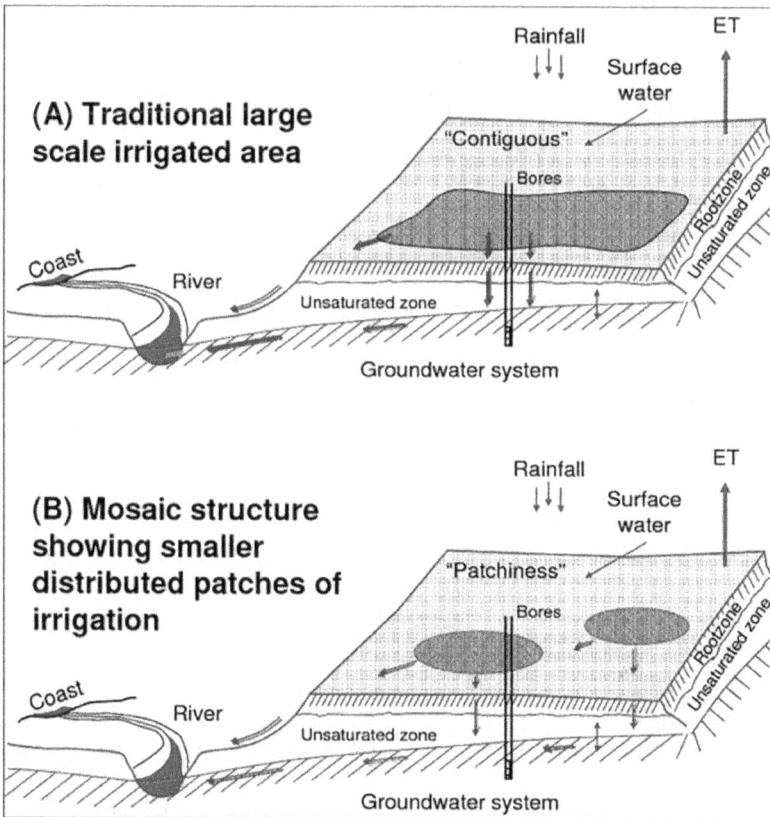

FIGURE 1.16 Schematic showing basic features of large contiguous irrigation compared with irrigation mosaics involving smaller patches of irrigation distributed across the landscape (Paydar et al., 2011).

By 1985, the development of available surface waters approached 51% of capacity, increasing to 60% just 10 years later (Batten et al., 2003; Murray Darling Basin Commission, 2000). There were increasing concerns that this level of water use was becoming unsustainable, such that by 1995 a cap on the extraction of water was agreed by the Council of Australian Governments (COAG).[6] This agreement stated that any irrigation expansion could only proceed with water gained through improved water use efficiency or the purchase of water from existing developments. Later with the introduction of the Water Act 2007 (Cth), water rights were now tradeable, and it also provided for the creation of the Murray–Darling Basin Authority. The Authority provides oversight for the management of water within the Basin; however, under the Australian Constitution (1901), water rights remained vested in the individual states (Coles, 2018).

1.6.2 Managing Irrigation as a Complex System

Decisions about irrigation are extremely complex, and there are many uncertainties. Experiences in the MDB and elsewhere have increased awareness of the risks and potential consequences of water use decisions and dealing with that complexity and uncertainty is a shared need and responsibility for catchment communities, development proponents and decision makers. At the same time, the community now expects that decisions about proposals for irrigation development will not only deliver economic benefits to proponents but will also deliver social and economic benefits to the broader community, with acceptable environmental impacts (Camkin et al., 2008).

Non-government organisations and community individuals are now better trained, better connected and generally better equipped to play a 'watchdog' role on government and other decision making, and on decision makers (Camkin et al., 2008). Due to restrictions on water allocation in existing irrigation areas, there is increasing pressure for the allocation of new and alternative water resources, despite the inherent complexity and uncertainty surrounding sustainable use of these resources. Government decision makers now find themselves in a position where they are required to make decisions within timeframes set by legislation, policy or political needs that can result in cautious or risk-averse decision making.

1.6.3 Irrigation Mosaics

In making changes to existing irrigation systems, or designing new irrigation areas, it is necessary to determine what unwanted impacts and what compensatory benefits are likely to accrue. To assist this evaluation process, the Northern Australia Irrigation Futures project introduced the concept of 'irrigation mosaics' as one possible mode of development for northern Australia. Irrigation mosaics – smaller discrete patches of irrigated land dispersed across the landscape (Figure 1.16) – may offer an alternative to traditional large-scale contiguous irrigation systems and have the potential to deliver improved social and economic opportunities for rural and remote (often indigenous) communities in northern tropical Australia (Paydar et al., 2011).

A review by Paydar et al. (2011) drew findings and lessons learned from studies of other systems dealing with spatial patterns in the landscape. Within this framework, while irrigation mosaics show some promise, a need was identified for data and better understanding of:

- patch number, size, shape and connectivity on evapotranspiration from irrigated patches of land within a mosaic structure;
- the fate of solutes within the mosaic and recharge to groundwater and the surrounding land and its impact on groundwater quality;
- soil and water salinisation; and
- system losses and biodiversity (Paydar et al., 2011).

1.7 Channelling Irrigation Research – Irrigation System Harmonisation

Prior to the release of the National Water Security Plan (NWSP), the Cooperative Research Centre for Irrigation Futures (CRC IF)[7] was established in 2003. The CRC IF was focussed on delivering research, education and training that gave confidence to growers, industry, government and communities to invest in better irrigation, a better environment and a better future (Irrigation Australia, 2020). Between 2006 and 2010, the CRC IF ran two successful research programmes *Irrigation Toolkits to Improve Enterprise Performance (Toolkits)* and *System Harmonisation through Regional Irrigation Partnerships (System Harmonisation™).*

The 'Irrigation Toolkits' programme delivered applications based on research and innovation to the on-farm irrigation sector to improve production, profitability and sustainability of irrigation enterprises. *System Harmonisation* was developed as a strategy to improve cross-organisational communication and system-wide management and improve production and environmental outcomes (Khan, 2010). Bridging the gap between water policy, water resource management and scientific communities was a key outcome; from the setting of research agendas to the free flow of information for use in management and policy making to enhance the multifunctional productivity of water resources (Khan, 2010).

Historically irrigation and environmental sustainability have been managed as two competing enterprises under separate and divergent control. This often translated into polarised approaches to resource management, to the detriment of both production and environmental sustainability. As its key objective, the System Harmonisation approach sought to

> *identify business opportunities for irrigators to become an integral part of an expanding environmental services industry and in doing so support a truly sustainable and diversified irrigation business environment.*

> *(Khan, 2010)*

System Harmonisation establishes the base physical, economic and social position of the region, and identifies the key pressure points in the system and the system constraints. Key pressure points in a water system can be of a biophysical, economic, social, environmental or institutional nature, and it is the changes in these key pressure points than need to be assessed, in a comprehensive and systematic way, to enhance the multifunctional productivity of irrigation systems (Khan, 2010).

This approach also recognised that in Australia, as elsewhere, there is a major lag between research and management policy since most water management policy is based on outdated knowledge and technology, and there is a paradigm lock between water scientists, policy makers and users. In many catchments, often stakeholders are unaware of what technical facilities are available and scientists do not appreciate how to become part of real solutions. System Harmonisation provides opportunity for the sharing of key questions and answers, and for participatory planning and policy development (Box 1.1).

By recognising that, irrigation is part of a dynamic, complex social-ecological system that requires new approaches based on integrating science, policy, planning, management and communities to ensure irrigation continues, there is a requirement to:

- work with regional irrigation partners to increase profitability and reduce the environmental footprint of improved irrigation systems.
- collaborate with researchers from different disciplines working together to find new ways to increase understanding to support implementation of new, more resilient land and water management strategies.
- improve communication between organisations and individuals as well as more transparent and objective decision making to support irrigation planning and management at all levels within a catchment and beyond (Bristow and Stubbs, 2010).

Box 1.1   System Harmonisation Provides Opportunity for Q&A and Participatory Planning and Policy Development

- Do we have the right policies and institutions in place?
- Are we using appropriate economic models that account for impacts on the environment to support long-term decision making?
- If current trends in population, water demand and energy usage persist, where will we get our water and food from?
- Should we encourage more local food production through careful design of urban and peri-urban zones and set aside particular areas as 'horticultural precincts'?
- Are we properly considering systems and long-term thinking for the benefit of future generations?
- Can we design improved allocation and use of water resources via collective discussion and negotiation as opposed to an adversarial or prior right, precedent or entitlement, based approach?
- Are irrigation communities demonstrating environmental, social and economic responsibility?
- What if governments introduce changes that mean irrigation is no longer viable in a particular catchment; what will the people do?
- How do you decide where and why you establish new irrigation schemes?
- Where should irrigation be located within a catchment, what should it look like and how should it be managed?
- What role could mosaics play in helping build more resilient and regenerative irrigated catchments?
- How do we redesign irrigation schemes and irrigation businesses, so they remain viable and profitable in highly uncertain and changing climatic and economic environments?
- And many more equally difficult questions.

Source: After Bristow and Stubbs (2010).

Above all, System Harmonisation relies on collaboration, and collaboration requires a change in attitude, a high level of commitment and, ultimately, a lot of hard work. It takes time to build the trust necessary for effective collaboration. Key findings from the CRC IF on Irrigation System Harmonisation are presented in Box 1.2 – they have relevance to the future of irrigation in Australia and beyond.

System Harmonisation was an experiment in itself, one that sought to deliver a whole new approach to better integrating and improving the research and practice of irrigation within a catchment context. As an experiment, the programme was particularly successful in demonstrating that many different stakeholders can come together to tackle difficult and complex problems. These were problems that they could not tackle on their own but by creating partnerships and collaborations with other stakeholders

Box 1.2   Key Findings from the System Harmonisation Programme Workshops and Region Community Engagement Focus Groups

The Environment
Integrating irrigation systems into the broader environment

- Progress of the region is closely linked to progress in understanding and improving the environmental condition of the region.
- The environment was not viewed as critical infrastructure for the irrigators and others in the community.
- The environment is the cornerstone which underlies irrigation and its future sustainability and profitability.

Science
It's all about science, but not science on its own

- Bringing all different types of science together with economics, environment and communities to solve catchment-scale problems.
- Requires strong disciplinary skills working collaboratively and the science needs to be flexible, not in terms of the rigour of the process or the statement of the findings, but rather in the way the issues are approached.
- The science evolves as the community knowledge grows in their understanding both the system and the science underpinning it.
- Leadership is a key component with communities and irrigators struggling to understand the science and the need for interventions.
- Bridging this divide would appear to be one of the biggest hurdles to Australia having a resilient environment and a profitable irrigation future.

Irrigators
Irrigators are people at the centre of change

- Irrigation and irrigators in Australia are at a critical fork in the road.
- People are now more aware of importance and centrality of the health of the environment.
- With population and demand for food and fibre, the products of irrigation continue to grow.
- Irrigators find themselves in the middle of conflicting needs to produce more food with less impact.
- System harmonisation has potentially offered a way forward, to address both needs and promote productive and sustainable irrigation futures.
- Irrigators failure to engage in this process, with their main focus on today, rather than tomorrow, next year and the next 50 years.
- This attitude is not dissimilar to businesses and communities caught in this situation; however, if this approach does not change, the future of irrigation will be stunted and difficult.

Leadership and Collaboration Challenge of people and how we work and how we are involved in learning about leadership and collaboration	• Effective communication and collaborative relationships are key elements to evoke change. • Leadership is a critical ingredient when dealing with the disparate range of stakeholders associated with irrigation schemes and regional communities. • Good leadership provides an environment where relationships can develop and grow and create collaborative opportunities. • These are extremely valuable asset to underpin the sustainable future for irrigation.
Business and Economics The future of irrigation is about economics, but it is also about the environment.	• Irrigation regions and individual irrigators need to make good economic decisions to ensure they have an irrigation future. • Regional Irrigation Business Partnerships are viewed as critical to assist in decision making. • Communication of change economics and recognition of environmental values are important in order to articulate the development of a sustainable future for irrigation.

Source: Adapted from Bristow and Stubbs (2010).

they could harness the energy and commitment of individuals and organisations to the group to deliver local-, regional- and catchment-scale changes (Bristow and Stubbs, 2010).

1.8 Conclusions

In conclusion, it is acknowledged that satisfying the demand for food into the future, from a static or even decreasing land base, coupled with declining in situ water resources, poses a significant threat to water and food security in Australia. The subsequent chapter will present productivity data, and address some of the environmental, water management and governance issues that can move Australia towards having a more efficient, productive and sustainable irrigation industry.

Notes

1 In WA, this is managed for the purposes of public water supply and to reduce pressure on the aquifers. But the techniques can be applied elsewhere.

2 Note that in the Australian context a financial year is defined as a 12-month period commencing on July 1st and finishing on June 30th the following calendar year (e.g. 1 July 2017 to 30 June 2018).

3 We acknowledge that many aspects of cultural knowledge relating to these matters are confidential and restricted to adults of one gender or the other.

4 Water Irrigation Act Amendment 1889 (Vic) Part II. & Part III.

5 The Scheme transfers water from the Snowy River and some of its tributaries into the MDB and provides approximately 2,100 GL (7.4 × 1,010 cu ft) of water a year to the Basin. The additional water used for irrigation generates about A\$3 bn per annum (c2007) representing more than 40% of the gross value of the nation's agricultural production (https://web.archive.org/web/20070830103344/, http://www.cultureandrecreation.gov.au/articles/snowyscheme/).

6 COAG consists of representatives from each of the Australian States and Territories, and a representative of Local Government. It was brought together to initiate, develop land and water management at both State and National levels to enhance cooperation between the states and improve

environmental magnet as part of the Decade of Landcare and the need for Australia to meet its international treaty obligations.

7 The CRC IF was a collaboration of 15 partners from Australian and state governments, academic and research organisations, and private enterprise.

References

ABS, Australian Bureau of Statistics. 2020. Water Use on Australian Farms: Canberra: Australian Government. https://www.abs.gov.au/statistics/industry/agriculture/water-use-australian-farms/latest-release Accessed Oct 10 2020.

Alley, W.M., 2016. Drought-proofing groundwater. *Groundwater* 54 (3), 309.

Anon. 1983. *Rural Industry in Australia*. Australian Government Publishing Service, Canberra.

Batten, G., Katupitiva, A., and J. Pratley. 2003. Irrigation management. In *Principles of Field Crop Production*, edited by J. Pratley, pp. 418–462. Oxford University Press, Oxford.

Bristow, K.L. and T. Stubbs 2010. *Reinventing Irrigation Catchments. The System Harmonisation Story*. Cooperative Research Centre for Irrigation Futures. IF Technologies Pty Ltd., Townsville. ISBN: 9780980810950.

Bureau of Meteorology, 2019. Drought Tracker Archive. 8 October 2019. http://www.bom.gov.au/climate/drought/archive/20191008.archive.shtml. Accessed Oct 10 2020.

Bureau of Meteorology. 2020. Annual Climate Statement 2019. Australian Government. http://www.bom.gov.au/climate/current/annual/aus/. Accessed Oct. 10 2020.

Camkin, J.K., Bristow, K.L., Petheram, C. Paydar, Z. Cook, F. J., and J. Story 2008. Designs for the future: The role of sustainable irrigation in Northern Australia. *WIT Transactions on Ecology and the Environment* 112:293–302. doi: 10.2495/SI080291

Clarkson, C., Jacobs, Z., Marwick, B., Fullagar, R., Wallis, L., Smith, M., Roberts, R.G., Hayes, E., Lowe, K., Carah, X., Florin, S.A., McNeil, J., Cox, D., Arnold, L.J., Hua, Q., Huntley, J., Brand, H.E.A., Manne, T., Fairbairn, A., Shulmeister, J., Lyle, L., Salinas, M., Page, M., Connell, K., Park, G., Norman, K., Murphy, T., and C. Pardoe (2017). Human occupation of northern Australia by 65,000 years ago. *Nature* 547 (7663), 306–310. doi: 10.1038/nature22968

Coles, N.A. 2017. Fracking Australia, from nirvana to crisis management? The unconventional gas industry in eastern Australia and the development of the associated Federal and State water and environmental policy and legislation. LLM, Thesis University of Dundee. Dundee Scotland. UK.

Coles, N.A. 2018. Using an adaptive environmental management framework to regulate the unconventional gas industry: Queensland a case study. *New Water Policy and Practice Journal* 4 (2), 57–73.

Coles, N.A. 2019. Unconventional gas: New energy nirvana or environmental menace? Centre for Energy, Petroleum and Mineral Law and Policy, CAR 18, 2019. University of Dundee. Dundee. Scotland UK.

Coles, N.A., Stanton, D. and C. W. Baek 2021. A review of surface enhanced experimental catchments to improve farm water security and resilience in a drying climate in southwestern Australia. *Water Productivity Journal* 1(3), 11–23.

Creighton, C, Meyer, W.S., and S. Khan 2004. Farming and land stewardship. Case study- Australia's innovations in sustainable irrigation. In: *4th International Crop Science Congress; September 2004*; the Regional Institute. http://hdl.handle.net/102.100.100/187979?index=1.

CTHS (2006). Centre for Tasmanian Historical Studies. www.utas.edu.au/library/companion_to_tasmanian_history Accessed Nov 10 2020.

Department of Primary Industries and Regions. 2019. Water storage project to provide a $33 million boost for McLaren Vale wine region. Government of South Australia Media Release, 27 August 2019. https://roadsonline.com.au/water-storage-infrastructure-33m-boost-for-mclaren-vale-wine-region/ Accessed Oct 30 2020.

Department of Sustainability and Environment. 2006. *Stream Flow Management Plans*. Victorian Government, Melbourne, VIC.

Department of Water Resources Victoria. 1989. *Water Victoria: A Resource Handbook*. VGPO, Victorian Government, Melbourne, VIC.

Department of Water. 2010. *Operational Policy 1.01 – Managed Aquifer Recharge in Western Australia*. Department of Water, Perth, WA.

Dillon P. and M. Arshad (2016) Managed aquifer recharge in integrated water resource management. In *Integrated Groundwater Management*, edited by Jakeman A.J., Barreteau O., Hunt R.J., Rinaudo J.D., and A. Ross. Springer, Cham. doi: 10.1007/978-3-319-23576-9_17

Eslamian, S.S. 1995, Regional Flood Frequency Analysis Using a New Region of Influence Approach, Ph.D. Thesis, Univ. of New South Wales, School of Civil Engineering, Dept. of Water Engineering, Sydney, NSW, Australia, 380 P.

Foley, J.C. 1957. *Droughts in Australia: Review of Records from Earliest Years of Settlement to 1955*. Australian Bureau of Meteorology.

Goss, K. 2003. Comprehensive water management in the Murray-Darling basin. In *Proceedings of River Engineering*, JSCE 8, February, pp. 1–6.

Groundwater Governance Global Framework for Action 2015. Global Framework for Action. http://www.groundwatergovernance.org/fileadmin/user_upload/groundwatergovernance/docs/GWG_FRAMEWORK_FR.pdf Accessed Oct 27 2020.

Hill, R. and L. Williams 2009. Indigenous natural resource management: Overcoming marginalisation produced in Australia's current NRM model. In *Contested Country: Local and Regional Natural Resources Management in Australia*, edited by M. Lane, C. Robinson, and B. Taylor, pp. 161–178. CSIRO. Canberra, Australia.

Irrigation Australia. 2020. *Irrigation Futures Publications/CRC IF*. Accessed on Dec 7 2020 at https://www.irrigationaustralia.com.au/publications/irrigation-futures Accessed Sep 10 2020.

Jackson, S., Moggridge, B., and R. Robinson 2010. Summary of the scoping study: Effects of change in water availability on Indigenous people of the Murray-Darling Basin. Report prepared by CSIRO for the Murray Darling Basin Authority, Australia, October 2010.

Khan, S. 2010. System Harmonisation: Concept, Issues and Progress. Cooperative Research Centre for Irrigation Futures Technical Report No. 16/10. CRC IF Technologies Pty. Ltd, Australia.

Khan, S. and M.A. Hanjra 2008. Sustainable land and water management policies and practices: A pathway to environmental sustainability in large irrigation systems. *Land Degradation & Development* 19 (5), 469–487.

Kingsford, R.T. 2000. Ecological impacts of dams, water diversions and river management on floodplain wetlands in Australia. *Austral Ecology* 25 (2), 109–127. doi: 10.1046/j.1442-9993.2000.01036.x

Kirby, M. 2011. Irrigation. In *Water*, edited by I. Prosser, p. 192. CSIRO Publishing, Canberra, ACT.

Kretschmer, P. 2017. *Managed Aquifer Recharge Schemes in the Adelaide Metropolitan Area*. Government of South Australia, Department of Environment, Water and Natural Resources, Adelaide, SA.

Langford-Smith, T. and J. Rutherford 1966. *Water and Land. Two Case Studies in Irrigation*. Australian National University Press, Canberra, ACT.

Langton, M. 2002. Freshwater. In *Background Briefing Papers: Indigenous Rights to Waters Lingiari Foundation*, edited by Lingiari Foundation, Lingiari Foundation, Broome, WA.

Mason-Cox, M. 1994, *Rivers and Water Supply Commission Lifeblood of a colony: A history of irrigation in Tasmania*. Rivers and Water Supply Commission, Hobart, TAS.

McCrea, A.F. and M.R. Rivers. 2003. "Sustainable Irrigation - a collective effort for regional development." Proceedings of the International Conference of the Network of Regional Governments for Sustainable Development, Fremantle, WA.

MLDRIN (Murray Lower Darling Rivers Indigenous Nations), NBAN (Northern Basin Aboriginal Nations) and NAILSMA (North Australian Indigenous Land and Sea Management Alliance). 2014. Cultural flows; Literature Review. A guide prepared for the Cultural Flows Planning and Research

Committee as part of the National Cultural Flows Research Project. http//culturalflows.com.au/. Accessed Nov 10 2020

MLDRIN (Murray Lower Darling Rivers Indigenous Nations), NBAN (Northern Basin Aboriginal Nations) and NAILSMA (North Australian Indigenous Land and Sea Management Alliance). 2020. National Cultural Flows Research Project. http://culturalflows.com.au/ Accessed Nov 10 2020

Murray Darling Basin Commission, MDBC 2000. *Basin Review of the Operation of the Cap-Overview Report of the Murray Darling Basin Commission*. Murray-Darling Basin Commission, Canberra, ACT.

Murray–Darling Basin Authority. 2019. Climate change and the Murray–Darling Basin Plan. Canberra. CC BY 4.0, Australia.

Paydar, Z., Cook, F., Xevi, E., and K.L. Bristow 2011. An overview of irrigation mosaics. *Irrigation and Drainage* 60 (4), 454–463. doi: 10.1002/ird.600

Qureshi, M.E., Wegener, M. K., Bristow, K.L., and S.R. Harrison 2001. Economic evaluation of alternative irrigation systems for sugarcane in the Burdekin delta in north Queensland, Australia. *WIT Transactions on Ecology and the Environment* 48, 11. doi: 10.2495/WRM010041.

Ritchie, H. 2017. Water Use and Stress. https://ourworldindata.org/. Accessed Oct 20 2020.

Roberts, J. and M. Henneveld. 2003. Grower takeover of irrigation schemes in Western Australia: A quiet revolution. In *Proceedings of the Australian Water Association. 20th Convention-Ozwater Innovations in Water Conference, Perth, WA, 6th–10th April*, Australia.

Rose, D.B. and Australian Heritage Commission. 1996. *Nourishing Terrains: Australian Aboriginal Views of Landscape and Wilderness*. Australian Heritage Commission, Canberra, ACT.

Ross, A. 2018. Speeding the transition towards integrated groundwater and surface water management in Australia. *Journal of Hydrology* 567, e1–e10. doi: 10.1016/j.jhydrol.2017.01.037.

Skinner, D. and J. Langford 2013. Legislating for sustainable basin management: The story of Australia's Water Act (2007). *Water Policy* 15, 871–894.

Sunwater. 2020. Our History: Delivering Water for Prosperity. Sunwater. https://www.sunwater.com.au/. Accessed Oct. 30 2020.

Sustainable Australia Winegrowing, 2019. Webpage. https://mclarenvale.info/industry-development/sustainable-australia-winegrowing. Accessed Sept. 2020.

Tan, P. and S. Jackson 2013. Impossible dreaming - Does Australia's water law and policy fulfil Indigenous aspirations? *Environmental and Planning Law Journal* 30, 132–149.

The Lead. 2019. Dam Big Boost for McLaren Vale irrigators. Adelaide. August 27, 2019 http://theleadsouthaustralia.com.au/industries/primary-industries/dam-big-boost-for-mclaren-vale-irrigators/. Accessed Nov. 2020.

Thomas, R. 2021 Photo. https://en.wikipedia.org/wiki/Gascoyne_River. Accessed Nov 10 2020.

Ticehurst J.L. and A.L. Curtis 2019. Assessing Conjunctive Use Opportunities with Stakeholders in the Murray-Darling Basin, Australia. *Journal of Water Resources Planning and Management* 145 (5), 05019008. doi: 10.1061/(ASCE)WR.1943-5452.0001069.

Water and Rivers Commission. 2000. *Western Australia Water Assessment 2000- Water Availability and Use*. Water and Rivers Commission Policy and Planning Division, Perth, WA.

Water Education Foundation. 2021. Conjunctive Use. Aquapedia Background. Sacramento, California. USA. https://www.watereducation.org/aquapedia/conjunctive-use. Accessed Jan. 2021.

World Bank. 2021. Agricultural Irrigated Land (% of Total Agricultural Lands). https://tradingeconomics.com/australia/agricultural-irrigated-land-percent-of-total-agricultural-land-wb-data.html. Accessed Feb 2021.

https://web.archive.org/web/20061123113332/http://www.watercorporation.com.au/D/dams_all.cfm Accessed Oct 2020.

https://en.wikipedia.org/wiki/List_of_dams_and_reservoirs_in_Australia Accessed Oct 2020

2

Irrigation Developments in Australia: Irrigation and Agricultural Production

Neil A. Coles
University of Leeds
The University of
Western Australia.

Jeff Camkin
The University of
Western Australia

The words of former Prime Minister Alfred Deakin, spoken at a conference of 'irrigationists' in 1890, are still, if not more, relevant to irrigation in Australia today:

DOI: 10.1201/9781003353928-3

It is not the quantity of water applied to a crop; it is the quantity of intelligence applied which determines the result – there is more due to intelligence than water in every case.

Alfred Deakin (1856–1919)

2.1 Introduction

Water availability in Australia, while always subject to climate variability, has in recent decades become increasingly unreliable as global heating has altered rainfall distribution patterns, delivery frequency and quantity, which has affected water security, cost and supply reliability, and promoted the need to rethink the allocation of this scarce resource (Kirby, 2011; Coles, 2017). Shifts in rainfall patterns now being experienced are having an increasing impact on the quantity and quality of water available within the established irrigation regions. Tighter government controls, new water legislation, improved monitoring and the creation of markets for trading in water entitlements ('water markets'), since the early 1990s, have pushed irrigators and the water sector to reform and rationalise, thus promoting more intensive competition for available water resources (Coles, 2015). A food production system worth more than A$17.7B (Australian Bureau of Statistics, 2020a) now finds itself without the luxury of new water sources, which it has relied upon in the past, and is also faced with growing pressure from urban, cultural and environment demands for this diminishing resource.

While irrigation can bring many benefits to individuals, communities and regions, it can also lead to unwanted and often unintended environmental consequences. The impacts may be local, such as impacts on local water availability or quality, or regional, such as impacts on downstream river basins and coastal zones. As a specific example, the intensification of agriculture often results in increased use of pesticides and fertilisers, which runoff with rainfall or over-irrigation, or percolate through the soil, or move with drainage water, and can result in pollution of groundwater, surface waters and the near-shore marine environment. Runoff water can carry sediment, animal waste and other soil surface pollutants into surface water, which may be used further downstream for irrigation or other purposes, or it impacts upon ecosystem functions and services. The cumulative effect of these impacts may impair the long-term sustainability of an irrigation project and other economic and non-economic activities in the surrounding area, reducing its acceptability to local and regional communities. Addressing these issues will require Australia and many other countries where similar problems occur, to implement improvements in: irrigation practices, innovative water management and governance, and enhanced automated soil, water and environmental monitoring techniques (Rivers et al., 2015; Zia et al., 2013).

As with all water users in Australia, irrigated agriculture must inevitably focus on becoming more water use efficient productively, requiring targeted soil and plant research, technological solutions, and their relatively rapid adoption by the industry and the communities they support. In this chapter, agricultural produce delivered by irrigated agriculture in Australia is examined and the relationship with the governance, management and protection of Australia's water resources is discussed.

2.2 Irrigation and Agricultural Production

Given Australia's large deserts and irregular rainfall, irrigation is a necessary requirement to sustain large-scale agricultural production. There has been a long history of using irrigation to augment production and deliver high-value crops or pastures that significantly contribute to the economy, although the management of the Australian environment, soils and water resources that deliver this bounty has not always been beneficial for these resources (Batten, 2003; Kirby, 2011).

2.2.1 Agricultural Production

As the land used for irrigation rapidly increased through the 20th century, there was a commensurate increase in the Gross Value of Irrigated Agricultural Production (GVIAP), coupled with a diversification

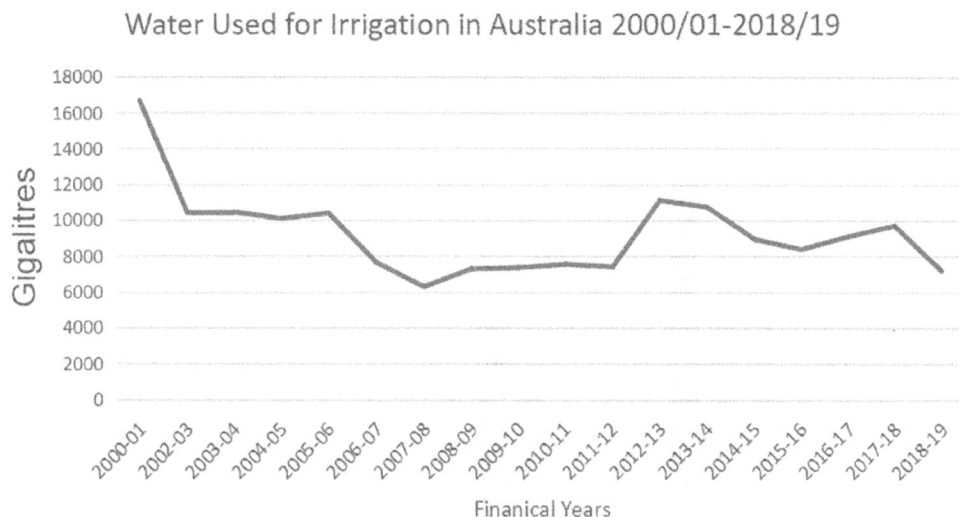

FIGURE 2.1 The volume of water used for irrigation across Australia from 2000–2001 to 2018–2019 financial years. Note the steep decline for drought years 2001–2002 to 2008–2008. With a similar decline experienced in 2017–2018 to 2018–2019 as the next drought phase became more widespread across the eastern States (Australian Bureau of Statistics, 2020b).

in the crops, animal production, horticulture and viticulture associated with these areas. The disparity between rainfed agriculture and irrigated production is amply demonstrated in the value of the produce; using ~0.5% of Australia's agricultural lands, irrigation delivers 23% of the total gross value of agricultural production (GVAP) and 50% of the profit at full equity (Creighton et al., 2004; Kirby, 2011).

These trends continued into the 21st century, with irrigated agriculture contributing 27%–35% of GVAP between 2000 and 2019 (Australian Bureau of Statistics, 2020a). Most notably, the amount of water used in irrigated agriculture has declined over the same period (from 16,600 GL in 2000–2001 to 7,200 GL in 2018–2019), averaging 8,822 GL (Figure 2.1), while the value in real terms increased from A\$9.6B (2000–2001) to A\$17.7B (2017–2018)[1] (Australian Bureau of Statistics, 2020b). The fall in water use has been attributed to several factors: the millennium drought (1996–2009) followed by the drought years 2016–2019, reduced water allocations under revised licencing provisions and the reallocation of water to meet new environmental flow requirements since 2007. Irrigators have also increased water efficiency in line with new legislation, have responded to water scarcity and have modified their crop selection, to maintain productivity (Figure 2.2) (Australian Bureau of Statistics, 2020b). Water use (Figure 2.3) and GVIAP for each state is given for the financial years 2009–2010 to 2011–2012 (Figure 2.4).

Table 2.1 provides a summary of water use on Australian farms for the 2015–2016 financial year. Irrigation is shown as the major water user across Australia. The four highest water using states, NSW, Victoria and QLD, followed by South Australia – all Murray–Darling Basin (MDB) states – consumed nearly 60% of all waters allocated for irrigation in Australia. As noted earlier, water allocation can vary quite markedly. A total of 16,000 GL was made available for irrigation in 2000, prior to the beginning of the millennium drought, the introduction of new water legislation to enforce the CAP, specific provisions for environmental flows and expansion of the water trading market. Water allocations have been maintained at between 6 and 11,000 GL since that time, in part due to a reduction in water allocations and a national government buy-back scheme, and in part due to climate change and recurring droughts or low rainfall years within the MDB.

The type and value of crops grown in the irrigated areas of the different states of Australia, and the volume of water used to grow them, are given in Figures 2.5 and 2.6 for two financial years (2017–2019). The highest water users are rice and cotton, which are mainly flood irrigated, along with sugar cane, and this is reflected

GVIAP of selected commodities

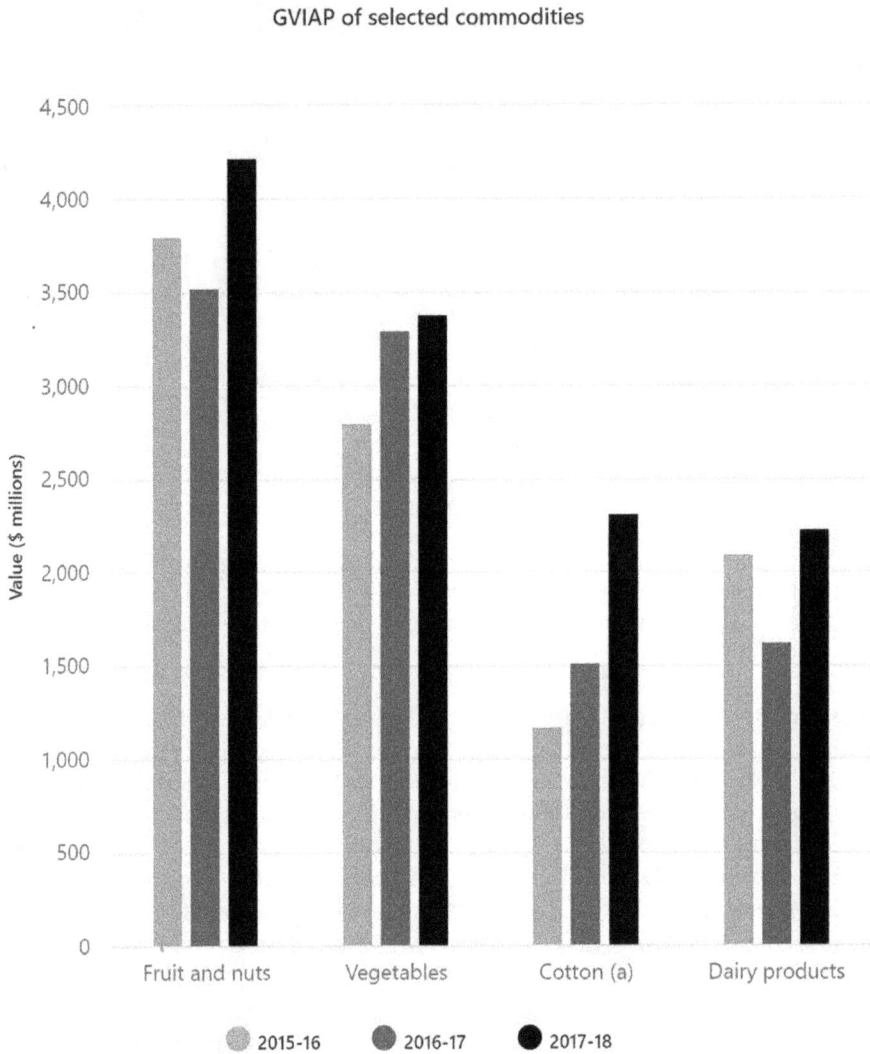

FIGURE 2.2 Value of key irrigated crops and industries produced in Australia for the three financial years between 2015–2016 and 2017–2018. Value varies year on year and reflects market and water scarcity conditions (Australian Bureau of Statistics, 2020b).

in the average amount of water applied per hectare, ranging from 9 to 12.5ML/ha for rice and around 4–4.3 ML/ha for grapevines and vegetables. Low-water-demand horticulture and viticulture crops such as fruits, nuts, grapes and vegetables are typically watered with sprinklers, drip irrigation and other lower water use methods, which accounts for their relative low water use to crop value ratio (Figures 2.2 and 2.6).

2.3 Managing the Water – Governance and Legislation

Water supply scarcity issues in Australia are historically linked to its arid climate and were managed for thousands of years by indigenous communities in harmony with the natural cycles of wet and dry. However, since European settlement, water reliability problems have been a significant driver of change and adaption, if not exploitation. The first colonial instance of 'a lack of reliable water supply' traced back to January 1788, when the new governor of the colony of Australia, Governor Philip, relocated

Volume of Water Applied, 2009-10 to 2011-12

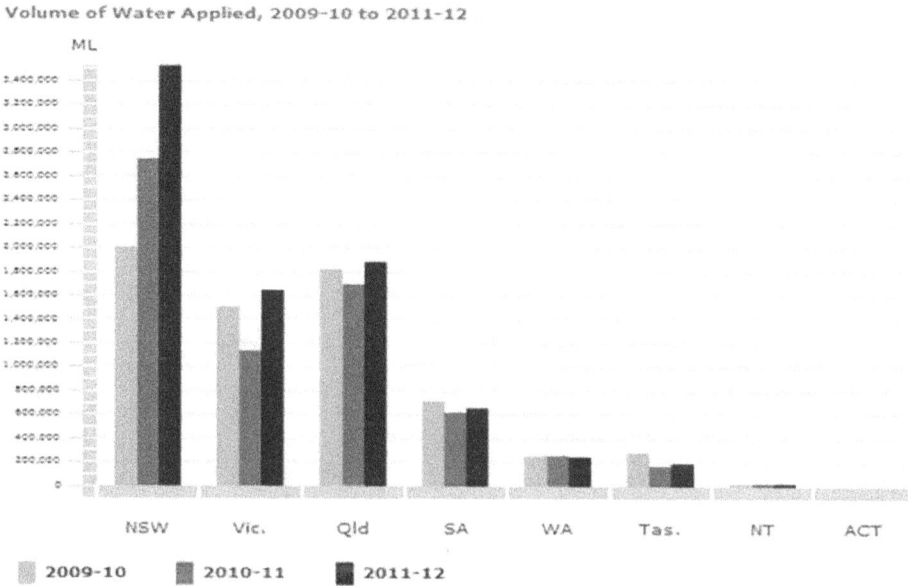

FIGURE 2.3 Volume of water applied for irrigated agriculture for each state and territory between 2009 and 2012. This shows a significant increase in water availability in NSW as the drought eased in 2009.

Gross Value of Irrigated Agricultural Production, 2009-10 to 2011-12

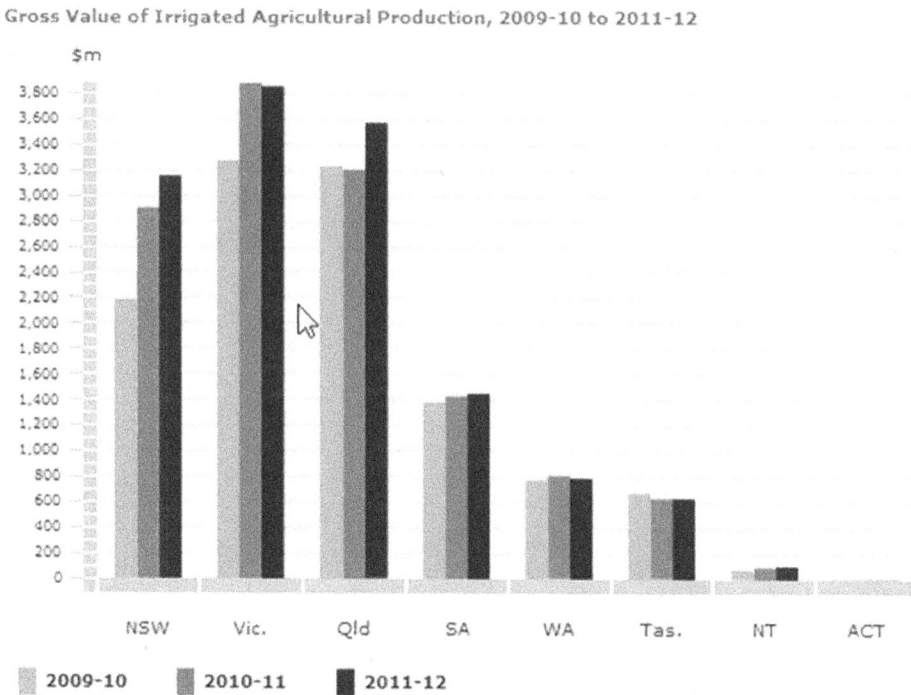

FIGURE 2.4 Variation in GVIAP for each state for the period 2009–2012. It shows a generally increasing trend in value, although the water applied for irrigation in each state and territory varies year on year for the same period. Values reflect shifts in crop types, livestock and total area irrigated (Figure 2.6).

TABLE 2.1　Water Use on Australian Farms for the Financial Year Ended 30 June 2016

	Aust.	NSW	VIC	QLD	SA	WA	TAS	NT	ACT	MDB[a]
				Agricultural Water Use						
Agricultural businesses (× 100 ML)	85.7	26.1	20.8	18.2	9.5	8.4	2.3	0.4	0.0	35.5
Agricultural businesses irrigating (`000)	22.7	5.3	6.0	5.4	3.1	1.5	1.2	0.2	0.0	9.2
Total water use (`000 ML)[b]	9,157.3	2,805.3	2,095.0	2,646.1	858.8	372.6	332.1	47.0	0.3	5,209.9
Water applied for irrigation (`000 ML)[c]	8,381.4	2,610.9	1,946.1	2,433.5	777.8	287.5	308.7	16.9	0.1	4,938.4
Water applied for other agricultural purposes (`000 ML)[d]	775.9	194.5	148.8	212.6	81.0	85.1	23.5	30.1	0.3	271.6
Change in total water use from 2014 to 2015 (%)	−2.5	−15.5	−10.2	11.3	15.8	16.7	36.6	−21.9	32.5	−12.6
				Sources of Agricultural Water						
Irrigation channels or pipelines (`000 ML)	3,096.2	965.7	1,067.7	789.2	117.9	101.2	53.2	1.3	0.0	2,087.6
On-farm dams or tanks (`000 ML)	980.1	228.6	103.1	410.5	20.1	87.6	126.4	3.6	0.2	438.5
Rivers, creeks or lakes (`000 ML)	2,412.5	885.2	540.0	610.1	231.4	14.1	124.0	7.5	0.1	1,656.4
Groundwater (`000 ML)	2,357.2	675.5	292.3	761.7	432.3	141.6	19.3	34.4	0.0	926.2
Recycled/re-used from off-farm (`000 ML)	161.0	23.3	49.9	57.3	11.3	15.6	3.6	0.0	0.0	56.2
Town or country reticulated mains supply (`000 ML)	126.5	21.3	37.5	8.8	43.6	11.8	3.4	0.1	0.0	35.6
Other water sources (`000 ML)	23.8	5.7	4.5	8.4	2.2	0.7	2.2	0.0	0.0	9.5

Source: Australian Bureau of Statistics (2020b).

[a] Murray–Darling Basin (MDB).

[b] Includes water applied for irrigation and other agricultural purposes.

[c] Includes water applied to pastures and crops.

[d] Includes livestock drinking water, dairy or piggery cleaning, etc.

Value of crops

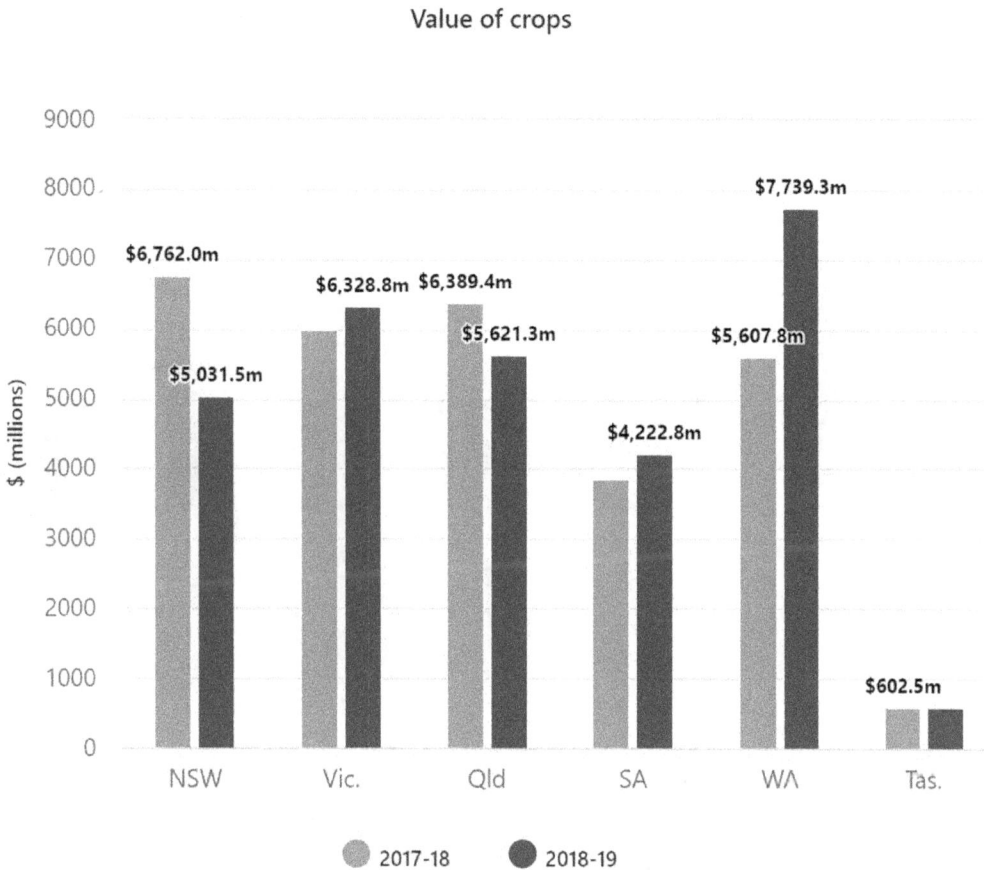

FIGURE 2.5 Comparison of the gross value of irrigated crops grown in each Australian state in 2017–2018 and 2018–2019 financial years.

the first settlement from Botany Bay to Port Jackson within the first week of his arrival from England (Hubert and Patrick, 2000).

Since then, issues surrounding the supply, management and equitable distribution of water have been a recurring theme. In the 100 years or so between settlement and 1896, access to water in NSW (as in the rest of Australia at the time) was based on British common law riparian rights. Wade, in his 1909 speech before the Sydney University Engineering Society, proclaimed that:

"it is one of the ironies of fate" that the "(riparian rights) Common law of Great Britain"

was (by default) the instrument of choice for the water law framework for a country as arid and ephemeral as Australia[2] (Wade, 1909). The difference in water availability, reliability, quality and management between the two countries was and remains stark (Coles, 2016). Consequently, water resource issues in Australia remains a priority for all governments, is increasingly a concern for the Australian community, and has continued to be at the forefront of environmental protection and sustainability legislation to enhance long-term security in the face of increasing demand, reducing water availability and quality, and changing community values (Coles, 2016).

2.3.1 Parallel Developments in Water Governance

Since the 1880s, when large-scale irrigation began at Renmark and Mildura, governments have stimulated the development of water delivery schemes, and have developed water law in parallel to these developments

Water application rate for selected crops and pastures

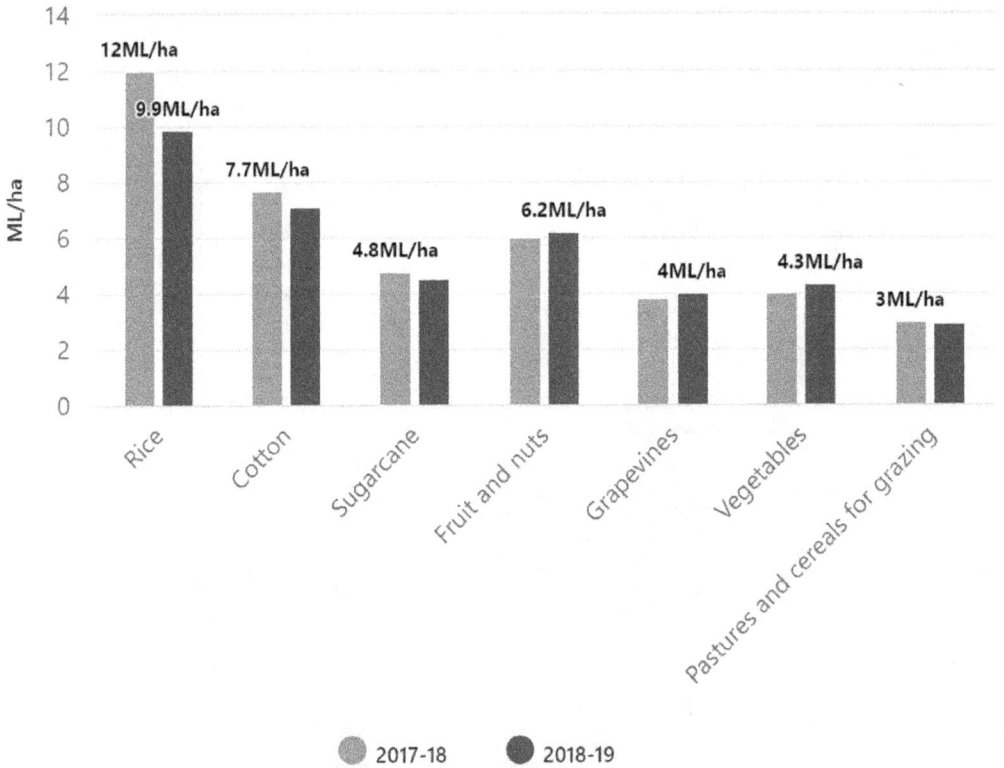

FIGURE 2.6 Comparison of the water use per ha for irrigated crops grown in each Australian state in 2017–2018 and 2018–2019 financial years.

(see Table 2.2). The Federation of States that formed the Commonwealth of Australia in 1901 created a federal government in which states are represented but retain substantial powers of self-governance. This continues to be a challenging environment in which to devise and implement coordinated policy to successfully manage the nation's water resources and thus guarantee water security for irrigators, communities and the environment (Coles, 2016). The River Murray Waters Agreement of 1914,[3] which aimed to address some of the issues raised by downstream states regarding river flows, was operational for 70 years and was notable in that it was the first intergovernmental water agreement in Australia (Kneebone and Brakell, 2014).

There were several other intergovernmental agreements on water management[4] during the mid-1900s (Table 2.2), but they tended to be bilateral or trilateral agreements between two or three basin states and/or the Commonwealth Government (Kildea and Williams, 2010). However, interjurisdictional management and allocation of the waters for the environment, agricultural development and human consumption within the MDB remained largely unresolved for the next 120 years (Coles, 2016).

This was often complicated by extended periods of low rainfall (droughts) leading inevitably to low flows, soil moisture deficits and loss of crops and associated incomes. Over the last 170 years these have been exacerbated by irrigation expansion, over-allocation, population growth and climate change (Coles, 2019). From the 1970s, the declining health of the rivers in the MDB has been a cause for increasing concern for both state and national governments, reflecting widespread community concern across Australia and even internationally.

TABLE 2.2 Timeline for the Introduction of Water Law in Each State and Federally to Manage Water Allocation for Irrigation and the Environment in Australia

Year	State	Legislation
1886–1887	Victoria	*Irrigation Act 1886 (Vic) – Amended 1889* Allowed for the creation of irrigation trusts rationalisation of all surface water resources. 90 trusts in existence by year of Federal 1901
1886	Victoria	*The River Goulburn Weir Act 1886[2]*
1887	Victoria	*The Water Supply Loans Act 1887*
1898	TAS	*Clyde Water Act (1898)*
1900	Australia	*The Constitution of the Commonwealth of Australia* (Section 100 – The Commonwealth shall not, by any law or regulation of trade or commerce, abridge the right of a State or of the residents therein to the reasonable use of waters of rivers for conservation or irrigation.)
1905	Victoria	*Water Act 1905* Established the Victorian State Rivers and Water Supply Commission (SRWSC) with oversight for surveys of available water resources and planning for their storage and utilisation. Oversight of local water, sewerage, and irrigation authorities
1906	NSW	*Burrinjuck and Murrumbidgee Construction Act 1906* *Barren Jack and Murrumbidgee Canals Construction Act 1906 (Cth)* Authorised the construction of the Burrinjuck Dam on the Murrumbidgee River near Yass (MDB)
1909	Victoria	*Water Act of 1909* Statutory responsibility for construction of 'national work' transferred to SRWSC
1912–1941	NSW	*Irrigation Act, 1912* Abolished the Murrumbidgee Irrigation Trust and create post of the Commissioner for Water Conservation and Irrigation with broad reaching powers under the Act. Later amended in 1916 and 1926 to account for expanded developments, soldier settlers' programme, larger farm sizes and financial considerations *Irrigation (Amendment) Act, 1916* *Irrigation (Amendment) Act, 1926* *Irrigation Amendment Act, 1931* Previous legislation provided for remissions of rent or water rates, but not of other forms of indebtedness *Irrigation (Amendment) Act, 1941*
1913	NSW	*Water Conservation and Irrigation Commission (NSW)* **created**
1914	WA	*Rights in Water and Irrigation Act 1914*
1914	Australia	*River Murray Waters Agreement (RMWA) 1914* RMWA (1914) was negotiated by the Commonwealth, New South Wales, South Australia and Victoria, and implemented by legislation in each jurisdiction in 1915. The RMWA addressed issues raised by downstream States in regard to river flows, was operational for 70 years, and is notable in that it was the first intergovernmental water agreement in Australia
1924	NSW	*Irrigation Holdings (Freehold) Act, 1924*
1932	NSW	*Water Conservation and Irrigation Advisory Committee Act, 1932* Create the WCIA committee to oversee irrigation in NW and in the Murrumbidgee Irrigation Area

(Continued)

TABLE 2.2 Continued Timeline for the Introduction of Water Law in Each State and Federally to Manage Water Allocation for Irrigation and the Environment in Australia

Year	State	Legislation
1942	VIC	*Water Act 1942* *Enabled the Commission to control cultivation of land subject to drift near water channels*
1957	TAS	*Water Act (1957)* *Established the rivers and water supply commission of Tasmania giving it extensive powers over the regulation and use of Tasmanian water resources*
1960	WA	*Northern Development (Ord River) Act 1960*
1969	VIC	*Groundwater Act 1969* *Invested in the SRWSC responsibility for the utilisation and conservation of groundwater in Victoria*
1970	NSW	*Clean Waters Act 1970 (NSW)* *Directed at the prohibition of pollution of waters except where such pollution is managed by way of a licence, although ultimately it creates the offence of the pollution of waters*
1973	NSW	*Private Irrigation Districts Act 1973 (NSW)*
1974	NSW	*Irrigation (Amendment) Act 1974 (NSW)*
1975	VIC	*Water Resources Act 1975* *Established a Ministry of Water Resources and Water Supply*
1976	NSW	*Water Resources Commission (NSW)* **replaces the Water** *Conservation and Irrigation Commission (NSW)* **Granted additional powers to manage groundwater resources**
1993	Australia	*Murray–Darling Basin Act 1993 (Cth)* *Enacted the Murray–Darling Basin Agreement – The MBDA replaced the 1915 River Murray Waters Agreement (RMWA)*
1994	NSW	*Irrigation Corporations Act 1994 (NSW)* *A consequence of this Act was the creation of privatised irrigation corporations through the transfer of ownership and operations of the irrigation areas and irrigation districts in NSW*
1995	NSW	*Department of Land and Water Conservation (DLWC)* **Authorised to manage irrigation schemes and water markets**
1999	TAS	*Water Management Act (1999)* *Retained the Water Commission as a government agency, giving it ownership, management and control over the operation of public irrigation schemes in Tasmania*
2000	NSW	*Water Management Act 2000 (NSW)*
	QLD	*Water Act 2000 (Qld)*
2007	WA	*Water Resources Legislation Amendment Act 2007 (WA)* *Implementation of water law reforms in WA to meet the state's obligations the Implementation Plan for the National Water Initiative signed in April 2006*
2007	Australia	*Water Act 2007 (Cth)* *Promote and coordinate effective planning and management for the equitable, efficient and sustainable use of the water and other natural resources of the Murray–Darling Basin, including by implementing arrangements agreed between the Contracting Governments to give effect to the Basin Plan, the Water Act and state water entitlements* *Also paved the way for the proclamation of the Murray–Darling Basin Plan (MDBP) 2012*

Note this is not a comprehensive list but provides some key examples of key legislation.

2.3.2 Realisation of a National Water Policy

Options were put forward to introduce new legislation to manage water allocation, quality and flow in the mid- to late 1970s, although at this time environmental flow[5] was not given due consideration (Coles, 2017). Most of the water and environmental policy reforms of note for the MDB in the last 50 years began with the implementation of the NSW Clean Waters[6] Act[7] in 1970, along with the introduction of other legislation, for example, Water *(Amendment)* Act[8] 1971.

While there had been changes in water and environmental laws across all states (Table 2.2), towards the end of the 20th century there was emerging weaknesses in the regulatory regime (Coles, 2017). While attempting reforms, state governments were still granting a substantial number of new water licences to irrigators and others – in spite of the decline in available water flows and rainfall – with generous extraction allocations attached (Holley and Sinclair, 2013). The expansion of irrigation, discussed in the previous sections, led to the situation where 'waters' in the MDB were considered over-allocated, and water was required to be re-allocated to the environment to recover the ecological functioning of the river system. In response, The Council of Australian Governments (COAG) ushered in major reforms in 1994[9] focused on the environment and water quality, pricing and cost recovery, water allocations and trading, public consultation and institutional reform. These initiatives and reforms have accelerated, and refocused, in the last two decades as Australia struggled with a decade-long drought, climate change and socio-economic structural shifts, particularly in the eastern states (Coles, 2015).

Further water reforms were enacted through the Intergovernmental Agreement on a National Water Initiative (NWI) 2004.[10] These initiatives aimed at improving the economic efficiency of Australia's water management while protecting water resources and the environment. Overlaying these physical aspects of change were changing socio-economic considerations and the wide-ranging political views that were to shape the government's response and water policy in Australia. This approach has driven important changes in water law in Australia, increased the level of monitoring, management, licencing and trading in water entitlements, with all irrigators requiring authorisation to extract specified amounts of water from rivers or bores (groundwater) or from irrigation supply systems.

2.3.3 A National Plan for Water Security (NPWS)

In January 2007, the Australian Government announced a 10-year, Aus\$10 billion NPWS (Australian Government, 2007), followed by the National Partnership Agreement on Water for the Future,[11] largely in response to the decade-long millennium drought. Furthermore, in 2007, the Commonwealth took overall leadership for water management in the MDB through a new *Water Act 2007* (Cth), replacing the MDB Agreement and repealing the *Murray–Darling Basin Act 1993 (Cth)*. The *Water Act* introduced new powers that help the Australian Government coordinate a national approach to water management outlined in the National Partnership Agreement, and also meet the challenges facing water management in the MDB (Murray–Darling Basin Authority, 2011) These actions also addressed the findings of the National Land and Water Resources Audit (2001), which revealed that less than one-third of the river systems in Australia were being monitored and the *Murray–Darling Basin Act* was failing to deliver promised benefits and reforms within and between the basin states, particularly in South Australia and eastern Australia, where salinity problems became cause for increasing concern (Coles, 2016).

Another 5 years of intense, basin-wide discussion resulted in amendments to the *Water Act 2007* with the proclamation of the Murray–Darling Basin Plan (MDBP) 2012 (Table 2.1). As part of the implementation of the MDBP, billions of dollars have been spent over the past decade on irrigation modernisation within the basin. The purpose of the modernisation process was to increase water use efficiency and improve productivity. However, the states still retained the right to the control, use and flow of all water in rivers, lakes and aquifers, as well as all water occurring naturally on or below the surface of the ground within the state, provided it meets the national objectives set out in the MDBP (Coles, 2017).

This management and control were promoted through the establishment of the various Water and Irrigation Commissions or Departments under pre-existing legislation in each state and through amendments to state and territory water legislation, as required under the NWI (Table 2.1). These actions stemmed from the realisation that there was a need for a single, powerful and independent authority to coordinate and manage each state's water resources.

State authorities dominate the assessment and control of water resources because, under the Australian Constitution, primary responsibility for water management rests with individual state governments (Coles, 2015). Each state government has, consistent with the NWI, the Water Act, 2009 (Cth) and National Partnership Agreements (NPAs), revised and strengthened its legislation and introduce management reforms to provide greater security for increasingly scarce water resources (Table 2.2).

Through the MDBP (2012), the Commonwealth and basin states have been able to raise the awareness of the public, irrigators and industry on the value of water and its links to the environment from which it is obtained, and they continue to promote the need to sustainably manage Australia's scarce water resources and fragile environment. The loss of these resources through ineffective regulation would be inestimable and the impact multi-generational (Coles, 2019).

2.4 Northern Australia: Is this the Last Large-Scale New Irrigation Opportunity?

The MDB makes a critical contribution to Australia's food supply and to the nation's export earnings. However, while two-thirds of Australia's irrigation occurs in the basin, its future is challenged by climate change, return of water for the environment and an increasingly open water market (CSIRO, 2011). The development of northern Australia has been a policy ambition for over a century, and the region now faces some of the same problems as the MDB and other irrigated regions of Australia, with greater demands placed on water security, for both productivity and the environment.

The Northern Australian Irrigation Futures[12] project highlighted the importance of developing a system-wide understanding of the role and impact of irrigation in northern Australia. Conclusions drawn from the project (Box 2.1) suggested that decisions on the future of irrigation in northern Australia are about people and relationships, and no single 'sustainability framework' can hope to ensure sustainability. Dealing successfully with the complexity of developing irrigation in northern Australia will require decision-making and irrigation management systems that better utilise existing and emerging technologies and approaches to deliver long-term ecologically sustainable development (Land and Water Australia, 2008).

2.4.1 Northern Australia Sustainable Yields (NASY) Assessments

Northern Australia is renowned for its high rainfall, intact tropical environments and relatively low levels of development, with around two-thirds of Australia's runoff occurring in this region. However, the rainfall is highly seasonal, with intense monsoonal rains in 'summer' and little rain through the 'winter' (or dry season). The Water Resources programme of the northern Australia Water Futures Assessment was established to develop a better understanding of water availability in northern Australia to determine its development potential. Their findings were released in the final reports from the NASY project.

The NASY project provided the first consistent, robust and transparent assessment of current and likely future water availability across the three jurisdictions of northern Australia (i.e., WA, NT and QLD), including an assessment of possible future climate implications. The NASY project provided the science to help governments, industry and communities consider the environmental, social and economic aspects of the sustainable use and management of the water resources of northern Australia, under four different climate and development scenarios (Australian Government, 2020a) with the key findings given in Box 2.2.

Box 2.1   Twenty Take-Home Messages from the Northern Australia Irrigation Futures Project

1. Research processes which effectively contribute to the integration of science, policy and stakeholders are valued highly by a wide range of stakeholders.
2. The land and water resources of northern Australia are already being used and decisions are about redirecting these resources to different uses.
3. Generating localised short-term benefits from irrigation are 'easy'; delivering catchment-scale long-term sustainability is the challenge.
4. We need to develop the capacity to view, understand and manage northern Australia through a 'northern lens' which takes account of the national and international context.
5. Groundwater can be critical to base flow and maintenance of ecological function.
6. Water quality is as important as quantity, especially in meeting ecological needs.
7. Irrigated systems are complex systems, and we need to accept, understand and manage that complexity.
8. Water availability and storage needs for irrigation in event-driven tropical systems are poorly understood.
9. We need to ensure policies and management strategies make sense for event-driven tropical systems.
10. Irrigation must be preceded by catchment-scale salt and nutrient management plans to deliver on long-term sustainability objectives.
11. We must set and meet groundwater quantity (level) and quality targets in irrigated systems and adjust management practices to meet those targets.
12. 'Efficiency' is not the answer to everything; the aim is to build and maintain resilience in irrigated systems.
13. Irrigation and water management is an individual and collective responsibility.
14. There is growing interest in irrigation mosaics as an alternative approach to traditional large-scale contiguous irrigation systems.
15. Irrigation mosaics may have both negative and positive biophysical effects compared with more traditional systems, with a possible net positive impact.
16. Further research is required on the biophysical, ecological, social and economic advantages and disadvantages of irrigation mosaics.
17. Dealing with complexity, uncertainty and risk in irrigation decision-making emerges as a shared need and responsibility for catchment communities, proponents and governments.
18. Dealing successfully with the complexity of irrigation in northern Australia to achieve long-term ecologically sustainable development will require decision-making and irrigation management systems that better utilise existing and emerging technologies and approaches.

19. Implementing frameworks (including catchment knowledge platforms and ecologically sustainable development (ESD) component tree systems) which effectively integrate science, policy and stakeholders will support more comprehensive, transparent and consistent planning and decision-making.
20. Above all else, decisions about the future of irrigation in northern Australia are about people and their relationships with the environment.

Source: Land and Water Australia (2008).

Box 2.2   Key Findings of the NASY Project

- Despite popular perceptions that northern Australia has a surplus of water, the climate is extremely seasonal, and the landscape may be described as annually water limited.
- Northern Australia experiences high rainfall during the wet season, but most of this rain falls near the coast, where there are generally fewer opportunities for engineered storage or diversion. Annual variation in rainfall is high.
- Runoff follows a similar pattern to rainfall; potential dam sites located inland receive less water than coastal areas and suffer very high evaporation rates.
- In the near future, potential evapotranspiration is likely to increase while rainfall is likely to be similar to historical levels, which were generally drier than the last decade, especially in the west.
- Groundwater reserves may provide alternative water source; however, more information is needed to access future groundwater availability.
- Shallow aquifers rapidly fill during the wet season and drain through the dry season. Consequently, there is limited opportunity for enhanced aquifer recharge. There are few river reaches that flow year-round. Those that do are generally sustained by localised groundwater discharge, and have high cultural, social and ecological value.

Source: Australian Government (2020a).

2.4.2 Developing Northern Australia: A Policy Perspective

The 2015 'Our North Our Future: White Paper on Developing Northern Australia' laid out the policy framework to achieve the government's vision for the north. The White Paper considers that:

"With the right policies, success in the north will mean that within a few generations we can expect that there will be a sharp increase in the scale and breadth of activity" and that *"...the north will be an exemplar of sustainable development."*

Australian Government (2015)

This White Paper continues to set the priorities for unlocking the potential over 20 years. One aspect focuses on developing the north's water resources to contribute to building the northern workforce and improving the governance of northern Australia (Australian Government, 2020b). The Australian Government's Office of Northern Australia leads implementation of the northern agenda, in collaboration with 14 government departments and agencies of the three northern jurisdictions; with 45 of the 51 White Paper measures already delivered, the plan was considered to be on track towards full implementation by August 2020 (Australian Government, 2020b).

While many of the measures support the potential for irrigation development in northern Australia in some way, it is instructive to note that a summary of the measures implemented does not directly mention irrigation. This reflects an important shift from the historic development agenda when irrigation expansion was the over-riding focus of visions for 'nation-building' projects in northern Australia to a much broader vision of what is possible or probable in northern Australia's future.

But while recent progress reports send positive signals about the new, broader vision for northern Australia, it was only 15 months earlier, in May 2019, that the Australian Government issued a press release *CSIRO investigates 'Super food bowl' in the NT*, as part of the $15 million package of Northern Australia Water Resource Assessments. According to the press release:

> *The Liberal and Nationals government understands if you just add water it can help to transform rural communities and economies by expanding agricultural production …*
>
> *(Australian Government, 2019)*

Not necessarily a defining moment in Australian Politics, but it is reflective of the journey that successive governments have made in efforts to try to justify irrigated expansion in the north, despite evidence to the contrary.

2.4.3 Assessment of Irrigation Potential in the North

Ash and Watson (2018) explored past agricultural developments, including some that persist today and those that have failed, to determine critical factors in success or failure. Their aim was to identify where most effort should focus in supporting contemporary agricultural developments, albeit with a broadening understanding of the complexity of the issues facing this realisation. Among their key findings, the authors noted that aside from the climate and environmental associated with farming developments in northern Australia, the following challenges persisted:

- financing and investment planning,
- land tenure and property rights,
- land and operational management,
- skills development and training, and
- supply chains (Ash and Watson, 2018).

Mirroring the previous work carried out under the Northern Australia Irrigation Futures project, Ash and Watson (2018) noted that rapid growth, insufficient investment (both time and money) and inappropriate farming systems applied to the land and water resources resulted in unsustainable and uneconomic developments. They also pointed to two modern departures from the historical focus on developing northern Australia:

(i) the acknowledgement that development should not disadvantage indigenous people, that indigenous people have strong interests and rights in land and water resources and that these resources will be deployed to further indigenous economic development; and

(ii) assessing environmental impacts of more intensive development is more rigorous than in the past and the resources and timeframes required for these processes are often underestimated (Ash and Watson, 2018).

2.4.4 Cooperative Research and Investment

The Cooperative Research Centre for Developing Northern Australia (CRCNA) was established in 2015, with the aim to bring together industry, universities and other research bodies, regional development organisations, all northern jurisdictions, and international partners in a collaborative industry-led R&D venture. The CRCNA will assist businesses, governments and researchers identify opportunities for business and growth in the north to deliver evidence-based information to support and inform the development of northern Australia (CRCNA, 2020a). Key aims are to:

- De-risk the Northern Australian investment landscape.
- Deliver a coordinated approach to sector development.
- Inform appropriate supply-chain development and infrastructure planning across northern Australia.
- Deliver research, development and extension solutions with impact.
- Build the strategic research capability and develop the workforce skills of Northern Australians.

The key opportunities and challenges for irrigation in northern Australia identified in a recent CRCNA report are consistent with those identified through previous studies, including:

- water resources exist, but there are hydrological, ecological, social, cultural and practical constraints to their use;
- investment in new and existing infrastructure (irrigation, water, roads and rail, processing and storage, etc.) will be essential to maximising value, but that investment requires reduced risk; and
- high input costs are prohibitive for farming intensification through irrigation in at least some areas (CRCNA, 2020b).

2.5 The Future of Irrigation in Australia

While clearly focused on the MDB, the NPWS also gave emphasis to the future of irrigation in northern Australia. The plan noted that:

> *While we go about repairing damage and adapting to new conditions in the south, we must look to the north. We have important water resources and environmental assets there which must be sustained. However, there is also opportunity for further development of northern land and water resources, and we must understand how to do that wisely.*

The 2007 NPWS was not the first to pay attention to opportunities for further irrigation development, particularly in northern Australia – grand visions of 'nation-building' developments have often centred around the development of large irrigation schemes, with irrigation developments in the north variously described as the last frontier, the new frontier and the next frontier (Ash et al., 2017).

However, there is a note of caution provided by previous attempts to 'irrigate the north'. More than 100 years ago, for example, Australia's leaders envisaged a large dam in the vast, remote and largely unpopulated Kimberley region of north-western Australia. Seventy years later, the Argyle dam was built on the Ord River, creating the largest man-made reservoir on the dry Australian continent.

But visions of unlimited water for irrigation creating a new food bowl for Australia and the world have not eventuated. Instead, a different set of community values have emerged, based on maintaining a new hydrologic regime, new ecosystems (including Ramsar wetlands) and new opportunities such as recreational fishing and tourism, while still including an expansion of irrigated agriculture. The paradigm of water for development that existed over 100 years ago is now more balanced with other objectives, such as environmental conservation, recognising indigenous cultural heritage, eco-tourism and social equity (Camkin, 2011).

Increasing interest in agricultural development in northern Australia continues to be driven by identified opportunities such as the proximity to large, growing and increasingly prosperous Asian markets, increasing global demand for food and fibre, and the development of economically sustainable and vibrant communities as a policy objective. However, there are several key challenges facing expanding agricultural development in northern Australia, including:

- accessing suitable land and water resources to underpin expanded agricultural production.
- navigating the various approval processes associated with land tenure, Native Title, water resource plans, environmental impact, etc.;
- sourcing the significant capital investment required to support the high cost of 'greenfield' agricultural development;
- cost-effectively, reliably and sustainably growing agricultural products in the northern tropical environment and getting them to market via efficient supply chains; and
- establishing new and viable export markets for high-value, perishable fruit and vegetable products with high seasonality of supplies, and maintaining the ecological values of northern Australia (Ash et al., 2017).

As both the Northern Australia Irrigation Futures project and the NPWS noted, northern Australia holds an iconic status for many Australians. The interplay between the landscapes, rivers and strongly monsoonal weather patterns has resulted in unique and diverse ecological systems that will need special care to retain their integrity. At the same time, with some 60%–70% of Australia's freshwater discharging from tropical rivers, there are pressures from various quarters to extract some of the water, including for irrigated agriculture.

A unique and historic opportunity exists to ensure that management of Australia's northern water resources takes place within a strategic, ecologically, culturally and economically sustainable framework. Such a framework can help to build the basis for developing sustainable irrigation across tropical Australia, through the provision of new knowledge, tools and processes to support debate and decision-making regarding irrigation in northern Australia by understanding of river and catchment attributes and the risks and benefits associated with irrigation (Camkin et al., 2007).

Petheram et al. (2008) identified a range of implications for future design and management of tropical irrigation. The authors found that a significant challenge for irrigation development in northern Australia is to find crop varieties suited to the tropical environment. Factors compounding the question of 'what to grow' extend beyond plant physiology and include markets, weeds, pests and production costs.[13] Recent experience suggests that direct extrapolation of yield, physiology and phenology data from other regions is often unreliable. Crops and farming systems need to be evaluated at specific locations to assess true yield potential and risks, and to determine appropriate management practices before these are applied on a large commercial scale (Petheram et al., 2008).

2.6 Learning from the Past to Create a Sustainable Irrigation Future

In Australia, there now exists a widespread awareness to learn from the past, and from previous decisions that have degraded or degrading irrigation schemes and environments in which they operate. This increasing recognition of the need to view and manage these systems sustainably and to understand that many new developments, such as those proposed for the northern Australia, should be evaluated within the context of the complexity of the specific environment and cultural needs. A different approach to irrigation development is required than the previous 'nation building – development at all cost' view.

More importantly, irrigation planning in northern Australia, for example, must consider lessons from past experiences, and use data available from the north and other tropical regions of the world, in addition to the lessons learnt and ideas from southern Australian experiences. This new approach requires transparency and accountability to local communities, and resources need to be made available to deal with unexpected and/or unwanted problems that invariably occur some years after irrigation has been implemented. Early identification and rectification of emerging problems require vigilant and ongoing monitoring upstream, within and downstream of irrigation areas (Petheram et al., 2008).

As the Northern Australia Irrigation Futures project highlighted over a decade ago, the central message from rethinking irrigation development in Australia's north is clear – the development of irrigation in northern Australia, and of northern Australia itself, must be viewed through a 'northern lens'. The key question is not "How can we develop irrigation in northern Australia?" The key questions are: What is our vision for northern Australia? What role does irrigation have, if any, in that vision? And, if irrigation is part of the vision for northern Australia, what is the best way to go about it? (Camkin et al., 2008).

Only by looking through such a 'northern lens' can we hope to ensure that further irrigation development in Australia, particularly the north, is suited to the ecological, social, cultural and economic context to deliver the widest possible benefits. By adopting more progressive and inclusive attitudes towards cultural diversity, recognising and learning from indigenous culture in Australia, and acknowledging the prior knowledge of country and strong cultural relationship to water, we can create better consultative and management strategies for Australian landscapes (Tan and Jackson, 2013; Hemming et al., 2019).

2.7 Conclusions

Based on the discussions presented in the last two chapters, it is clear that, the future of irrigation in Australia will need one unifying thing more than everything else. Not data, not information, not knowledge – it will need wisdom. Australia needs the wisdom to confront difficult, sometimes unpalatable challenges. This endeavour and commitment to seek out and apply all the data, information and knowledge that is available, in all its forms, is commendable. In using this approach, it furthers the perceived need to use wisdom to accept and address even greater levels of certainty in regard to sustainable irrigation practices and food production. By adopting this pathway, then future generations can be assured that the important contribution irrigation makes to Australia can continue, and that it is delivered through a sustainable, cultural and environmentally equitable process and imbues the necessary climate resilience. By taking this broader perspective to further irrigation developments in Australia, we ensure that the lessons from previous experiences in the MDB and elsewhere are heeded.

Notes

1 Latest available figures from ABS in 2020.
2 NSW inherited the common law of riparian rights by virtue of the Australian Courts Act, 1828. This act established July 25, 1828 as the date for the introduction of British common law, including riparian rights, to the Colony of NSW.
3 RMWA (1914) was negotiated by the Commonwealth, New South Wales, South Australia and Victoria, and implemented by legislation in each jurisdiction in 1915.
4 These tended to cover the associated issues of water conservation, irrigation, water rights and water supply as these jurisdictions are at times inextricably linked.
5
  The concept of an environmental flow includes (but is not limited to): volume or water over some time base; velocity of water in channel; duration of flow events; water-level; natural and

human induced variation flows on an annual and longer time scale; need for pulses of high flows (e.g., to stimulate fish breeding); and the rate, of change of flow

Cullen, P. 1994.

A rational for environmental flows. Water and the Environment Newsletter No., 318 (Sep–Oct). www.nccnsw.org.au/glossary.

6 "Waters" means any river, stream, lake, lagoon, swamp, wetlands, unconfined surface water, natural or artificial watercourse, dam or tidal waters (including the sea), or part thereof, and includes water stored in artificial works, water in water mains, water pipes and water channels, and any underground or artesian water, or any part thereof.

7 NSW Clean Water Act 1970, No 78. An Act to make provisions with respect to the prevention or the reduction of pollution of certain waters; and for purposes connected therewith. The Act was repealed by Sch 3 to the Protection of the Environment Operations Act 1997, No 156 with effect from 01-07-1999. http://www.austlii.edu.au/au/legis/nsw/repealed_act/cwa1970131/.

8 Along with further Amendments to Act in each year between 1976–1981, 1983–1986 and 1988. http://www.legislation.nsw.gov.au/#/view/act/1912/44/history.

9 These were outlined in the *Water Reform Framework Agreement*. COAG meeting, February 1994, Communiqué, Attachment A.

10 The NWI included the following statements aimed at providing an approach to improving water resource management and allocation by: (i) improving the security of water rights – giving them effectively the same legal status as property rights – by creating a nationally compatible system of water entitlements providing perpetual access to a share of water resources available to irrigators (as opposed to a fixed volume); (ii) ensuring water is put to best use by creating, and encouraging trading in, a water market encompassing the' entirety of the MDB that allows participants to trade water rights both intrastate and interstate; (iii) restoring over-allocated river systems to environmentally sustainable levels; and (iv) encouraging water conservation in our cities, including better use of storm water and recycled water.

11 The National Partnership Agreement on Water for the Future will contribute to increased efficiency of rural water use, protect and improve the environmental health of freshwater and freshwater-dependent ecosystems, prepare communities for climate change, help secure water supplies for towns and cities and assist households and businesses to use water more efficiently. The agreement did not cover projects outlined in the Inter-Government Agreement on Murray Darling Basin Reform or the National Partnership Agreement on the Great Artesian Basin Sustainability Initiative.

12 The Northern Australia Irrigation Futures project was a collaboration between the Cooperative Research Centre for Irrigation Futures, CSIRO, Australian Government and the Governments of Queensland, Northern Territory and Western Australia between 2003 and 2010. It aimed to deliver a framework for use by policy makers, regulators, managers and investors to ensure irrigation is developed in a sustainable manner across northern Australia.

13 'Costs' included operational elements such as: fertilizer, energy for pumping water, transport, labour, distance to interstate and export markets and vulnerability to external factors (e.g. airline strike in the case of the Ord River Irrigation Area (ORIA)).

References

Ash, A. and Watson, I. 2018. Developing the north: Learning from the past to guide future plans and policies. *The Rangeland Journal*, 40, 301–314. https://doi.org/10.1071/RJ18034

Ash, A., Gleeson, T., Hall, M., Higgins, A., Hopwood, G., MacLeod, N., Paini, D., Poulton, P., Prestwidge, D., Webster, T., and Wilson, P. 2017. Irrigated agricultural development in northern Australia: Value-chain challenges and opportunities. *Agricultural Systems*, 155, 116–125.

Australian Bureau of Statistics, ABS. 2020a. *Value of Agricultural Commodities Produced, Australia*. Australian Government. Available at https://www.abs.gov.au/statistics/industry/agriculture/value-agricultural-commodities-produced-australia/2018-19. Accessed on 9 September 2020.

Australian Bureau of Statistics, ABS. 2020b. *Water Use on Australian Farms*: Canberra: Australian Government. Available at https://www.abs.gov.au/statistics/industry/agriculture/water-use-australian-farms/latest-release. Accessed on 9 September 2020.

Australian Government. 2007. A National Plan for Water Security. The Hon John Howard MP, Prime Minister of Australia. 25 January 2007.

Australian Government. 2015. Our North, Our Future: White Paper on Developing Northern Australia. Commonwealth of Australia.

Australian Government. 2019. CSIRO investigates 'Super food bowl' in the NT. The Hon Michael McCormack MP, Deputy Prime Minister, Minister for Infrastructure, Transport and Regional Development. Media release. 29 March 2019, Australia.

Australian Government. 2020a. Northern Australia Water Futures Assessment – Water Resources Program. Available at https://www.agriculture.gov.au/water/national/northern-australia/northern-australia-water-futures-assessment/water-resources-program. Accessed on 9 December 2020.

Australian Government. 2020b. Progress update: meeting measures on the Our North Our Future White Paper. Office of Northern Australia, Government of Australia. Available at https://www.industry.gov.au/policies-and-initiatives/developing-northern-australia. Accessed on 9 December 2020.

Batten, G., Katupitiva, A., and J. Pratley. 2003. *Irrigation management. In Principles of Field Crop Production*, edited by J. Pratley, pp. 418–462. Oxford University Press, Oxford.

Camkin, J.K. 2011, Adapting to changing hydrology, ecology and community attitudes to water at the Ord River, Northwestern Australia. *Journal of Hydrologic Environment*, 17(11), 17–30.

Camkin, J.K., Bristow, K.L., Petheram, C., Paydar, Z., Cook, F.J., and Story, J. 2008. Designs for the future: The role of sustainable irrigation in Northern Australia. *WIT Transactions on Ecology and the Environment*, 112, 293–302. https://doi.org/10.2495/SI080291

Camkin, J.K., Kellett, B.M., and Bristow, K.L. 2007. *Northern Australia Irrigation Futures: Origin, Evolution and Future Directions for the Development of a Sustainability Framework*. CSIRO Land and Water Science Report No.73/07, CRC for Irrigation Futures Tech. Report No. 11/07. 48 pp.

Coles, N. 2015. Water industry (law) reforms: The adoption of Australian drinking water guidelines in Western Australia — from targets to aspirations. *New Water Policy & Practice*, 1(2), 68–83. https://doi.org/10.18278/nwpp.1.2.6

Coles, N.A. 2016. From Development to Sustainability: The Shifting Focus of Water Law in NSW. Upublished Essay, Australia.

Coles, N.A. 2017. Fracking Australia, from Nirvana to Crisis Management? The Unconventional Gas Industry in Eastern Australia and the Development of the Associated Federal and State Water and Environmental Policy and Legislation. LLM, Thesis University of Dundee. Dundee, Scotland.

Coles, N.A. 2019. *Unconventional Gas: New Energy Nirvana or Environmental Menace?* Centre for Energy, Petroleum and Mineral Law and Policy, University of Dundee, Dundee, Scotland.

CRCNA. 2020a. Cooperative Research Centre for Developing Northern Australia Home Page. Available at https://crcna.com.au/. Accessed on 9 December 2020.

CRCNA. 2020b. *State of the North 2020*. Cooperative Research Centre for Developing Northern Australia. Available at https://crcna.com.au/. Accessed on 9 December 2020

Creighton, C., Meyer, W., and Khan, S. 2004. Farming and land stewardship: Case study - Australia's innovations in sustainable irrigation. In *Agronomy Australia Proceedings*, Australia.

CSIRO. 2011. Irrigation. In: *Water: Science and Solutions for Australia*, edited by I.P. Prosser. CSIRO Publishing, Collingwood, VIC.

Hemming, S, Rigney, D., Bignall, S., Berg, S., and Rigney, G. 2019. Indigenous nation building for environmental futures: Murrundi flows through Ngarrindjeri country. *Australasian Journal of Environmental Management*, 26(3), 216–235. https://doi.org/10.1080/14486563.2019.1651227

Holley, C. and Sinclair, D. 2013. Non-urban water metering policy: Water users' views on metering and metering upgrades in New South Wales, Australia. *Australasian Journal of Natural Resources Law and Policy*, 16(2), 101–131.

Hubert, C. and Patrick, D.J. 2000. One hundred years+of riparian legislation in New South Wales. *Australian Environmental Law News*, 3, 39–45.

Kildea, P. and Williams G. 2010. The constitution and the management of water in Australia's rivers. *Sydney Law Review*, 32, 595–616.

Kirby, M. 2011. Irrigation. In *Water*, edited by I. Prosser, p. 192. CSIRO Publishing, Canberra.

Kneebone, J. and Brakell, D. 2014. Water law: Water law in NSW: An overview. *LSJ: Law Society of NSW Journal*, 1, 79.

Land and Water Australia. 2008. Northern Australia Irrigation Futures – Final Report. Providing New Knowledge, Tools and Processes to Support Debate and Decision Making Regarding Irrigation in Northern Australia. National Program for Sustainable Irrigation CDS23- Final Report. CSIRO, LWA, NPSI, CRC IF, Australian Government, Queensland Government, Northern Territory Government and Government of Western Australia.

Murray–Darling Basin Authority. 2011. An overview of Australian An overview of Australian Government agencies, programs, roles and responsibilities Government agencies, programs, roles and responsibilities. Managing Australia'swater resources. Factsheet. 0./11. Available at https://www.mdba.gov.au/sites/default/files/pubs/FS_water_resources.pdf. Accessed 11 November 2020.

Murray-Darling Basin Plan (MDBP). 2012. Basin Plan 2012. *Water Act 2007* 44(3)(b)(i). https://www.legislation.gov.au/Details/F2021C01067 Accessed 11 November 2020.

National Land and Water Resources Audit, NLWRA. 2001. Australian Water resources Assessment 2000. *Surface water and groundwater - availability and quality*. Land & Water Australia. Commonwealth of Australia. Canberra.

Petheram, C., Tickell, S., O'Gara, F., Bristow, K.L., Smith, A., and Jolly, P. 2008. CRC for Irrigation Futures Technical Report No. 05/08. CSIRO Land and Water Science Report 19/08. February, Australia.

Rahman, A., Hajani, E., and Eslamian, S., 2017, Rainwater harvesting in arid regions of Australia. In *Handbook of Drought and Water Scarcity: Environmental Impacts and Analysis of Drought and Water Scarcity*, edited by S. Eslamian and F. Eslamian, pp. 489–500, Taylor and Francis, CRC Press, Boca Raton, FL.

Rivers, M., Coles, N.A., Zia, H., Harris, N. R., and Yates, R. 2015. How could sensor networks help with agricultural water management issues? Optimizing irrigation scheduling through networked soil-moisture sensors. In *2015 IEEE Sensors Applications Symposium (SAS)*, 13–15 April 2015.

Tan, P-L., and Jackson, S. 2013. Impossible dreaming - Does Australia's water law and policy fulfil Indigenous aspirations? *Environmental and Planning Law Journal* 30:132–149.

Wade, L.A.B. 1909. Irrigation from the Murrumbidgee River; New South Wales, Paper Read before the Sydney University Engineering Society. 17 November 1909, Australia.

Zia, H., Harris, N., Merrett, G., Cranny, A., Rivers, M., and Coles, N. 2013. A review on the impact of catchment-scale activities on water quality: A case for collaborative wireless sensor networks. *Journal of Computers and Electronics in Agriculture*, 96(96):126–138.

3

Optimisation of Twinged Stormwater Storage for Golf Course Irrigation: A Case Study in Sydney

Monzur A. Imteaz
Swinburne University of Technology

Abdallah Shanableh
University of Sharjah

Md. Maniruzzaman Bin A. Aziz
University of Technology Malaysia (UTM)

3.1 Introduction

Due to population growth as well as ever-increasing anthropogenic activities, demand for water has been increasing day by day. On the other hand, existing water sources are depleting day by day. Moreover, with the impact of climate change rainfalls in many regions are experiencing higher degree of variability. As a result, prime source of water through rainfall has become uncertain and often causes tremendous water stress to the community. In many cities, government authorities have been struggling to maintain sufficient water for the residents and other users. To overcome the stress, many authorities offer incentives and grants for different water saving ideas and innovations. One of several water conserving techniques is on-site stormwater harvesting for non-drinking purposes. However, there is a lack of knowledge on the actual cost-effectiveness and performance optimisation of any stormwater-harvesting system; in particular, the proposed design storage volume could be overestimated or underestimated.

The biggest limitation of stormwater-harvesting schemes is the rainfall variability, which will control the size of the storage needed and can't be based on mean annual rainfall data. In addition, it has been predicted that due to the effect of global warming, Australia's rainfall pattern will be changed significantly, which adds further complexity (Imteaz et al. 2021). A range of approaches is needed to address the water shortage including stormwater harvesting, which has become a viable additional source of water with government authorities, large institutions, community groups and industries. However, a proper in-depth understanding of the actual effectiveness of any proposed on-site stormwater-harvesting system including a life cycle costing analysis is often lacking. It is necessary to quantify the expected

DOI: 10.1201/9781003353928-4

amount of water that can be saved and used through any particular harvesting technique based on contributing catchment size, detention pool volume, geographic location, weather conditions and purpose of water use (Rahman et al., 2017).

To encourage more and more implementations of stormwater harvesting, different government authorities have been promoting and providing financial incentives to the end-users for installations of stormwater-harvesting systems, which will reuse stored stormwater and consequently will reduce the potable water demand. However, even with several educational and awareness campaigns and financial incentives, there is a general reluctance to adopt any potential stormwater-harvesting measure. The main reasons behind this are that people are not aware of the effectiveness of any potential stormwater-harvesting system and the optimum size of the storage required to satisfy their performance requirements. There have been several studies on the effectiveness of rainwater harvesting for domestic as well as commercial buildings around the world. Among the published studies on potential water savings, Shanableh et al. (2018) presented for the city of Sharjah (UAE), Moniruzzaman and Imteaz (2017) presented for the city of Sydney (Australia), Imteaz et al. (2016a) presented for the city of Adelaide (Australia), Imteaz et al. (2016b) presented for the city of Melbourne (Australia), Imteaz et al. (2014) presented for the city of Canberra (Australia) and Imteaz et al. (2012) presented for Nigeria. Some researchers converted water saving potentials to monetary values through economic analysis considering costs and potable water saving benefits. Imteaz and Moniruzzaman (2018) presented economic analysis and reasonable government rebate through single-household rainwater tank for the city of Sydney (Australia), Imteaz et al. (2017) presented economic benefits of residential rainwater tanks for the city of Kathmandu (Nepal), Gomez and Teixeira (2017) presented effect of incentives on economic feasibility of residential rainwater harvesting in Brazil, Karim et al. (2015) presented reliability and economic analysis of residential rainwater tanks for different cities of Bangladesh, Matos et al. (2015) presented economic analysis of large-scale commercial rainwater tank in Portugal, Rahman et al. (2012) presented economic benefits of rainwater tanks for multi-story residential buildings in Sydney (Australia) and Imteaz et al. (2011) presented detailed economic analysis with varying water prices through large-scale rainwater tanks from a university buildings in Melbourne (Australia). While doing costing analysis, some researchers calculated payback periods of rainwater tank under different scenarios (Karim et al., 2021; Paudel and Imteaz, 2019). Imteaz et al. (2021) presented a further study on payback periods through developing a generalised equation of payback periods for any roof and tank sizes for the city of Dhaka (Bangladesh).

Earlier, this sort of analysis was performed based on average annual rainfall amount. However, analysis using long-term average rainfall will not depict the real situation. In reality, the total annual rainfall is not evenly distributed over the year; rather, it is the sum of several random events of different magnitudes. An analysis using only average rainfall magnitude is unable to produce the effects of random rainfall events and magnitude. To overcome this drawback, it is proposed that an analysis using real daily rainfall data should be used. Imteaz et al. (2012) through a case study in Nigeria using daily rainfall data showed that analysis using monthly rainfall data provides erroneous results. A daily water balance modelling is most appropriate for such analysis, where daily rainfall amount is considered as input, whereas daily water usage and losses are considered as output. The difference between the input and output is the amount stored in the tank, until the tank becomes full. Once the tank is full, the subsequent runoff from the roof will be diverted (i.e. lost) as overflow. The calculations are performed with the constraints of tank volume and roof area.

All the above-mentioned studies and many other studies were conducted on rainwater harvesting from residential and/or commercial buildings (i.e. from building roofs). Study related to on-site rainwater harvesting from farms to facilitate irrigation of the crops/grass is scarce. However, it is obvious that through such on-site rainwater harvesting, crop water demand can be augmented to some extent. This paper investigates effectiveness of on-site stormwater-harvesting pond for the purpose of augmenting irrigation demand for golf course grasses with a case study for a golf course located in Guildford, Sydney (Australia).

3.2 Methodology

Basic methodology utilised in this study was a daily water balance modelling considering actual daily rainfall and evaporation data. A rainfall-runoff model based on the Australian Water Balance Model concept (Boughton, 1993) was developed. In the model, input values were the daily rainfall amount for a whole year. From the model, daily runoff from the selected contributing catchment was calculated considering all the significant losses. Generated runoff was diverted to selected storage volume for storage/use. Once available storage became full, excess runoff was diverted via spillway (which in fact lost from the available storage). The model was used for three different years (dry year, average year and wet year) to visualise storage situations under different climatic scenarios. Also, the model was used for a particular area in Western Sydney. Utilising developed model, sensitivity analysis was performed to predict availability of stormwater for grass irrigation for incremental catchment sizes and incremental storage volumes.

3.3 Study Area and Data

As a case study, a golf course located in Guildford (Western Sydney) was considered. Total catchment area (which was a variable for the sensitivity analysis) was assumed to be flowing to the pond/storage. While performing sensitivity of contributing catchment size, storage volume was assumed to be 1,800 m³. While performing sensitivity of storage volume, contributing catchment area was assumed to be 20 ha.

Historical (1945–2004) rainfall and evaporation data for the area were collected from the Bureau of Meteorology website (www.bom.gov.au). From historical rainfall data, three different calendar years were chosen as dry year, average year and wet year. These were selected based on first decile, fifth decile and ninth decile of annual rainfalls for the entire historical period. Table 3.1 shows the selected rainfall amounts and corresponding years. Table 3.2 shows the monthly evaporation losses used in the model for the area (Figure 3.1).

TABLE 3.1   Selected Rainfalls and Years

Decile	Annual Rainfall (mm)
First Decile (dry year)	630 (1980)
Fifth Decile (average year)	967.6 (1954)
Ninth Decile (wet year)	1,351.4 (1952)

TABLE 3.2   Monthly Evaporation Data

Month	Evaporation (mm)
January	171
February	140
March	124
April	93
May	65
June	54
July	59
August	84
September	111
October	143
November	159
December	185

FIGURE 3.1   Aerial photo of the study area.

3.4 Water Balance Modelling

The three annual sets of historic daily rainfalls (i.e. the 1952, 1954 and 1980) were separately used to calculate the daily runoff from the catchment. The daily water balance model accounted for infiltration loss, soil storage, evapo-transpiration loss and potential water harvesting. From the daily rainfall amount, daily losses were subtracted and the remaining rainfall excess value was multiplied with catchment area to calculate runoff generated from the contributing catchment area. Generated runoff was diverted to the available storage. Available storage capacity was compared with the accumulated runoff. If the accumulated runoff becomes bigger than available storage volume, excess water will be lost from the system. Evaporation loss from the water storage was also considered. The model calculated the daily water demands of different types of grasses (for tees, greens and fairways) within the golf course. Monthly crop factors are shown in Table 3.3. From calculated water demand and simulated available water, daily water deficits were calculated. If in a specific day, available water is less than the water demand, then daily water deficit is simply the difference between the water demand and available water. If available water is more than or equal to water demand, then daily water deficit is zero. After simulated runoff is directed towards the pond and storage builds up, any excess runoff in addition to pond volume will be diverted (spilled) and escaped from the system.

3.5 Results Analysis

The model was simulated for the whole calendar year (for three different years). Figure 3.2 shows the effect of storage volume on average daily available water volume in a dry year for a contributing catchment area of 20 ha. It is found that effect of storage volume becomes insignificant for storage volumes larger than 2,000 m^3. It is because in a dry year, even storage volume is higher; rainfall amounts are not enough to fill the storage. However, average available water volume increases with the increase of storage volume up to a volume of 2,000 m^3. However, this threshold storage volume will be different for different locations.

Figure 3.3 shows the effect of storage volume on average daily available water volume in an average year for the same contributing catchment area. It is found that for an average year, average daily

TABLE 3.3 Monthly Crop Factors

Month	Crop Factor
January	100
February	93
March	93
April	87
May	80
June	67
July	67
August	73
September	80
October	87
December	100

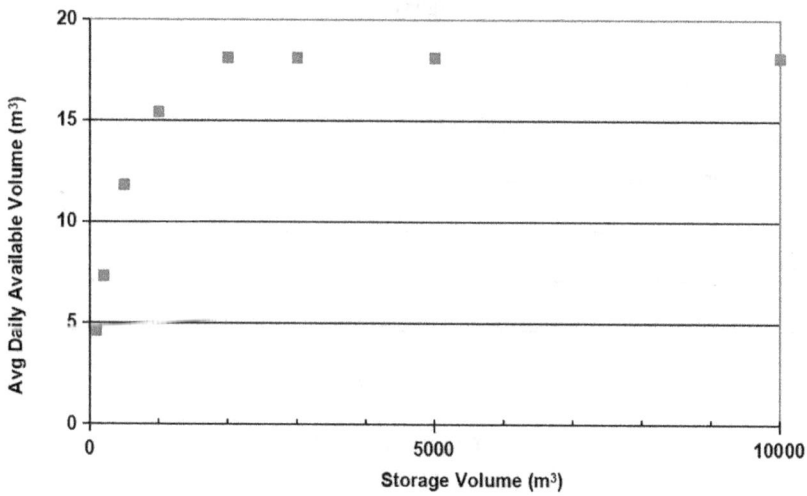

FIGURE 3.2 Effect of storage volume in a dry year.

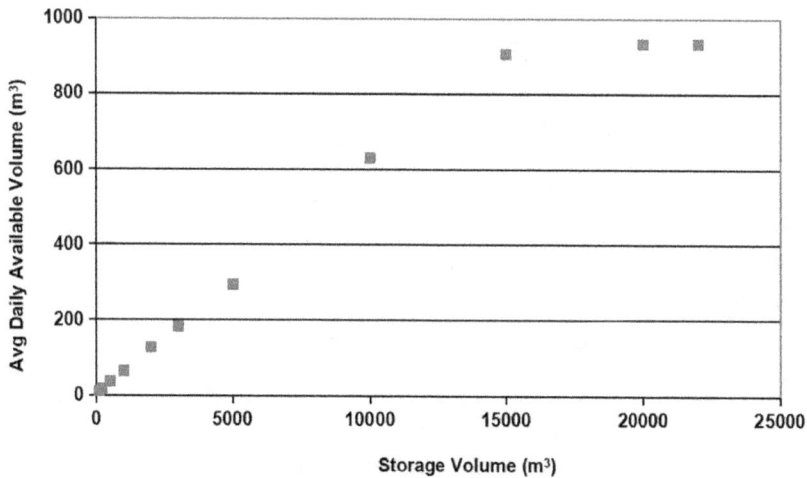

FIGURE 3.3 Effect of storage volume in an average year.

available water volume increases with the increase of storage volume, up to a volume of 15,000 m³, beyond which effect of storage volume becomes insignificant. However, in a wet year, effect of storage volume is significant up to a storage volume of 50,000 m³ and in a wet year up to about 29,000 m³ of water is available per day for irrigation. Figure 3.4 shows the effect of storage volume on daily average available water volume.

Figure 3.5 shows the comparison of three different years in regard to effectiveness of storage volume on total yearly available water volume. It is clear from the figure that effectiveness of storage volume is insignificant in dry and average years compared to the effectiveness in wet year.

Figure 3.6 shows the effect of storage volume on average daily water deficit in a dry year. Average water deficit amount is decreasing with the increase of storage volume; however, it becomes insignificant beyond a storage volume of 2,000 m³. Figure 3.7 shows the effect of storage volume on average daily water deficit in an average year. Obviously a similar trend was observed; however, the threshold storage volume for which effect of storage volume becomes insignificant was 15,000 m³. Figure 3.8 shows the effect of storage volume on average daily water deficit in a wet year. It is found that in a wet year the effect of storage volume in compensating daily water deficit is highly significant, i.e. average daily water deficit decreases significantly with the increase of storage volume.

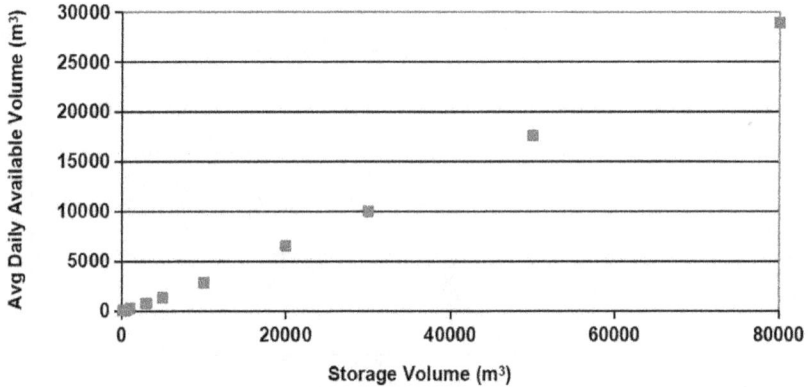

FIGURE 3.4   Effect of storage volume in a wet year.

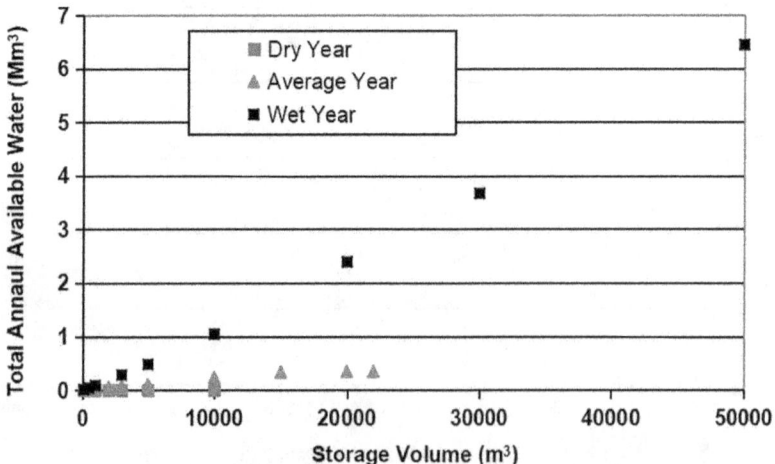

FIGURE 3.5   Effect of storage volume on total available water volume in different years.

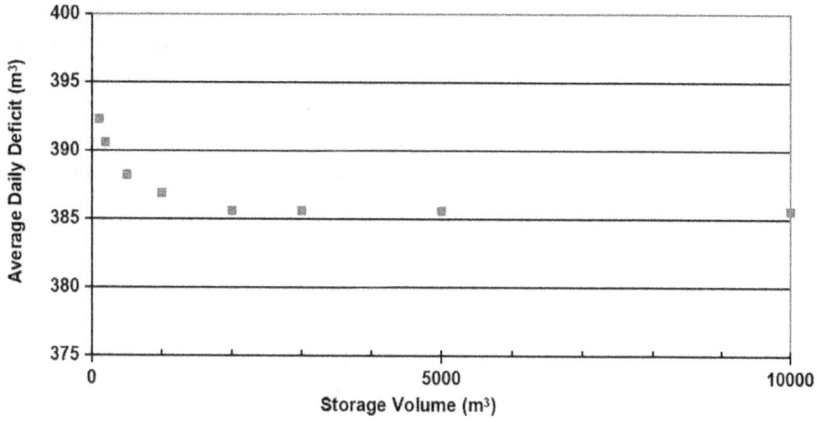

FIGURE 3.6 Effect of storage volume on average daily water deficit in a dry year.

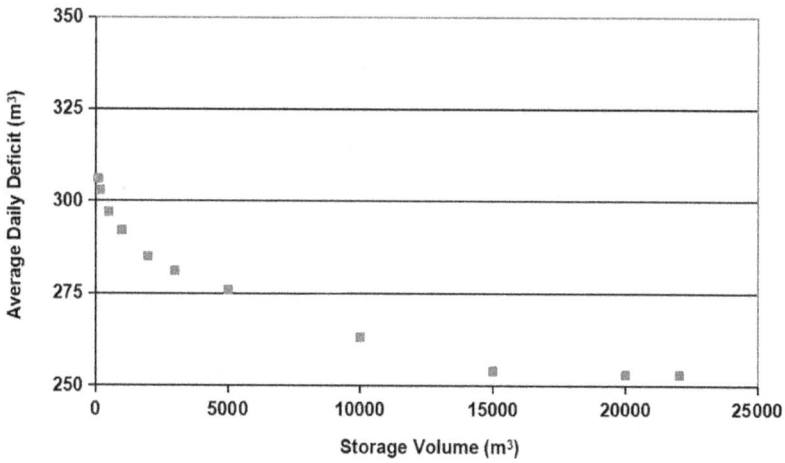

FIGURE 3.7 Effect of storage volume on average daily water deficit in an average year.

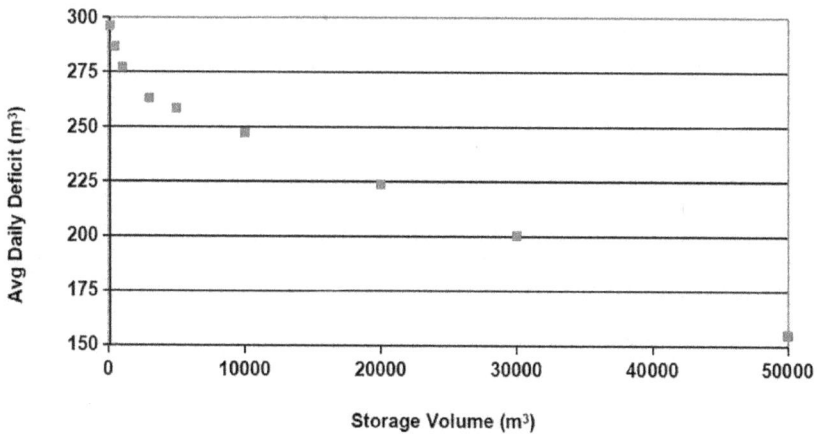

FIGURE 3.8 Effect of storage volume on average daily water deficit in a wet year.

Figure 3.9 shows the comparison of 3 years in regard to effectiveness of storage volume on total yearly water deficit. It is found that compared to a wet year, effectiveness of storage volume in a dry year to compensate water deficit is very insignificant. However, in an average year storage volume is significant up to a certain value.

Again the model was simulated to see the effect of contributing catchment area on available water and water deficit scenarios. For these scenarios, the storage volume was kept constant at 1,800 m³. Figures 3.10–3.12 show the effect of contributing catchment area on average daily water deficit for dry year, average year and wet year, respectively. In the x-axis of these figures, 'catchment area multiplication' means the original catchment area (20 ha) has been multiplied with the specified factors and water deficits were calculated for the new catchment area. It is to be noted that when contributing catchment area increases, not only runoff from rainfalls increase, irrigation demand for the increased grassed area also increases. From these figures it is clear that with the increase of catchment area, average daily water deficit increases linearly in all the three years. This is because of linear increment of area to be irrigated, hence the irrigation demand. Increment of irrigation demand was higher than the increment in runoff. In average and wet years, the deficits were lower than the dry year deficit, which is reasonable. However, there are minimal differences between an average-year and wet-year deficit scenarios. This is because

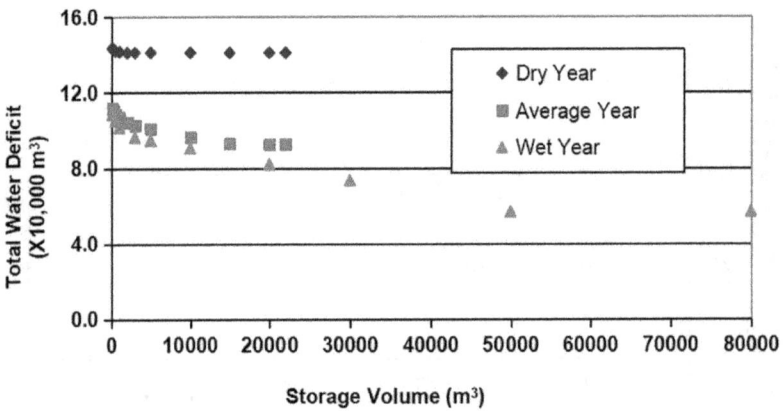

FIGURE 3.9   Effect of storage volume on total annual water deficit in different years.

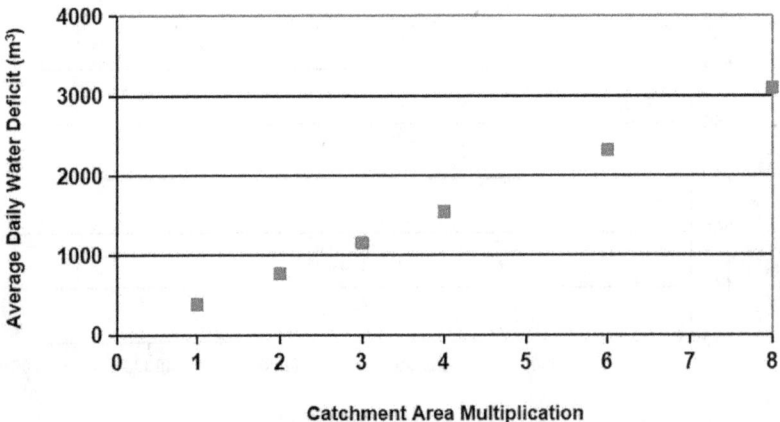

FIGURE 3.10   Effect of catchment area on average daily water deficit in a dry year.

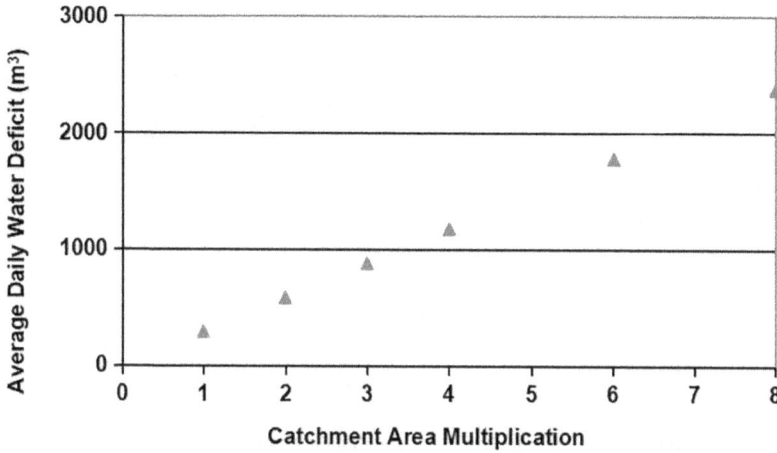

FIGURE 3.11 Effect of catchment area on average daily water deficit in an average year.

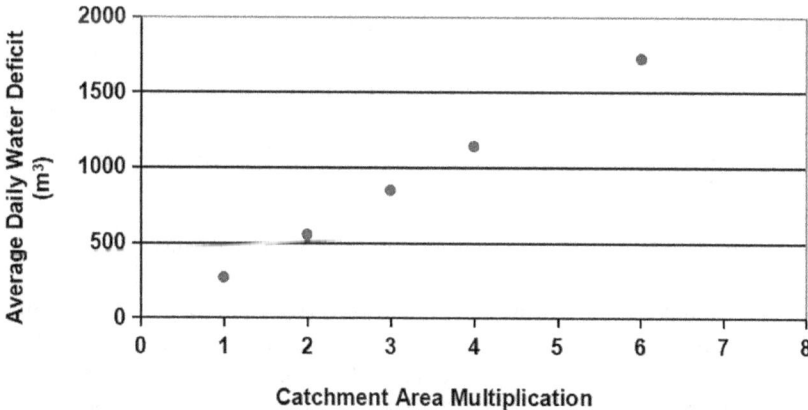

FIGURE 3.12 Effect of catchment area on average daily water deficit in a dry year.

with increased catchment area wet year will not produce any additional benefit due to limited storage area. Assumed storage area is not enough to hold additional runoff from increased catchment area.

Figure 3.13 shows the effect of contributing catchment area on average daily available water in three different years. It is found that for both the dry- and average-year amount of daily average available water is increasing with the increase of catchment area up to a certain stage, after which the effect of contributing catchment area becomes insignificant. This is because with the very big area, available storage is not enough to hold all the generated runoff. In a wet year, although magnitude of daily average available water was higher than that in the dry and average years, it was almost independent of increasing catchment area. Again, this is because of limited storage volume of the assumed storage.

3.6 Conclusions

This study investigated the effectiveness of storage volumes and contributing catchment areas on available water for irrigation and deficit in water volume. This study considered actual daily rainfall data instead of traditional average rainfall data. Analysis using actual rainfall data reveals actual possible scenario, instead of a hypothetical average scenario. Analysis was performed for three different years

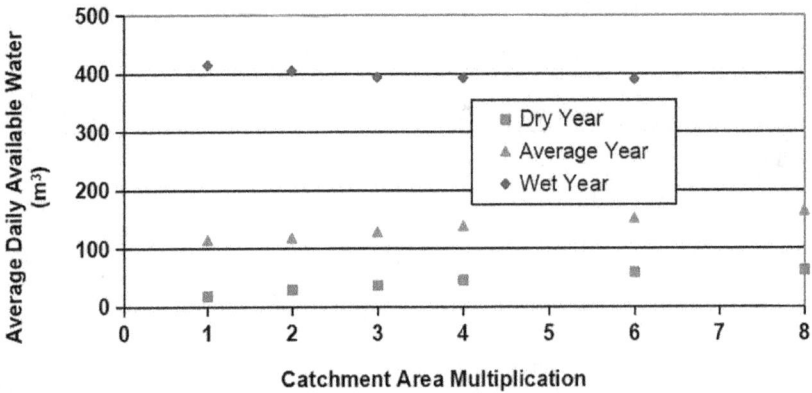

FIGURE 3.13   Effect of catchment area on average daily available water in different years.

having significant differences in rainfall amounts. However, this study is based on a particular geographical area in Western Sydney. Results of this sort of studies will vary with geographical locations, i.e. with different climatic conditions or, in general, with different rainfall intensities and patterns. Also, outcomes will be dependent on different soil conditions and grass types. Scenarios were produced by changing one parameter while keeping other parameter constant, i.e. when the effect of catchment area was investigated, storage volume was assumed unchanged and when effect of storage volume was investigated, catchment area was assumed unchanged. In reality, there might be numerous different types of combinations with storage volumes and catchment areas. However, for this primary study this was out of scope to consider numerous different types of combinations. A comprehensive decision support system would be able to produce effectiveness and water saving for any geographical location with any storage volume and any catchment area.

This study was performed for stormwater storage effectiveness in a golf course only. However, this sort of study can be performed for any potential stormwater storage system/facility applying similar daily water balance modelling concept. A comprehensive decision support tool can incorporate analysis of any potential storage system. Also, it is recommended that a life-cycle-cost analysis has to be performed to evaluate actual benefit of any potential storage volume to be constructed/installed.

References

Boughton, W. C. (1993) A hydrograph based model for estimating the water yield of ungauged catchments, in *Hydrology and Water Resources Symposium*, Newcastle, Australia, 30 June – 2 July.

Gomez, Y. G. and Teixeira, L.G. (2017) Residential rainwater harvesting: Effects of incentive policies and water consumption over economic feasibility. *Resources, Conservation & Recycling*, 127, 56–67.

Imteaz, M.A. and Moniruzzaman, M. (2018) Spatial variability of reasonable government rebates for rainwater tank installations: A case study for Sydney. *Resources, Conservation & Recycling*, 133, 112–119.

Imteaz, M.A., Adeboye, O., Rayburg, S. and Shanableh, A. (2012) Rainwater harvesting potential for southwest Nigeria using daily water balance model. *Resources, Conservation & Recycling*, 62, 51–55.

Imteaz, M.A., Bayatvarkeshi, M. and Karim, M.R. (2021) Developing generalised equation for the calculation of payback period for rainwater harvesting systems. *Sustainability*, 13, 4266. DOI:10.3390/su13084266

Imteaz, M.A., Karki, R., Hossain, I. and Karim, M.R. (2017) Climatic and spatial variabilities of potential rainwater savings and economic benefits for Kathmandu valley. *International Journal of Hydrology Science and Technology*, 7(3), 213–227.

Imteaz, M.A., Matos, C. and Shanableh, A. (2014) Impacts of climatic variability on rainwater tank outcomes for an inland city, Canberra. *International Journal of Hydrology Science and Technology*, 4(3), 177–191.

Imteaz, M.A., Paudel, U., Matos, C. and Ahsan, A. (2016a) Generalized equations for rainwater tank outcomes under different climatic conditions: A case study for Adelaide. *International Journal of Water*, 10(4), 301–314.

Imteaz, M.A., Paudel, U. and Santos, C. (2021) Impacts of climate change on weather and spatial variabilities of potential water savings from rainwater tanks, *Journal of Cleaner Production*, 311, 127491, DOI: 10.1016/j.jclepro.2021.127491.

Imteaz, M.A., Sagar, K., Santos, C. and Ahsan, A. (2016b) Climatic and spatial variations of potential rainwater savings for Melbourne. *International Journal of Hydrology Science and Technology*, 6(1), 45–61.

Imteaz, M.A., Shanableh, A., Rahman, A. and Ahsan, A. (2011) Optimisation of rainwater tank design from large roofs: A case study in Melbourne, Australia. *Resources, Conservation & Recycling*, 55(11), 1022–1029. DOI: 10.1016/j.rescorec.2011.05.013.

Karim, M.R., Bashar, M.Z.I. and Imteaz, M.A. (2015) Reliability and economic analysis of urban rainwater harvesting in a megacity in Bangladesh. *Resources, Conservation & Recycling*, 104, 61–67. DOI: 10.1016/j.resconrec.2015.09.010

Karim, M.R., Sakib, B.M.S., Sakib, S.S. and Imteaz, M.A. (2021) Rainwater harvesting potentials in commercial buildings in Dhaka: Reliability and economic analysis, *Hydrology*, 8, 9. DOI: 10.3390/hydrology8010009.

Matos, C., Santos, C., Bentes, I. and Imteaz, M.A. and Pereira, S. (2015) Economic analysis of a rainwater harvesting system in a commercial building. *Water Resources Management*, 29(11), 3971–3986.

Moniruzzaman, M. and Imteaz, M.A. (2017) Generalized equations, climatic and spatial variabilities of potential rainwater savings: A case study for Sydney. *Resources, Conservation & Recycling*, 125, 139–156.

Paudel, U. and Imteaz, M.A. (2019) Spatial variability of reasonable government rebates for rainwater tank installations: A case study for Adelaide, in *Sustainability Perspectives: Science, Policy and Practice*, Eds.: P.A. Khaiter & M.G. Erechtchoukova, Springer. DOI: 10.1007/978-3-030-19550-2_13

Rahman, A., Hajani, E., Eslamian, S., (2017) Rainwater harvesting in arid regions of Australia, in *Handbook of Drought and Water Scarcity, Vol. 2: Environmental Impacts and Analysis of Drought and Water Scarcity*, Eds.: S. Eslamian & F. Eslamian, Taylor and Francis, CRC Press, Boca Raton, FL, pp. 489–500.

Rahman, A., Keane, J. and Imteaz, M.A. (2012) Rainwater harvesting in greater Sydney: Water savings, reliability and economic benefits. *Resources, Conservation & Recycling*, 61, 16–21.

Shanableh, A., Al-Ruzouq, R., Yilmaz, A., Siddique, M., Merabtene, T. and Imteaz, M.A. (2018) Effects of land cover change on urban floods and rainwater harvesting in Sharjah, UAE, *Water*, 10(631), 1–17. DOI:10.3390/w10050631

II

African Case
Studies

4

Deficit Irrigation: A Review from Ethiopia

Desale Kidane
Asmamaw
Bahir Dar University
Ghent University

Jan Nyssen, Kristine
Walraevens
Ghent University

Pieter Janssens
Ghent University
Soil Service of Belgium

Wim M. Cornelis
Ghent University

Mekete Desse, Seifu
Admasu Tilahun,
and Enyew Adgo
Bahir Dar University

4.1 Introduction

Irrigation is a very old practice and dates back to the earliest civilizations of humankind. It makes a major contribution to food security, producing greater than 40% of the world's food and agricultural commodities (Kilob, 2015; Rockström et al., 2010; Fereres and Connor, 2004). Irrigated areas have doubled in recent decades and contributed much to the growth of agricultural productivity over the last 50 years (Haileslassie et al., 2016; Hagos et al., 2009; Tsegay, 2012; Yihun et al., 2013). It served as one of the key drivers behind the growth in agricultural productivity, increasing household income and alleviation of rural poverty, thereby highlighting the various ways that irrigation can alleviate poverty (Hagos et al., 2009; Dananto and Alemu, 2014). To meet food requirements by 2020, FAO (2002) estimated that food production from irrigated areas will need to increase from 35% in 1995 to 45% in 2020. This means that access to water for irrigation will become an issue of global concern and competition in the future, especially in the arid and semi-arid regions of the world (Fereres and Connor, 2004). Moreover, irrigation reduces the risk of expensive inputs being wasted by crop failure resulting from moisture stress (FAO, 2002). In the Comprehensive Assessment of Water Management in Agriculture, Molden (2007) stated a similar finding.

Likewise, traditional irrigation practices in Ethiopia dates back several centuries and continues to be an integral part of agriculture. It has been practiced to produce subsistence food crops (Awulachew et al., 2007; Asres, 2016; Mengist and Tilahun, 2009; Yohannes et al., 2017). Besides, modern irrigation systems were started in the 1960s to produce industrial crops in all parts of the Awash Valley and

DOI: 10.1201/9781003353928-6

the lower Rift Valley (Figure 4.1). Given the extreme meteorological and hydrological variability of Ethiopia, it is important that attention has to be given to enhance improved water management strategies to increase agricultural production through irrigated agriculture (Kidane et al., 2014). Hagos et al. (2009) suggested that improving irrigation development can be considered as one of the cornerstones of food security and a poverty reduction tool as it has the power to stimulate economic growth and rural food security improvement. Nationally, the area under irrigation has been increasing year after year (Table 4.1), although the contribution of irrigation development is limited (Awulachew et al., 2010). In arid and semi-arid areas of Ethiopia, water is the most limiting factor for crop production (Mengist and Tilahun, 2009). The amount and distribution of rainfall are not sufficient to sustain crop growth, and an alternative approach to use the accessible water for irrigation is needed (Rockström et al., 2009), apart from further investments in rain-fed agriculture (Cornelis et al., 2019). Though satisfying crop water requirements maximizes production from the land, it does not necessarily maximize the return per unit volume of water (Pereira et al., 2012). Improved irrigation technologies could increase agricultural productivity and enhance socioeconomic growth through increasing production (Mengist and Tilahun, 2009). Similarly, irrigation has been reported to increase yield by 100%–400% (English, 1990) in comparison with conventional rain-fed farming where no specific soil-water management practices are adopted.

Conversely, in Ethiopia, farmers' irrigation application is often either more or less than the crop water requirements (Beyene et al., 2018; Haileslassie et al., 2016). Over-irrigation increases the cost of production and might leach nutrients out of the root zone (Agide et al., 2016). However, efficient irrigation management strategies such as deficit irrigation (DI) are important for decision-makers, growers, and irrigation experts. Hence, to improve water productivity (WP), there is a growing interest in DI, an irrigation practice whereby water supply is reduced below maximum and mild stress is allowed with minimal effects on yield (Pereira et al., 2012). Under the conditions of scarce water supply and drought, DI can lead to greater economic gains by maximizing yield per unit of water (Rockström and Falkenmark, 2015). Therefore, in areas with water shortage, it is important to understand which stress level at which growth stage results in high WP (Kirda et al., 2005). This enables irrigators to know not only the critical growth stage but also the optimum stress level that can be imposed (Admasu et al., 2017).

The possibility for further irrigation development to meet food requirements in the coming years is, yet, constrained by decreasing water supply and growing competition for water (FAO, 2002). The situation is worsened by declining soil quality and soil fertility (Derib, 2013). The reliance on water has become a serious limitation on further development and threatens to slow down development and endangering food supplies. The big challenge for the coming decades will be to enhance food production with less water (Asmamaw, 2017), thus producing more crops per drop, particularly in arid and semi-arid regions with limited water resources (Molden et al., 2010). To enhance WP in arid and semi-arid regions, adequate knowledge about DI effects on WP and crop yield is critical (Hagos et al., 2009; Haileslassie et al., 2016; Yihun, 2015; Awulachew et al., 2010). This chapter intends to document some basic information about the water use in Ethiopia and the impact of DI strategies on WP and yield, and the need and adoption of DI for decision-makers and growers. Besides, barriers that prevent fast implementing of DI are addressed.

4.2 Water Resources Development in Ethiopia

Ethiopia is endowed with huge surface and groundwater resources that have given it a prestige of being called the "water tower" of Northeast Africa (Table 4.1). There are 12 river basins with an annual runoff volume of 122 billion cubic meters and an estimated 2.6–6.5 billion cubic meters of groundwater potential (Awulachew et al., 2010; Figure 4.1). This makes an average of 1,575 m^3 of physically available water per person per year, which is a relatively large volume (Yihun, 2015). However, rain-fed-based farmers in the dry land regions don't receive sufficient water at their farms in the dry season. This makes it impossible for farmers to harvest more than once a year (Hagos et al., 2009; Awulachew et al., 2005).

TABLE 4.1 Annual Runoff, Total Area, and Irrigation Potential by River Basin

Drainage System	River Basin	Total Area (ha)	Total Area (%)	Annual Runoff (×10⁹ m³)	Total Runoff (%)	Irrigation Potential (ha)
Nile		**36,881,200**	**32.4**	**84.5**	**69**	
	Abbay (blue Nile)	19,981,200	17.6	52.6	42.9	815,581
	Baro-Akobo	74,10,000	6.5	23.6	19.3	1,019,523
	Tekeze	8,900,000	7.8	7.63	6.2	83,368
	Mereb	5,700, 000	0.5	0.26	0.6	NA
Rift Valley		**31,764,000**	**27.9**	**29.0**	**23.7**	139,300
	Awash	11,270,000	9.9	4.6	3.7	134,121
	Afar-Denakil	7,400,000	6.5	0.86	0.7	158,776
	Omo-Gibe	7,820,000	6.9	17.96	14.7	67,928
	Central lakes	5,274,000	4.6	5.6	4.6	
Shebelle-Juba		**37,0126,400**	**32.7**	**8.95**	**7.3**	
	Wabi-Shebelle	20,021,400	17.6	3.15	2.6	
	Genale-Dawa	17,105,000	15.1	5.8	4.7	1,074,720
Northeast coast		**7,930,000**	**7**	**0**	**0**	237,905
	Ogaden	7,710,000	6.8	**0**	0	0
	Gulf of Aden	220,000	0.2	**0**	0	0
Total		**113,681,600**	**100**	**122**	**100**	3,731,222

Source: Awulachew et al. (2010).

NA = not available.

Bold values differentiate the drainage system contributions with their component river basin contribution.

Although Ethiopia has abundant rainfall and water resources, its agricultural system has not yet fully benefited from the technologies of irrigation water management (Hagos et al., 2009; Kidane et al., 2014; Yihun et al., 2013). Most of the rural dwellers are among the poorest in the country with limited access to agricultural technologies and little to no access to agricultural markets and technological innovations (Awulachew et al., 2010).

Agide et al. (2016) and Schmitter et al. (2017) also suggested that Ethiopia's irrigation potential has been estimated at up to 3.7 million hectares and could contribute a lot in the overall development of the country, provided that improved irrigation strategies are applied. Using dynamic vegetation and water balance model, Rockström and Falkenmark (2015) categorized Ethiopia among the countries with green water freedom under blue water shortage. This means that though the country suffers from a blue water shortage, it should be able to produce enough food if green water is considered and is managed well. Blue water refers to a stored runoff in dams, lakes, rivers, and recharged water in aquifers, while green water is rainwater stored in the soil as soil moisture and generated from infiltrated rainwater (Falkenmark and Lundqvist, 1995). Key adaptation and resilience-building strategies suggested by these authors for countries within the category of green water freedom under blue water shortage are "rainwater management and soil moisture conservation, runoff water harvesting, spatial catchment planning, adaptive water governance at micro-catchments and meso-catchment scales", similar to countries falling in the category of green water freedom under chronic blue water shortage. Since in countries like Ethiopia, blue water shortage is not chronic (though limited), these authors suggested supplemental irrigation (SI) as well as micro-catchment water harvesting, as an additional strategy (Rockström and Falkenmark, 2015). Likewise, Molden (2007) also summarized agricultural water management practices that comprise a range of possibilities from producing under fully irrigated to entirely rain-fed conditions (Figure 4.2).

SI is different from DI, in that SI is the addition of small amounts of water to essentially rain-fed crops during the wet season when a shortage of rainfall occurs because of quickly offset of the rainfall and thus to bridge dry spells as well as to increase and stabilize yields, whereas DI is applied during the long dry season as the main source of water (Geerts and Raes, 2009; Oweis and Hachum, 2009). Unlike DI,

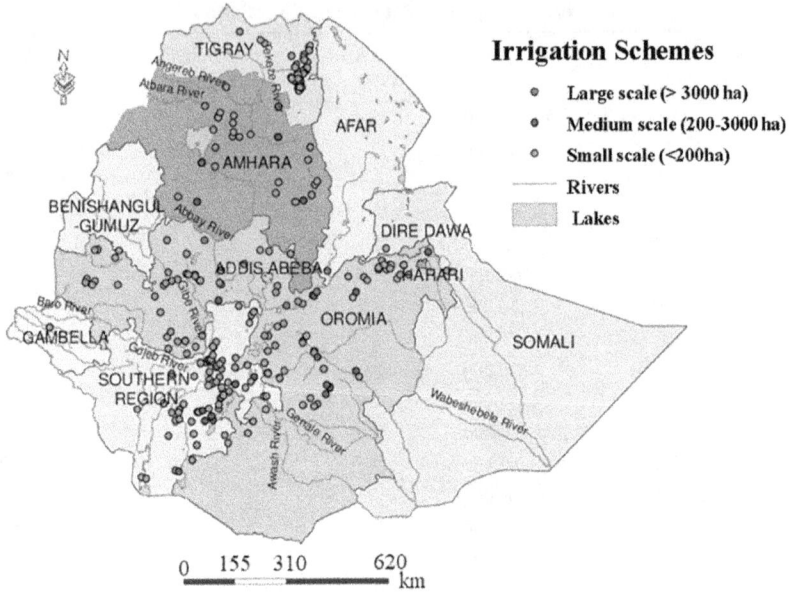

FIGURE 4.1 The distribution of irrigation infrastructures in Ethiopia (Awulachew et al., 2010). The mean annual specific runoff varies from zero to 35 L/sec/km^2 (Awulachew et al., 2007). In the majority of the river basins, minimum flows occur in the period from December to March. Apart from the big rivers and their major tributaries, there is hardly any perennial flow in areas below 1,500 m (Awulachew and Merrey, 2008). Mainly perennial streams and springs exist only in the vicinity of mountains with an annual rainfall of more than 1,000 mm (Awulachew and Merrey, 2008). About 70% of the total runoff takes place during the period from June to September. Dry season flow originates from springs that provide base flows for SSI (Awulachew et al., 2007). The current actual consumption from surface water resources is less than 5%, and only very little attempts from groundwater. Awulachew et al. (2010) stated that irrigation could represent a basis of the agricultural development, which may contribute up to ETB 140 billion ($US 5 billion) per year to the economy and potentially moving up to 6 million households into food security.

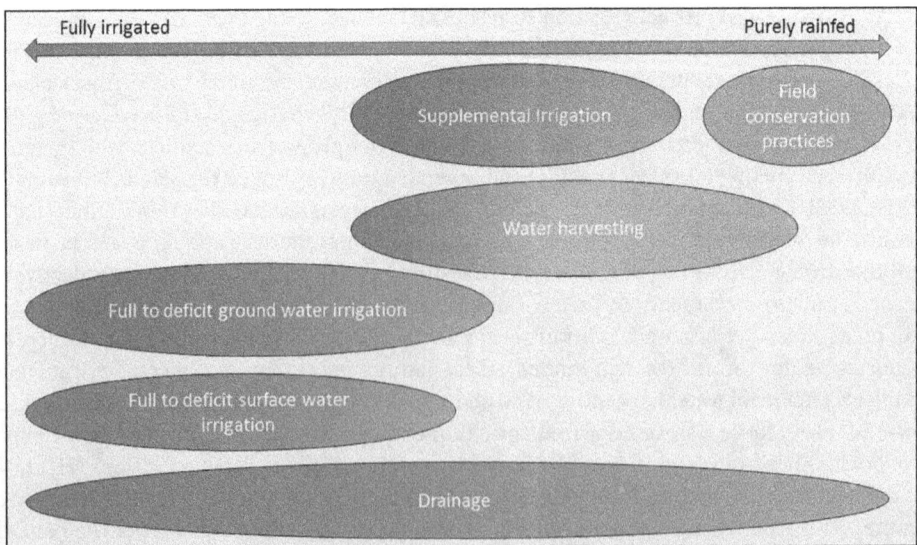

FIGURE 4.2 Diverse options for agricultural water management along the spectrum (adapted from Molden, 2007).

the timing and amount of SI cannot be determined in advance given the natural variability in season-to-season and within-season rainfall levels (Oweis and Hachum, 2012). Water for SI comes largely from surface sources, but shallow groundwater aquifers increasingly are being used. In the case of DI, the principal source of moisture is fully controlled irrigation water (groundwater and surface water), and highly variable limited precipitation is only supplementary (Oweis and Hachum, 2009). SI is dependent on the precipitation of a basic source of water for the crop.

4.3 Common Irrigation Systems in Ethiopia

Appropriate irrigation water management seeks to optimize and efficiently use water for the highest possible production of crops and minimizing water loss (Belete et al., 2011). Several different irrigation methods apply water to the soil surface. These irrigation methods are employed to irrigate plants with the primary goal of uniformly storing water in the efficient root zone in an optimal amount while minimizing water losses and maintaining crop output with the desired product quality.

The choice of an irrigation system relies on the area's or country's irrigation tradition. In agriculture, irrigation methods can be categorized into three major groups such as drip, sprinkler, and surface irrigation. For thousands of years, traditional flood irrigation has been applied in Ethiopia by smallholder farmers though few investors introduced sprinkler and drip irrigation methods (Awulachew et al., 2007). These new systems have been mainly applied for horticultural crops and sugarcane plantations around the Rift Valley Region and upper and middle Awash Valley. However, in all state-owned large-scale sugarcane farms such as Tendaho, Kesem, Metehara, Omo, Wenji, and other medium- and small-scale irrigation schemes, furrow irrigation is widely practiced (Awulachew et al., 2007).

Most of the traditional small-scale irrigation (SSI) infrastructures are using temporary river diversion structures made of locally available materials. The performance of SSI infrastructures in Ethiopia is very poor (Haileslassie et al., 2016). The area under traditional irrigation is bigger than the area under modern irrigation infrastructure, and the possibilities for obtaining increased food production from these infrastructures are significant. Moreover, the canals are unlined with inadequate controlling, discharge, and water depth measuring structures. Siltation in canals and inappropriate drainage are serious problems in most SSI infrastructures (Hagos et al., 2009; Dananto and Alemu, 2014). Effective agronomic practices, soil fertility management, appropriate crop selection, and rotation are not properly implemented in most SSI structures (Agide et al., 2016). The efficiency of surface irrigation ranges from 40% to 60%, and during water-scarce seasons, this system does not support the community (Agide et al., 2016; Haileslassie et al., 2016). Therefore, adopting as well as properly applying DI which reduces the non-productive water loss (evaporation, runoff, deep percolation) and improves productive water (transpiration) can contribute to securing water availability and enhance WP.

4.4 Economic Benefits of Irrigation

The expansion of irrigation agriculture has a significant potential to improve productivity and reduce vulnerability to climate variability (Awulachew and Ayana, 2011; Erkossa et al., 2014). As stated earlier, SI reduces the risk of crop failure in the wet season, whereas deficit/fully irrigation served as the main source of economic crops in the dry season. SSI can thus be a promising vehicle for rural development. It can offer the farmer increased security of crop production while avoiding many of the problems which have been experienced by a shortage of rainfall. However, many existing irrigation systems perform below potential (Agide et al., 2016; Schmitter et al., 2017).

Mengistie and Kidane (2016) found that the annual average income of growers before and after irrigation was US$ 141.29 and US$ 721.36, respectively. This approved that irrigation-based farming has a big contribution to increasing the income of the growers than the households that abstained from rain-fed farming. Thus, increases in irrigation efficiency have been found to increase the role of irrigation in the national economy.

The economic analysis of DI on tomato was conducted by Kifle (2018) in southern Ethiopia. The highest net income (US$ 8,067 ha^{-1}) was achieved at 100% ETc applied all over the growth stages and the least net income (US$ 5,794.37 ha^{-1}) was found at a 50% deficit at mid-stage. The largest marginal rate of return (1,396%) was acquired at 100% ETc and the smallest marginal rate of return (247%) was obtained at 85% ETc. The marginal rate of return tells us that the amount of additional income obtained for every 1 US$ spent. Hence, optimum irrigation acquired an additional 13.96 US$ for every 1 US$ spent.

Hagos et al. (2009) studied the importance of irrigated agriculture to the Ethiopian economy, and the results revealed that irrigation generates an average income of about 323 US$/ha under farmers-managed irrigation systems compared to an average income of 147 US$/ha for rain-fed systems. This shows that, after accounting for annual investment replacement cost, the adjusted gross margin from irrigation was 219.7% higher than the gross margin from rain-fed agriculture. According to their findings, irrigation contributed roughly 5.7% and 2.5% to agricultural Gross Domestic Product (GDP) and the overall GDP, respectively, during the 2005/2006 cropping season. By the year 2009/2010, the contribution of irrigation to agricultural GDP and overall GDP was estimated to be about 9% and 3.7%, respectively. They further explained that after relaxing some of the basic assumptions, the future contribution of irrigation to agricultural GDP rises to almost 12%, while the effect on overall GDP will be around 4%. To achieve these results, there is a need to improve irrigation water management strategies, enhance the provision of agricultural inputs, increase the performance of agricultural extension works, expand market access conditions, and advance the management of SSI infrastructures to increase efficiency at all levels. In line with this finding, Ali et al. (2007) also reported.

4.5 Irrigation Scheduling

Irrigation scheduling refers to when and how much water to apply to an irrigated crop to maximize net returns (Geerts and Raes, 2009). The maximization of net return requires a high irrigation efficiency. The knowledge of crop water requirements, expressed in terms of crop evapotranspiration (ETc), is an important practical consideration to improve WP in irrigated agriculture. WP can be improved by using proper irrigation scheduling, which is essentially governed by ETc (Geerts and Raes, 2009). A wisely designed irrigation schedule can ensure the optimal use of allocated water and enhance WP in the water resource-limited region. The risk with DI can be minimized through proper irrigation scheduling and avoiding water stress in water-stress-sensitive stages (Geerts and Raes, 2009; Pereira et al., 2012). It is also imperative to achieve a uniform water distribution to maximize the benefits of irrigation scheduling. Accurate water application prevents over- or under-irrigation. Over-irrigation wastes water, energy, and labor, as well as leaches nutrients below the root zone which reduces crop yields (Hagos et al., 2009; Furgassa, 2017; Grimes and Yamada, 1982). It should be noted that over-irrigation might be needed to meet the leaching requirements in case of excessive salt concentrations in the root zone. Under-irrigation stresses the plant, resulting in yield reductions. Irrigation scheduling has tremendous advantages when environmental, crop production, and water use issues are considered (Yihun et al., 2013; Admasu et al., 2019). The benefits of irrigation scheduling can be summarized as a reduction in energy, water, and labor costs through less irrigation; a lowering of fertilizer costs by reduced surface runoff and deep drainage; enhanced net returns through increased yields and improved crop quality; a minimization of waterlogging problems; assisting control of root-zone salinity problems through controlled leaching; and additional crops through savings in irrigation water (Furgassa, 2017).

The improvement of irrigation schemes requires all-round interventions, starting with involving the community in the system management up to improving water-measuring structures such as Parshall flumes and supplying options for when and how much water to apply. Irrigation scheduling is directly related to profitable production and sustainable agricultural practices (Jensen, 2007). Due attention to irrigation scheduling can help to assure high yields and better-quality products. It is one of the most important tools for developing the best management practices for irrigated areas in any region.

How much water to apply and when to irrigate depends on the soil's available moisture storage capacity and the amount of available water depleted from the soil profile by crop water use (Hook et al., 2000). Hence, one can use the "feel and appearance method" to determine the amount of soil moisture. As water content changes, the "feel" of soil changes. Besides, an effective and precise irrigation schedule can be designed using integrated information from the soil, plant, and atmosphere (Janssens et al., 2011). Soil-water status is mostly acquired using soil sensors which can measure soil-water content (θsoil) or soil-water potential (Ψsoil). Information from soil moisture sensing instruments can support informed decisions about when and how much to irrigate (Janssens et al., 2011). Irrigation scheduling can be easily designed using CropWat, AquaCrop, or other user-friendly freely available software packages for different watering regimes. CropWat and AquaCrop are empirical decision support systems developed by the Land and Water Division of the FAO and are well known for their easy estimation of crop water demands under different irrigation practices (Rodrigues et al., 2013). The models consider climate, soil, soil moisture, and crop data. CropWat is developed based on the FAO Irrigation and Drainage papers No. 56 "Crop Evapotranspiration" and No. 33 "Yield response to water". A fundamental equation as stated by Jensen (2007) explained that the relative loss in yield is proportional to the relative reduction in ET. The Penman–Monteith equation and respective crop coefficients are used to calculate crop ET rates. Crop growth is simulated by a linear model where gross dry matter production of a standard crop is empirically calculated and crop-dependent correction factors for climate, growth, and yield are applied (Doorenbos and Kassam, 1986). Unlike CropWat, AquaCrop calculates biomass from crop transpiration and a biomass WP coefficient. Yield is then derived from biomass by considering a harvest index. As with CropWat, the Penman–Monteith equation and crop coefficients are used to calculate crop ET and separate the non-productive consumption of water (soil evaporation) from the productive consumption of water (transpiration). Though FAO published a new Irrigation and Drainage paper No. 66 "Crop yield response to water" with the release of AquaCrop, the ideas behind the model are outlined in Steduto et al. (2009). In Ethiopia, meteorological stations are distributed mainly in the north, southwest, central, and northwest parts. Most of these stations are not operating well and some of the stations record only rainfall and temperature. This affects the design and preparation of irrigation scheduling for crops especially for remote arid regions where adequate data are not available. Moreover, there are no well-organized soil database systems including field capacity, wilting point, bulk density, texture, hydraulic conductivity, and other soil properties which could be used for irrigation scheduling. Yet, pedotransfer functions have been developed for a range of Ethiopian soils (Kibebew, 2003). In general, irrigation scheduling with estimated ETc and a target soil-water potential level can provide feedback for managing irrigation strategies. Precise irrigation scheduling has resulted in optimum yield and quality. Researchers conclude that DI management requires optimizing the timing, amount of water to be applied, and the degree of plant stress within the restriction of available water (Geerts and Raes, 2009; Jensen, 2007). Besides, it significantly improves the complexity of the decision process (Pereira et al., 2012; Rodrigues et al., 2013; Rusere et al., 2012).

4.6 Deficit Irrigation

4.6.1 Principles

DI is a strategy in which the crop is intentionally deprived of water (Geerts and Raes, 2009). It is a strategy dealing with a severely limited water supply. Besides, other approaches such as water harvesting or precision irrigation can be employed to help save water and improve the irrigation efficiency of low productive systems (Fereres and Soriano, 2007). Water restriction is limited to drought-tolerant stages, commonly the vegetative stages and the late-ripening period (Fereres and Soriano, 2007). DI intends to improve WP and to bring more area under irrigation with the available water. This concept of irrigation scheduling has different names, such as regulated DI, pre-planned deficit evapotranspiration, and DI (English, 1990) but all mean the same, i.e. irrigating at a lower rate than 100% ETc, assuming that capillary rise from shallow groundwater tables is not an issue. DI is an optimization strategy in which irrigation can be applied stage-wise during drought-sensitive growth stages of a crop only or the whole crop-growing season (Zahraei et al., 2017). Furthermore, based on the objectives of water application, DI can be applied through regulated DI and partial root-zone drying. To estimate the role of DI, recognizing the concept of WP expressed in kg/m³ is essential. WP is an efficiency term, expressing the amount of marketable product (e.g. kilograms of grain or fruit) to the quantity of input (cubic meters of water) needed to produce that output (Kijne et al., 2009). It is a synonym to the widely used water use efficiency (WUE). Moreover, the term WP also refers to the ratio of "economic value" of production (in terms of the gross or net value of the product) relative to water use (Eqs. 4.1 and 4.2).

$$WP = \frac{\text{Yield (kg)}}{\text{Evapotranspiration}\left(\text{m}^3\right)} \tag{4.1}$$

$$WP = \frac{\text{Monetary value}\left(\text{USD\$}\right)}{\text{Water applied to produce}\left(\text{m}^3\right)} \tag{4.2}$$

DI especially allows economic optimization of water use regarding crop output where water is scarce.

4.6.2 The Need for Improving WP in Ethiopia

Agriculture water withdrawal accounts for about 75% of all withdrawals in developing countries, and FAO predicts a 14% net expansion of use between the years 2000 and 2030 to meet food demands (Steduto et al., 2012). The growing demand for water and the increasing competition between water users evoke the need to manage water resources effectively and to progress economic performance to assure sufficient food for future generations with the same or less water and land than that which is currently available for agriculture (Kilob, 2015). Improving WP can be an important pathway for poverty reduction, where WP is very often relatively low (Meskelu et al., 2017). Enhancing WP requires, among other things, improved irrigation technologies and better knowledge. Hence, new approaches to irrigation are needed for irrigated agriculture to ensure viable water management.

Irrigation is widely criticized as a wasteful user of water, especially in water-deprived regions. Thus, DI is not only of high relevance in water-scarce areas; it also possibly enhances WP and decreases water use in irrigated systems (Fereres and Soriano, 2007). For arid and semi-arid regions, DI strategies are very exciting for an efficient allocation of scarce water resources. DI maximizes WP with better harvest quality (FAO, 1979). Moreover, since water use is reduced, the irrigated area can be increased and additional crops can be irrigated augmenting the diversity of the household production. DI reduces the leaching effects of nutrients from the root zone and agrochemicals and preserves the groundwater quality (Furgassa, 2017). Moreover, it reduces the risk of the development of certain diseases linked with high humidity (e.g. fungi) that are common in optimal irrigation systems (Fereres and Soriano, 2007). Researchers have evaluated the feasibility of DI to significantly save irrigation water without significant yield decrease (Fereres and Soriano, 2007; Meskelu et al., 2017). For instance, cotton shows complex responses to DI because of its deep root system, its ability to sustain low leaf water potential, and osmotically regulate leaf-turgor pressure (Furgassa, 2017). Hence, understanding crop response to water stress can be the key strategy for realizing a sustainable agricultural production under such circumstances.

4.6.3 Management and Adoption

Water availability, water use, and nutrient supply to plants are closely interacting factors influencing plant growth and yield production. Integrated soil fertility management (ISFM) is a collection of farming methods tailored to local circumstances to maximize nutrient and WUE and enhance agricultural productivity. ISFM strategies focus on the combined use of mineral fertilizers and locally available soil amendments (such as lime and phosphate rock) and organic matter (crop residues, compost, and green manure) to replenish the lost soil nutrients. This improves both soil quality and the efficiency of fertilizers, water holding capacity of soils and other agro-inputs. Also, ISFM promotes improved germplasm, agroforestry and the use of crop rotation, and/or intercropping with legumes.

In Ethiopia, where water-induced soil erosion is common, flood and furrow irrigation practices are widely used (Admasu et al., 2017; Erkossa et al., 2011; Haileslassie et al., 2016). Thus, improved soil and water management practices are needed because the erratic rainfall and improper irrigation water erode nitrates in solution and phosphate adsorbed to sediments, as well as other soluble chemicals. In the same way, irrigation and fertilization have to be applied in the right amounts at the right time to attain better yield (Bekele, 2017; Dananto and Alemu, 2014). However, in Ethiopia, farmers do not optimally practice fertilization and irrigation in an integrated way probably due to lack of appropriate research recommendations, absence of extension services and irrigation methods. Under drip and sprinkler irrigation techniques, it is possible to apply fertigation precisely, but in furrow irrigation systems, the integration of fertilizers and irrigation water cannot be applied at the same time; fertilizers are rather broadcasted traditionally. As regards DI strategies, the manager and the growers need to know the level of transpiration deficiency allowable without significant yields reduction. Given scarce water management, DI is intended to increase the WP of a crop by eliminating irrigations that have little impact on yield (Molden et al., 2010). Results from various experiments indicated that the application of DI may have small yield reduction compared with optimal irrigation (Ali and Talukder, 2008; Bekele, 2017). However, these reductions are minimal compared with the benefits gained through diverting the saved water to irrigate other fields for which water would normally be insufficient under traditional irrigation practices. Moreover, before implementing a DI strategy, it is necessary to know crop yield responses to water stress (Kirda et al., 1999). High-yielding varieties are more sensitive to water stress than low-yielding varieties; for example, DI had a more adverse effect on the yields of improved maize varieties than on those of traditional varieties (Ali and Talukder, 2008; Doorenbos and Kassam, 1986). Crops or crop varieties that are most suitable for DI are those with a short growing season and tolerant to drought. To ensure a successful DI strategy, it is necessary to consider the water retention capacity of the soil. In sandy soils, plants may undergo water stress quickly under DI, whereas plants in deep soils of fine texture may have a lot of water at low matric potential (Kijne et al., 2009). Therefore, success with DI is more probable in fine-textured soils. Under DI strategy, agronomic practices may require modification, e.g. decrease plant population, apply less fertilizer, adopt flexible planting dates, and select shorter-season varieties. The proper application of DI strategy can generate significant savings in irrigation water for re-allocation. The water demand has been growing globally at more than twice the rate of the population increase, and many regions are reaching the limit at which reliable water services can be delivered (Seid and Narayanan, 2015). Furthermore, water use in irrigation schemes often exceeds the water supply. However, water management should be developed to increase the economic return of irrigation water. Optimizing the economic return of water can be done by applying water-saving strategies like DI and growing high-value crops (Haileslassie et al., 2016). Modern and flexible irrigation systems with reliable irrigation water delivery services give farmers and water managers more options to reduce water losses and invest in modern irrigation techniques. The adoption of new technology, such as the method of production by farmers, undergoes many steps. The initial step is awarding the growers to recognize the existence of the new technology. Secondly, farmers who are aware become motivated and search for details on the new method. Thirdly, the cultivators wisely evaluate alternative water management techniques. At this step, the new strategies are accepted or rejected. The final step is the adoption of new technology. Similarly, the adoption of DI will proceed with these procedures. Based on this review and to our knowledge, there are no studies carried out about the adoption of DI strategy in Ethiopia, even in Africa. Hence, this review chapter does not include more information regarding DI works done in Ethiopia and its adoption but describes the author's understanding and some field observations. According to this review and some field observations, farmers do not know even information concerning DI indicating a need for intensive promotion of the adoption of DI through scientific knowledge transfer and extension work. The adoption of irrigation technology is probably constrained by lack of extension works, absence of farmer's involvement during the experiment, and the restriction of DI experiments on pot or plot level. Under these conditions, farmers would probably be motivated to adopt DI if better information flows are assured. Similarly, FAO (2002), Fereres and Soriano (2007), and Geerts and Raes (2009) stated that better knowledge of crop response to drought stress about when introducing DI is desirable. Likewise, Foltz (2003) studied the adoption of drip irrigation in Tunisia and witnessed that information and credit limits had the biggest impact on technology adoption. The same author confirmed that better-informed farmers had a higher probability of adopting drip irrigation technology. Alcon et al. (2014) also conclude that the move to water-saving technologies was slow in the absence of policy-supported changes and farmers' advises. Similarly, Ward and Pulido-Velazquez (2008) indicated that the adoption of efficient water application strategies can enhance WUE. The absence of organized databases on actual and potential irrigation, use of over- or under- irrigation, inappropriate irrigation infrastructures such as canals, insufficient inputs provision, lack of farmers supervisions and poor market access or linkages, low awareness of farmers regarding crop water requirement, irrigation schedule and DI, absence of farmers involvement during DI trials, and knowledge sharing mechanisms were identified as the major bottlenecks for further developing the irrigation sector (Derib, 2013; Amede, 2015; Mengistie and Kidane, 2016). Directly or indirectly, these affect the adoption of DI. Irrigation researchers, experts, government authorities, and cultivators must take the lead in the adoption of improved irrigation strategies to water use and management as well as documenting the claims on the resources in a worthy way, and pursuing excellence in the practice of DI.

4.7 Water Productivity and Yield of Major Irrigated Crops in Ethiopia

In Ethiopia, DI experiments have been mainly conducted at or near agricultural research centers and a few universities. Cereal crops and maize were mainly grown during the rainy season (from June to September). During the long dry season farmers irrigated commonly maize, potato, tomato, and onion crops. Table 4.2 gives an overview of yield, WP, and relative yield reduction for four of these crops. The findings shown are discussed in the next sub-sections.

4.7.1 Maize

The effect of stage-wise moisture stress on maize (*Zea mays* L.) yield and WP was studied in western Ethiopia under clay loam soil by Admasu et al. (2017). They found that irrigating at all growth stages resulted in maximum grain yield (8.36 ton/ha), followed by irrigating all stages except the initial stage (6.89 ton/ha). The minimum grain yield (1.02 ton/ha) was obtained under rain-fed conditions (without water conservation practices), followed by irrigating only at the initial stage (1.83 ton/ha), though differences were statistically not significant. When moisture stress happened both at the development and mid-season stages, the yield was highly influenced (3.01 ton/ha). Moisture stress at flowering and pollination resulted in unfilled kernels on the cob, which reduced grain yield by 6%–8% per day of stress. If the plant was stressed after flowering, kernel size was reduced. The lowest WP ($0.50 \, \text{kg/m}^3$) was found at optimum irrigation, whereas the highest ($2.65 \, \text{kg/m}^3$) was observed at stress during development, mid-season and late-season stages. Meskelu et al. (2014) also found similar results. In another study, Seid and Narayanan (2015) tried to find suitable furrow irrigation systems and levels of DI which allow achieving optimal crop yield, quality, and WP of maize in Melkassa, central Ethiopia. They found that the highest WP value ($2.06 \, \text{kg/m}^3$) was gained from an alternate furrow irrigation (AFI) system at 70% ETc. However, the highest grain yield (8.4 ton/ha) was obtained from conventional furrow irrigation (CFI) at 100% ETc, though it was not significantly different compared with 85% ETc. AFI at 70% of ETc indicated a 20% yield decrease and saved 65% water while 100% ETc resulted in 50% water saving for 5.5% yield

TABLE 4.2 Yield, WP, and Relative Yield Reduction for Four Major Irrigated Crops in Ethiopia

Applied Irrigation Water	Grain Yield (ton/ha)	WP (kg/m³)	Relative Saved Water (%)	Reduced Yield (%)	Crop	Sources
Irrigate all growth stages	8.58	0.50	0	0	Maize	Admasu et al. (2017)
Irrigate all stages except I	6.89	0.57	7.41	19.69		
Irrigate all stages except D	4.41	0.57	7.41	48.60		
Irrigate all stages except MS	4.58	0.95	10.19	46.62		
Irrigate all stages except M	5.71	0.94	14.93	33.45		
Irrigate all stages except I & D	4.67	0.97	20.36	45.57		
Irrigate all stages except I & MS	5.38	1.28	10.56	32.30		
Irrigate all stages except I & M	5.39	0.84	17.58	32.18		
Irrigate all stages except D & MS	3.01	0.72	14.38	64.91		
Irrigate all stages except D & M	3.46	0.82	21.15	59.67		
Irrigate all stages except MS & M	4.06	1.84	35.28	52.68		
Irrigate only at M	3.21	1.35	28.86	62.59		
Irrigate only MS	3.87	1.22	17.53	54.89		
Irrigate only D	3.77	2.29	14.78	56.06		
Irrigate only I	1.83	2.65	29.66	78.67		
No irrigation	1.02	–	35.09	88.11		
100% ETc	5.52	0.67	0	0	Maize	Admasu et al. (2019)
85% ETc	5.21	0.78	15	5.62		
75% ETc	4.85	0.78	25	12.14		
65% ETc	4.06	0.75	35	24.45		
55% ETc	3.53	0.77	45	36.05		
45% ETc	2.58	0.75	55	53.26		
35% ETc	2.1	0.81	65	61.96		
25% ETc	1.46	0.98	75	73.55		
100% ETc	4.5	0.54	0	0	Maize	Rodrigues et al. (2013)
85% ETc	4.2	0.66	15	6.67		
70% ETc	4.1	0.78	30	8.89		
50% ETc	3.7	0.96	50	17.78		
100% ETc	31.9	0.82	0	0	Tomato	Kifle (2018)
85% ETc	25.0	12	15	21.63		
70% ETc	23.4	10.9	30	26.65		
50% ETc	22.8	11	50	28.53		
100% ETc	40.38	10.1	0	0	Onion	Dirirsa et al. (2017)
25% deficit at D	39.85	10.67	7.41	1.35		
25% deficit at bulb formation (B)	35.00	9.66	10.19	13.49		
50% deficit at D	34.50	10.06	14.93	14.76		
50% deficit at B	25.93	8.07	20.36	36.31		
25% deficit at I and D	39.00	10.81	10.56	3.49		
25% deficit at B and D	32.90	9.90	17.58	18.79		
25% deficit at B and M	34.78	10.07	14.38	14.06		
50% deficit at I and D	31.67	9.96	21.15	21.87		
50% deficit at D and B	20.82	7.98	35.28	49.16		
50% deficit at B and M	23.35	8.14	28.86	42.80		
25% deficit at I, D and M	39.05	11.74	17.53	3.34		
25% deficit at I, B and M	30.60	8.90	14.78	24.59		
25% deficit at I, D and M	24.19	8.53	29.66	40.70		
25% deficit at I, B and M	21.31	8.14	35.09	47.90		

(*Continued*)

TABLE 4.2 (*Continued*)   Yield, WP, and Relative Yield Reduction for Four Major Irrigated Crops in Ethiopia

Applied Irrigation Water	Grain Yield (ton/ha)	WP (kg/m³)	Relative Saved Water (%)	Reduced Yield (%)	Crop	Sources
100% ETc	18.77	2.79	0	0	Tomato	Kifle and Gebretsadikan (2016)
Stress during I	11.11	1.81	9	41		
Stress during D	13.52	2.71	26	28		
Stress during MS	83.33	1.6	23	56		
Stress during MS	10.43	1.88	18	44		
25% deficit via the growing season	14.44	2.86	25	23		
35% deficit via the growing season	10.06	2.3	35	46		
50% deficit via the growing season	7.03	2.09	50	63		
AFI irrigated at 100% ETc	49.71	15.27	0	0	Tomato	Bekele (2017)
AFI irrigated at 75% ETc	48.06	19.69	25	3.32		
AFI irrigated at 50% ETc	43.72	26.86	50	12.05		
FFI irrigated at 100% ETc	46.62	11.93	0	6.22		
FFI irrigated at 75% ETc	42.37	14.46	25	14.77		
FFI irrigated at 50% ETc	39.56	20.26	50	20.42		
CFI irrigated at 50% ETc	47.08	14.46	50	5.29		
CFI irrigated at 75% ETc	52.60	10.77	25	–		
CFI irrigated at 100% ETc	57.38	8.82	0	0		
60% ETc	36.21	14.62	40	–	Potato	Hilemicael, and Tibebe (2018)
80% ETc	29.91	9.08	20	15.05		
100% ETc	35.22	8.52	0	0		
140% ETc	34.69	6.99	−40	1.50		
120% ETc	38.24	6.61	−20	–		
100% ETc	46.9	7.40	0	0	Tomato	Bisa (2018)
Irrigate all except I	34.64	6.20	12	26.14		
Irrigate all except D	26.53	5.92	29.27	43.43		
Irrigate all except M	39.69	9.65	35.14	15.37		
Irrigate all except MS	39.26	9.54	35.14	16.29		
Irrigate all except I & D	19.39	4.96	38.29	58.66		
Irrigate all except I & MS	36.77	8.23	29.51	21.60		
Irrigate all except I & M	32.16	7.19	29.51	31.43		
Irrigate all except D & MS	28.94	12.82	64.42	38.29		
Irrigate all except D & M	25.89	7.68	46.85	44.80		
Stress at MS and M	42.38	14.13	52.71	9.64		
Stress at I, D, & MS	15.15	8.98	73.43	67.70		
Irrigate all except MS	20.11	7.18	55.86	57.12		
Irrigate all except D	26.47	7.88	47.08	43.56		
Irrigate all except I	18.81	16.46	81.98	59.89		
Stress at all stages	17.59	30.79	90.99	62.49		

AFI, Alternative furrow irrigation; B, Bulb formation stage; CFI, Conventional furrow irrigation; D, Development stage; FFI, Fixed furrow irrigation; I, Initial stage; M, Maturity (stage); MS, Mid-season stage.

reduction. Moisture stresses lead to a decrease of chlorophyll content which will reduce the amount of food produced in the plant (Dirirsa et al., 2017). The ability of crops to partially recover from the effect of early water stress has also been observed in other studies (Kirda et al., 1999). The same authors explained that, under limited water conditions, it is better to start by subjecting the crops to stress early in the season. By doing so, the crop adapts to limited watering conditions with the stress not being severely concentrated at any one time. The effect of moisture stress in all crop growth stages on maize yield and

WP was evaluated by Admasu et al. (2019) in Awash Melkassa, central Ethiopia. They showed that grain yield was proportional to the availability of water; as stress intensity increased, grain yield decreased (Table 4.2). Rather than stressing crops during a specific growth stage, they reduced the irrigation dose at each irrigation time. In comparison with optimal irrigation (with irrigation dose equaling 100% ETc) with a yield of 5.53 ton/ha, yield under a moisture stress treatment with irrigation dose equaling 85% ETc was not significantly lower (5.20 ton/ha). Minimum grain yield was noted when irrigation dose was 25% ETc (1.47 ton/ha) and this was statistically inferior to all other treatments. Applying water by 25%, 35%, 45%, 55%, 65%, and 75% of ETc led to declined grain yield of maize by 12%, 26%, 36%, 53%, 62%, and 73%, respectively. In another study carried out by Admasu et al. (2017) on maize yield and WP under clay loam soil in western Ethiopia, the maximum grain yield (8.36 ton/ha) was attained under an irrigation dose of 100% ETc, whereas the minimum grain yield was found from no irrigation (1.02 ton/ha) and irrigating only initial stage (1.83 ton/ha). Grain yield production and the availability of water were proportionally decreased. Nevertheless, the highest WP (2.65 kg/m) was obtained from treatments irrigated only initial stage, whereas the lowest WP (0.50 kg/m) was recorded from optimal irrigation treatments. In line with these findings, Rusere et al. (2012) on the effects of DI on winter silage maize production in Zimbabwe conclude that, with increasing moisture stress, the dry matter production of the crop decreases directly by decreasing cell division and enlargement and indirectly by reducing the rate of photosynthesis. In Adami Tulu, Central Rift Valley Region of Ethiopia a field experiment under loam soil indicated the highest maize yield (4.52 ton/ha) from optimal irrigation with 100% ETc which was not significantly different from the 85% of ETc irrigation level (Furgassa, 2017). In terms of irrigation and WP, 50% ETc DI application gave the highest WP which was significantly different from all other treatment combinations. Overall, it appears that with irrigation doses being reduced from 100% to 85% ETc, maize yields are not significantly affected, while WP is significantly larger and more water can be saved. The application of 85% ETc irrigation water appears to be a promising alternative for water and labor-saving with a negligible trade-off in yield. But, applying water during crown root initiation, flowering, tasselling, and grain filling stages while depriving water during initial, vegetative, and late repining periods resulted in higher WP with insignificant maize yield reduction. Therefore, considering phenological response to water stress sounds better.

4.7.2 Potato

The effect of DI on yield and WP of furrow-irrigated potato (*Solanum tuberosum* L.) was studied in Atsibi-Wemberta, Northern Ethiopia by Kifle and Gebretsadikan (2016). They found that the potato yield was significantly affected by water stress (Table 4.2). The highest yield (18.77 ton/ha) was found under optimum irrigation (100% ETc), whereas minimum yield (7.04 ton/ha) was obtained under partially stressed treatment (50% ETc). Stresses occurring at the middle stage or flowering affected yield more as compared to treatments with stresses at the initial and maturing stage. Giving irrigation doses of 65% of ETc throughout the growing season proved to be better than stressing the crop only at the middle stage. Though not significantly different, 65% ETc and 80% ETc had the highest (2.86 kg/m^3) and the lowest (1.60 kg/m^3) WP, respectively. This demonstrated that applying 75% of ETc has better WP than optimal irrigation (2.71 kg/m^3). In an agro-ecologically comparable study in Melkassa, central Ethiopia, similar findings were reported by Seid and Narayanan (2015) on loam and clay loam soil. Similarly, Yihun et al. (2013), Yihun (2015), and Mengist and Tilahun (2009) found comparable results on clay loam soils in central Ethiopia. Those studies have also stated that applying DI with irrigation doses of 75% of ETc did not affect yield significantly. A field experiment conducted at Kulumsa in central Ethiopia indicated that irrigating at 100% ETc resulted in significantly highest tuber yield (33.39 ton/ ha[1]) which was similar to that when irrigating at 80% ETc (31.91 ton/ha) (Admasu et al., 2016). However, the lowest tuber yield was found at 60% ETc irrigation with a value of 28.77 ton/ha, which was comparable to tuber yield found at 120% ETc irrigation. They argued that referring to the highest tuber yield at 100% ETc, tuber yield decreases more as irrigation water amount increased (120% ETc and 140% ETc) than irrigation water amount decreased (80% and 60% ETc, respectively). For two consecutive years (2013–2014), Hilemicael and Tibebe (2018) conducted a DI experiment on potato in Holetta Agricultural Research Center on a clay loam soil. Among the tested treatments, 140% ETc and 80% ETc gave the highest (38.24 ton/ha) and the lowest yield (29.91 ton/ha), respectively. They also found the highest WP (14.62 kg/m³) under 60% ETc, whereas the lowest WP (0.24 kg/m³) under 140% ETc. The overall results showed that applying 75% of ETc throughout the growing season of potato has better WP than applying optimal irrigation with 100% ETc.

4.7.3 Tomato

In a study on the response of tomato (*Solanum lycopersicum* L.) to DI in Ambo Plant Protection Research Center in central Ethiopia, Kifle (2018) recorded that maximum tomato fruit yield (43.4 ton/ha) was gained under optimal irrigation, with irrigation doses (applied during the complete growing season) of 70% and 85% of ETc showing moderate yield decreases (32.1 and 34.2 ton/ha), respectively. The latter treatments showed, in turn, significantly better fruit yield than the 50% ETc treatment during the growing season. Similarly, Birhanu and Tilahun (2011) observed a significant decrease in total yield and marketable yields of tomato as the deficit level increased in southeast Ethiopia. The marketable yield was 41.5 and 15.1 ton/ha for 100% and 25% ETc irrigation water amounts, respectively. The total and marketable yield of tomato was the lowest for 25% ETc. The fresh fruit yield of tomato was reduced under 75%, 50%, and 25% ETc treatments by 6.8%, 48.5%, and 71.0%, respectively. Moderate water deficits could significantly improve fruit quality of tomato without depressing marketable yields in relation to fully irrigated treatments: size, shape, juiciness, and color of the fruit were improved, while total solids (dry matter content) and acid content were reduced. The decrease in solids will lower the fruit quality for processing. Prolonged water deficit also led to fruit cracking. In choosing the irrigation practices attention must, therefore, be given to the type of essential end product. Also in Ambo, central Ethiopia, Firrissa and Bekele (2021) found that tomato fruit yield and WUE responded significantly different to the different furrow systems as well as to different DI levels. The maximum yield was recorded at CFI and 100% ETc which was significantly different compared to fixed furrow irrigation with 50% ETc but not compared to AFI with 75% ETc. CFI with 100% ETc showed the lowest WP. Implementation of AFI would lead to 38%–40% more water being available to irrigate more land. Moreover, Bisa (2018), at Debrezeit Agriculture Research Center, central Ethiopia, reported no significant difference in yield between irrigating at all growth stages and only the first two stages because of considerable rainfall contribution at a mid and late stage. By irrigating only the initial and development stages, 42.38 ton/ha total harvested yield was recorded and 334.28 mm irrigation water was saved. Under these rainfall conditions, either irrigating or not in the late and mid-stage did not show any significant difference in yield. The control treatment gave the highest WP (30.79 kg/m³). Irrigating only in the first two stages results in a WP of 14.13 kg/m³ compared with irrigating in all stages which yielded the lowest WP value (7.40 kg/m³). For tomato in water-scarce areas applying 85%–70% of ETc is suggested as it results in a minimum yield reduction, while significantly enhancing WP. It can thus be concluded that the level of irrigation water supply should in most cases be 60%–100% of ETc needs to improve WP and fruit quality with less yield reduction.

4.7.4 Onion

The effects of DI on yield and WP of onion (*Allium cepa* L.) was assessed through field experiments at Ambo Agricultural Research Center, central Ethiopia, by Temesgen et al. (2018). Results indicated that the highest total onion yield (46.7 ton/ha) was obtained from a full irrigated treatment, which was not statistically different from treatments that were not irrigated during initial growth stage and irrigated with 75% ETc during the rest of phenological stages, and a treatment with no irrigation during bulb maturity stage and irrigated with 75% ETc. WP varied from 7.7 kg/m³ for the control to 14.9 kg/m³ for the 50% stressed and not irrigated during maturity-stage treatment. However, such yield reduction may not be tolerable from the perspective of individual producers. In a DI field trial using drip irrigation on clay loam soil at Melkassa, central Ethiopia, Enchalew et al. (2016) observed that marketable bulb yield and total bulb yield of onion were significantly different among DI treatments. The highest total bulb yield (15.69 ton/ha) was observed from 100% ETc which was not significantly different from the treatment receiving 90% ETc. The maximum WP (25 kg/mm) was observed for the treatment receiving 70% ETc while the biggest onion bulb diameter was observed from treatments receiving 100% to 70% ETc. The yield response factor Ky ranged between 0.8 and 1.7. Thus, considering Ky as a limiting factor, 80% ETc application resulted in marginal yield loss and beyond that yield losses were not acceptable. In a study carried out at Melkassa Research Center, central Ethiopia, on clay loam soil, Dirirsa et al. (2017) observed the highest bulb yield (40.38 ton/ha) from the control treatment with no significant difference with the 75% ETc treatments except when the deficit occurred at the bulb formation stage. The highest WP (11.74 kg/m) was recorded when there was no deficit at the bulb formation stage and a 25% deficit at other stages. It was thus shown that onion bulb yield was most sensitive to water deficit at the bulb formation stage. This can guide irrigation scheduling to achieve optimal onion bulb production under water-scarce conditions. Similarly, Tibebe and Hailemichael (2018) reported a significant difference in onion yield and WP at Holetta Research Center, central Ethiopia. The results show that the marketable onion yield was 29.07, 25.7, 28.23, 30.41, and 25.31 ton/ha from 60%, 80%, 100%, 120%, and 140% ETc DI treatments, respectively. Besides, the highest WP (8.61 kg/m^3) was recorded at 100% ETc followed by 60% ETc (8.12 kg/m^3) and 120% ETc (6.06 kg/m^3), and the water application of 80% ETc recorded the least WP of 5.64 kg/m^3, respectively. Thus, optimal irrigation is the best WP application without significant yield reduction. Overall, it can be concluded that emergence, transplanting, and bulb formation stages are most sensitive to drought stress. Skipping irrigation during maturity stages and irrigating with 75% ETc otherwise has more water-saving and WP enhancement potential with a tolerable level of yield reduction.

4.8 Conclusions

Traditional SSI practices have been implemented for centuries in Ethiopia though their contribution to livelihood improvement is limited. However, the rapidly growing population (~110 million) coupled with water scarcity and increasing competition for scarce water requires adequate food production and related agricultural products. This can be achieved by enhancing the productivity of water in agriculture using more efficient irrigation strategies such as DI, which has proved a success story for some crops in various parts of Ethiopia. Many findings revealed that DI usually has higher WP than optimal irrigation, resulting in more crop per drop. Flowering and crown root initiation stages are the most drought-sensitive periods compared to the initial and late-season stages. The adoption of DI has been constrained by many factors such as the absence of extension works and farmers' involvement in the DI experiments. The wise use of DI can generate significantly adequate irrigation water which could be used to bring an additional area under irrigation and could be taken as a better solution for water-scarce areas. Absence of databases about actual and potential irrigation maps, application of over- or under-irrigation, improper irrigation infrastructures such as canals, inadequate provision inputs, and market linkage problems were identified as the significant bottlenecks. Developing databases about current and future irrigation potentials; awareness creation for farmers about ETc, irrigation scheduling, and DI; involving them in DI trials; and arranging seminars could scale up the adoption of DI and thus improve yield and WP.

References

Admasu, R., Michael, A.W., and Hordofa, T. (2019). Effect of moisture stress on maize (*Zea Mays* L.) yield and water productivity. *International Journal of Environmental Sciences & Natural Resources*, 16(4):555945.

Admasu, R., Tadesse, M., and Shimbir, T. (2017). Effect of growth stage moisture stress on maize (*Zea Mays* L.) yield and water use efficiency at West Wellaga, Ethiopia. *Journal of Biology, Agriculture and Healthcare*, 7(23):98–103.

Admasu, W., Tadesse, K., Hordofa, T., Deresse, Y., and Habte, D. (2016). Determining of optimal irrigation regimes and NP fertilizer rate for potato (*Solanum tuberosum* L.) at Kulumsa, Arsi Zone, Ethiopia. *Academia Journal of Agricultural Research*, 4:326–332.

Agide, Z., Haileslassie, A., Sally, H., Erkossa, T., Schmitter, P., Langan, S., and Hoekstra, D. (2016). Analysis of water delivery performance of smallholder irrigation schemes in Ethiopia: Diversity and lessons across schemes, typologies and reaches. LIVES Working Paper 15, ILRI, Nairobi, Kenya.

Alcon, F., Tapsuwan, S., Martínez-Paz, J.M., Brouwer, R., and Miguel, M.D. (2014). Forecasting deficit irrigation adoption using a mixed stakeholder assessment methodology. *Technological Forecasting and Social Change*, 83: 183–193.

Ali, M.H. and Talukder, M.S.U. (2008). Increasing water productivity in crop production. A synthesis. *Agricultural Water Management*, 95:1201–1213.

Ali, M.H., Hoque, M.R., Hassan, A.A., and Khair, A. (2007). Effects of deficit irrigation on yield, water productivity, and economic returns of wheat. *Agricultural Water Management*, 92:151–161.

Amede, T. (2015). Technical and institutional attributes constraining the performance of small-scale irrigation in Ethiopia. *Water Resources and Rural Development*, 6:78–91.

Asmamaw, D.K. (2017). A critical review of the water balance and agronomic effects of conservation tillage under rain-fed agriculture in Ethiopia. *Land Degradation and Development*, 28:843–855.

Asres, S.B. (2016). Evaluating and enhancing irrigation water management in the upper Blue Nile basin, Ethiopia: The case of Koga large scale irrigation scheme. *Agricultural Water Management*, 170:26–35.

Awulachew, S.B. and Ayana, M. (2011). Performance of irrigation: An assessment at different scales in Ethiopia. *Experimental Agriculture*, 47:57–69.

Awulachew, S.B. and Merrey, D.J. (2008). *Assessment of Small Scale Irrigation and Water Harvesting in Ethiopian Agricultural Development*. IWMI, Addis Ababa, Ethiopia.

Awulachew, S.B., Merrey, D., van Koopen, B., and Kamara, A. (2010). Roles, constraints and opportunities of small-scale irrigation and water harvesting in Ethiopian agricultural development: Assessment of existing situation. In *Proceedings of the ILRI Workshop*, Colombo, Srilanka, 14–16 March 2010; IWMI, Addis Ababa, Ethiopia.

Awulachew, S.B., Merrey, D.J., Kamara, A.B., van Koppen, B., de Vries, F.P., Boelee, E., and Makombe, G. (2005). *Experiences and Opportunities for Promoting Small-Scale/Micro Irrigation and Rainwater Harvesting for Food Security in Ethiopia*. IWMI, Addis Ababa, Ethiopia.

Awulachew, S.B., Yilma, A.D., Loulseged, M., Loiskandl, W., Ayana, M., and Alamirew, T. (2007). *Water Resources and Irrigation Development in Ethiopia*. IWMI, Colombo, Sri Lanka.

Bekele, S. (2017). Response of tomato to deficit irrigation at Ambo, Ethiopia. *Journal of Natural Sciences Research*, 7(23):48–52.

Belete, Y., Kebede, H., Birru, E., and Natea, S. (2011). *Small-Scale Irrigation Situation Analysis and Capacity Needs Assessment. A Tripartite Cooperation between Germany, Israel and Ethiopia*. Sustainable Land Management Programme, Ministry of Agriculture, Addis Ababa, Ethiopia.

Beyene, A., Cornelis, W., Verhoest, N.E.C., Tilahun, S., Alamirew, T., Adgo, E., Pue, J. D., and Nyssen, J. (2018). Estimating the actual evapotranspiration and deep percolation in irrigated soils of a tropical floodplain, northwest Ethiopia. *Agricultural Water Management*, 202:42–56.

Birhanu, K. and Tilahun, K. (2011). Fruit yield and quality of drip-irrigated tomato under deficit irrigation. *American Journal of Food Agriculture, Nutrition and Development*, 10:2139–2151.

Bisa, M. E. (2018). *Yield and Water Productivity Analysis of Tomato Crop under Water Stress Condition in Ethiopia*. Apple Academic Press, London, pp. 147–160.

Cornelis, W., Waweru, G., and Araya, T. (2019). Building resilience against drought and floods: The soil-water management perspective. In *Sustainable Agriculture Reviews*, Lichtfouse, E. (ed). Springer, Cham, Switzerland, vol. 29, pp. 125–142.

Dananto, M.U. and Alemu, E. (2014). Irrigation water management in small scale irrigation schemes: The case of the Ethiopian rift valley lake basin. *Environmental Research, Engineering and Management*, 1:5–15.

Derib, D.S. (2013). Balancing Water Availability and Water Demand in the Blue Nile: A Case Study of Gumara Watershed in Ethiopia. PhD Dissertation, Bonn, Germany.

Dirirsa, G., Woldemichael, A., and Hordofa, T. (2017). Effect of deficit irrigation at different growth stages on onion (*Allium cepa L.*) production and water productivity at Melkassa, Central Rift Valley of Ethiopia. *Resource Journal Agriculture Science. Resource*, 5(5):358–365.

Doorenbos, J. and Kassam, A.H. (1986). Yield response to water. FAO irrigation and drainage paper no. 33. FAO, Rome, Italy.

Enchalew, B., Gebre, S.L., Rabo, M., Hindaye, B., and Kedir, M. (2016). Effect of defcit irrigation on water productivity of onion (*Allium cepa L.*) under drip irrigation. *Irrigation & Drainage Systems Engineering*, 5(3):1–4.

English, M., (1990). Deficit irrigation. I: Analytical framework. *Journal of Irrigation and Drainage*, 116:399–412.

Erkossa, T., Awulachew, S.B., and Aster, D. (2011). Soil fertility effect on water productivity of maize in the upper Blue Nile basin, Ethiopia. *Agricultural Sciences*, 2(3):238–247.

Erkossa, T., Hagos, F., and Lefore, N. (eds.). (2014). *Proceedings of the Workshop on Flood-based Farming for Food Security and Adaption to Climate Change in Ethiopia: Potential and Challenge*s, Adama, Ethiopia, 30–31 October 2013. IWMI, Colombo, Sri Lanka.

Falkenmark, M. and Lundqvist, J. (1995). *World Freshwater Problems: Call for a New Realism. Background paper to the Comprehensive Global Freshwater Assessment*. Stockholm Environment Institute, Stockholm, Sweden.

FAO. (1979). Yield response to water by J. Doorenbos & A.H. Kassam. Irrigation and Drainage Paper No. 33. FAO, Rome, Italy.

FAO. (2002). Deficit Irrigation practices. Water Reports 22. Rome, Italy.

Fereres, E. and Connor, D.J. (2004). Sustainable water management in agriculture. In *Challenges of the New Water Policies for the XXI Century*, Cabrera, E. and Cobacho, R. (eds.), A.A. Balkema, Lisse, the Netherlands, pp. 157–170.

Fereres, E. and Soriano, M.A. (2007). Deficit irrigation for reducing agricultural water use. *Journal of Experimental Botany*, 58(2):147–159.

Firrissa, O, and Bekele, S. (2021). Validation and demonstration of alternate furrow and deficit irrigation on yield and water productivity of tomato in farmers field at Ambo, Western Shoa, Ethiopia. *American-Eurasian Journal of Agronomy*, 14(3):39–45.

Foltz, J.D. (2003). The economics of water-conserving technology adoption in Tunisia: An empirical estimation of farmer technology choice. *Economic Development and Cultural Change*, 5(2):359–373.

Furgassa, Z.S. (2017). The effect of deficit irrigation on maize crop under conventional furrow irrigation in Adami Tulu Central Rift Valley of Ethiopia. *Applied Engineering*, 1(1):1–12.

Geerts, S. and Raes, D. (2009). Review: Deficit irrigation as an on-farm strategy to maximize crop water productivity in dry areas. *Agricultural Water Management*, 96: 1275–1284.

Grimes, D.W. and Yamada, Y.H. (1982). Relation of cotton growth and yield to minimum leaf water potential. *Crop Science*, 22:134–139.

Hagos, F., Makombe, G., Namara, R.E., and Awulachew, S.B. (2009). Importance of irrigated agriculture to the Ethiopian economy. Capturing the direct net benefits of irrigation. IWMI Research Report 128. Colombo, Sri Lanka.

Haileslassie, A., Agide, Z., Erkossa, T., Hoekstra, D., Schmitter, P., and Langan, S. (2016). On-farm smallholder irrigation performance in Ethiopia: From water use efficiency to equity and sustainability. LIVES Working Paper 19, ILRI, Nairobi, Kenya.

Hilemicael, K. and Tibebe, M. (2018). Optimal irrigation scheduling for potato (*Solanum tuberosum*) at Holetta. *Journal of Natural Sciences Research*, 7(11):14–19.

Hook, J.E., Kincheloe, S., and Segar, W.I. (2000). Irrigation scheduling for corn - Why and how. In: *National Corn Handbook*, Rhoads, F. M. and Yonts, C. D. (eds). Iowa State University Extension, Iowa.

Janssens, P., Deckers, T., Elsena, F., Elsena, A., Schoofs, H., Verjans, W., and Vandendriessche, H. (2011). Sensitivity of root pruned 'Conference' pear to water deficit in a temperate climate. *Agricultural Water Management*, 99:58–66.

Jensen, M.E. (2007). Beyond irrigation efficiency. *Irrigation Science*, 25(3):233–245.

Kibebew, K. (2003). Estimating water retention for major soils in Ethiopia Hararghe region, Eastern Ethiopia. A PhD Dissertation, University of the Free State, South Africa.

Kidane, D., Mekonnen, A., and Teketay, D. (2014). Contributions of Tendaho Irrigation Project to the improvement of livelihoods of agropastoralists in the lower Awash Basin, Northeastern Ethiopia. *Ethiopian e-Journal for Research and Innovation Foresight*, 6(2):1–19.

Kifle, M. and Gebretsadikan, T.G. (2016). Yield and water use efficiency of furrow irrigated potato under regulated deficit irrigation, Atsibi-Wemberta, North Ethiopia. *Agricultural Water Management*, 170:133–139.

Kifle, T. (2018). Evaluation of tomato response to deficit irrigation at Humbo Woreda, Ethiopia. *International Journal of Research - Granthaalayah*, 6(8):57–68.

Kijne, J., Barron, J., Hoff, H., Rockstrom, J., Karlberg, L., Gowing, J., Wani, P.S., and Wichelns, D. (2009). Opportunities to increase water productivity in agriculture with special reference Africa and Asia. A paper prepared by SEI for the Swedish Ministry of Environment presentation at CSD 16, New York, USA, 14 May 2009.

Kilob, S. (2015). Simulation-Optimization of the Management of Sensor-Based Deficit Irrigation Systems. PhD Dissertation, Technische Universität Dresden, Germany.

Kirda, C., Moutonnet, P., Hera, C., and Nielsen, D.R. (eds.) (1999). Crop Yield Response to Deficit Irrigation: Report of an FAO/IAEA Co-ordinated Research Program by Using Nuclear Techniques, Italy.

Kirda, C., Topcu, S., Kaman, H., Ulger, A.C., Yazici, A., Cetin, M., and Derici, M.R. (2005). Grain yield response and N-fertiliser recovery of maize under deficit irrigation. *Field Crops Research*, 93:132–141.

Mengist, Y. and Tilahun, K. (2009). Yield and water use efficiency of deficit-irrigated maize in a semi-arid region of Ethiopia. *AJFAND Online*, 9:1635–1651.

Mengistie, D. and Kidane, D. (2016). Assessment of the impact of small-scale irrigation on household livelihood improvement at Gubalafto District, North Wollo, Ethiopia. *Agriculture*, 6:27.

Meskelu, E., Mohammed, M., and Hordofa, T. (2014). Response of Maize (*Zea mays* L.) for moisture stress condition at different growth Stages. *International Journal of Recent Research in Life Sciences*, 1(1):12–21.

Meskelu, E., Woldemichael, A., and Hordofa, T. (2017). Effect of moisture stress on yield and water use efficiency of irrigated wheat (*Triticum aestivum L.*) at Melkassa, Ethiopia. *Academic Research Journal of Agricultural Science and Research*, 5(2):90–97.

Molden, D. (ed.) (2007). *Water for Food, Water for Life: A Comprehensive Assessment of Water Management in Agriculture*. Earthscan/IWMI, London/Colombo, Sri lanka.

Molden, D., Oweis, T., Steduto, P., Bindraban, P., Hanjra, A.M., and Kijne, J. (2010). Improving agricultural water productivity: Between optimism and caution. *Agricultural Water Management*, 97:528–535.

Oweis, T. and Hachum, A. (2009). *Supplemental Irrigation for Improved Rain-fed Agriculture in WANA Region*. Integrated Water and Land Management Program, International Center for Agricultural Research in the Dry Areas (ICARDA), Aleppo, Syria.

Oweis, T. and Hachum, A. (2012). *Supplemental Irrigation, A Highly Efficient Water-Use Practice*. ICARDA, Aleppo, Syria.

Pereira, L.S., Cordery, I., and Iacovides, I. (2012). Improved indicators of water use performance and productivity for sustainable water conservation and saving. *Agricultural Water Management*, 108:39–51.

Rockström, J., and Falkenmark, M. (2015). Agriculture: Increase water harvesting in Africa. *Nature*, 519:283–285.

Rockström, J., Falkenmark, M., Karlberg, L., Hoff, H., Rost, S., and Gerten, D. (2009). Future water availability for global food production: The potential of green water for increasing resilience to global change. *Water Resources Research*, 45(7):1–16.

Rockström, J., Karlberg, L., Wani, S.P., Barron, J, Hatibu, N., Oweis, T., Bruggeman A., Farahani, J., and Qiang, Z. (2010). Managing water in rain-fed agriculture: The need for a paradigm shift. *Agricultural Water Management*, 97:543–550.

Rodrigues, G.C., Paredes, P., Gonçalves, J.M., Alves, I., and Pereira, L. S. (2013). Comparing sprinkler and drip irrigation systems for full and deficit irrigated maize using multicriteria analysis and simulation modelling: Ranking for water saving vs. farm economic returns. *Agricultural Water Management*, 126:85–96.

Rusere, F., Soropa, G., Svubure, O., Gwatibaya, S., Moyo, D., Ndeketeya, A., and Mavima, G.A. (2012). Effects of deficit irrigation on winter silage maize production in Zimbabwe. *International Research Journal of Plant Science*, 3(9):188–192.

Schmitter, P., Haileslassie, A., Desalegn, Y., Chali, A., Langan, S., and Barron, J. (2017). Improving on-farm water management by introducing wetting-front detector tools to smallholder farms in Ethiopia. LIVES Working Paper 28, ILRI, Nairobi, Kenya.

Seid, M.M. and Narayanan, K. (2015). Effect of deficit irrigation on maize under conventional, fixed and alternate furrow irrigation systems at Melkassa, Ethiopia. *International Journal of Engineering Research and Technology*, 4 (11):119–126.

Steduto, P., Hsiao, T.C., Raes, D., and Fereres, E. (2009). AquaCrop -the FAO crop model to simulate yield response to water: I. Concepts and underlying principles. *American Society of Agronomy*, 101:426–437.

Steduto, P., Hsiao, T.C., Fereres, E., and Raes, D. (2012). Crop yield response to water. FAO Irrigation and Drainage Paper No. 66. Rome, Italy.

Temesgen, T., Ayana, M., and Bedadi, B. (2018). Evaluating the effects of deficit irrigation on yield and water productivity of furrow irrigated onion (*Allium cepa* L.) in Ambo, Western Ethiopia. *Irrigation & Drainage Systems Engineering*, 7:203.

Tibebe, M. and Hailemichael, K. (2018). Water requirement and optimal irrigation on onion yield and productivity. In *Book of Irrigation and Watershed Management*, Atlabachew, W., Mohammed, M., and Abegaz F. (eds), Ethiopian Institute of Agricultural Research, Addis Ababa, Ethiopia.

Tsegay, A. (2012). Improving crop production by field management strategies using crop water productivity modeling: Case study of tef (*Eragrostis tef* (zucc.) Trotter) production in Tigray, Ethiopia. PhD thesis, KU Leuven, Belgium.

Ward, F.A. and Pulido-Velazquez, M. (2008). Water conservation in irrigation can increase water use. *Proceedings of the National Academy of Sciences*, 105:18215–18220.

Yihun, Y. M., Haile, A.M., Schultz, B., and Erkossa, T. (2013). Crop water productivity of irrigated teff in a water stressed region. *Water Resources Management*, 27:3115–3125.

Yihun, Y.M. (2015). Agricultural water productivity optimization for irrigated teff (Eragrostic tef) in a water scarce semi-arid region of Ethiopia. PhD thesis, UNESCO-IHE Institute for Water Education, the Netherlands.

Yohannes, D.F., Ritsema, C.J., Solomon, H., Froebrich, J., and van Dam, J.C. (2017). Irrigation water management: Farmers' practices, perceptions and adaptations at Gumselassa irrigation scheme, North Ethiopia. *Agricultural Water Management*, 191:16–28.

Zahraei, A., Saadati, S., and Eslamian, S., (2017). Irrigation deficit: Farmlands. In *Handbook of Drought and Water Scarcity, Vol. 3: Management of Drought and Water Scarcity*, Eslamian S. and Eslamian F. (eds), Taylor and Francis, CRC Press, Boca Raton, FL, pp. 343–358.

5

Investigating the Working Hydraulic Conditions of Gezira Two Main Canals in Sudan

Ageel I. Bushara
Zuhair Fayez Partnership Consultants

Younis A. Gismalla, AbdelNassir Osman, Y. Mohamed, and Ruaa A. Nasreldeen
Ministry of Irrigation and Water Resources

Saeid Eslamian
Isfahan University of Technology

5.1 Introduction

Gezira scheme is the largest irrigated scheme in Sudan. It is located in the central clay plain of Sudan and is irrigated from the Blue Nile. The scheme was constructed in 1925 and the irrigated area was doubled after the construction of Managil Extension in mid-1960s. The total irrigated area of the scheme is 0.9 ha. Gezira main canal was designed in 1920 and was widened twice before the construction of Managil main canal. When Managil main canal was designed in the 60s of the last century, Gezira main canal was almost 40 years of efficient operation in the heavily silted water of the Blue Nile. Thus, Managil main canal was constructed to form a second twin canal to the Gezira main canal, and it was supposed to be a duplicate of the Gezira canal. The same regime bed slope of Gezira, 6.5 cm/km, was adopted in the design of Managil canal (see Table 5.1). But, the bed level of Managil canal was 40 cm above that of Gezira, thus limiting the water depth to 3.95 m when keeping the working downstream water level at 416.98 m for both canals. The reason for raising the bed level of Managil canal is that Managil canal was designed to serve the area near Managil ridge which is higher than the area served by Gezira canal. Another reason is that Gezira main canal starts to branch from K57 with many regulators compared to Managil main canal.

The latest design section of Gezira main canal is shown in Table 5.1. This table also shows the hydraulic parameters of Managil main canal. The design of Managil canal was based on the Lacey's regime equations developed by Matthews for the Gezira canalization system in the early 50s assuming a single value for Lacey's silt factor ($f=0.63$), (Gismalla and Fadul, 2011). The design section of Managil canal is wider and shallower than that of Gezira. Gezira main canal is known to be stable, non-silting and non-scouring, while Managil main canal is not stable and the sediments deposit in the canal. The Gezira two main canals take off from Sennar dam on the Blue Nile. They run for 57 km before they discharge in a

DOI: 10.1201/9781003353928-7

TABLE 5.1   Different Hydraulic Parameters for Gezira and Managil Canals (Gismalla and Fadul, 2011)

Item	No. of Gates	BL (m)	B (m)	D (m)	Z	n	V (m/s)	S (10⁻⁵)	Q (m³/s)
Gezira	14	412.63	42.6	4.35	2:1	0.025	0.750	6.5	168
Managil	11	413.03	50.0	3.95	2:1	0.022	0.813	6.5	186

BL is the bed level above mean sea level (m.a.m.s.l), *B* is the bed width (m), *D* is the water depth (m), *Z* is the side slope (m/m), *n* is Manning's coefficient ($m^{-⅓}s$), *V* is the water velocity (m/s), *S* is the longitudinal slope (m/m), and *Q* is the discharge (m³/s).

common pool at K57. A group of head regulators at K57 distributes irrigation water between old Gezira and Managil Extension. The irrigation system in Gezira scheme comprises main, branch, major and minor canals.

The Blue Nile originates from Ethiopian highlands and brings high sediment loads during its flood season (July–October). This high sediment load has major influences on the design and the operation of the irrigated schemes. The reduced conveyance capacity of Managil main canal is one of the sedimentation impacts. The difference in the conveyance of the Gezira two main canals is always questioned, as Gezira is running its full design capacity while Managil main canal can only pass 60% of its design capacity.

The main objective of this study is to investigate the roughness and other hydraulic parameters of the Gezira two main canals under the prevailing sediment-laden flow conditions to better understand the hydraulic design parameters, and specifically, to work out the operational Manning's coefficients for the two main canals with high accuracies and correlate with bed slope, side slope, and water depths.

This chapter also analyses the existing conveyance capacity of the two canals, design sections, operating conditions, and sediment analysis to give conclusions on the different parameters.

5.2 Conveyance Capacity of Gezira Main Canals

Measurements of discharge, water depth and water width of the Gezira two main canals, viz. Gezira and Managil, were conducted on many occasions covering a complete hydrological year and different flow stages, low, medium and high. The measurements started in the July 2010 and ended in June 2011.

The maximum conveyance capacity for both Gezira and Managil main canals was calculated from these measurements is 243 m³/s. The maximum conveyance capacity of the two main canals now is the same, although the design capacity of Managil canal was 111% that of Gezira.

To study current hydraulic properties of the two canals, different parameters are investigated. Hydrographic surveys conducted on the two main canals have shown that the bed slope of both canals remained within the design value (6.5 cm/km), and side slopes are 2:1 (Gismalla et al., 2002). Using Manning's formula:

$$Q = \frac{AR^{\frac{2}{3}}\sqrt{s}}{n}, \tag{5.1}$$

where

  Q is the discharge (m³/s)
  A is the cross-sectional area (m²)
  R is the hydraulic radius, and is defined as follows:

$$R = \frac{A}{P}, \tag{5.2}$$

where

  A is the cross-section area (m²)
  P is the wetted perimeter (m).

Manning's formula can be linearized based on log transformation as follows:

$$\log Q = \log\left(AR^{\frac{2}{3}}\sqrt{S} \right) + \log\frac{1}{n},$$ (5.3)

where

    S is the longitudinal slope (m/m)
    n is the Manning's coefficient (m$^{-\frac{1}{3}}$s).

Equation (5.3) is the linearized form of Manning's formula and the formula can be represented graphically by plotting log Q in the x-axis and the log $(AR^{2/3} S^{1/2})$ on the y-axis as scatter plot and then can be fitted with linear line. The intersection of the linear line is the log $(1/n)$ and hence obtaining n.

For each canal and for each measurement occasion, the cross-sectional area and the hydraulic radii were calculated. Discharges were measured using Armfield current-meter type. The canals' cross-sections were divided into segments and the water velocity in each segment was measured using two-point method and the discharge was computed using mean section method (Boiten, 2003).

5.3 Sediment Entering Gezira Canals

The Blue Nile brings annually about 140 million tons of sediment during its flood season – July–October (Gismalla, 2009a). This sediment material originates mainly from heavy erosion in the upper catchments in Ethiopia. More than 97% of this transported sediment is very fine material composed of silt and clay (Gismalla, 2009a). The sediment concentration varies throughout the flood season having its maximum concentration in the second period of July.

Studies have shown that the sediment concentrations entering Gezira two main canals at Sennar are monotonically increasing from 700 ppm in the 1930s, to 3,800 ppm in the 1980s (HRL Wallingford, 1990) and more than 15,300 ppm in the last 10 years (Gismalla and Fadul, 2011). This fact is reflected on the trend of the total sediment load entering Gezira scheme which shows an increasing rate of 0.4 million tons/year (Gismalla, 2009b; Figure 5.1). Also, the sediment concentrations entering Managil main canal at Sennar are 10% lower than those entering Gezira main canal. This may be attributed to the flow curvature introduced by the intake configuration of Managil canal (Gismalla, 2005; Figure 5.2). On average 8.8 million tons of sediment enters Gezira scheme through Gezira and Managil main canals at Sennar. About 60% of this sediment enters the scheme through Gezira main canal (Gismalla, 2009b). Therefore, sediment concentrations entering Managil canal are not the cause of reduced water conveyance of the canal, as sediment concentrations entering Gezira canal are higher than those entering Managil canal.

5.4 Results and Discussions

Working Manning's coefficients (n) were determined based on all measurements and not on a single value. Manning's formula was linearized, Eq. (5.3), to obtain one value of n for each canal using all measurements. Figures 5.3 and 5.4 show the linearized form of Manning's formula for Gezira and Managil canals, respectively. Log $(AR^{2/3} S^{1/2})$ was plotted in the x-axis and log Q was plotted in y-axis. Very strong linear relations between log $(AR^{2/3} S^{1/2})$ and log Q were observed for both canals. The coefficients of determination, R^2, for Gezira and Managil were 0.98 and 0.97, respectively.

The derived working Manning's coefficients for Gezira and Managil were 0.013 and 0.021, respectively, which are different from the design values. The reduction of n coefficients for both canals indicates that the estimated discharges using design Manning's n values for Gezira and Managil are much less than the actual discharges. The value of n (Manning's coefficient) depends on the smoothness/roughness of the surface (Chow, 1959). The working Manning's coefficient for Gezira canal is almost half the design value. This indicates that the canal bottom and sides have become very smooth.

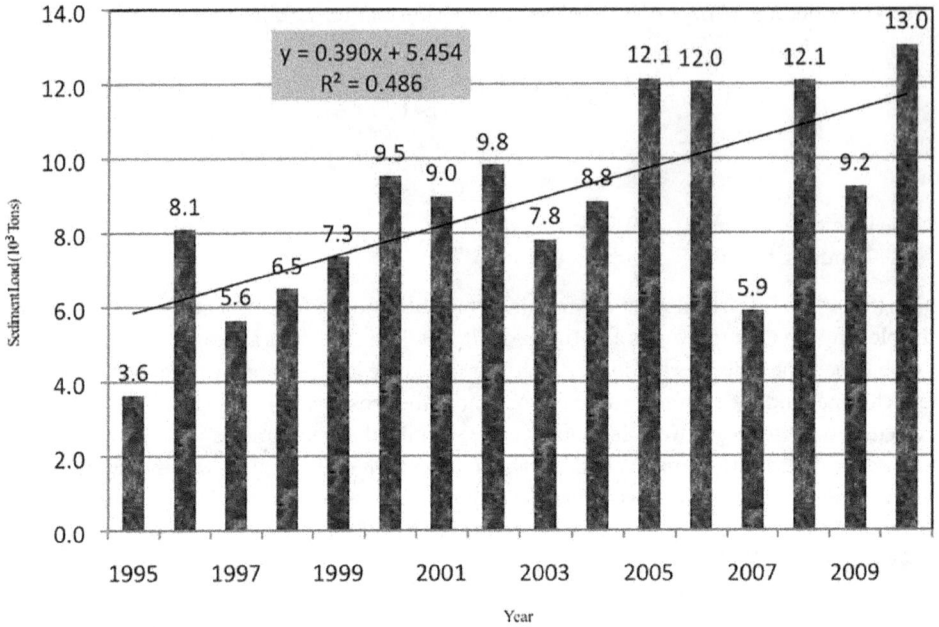

FIGURE 5.1   Annual sediment load entering Gezira scheme (1995–2010).

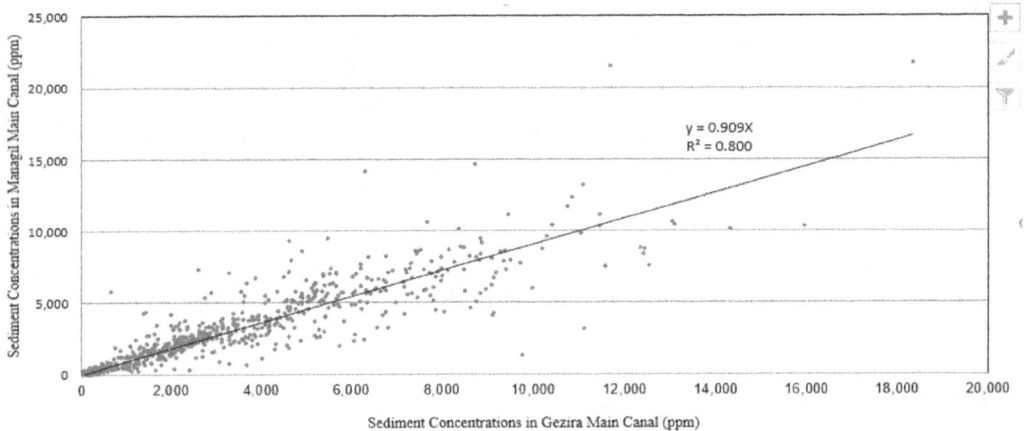

FIGURE 5.2   Sediment concentrations in the off-takes of the two main canals at Sennar (2002–2010).

The working Manning's n value for Managil has remained within the design value. The design values of n for Gezira and Managil canals were based on a certain roughness; however, with the operation of the two canals the parameters of the canals were changed. Surveys have shown that the effect of scouring for both canals is minimal (Gismalla et al., 2002), and no significant scouring was seen in the canals. Therefore, the differences between the design and the working Manning's coefficients are mainly due to sedimentation in the canals. These differences in hydraulic roughness in the two canals can be attributed partly to the incoming sediment concentrations, as the concentrations in Gezira off-take is slightly higher than those in Managil off-take (see Figure 5.2). From the Manning's formula it is noted that small change in n leads to a significant change in the estimated discharge. Therefore, the estimated discharges

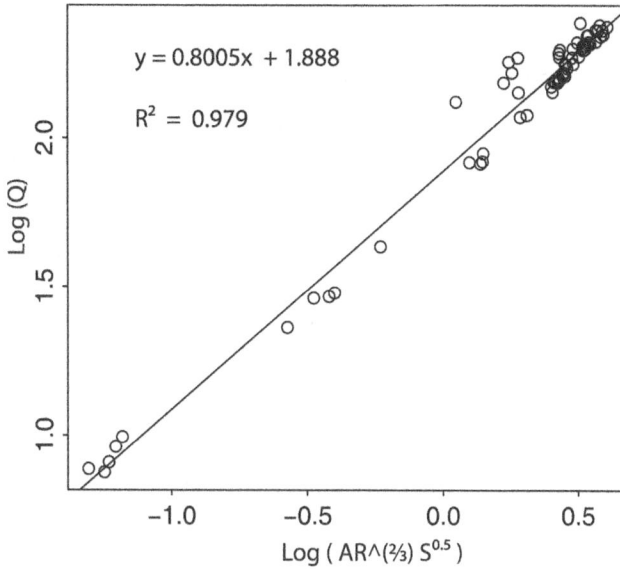

FIGURE 5.3 Linear relation between log ($AR^{2/3} S^{1/2}$) and log Q for Gezira canal. The derived Manning's coefficient is 0.013.

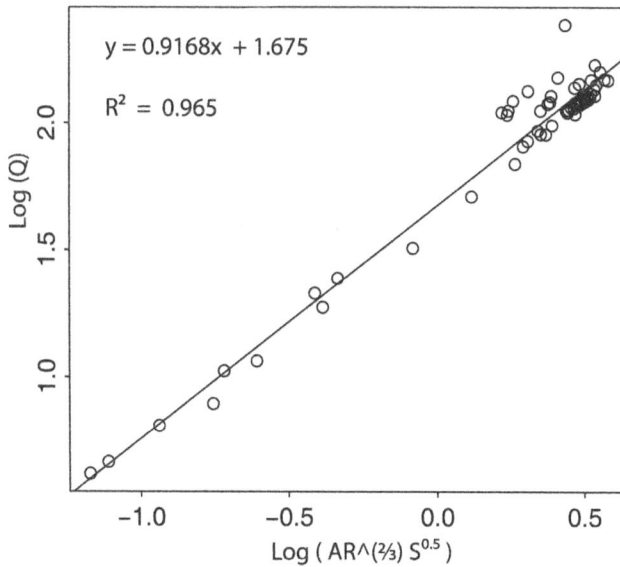

FIGURE 5.4 Linear relation between log ($AR^{2/3} S^{1/2}$) and log Q for Managil canal. The derived Manning's coefficient is 0.021.

with Manning's formula and using design n value for Gezira and Managil are much different from the discharge estimated using the same Manning's formula and the actual Manning's coefficients.

The design section of Managil canal is wider and shallower than that of Gezira; from the fact that stable canals, which carry fine sediment, have deeper and narrower sections it can be seen that Gezira section is more adequate than Managil section. The continuous human intervention to increase the carrying capacity of Managil canal has also contributed to the high roughness of the canal.

5.5 Conclusions

Measured discharges for a whole hydrological year in the Gezira two main canals were used for computing the working Manning's n values for Gezira and Managil canals. The following conclusion points can be drawn from the study:

- The working Manning's n values for Gezira and Managil canals were found to be 0.013 and 0.021 compared to the design values of 0.025 and 0.022, respectively.
- The estimated discharges using design Manning's n values for Gezira and Managil are much less than the actual discharges.
- No significant scouring was seen in the two canals, and the sediment concentration at the off-takes of the two canals is almost the same.
- The differences between the design and the working Manning's coefficients are mainly due to sedimentation in the canals.
- Gezira main canal became very smooth because of the adequate, deep and narrow, design section for the Blue Nile waters with high suspended sediment concentrations.
- Managil main canal has higher roughness because of the inadequate, flat and shallow, design section and human intervention to increase the carrying capacity of the canal.
- The dimensions of Gezira main canal can be considered in designing new stable canals in the similar hydraulic and soil conditions (Yousefi et al., 2016).

Acknowledgments

The authors would like to acknowledge the HRS-Sudan team who collected the field data.

References

Boiten, W., (2003). *Hydrometry, IHE Delft Lecture Note Series*. A.A. Balkema Publishers, Tokyo, Japan.

Chow, V.T. (1959), *Open Channel Hydraulics*, McGraw-Hill, New York.

Gismalla, Y. A. (2005). The Sediment Monitoring Programme for the Blue Nile System, Annual Report for the Year 2004. HRS-Sudan. Ministry of Irrigation and Water Resources - Sudan.

Gismalla, Y. A. (2009a). Sedimentation problems in the Blue Nile reservoirs and Gezira scheme canals: A review. *Gezira Journal of Engineering and Applied Sciences*, 4(2), 1–19.

Gismalla, Y. A. (2009b). *The Sediment Monitoring Programme for the Blue Nile System, Annual Report for the Year 2008*. HRS-Sudan. Ministry of Irrigation and Water Resources, Wad Medani, Sudan.

Gismalla, Y.A and Fadul E.M. (2011). Design and remodelling of stable canals in central Sudan. *Sudan Journal of Agricultural Research (Sudan)*, 17, 123–138.

Gismalla, Y.A., Bashir A.A., Ibrahim, E.A. and Adam, A.S. (2002). *Remodelling Managil Main Canal to Maintain a Stable Section*. Ministry of Irrigation and Water Resources, Wad Medanim, Sudan.

HRL Wallingford (1990). Research for Rehabilitation: Siltation monitoring study. Final Report. Ministry of Irrigation and Water Resources, Sudan.

Yousefi, N., Khodashenas, S. R., Eslamian, S. and Askari, Z. (2016). Estimating width of the stable channels using multivariable mathematical models, *Arabian Journal of Geosciences*, 9(321). DOI: 10.1007/s12517-016-2322-0

6

Plant Water Requirements and Evapotranspiration

Never Mujere
University of Zimbabw,
Department of Geography
Geoaspatial Sciences
and Earth Observation

Rajendra
Kumar Isaac
Deemed University

6.1 Introduction

Irrigation helps to avert major food crises in dry areas, a situation which most developing states seem to have become endemic. The activity is gaining prominence as about 20% of the world's 1500 million ha irrigated provide 40% and 70% of food and cereal production, respectively. In most developing countries is food produced from irrigated lands. For example, 50% of Indonesia, 70% of China and 80% of Pakistan's food requirements come from irrigated lands. Sudan's 85,000 ha Gezira Irrigation Scheme produces about 70% of Sudan's cotton, 50% of wheat, 30% of beans and 12% of sorghum. The scheme produces almost 50% of Sudan's total revenue (FAO, 2000). Given the importance of irrigation as a means of securing food in national economies, there is rising investment in irrigation projects by international donors and government departments. Some government agencies have been created primarily to spearhead irrigation development and monitor the performance of irrigation systems in most countries. Examples include the National Irrigation Administration in Philippines, International Water Management Institute in Sri Lanka and the Department of Agricultural Engineering in Zimbabwe.

DOI: 10.1201/9781003353928-8

The importance of adequate and reliable water supply in irrigated agriculture cannot be over-emphasised. However, information on the extent to which water supply meet crop water requirements at different phenological stages is critical for planning and management. This chapter, therefore, analyses the relationship between bean and wheat crop yields and water supply at needs at Nyanyadzi Irrigation Scheme in Zimbabwe.

6.2 Performance of Irrigation Agriculture

Despite the considerable potential of irrigation, several smallholder irrigation projects in most developing countries have been labelled socio-economic failures (Palmer-Jones, 1986; FAO, 2000). Low productivity is due to lack of inputs, low capacity of farmers to optimise production, lack of sufficient incentives, shortage of labour, water scarcity, little knowledge on field operation procedures, poor market access and lack of capital among other causes. Poor marketing strategies often result in most horticultural crops rotting in the fields (Meinzen-Dick, 1994; Rukuni et al., 1994; Mate, 1996; FAO, 2000; Tafesse, 2003).

Poor technical components of irrigation projects cause inefficient water use. Most smallholder schemes have irrigation efficiencies less than 60% due to weak water control measures (DFID, 2003). Thus, a lot of water is wasted and farmers have a general tendency to irrigate too soon and/or apply too much water. For example, FAO (2003) observed that farmers in Egypt were on average applying 50%–150% more water than was needed by the crops and for leaching requirements. Excess water supply caused water logging, thus reducing crop yields. Figure 6.1 shows causes and effects of poor irrigation performance.

The performance of an irrigation system is measured by the levels of its achievement in terms of one or several parameters that are chosen as indicators of the system's goals (Makadho, 1994). Notable indicators of irrigated agriculture performance are water delivery, agricultural production and socio-economic. Agricultural productivity is assessed in terms of crop yields (t) per unit irrigated area (e.g., t/ha) or per volume of water supplied (e.g. t/m^3 or t/mm).

Crop yields (t/ha) are estimated from crop output (t) and cropped area (ha). The ratio also indicates choices farmers make based upon the available of factors of production (Rukuni et al., 1994).. Poor agricultural production translates into poor financial and economic viability. Socio-economic performance indicators include irrigation gross margins and whether social objectives have been met (Pearce and Lewis, 1988; Pearce and Armstrong, 1990). Water delivery performance assesses the provision of water to the crop whether at the right time, at the right amount and at the right place. Indicators of water delivery performance include adequacy, reliability, equity and timeliness of water supply (Pearce and Lewis, 1988; Makadho, 1994; Chancellor and Hide, 1997). Adequacy measures the sufficiency of water in the irrigation system. It is an indirect measure of adequate soil moisture conditions in the root zone because it reflects the magnitude of reduction of water uptake by roots for evapotranspiration. To ascertain whether water supply was adequate, a comparison between water supply and demand was done. An index of comparison, known as relative water supply (RWS), provided data about the relative abundance or scarcity of water in the fields by matching water available to the farmers with that which was actually needed by them. RWS is estimated by dividing water supply and demand. Although RWS ≥ 1 implies that supply is adequate, in practice, RWS = 80% denotes the minimum requirements below which significant yield reduction occurs (Makadho, 1994).

Equity represents the uniformity or fairness in the spatial distribution of water use among fields by different farmers. The spatial coefficient of variation (CV) of evapotranspiration can be used as an indicator of equity of water use (Meinzen-Dick, 1994). Low CV indicates high level of equitable distribution of water in the system and vice versa.

Reliability of irrigation water supply denotes the degree to which water deliveries conform to some schedules or prior expectations of users (Nyamayevu et al., 2018). In Zimbabwe, a 90% reliable level is assigned for water supply in irrigation schemes.

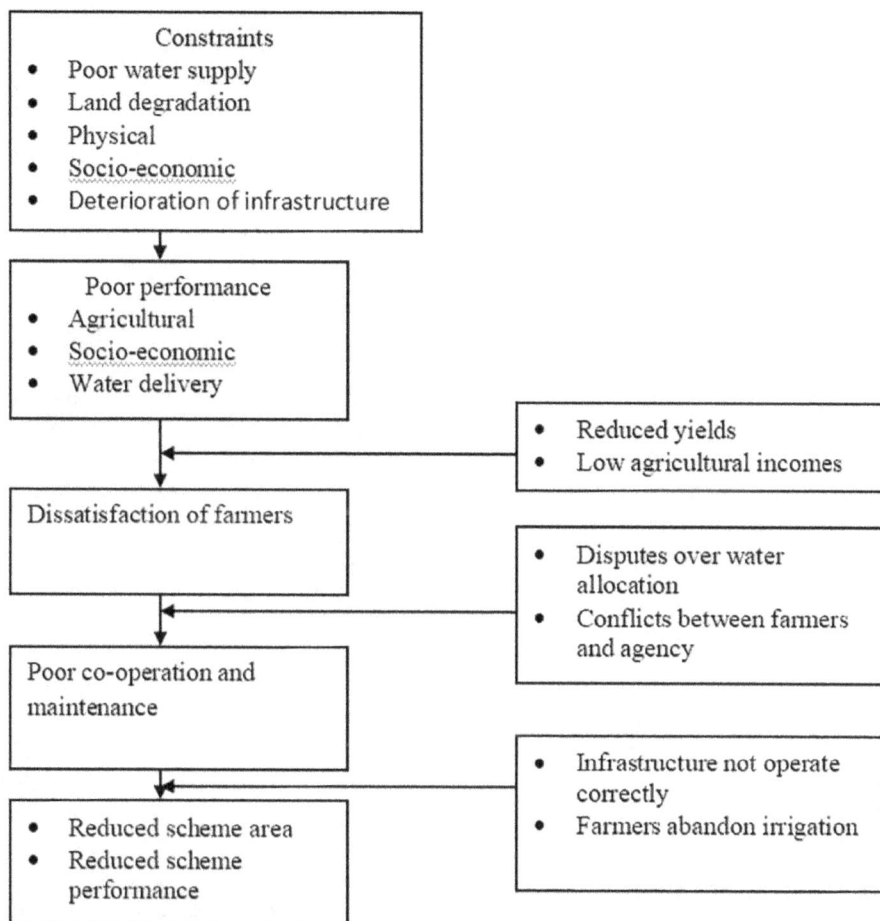

FIGURE 6.1 Causes and effects of poor irrigation scheme performance (Chancellor and Hide, 1997: p. 8).

Timeliness of water supply means correspondence of water deliveries to crop water or evapotranspiration needs (Makadho, 1996). Thus, water needs to be delivered to the crop timely to avoid crop water stress.

6.3 Water and Irrigation Agriculture

Irrigation agriculture though an important and, in many areas, indispensable activity to increase agricultural production in semi-arid regions of the world, is highly dependent on the availability of water. Irrigation water supply accounts for about 70% of world water abstraction and is delivered from dams or reservoirs, rivers and underground reserves. Rivers supply about 60% of the water used for irrigation (FAO, 2003). The growth of plant material is produced from a combination of water and other factors like air, soil, sunlight and human inputs. Water, as an essential component of all plant tissues, fulfils three primary functions of keeping plants erect when tissue cells are filled, cooling plants during evaporation from leaves, thus preventing overheating under hot conditions, and carrying nutrients in solution from the soil into the plants via roots (FAO, 2003). The delivery of water from the source to the irrigated field is critical since it affects the overall crop productivity in the irrigated plots.

The yield of a crop is determined by its genetic characteristics and the governing conditions in the growing environment. Such environmental factors include weather conditions, pests, diseases, soil fertility, salinity and soil water stress. Many studies have established that other factors being equal, the productivity of a given soil type is dependent on the supply of water during the crop growing season (Raes et al., 2003). Soil moisture is arguably the most significant parameter affecting crop yields and other variables such as albedo, canopy temperatures and evapotranspiration depend directly or indirectly on the soil moisture available to the crop (FAO, 2000).

The effects of water stress on yields vary according to stress duration, crop growth stage and crop variety. As compared to subsistence crops, horticultural crops are more sensitive to water supply (Lauraya and Sala; 1995). A sufficient quantity of water should be readily available to replenish soil water used by the crop during germination and flowering periods to protect yield potential and create a profitable return. FAO (2003) stated that for each additional water supply of 10 mm during the growing season, wheat production increases by 6.4%, while water supply during the harvesting season reduced the output by 39.6% of total production.

Rice variety IR20 has a potential yield of 6.2 t/ha which is achieved when the crop is flooded throughout the growing period. Early stress (43–81 days after seeding) reduces potential yield by 30%. Stress experienced when no irrigation is applied from 63 to 102 days after seeding (during the reproductive stage) reduces potential yield by 66%. Late stress (no irrigation from 63 days after seeding to harvest) causes yield reduction of 92%. Stress during the reproductive stage has drastic effects on yields because the potential recovery of the crop is very low. In the early vegetative period, crops have time to partially recover from stress, hence less drastic effect on yields. As compared to rice variety IR20, the rice variety IR5 recovers faster from low moisture stress (Chancellor and Hide 1997).

Poor spatial water distribution over the primary and secondary components of the irrigation systems results in major differences in water availability from one tertiary unit to the other, hence a key problem causing water shortage in irrigation schemes. This causes excess water to be provided in one area while others experience shortages. Head and middle irrigators often receive adequate or excess water which can be lost as surface drainage, while deficit water supply occurs at tail ends. Consequently, higher crop yields are obtained from fields closer to water sources than those far away.

Studies in Zimbabwe have shown how maize yields vary with respect to water supply from a night storage dam. Inequitable water distribution resulted in head irrigators achieving 20% higher yields than at the tail ends because they received on average twice the water supply. Maize yields decreased by 2.1 kg/ha for every 1 m distance from the plot closest to the night storage dam (Pearce and Lewis, 1988; Pearce and Armstrong, 1990; Pazvakavambwa and Van der Zaag, 2000). However, water distribution problems and conveyance losses (leakages and overtopping) along channels often cause water shortages at the head. In some instances, excess irrigation water taken by head irrigators led to loss of irrigated fields due to water logging, leaching of fertilisers and nutrients from the root zone, thus reducing crop yields. Flooding problems also cause post-flood humidity favouring the spread of crop diseases (Chancellor and Hide, 1997). In order to avoid over- and under-irrigating, it is prudent to provide water to the plants in amounts to satisfy the crop water needs. This can be achieved by objectively balancing water supply to the field and crop water demands.

No single factor can explain poor water supply in a given scheme. Low water supply from the sources to the irrigation schemes result from pump and engine breakdowns, low catchment rainfall especially in winter and during drought periods, poor groundwater and sand abstraction yields, siltation of rivers, storage weirs and dams, glaciation of water bodies, low river flows and dam levels. Inequitable irrigation water supply result from not changing spray-lines, poor infrastructure, canal leakage and corruption by administrators (Makadho, 1994; Lauraya and Sala, 1995; Tafesse, 2003). For example, a study by Manzungu (1996) at Chibuwe Irrigation Scheme in Zimbabwe showed that differential water supply resulted in relatively poor water supply to all the four irrigation blocks. Factors responsible for the variations in water supply included shifting of Save River course, thus reducing pump intake; accumulation of silt in the diversion canals; reduced Save River flows; poor condition of physical infrastructure; and frequent breakdown of electric pumps serving the irrigation blocks.

Besides affecting agricultural production, inadequate irrigation water supply often causes conflicts between stakeholders within irrigation schemes. Levine (1980) asserted that conflicts are inevitable because irrigation takes place in environments of significant variability and involves large numbers of people and groups. Conflicts can be internal or external. Internal conflicts occur within the bureaucracy, between the bureaucracy and farmers, between irrigators and other water users or among the farmers. Due to shortages of water in irrigation schemes, farmers and extension personnel clash on cropping programmes to be followed. In times of inadequate water supply farmers are less likely to follow the schedule of water delivery; hence, they clash over the priority of water use. Conflicts between head and tail irrigators had disturbed water distribution schedules (Mate, 1996). Upstream irrigators often cause the quality and quantity of water to diminish downstream (Smith, 1993; Mate, 1996; Manzungu, 1999). Such problems intensify during periods of water shortage when downstream users are left with less water for agricultural, domestic, recreational and industrial purposes.

6.4 Coping with Problems of Water Supply in Smallholder Irrigation Schemes

A number of strategies are employed in irrigation schemes when water supply is inadequate. Most interventions are short term and more than one strategy can be applied at a time. Measures Strategies to alleviate water shortage in irrigation schemes include; engineering works, cropping patterns and cropping methods. Engineering strategies comprise the use of irrigation systems giving maximum efficiency, conjunctive use of surface and ground water sources. Cropping patterns involve crop rotation, mixed cropping and spreading planting dates. Using different maturing varieties reduces the risk of complete crop production failure in times of water shortages.

Cropping methods include deep cultivation, cutting acreage, water harvesting, composting, mulching, drip irrigation, deficit irrigation and precision irrigation. Reducing the sizes of plots is done to spread the water scarcity equitably over all the plots. In Zimbabwe's Nyamaropa Irrigation Scheme, although winter season river flows and catchment rainfall decreased by 30 L/s and 158 mm between the 1994 and 1995 winter seasons, wheat and bean yields rose by 1.0 and 0.1 t/ha, respectively. This was due to farmers adopting the block irrigation system and reduced cultivated acreage. Wheat and bean acreages were reduced by 17 and 108 ha from the potential 130 and 158 ha, respectively. This allowed irrigation turns to be faster than when all cropland was cultivated in times of water scarcity (Magadlela, 1996). Nevertheless, Bolding (1996) observed that at Nyanyadzi Scheme, the proposed reduction of irrigated acreage during the 1995 season failed to take off since farmers demanded all their acres back.

However, some coping strategies employed by farmers in irrigation schemes may be illegal since they deviate from the set rules and regulations. Mate (1996) noted that during times of water shortages, farmers defy recommended water-use practices by breaking bunds, stealing water by inserting siphons in water courses away from their field intake and leaving water to run for longer periods once it is available. Such actions result in over-irrigation and water logging, thus depressing crop yields.

6.5 Irrigation Water Supply and Demand

In order for measures employed to cope with water shortages to be effective, adequate information on available crop water supply and demand is required. The total crop irrigation water demand in each cropping season is a sum of crop water requirements (ET_c less effective rainfall) and losses. Crop water demand for individual crops is influenced by crop type and growth stage (Allen et al., 1998). Generally, water supply becomes inadequate if it is less than 80% of ET_c. Thus, 80% of water demand denotes the minimum requirements below which significant yield reduction occurs (Makadho, 1994a, b). Water supplied in irrigated fields should satisfy plant or crop water requirements (ET_c) and seepage losses during conveyance, distribution and field application.

6.6 Modelling Crop Water Requirements Using the CROPWAT Model

The CROPWAT model version 5.7 is a simple empirical model for irrigation planning and evaluation developed by FAO (Raes et al., 2003). It is used to estimate ET_c using hydrological, meteorological and crop data (Figure 6.2). In addition, the model can estimate irrigation schedules using soil data. It oriented to simulate the response of crops to varying weather and irrigation conditions (Allen et al., 1998). The model is flexible; hence, it can be run at daily, monthly time step or entire growing season. It is appropriate where data on, for example, soil moisture and crop phenology are lacking. The model incorporates data for all

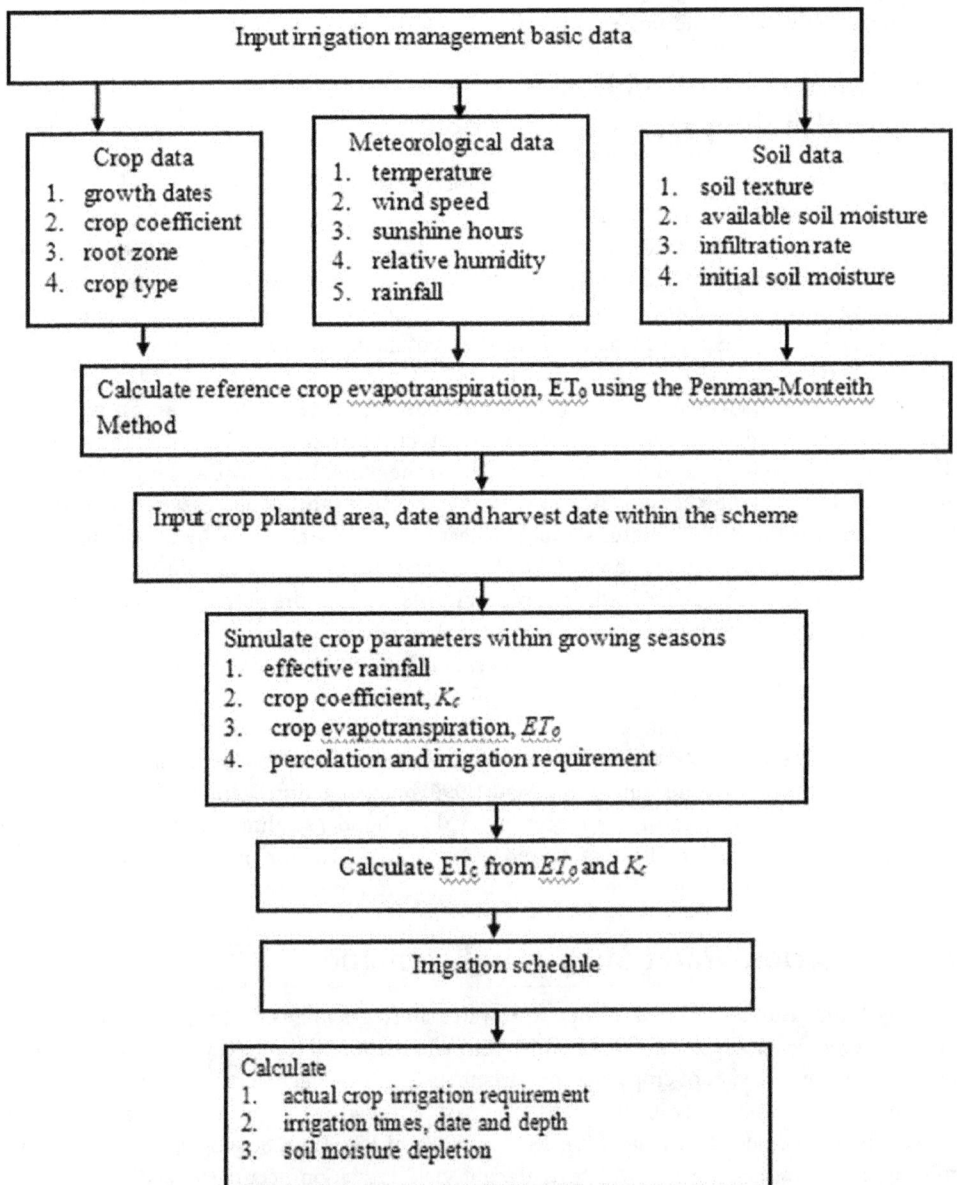

FIGURE 6.2   CROPWAT model flow chart (modified from Sheng-Feng et al., 1999; p. 193).

climate elements and provides values consistent with actual crop water-use data worldwide with minimum errors. It had been applied in Zimbabwe's all five natural farming regions to estimate reference crop evapotranspiration and reliable estimates of crop water requirements for a variety of crops were obtained (Smith et al., 2002, Kamali et al. 2015). Figure 6.2 shows the components of the CROPWAT model.

Crop water requirement, ET_c, is the amount of water which should be applied to the soil to replace that lost to evapotranspiration from planting to harvest for a given crop (Doorenbos and Pruitt, 1977; Allen et al., 1998). Different crops have different crop water requirements at various growth stages. Estimation of ET_c take into account the effect of climate (given as reference crop evapotranspiration, ET_o) and crop characteristics (given by crop coefficient, K_c). Thus, ET_c is estimated as a product of reference crop evapotranspiration, ET_o and crop coefficient, K_c (Allen et al., 1998):

$$ET_c = ET_0 * K_c \qquad (6.1)$$

The crop coefficient (K_c) is a crop-specific evapotranspiration value used with ET_o to estimate ET_c. It integrates the effect of both crop transpiration and soil evaporation on ET_c. Values of K_c vary with crop type, stage of growth, cropping season and cultural practices (Doorenbos and Kassam, 1979; Allen et al., 1998). Standard K_c values produced by FAO are used to estimate ET_c in CROPWAT model.

Reference crop evapotranspiration, also called potential evapotranspiration (ET_o), is the maximum water loss from a surface not short of water (Allen et al., 1998). In CROPWAT, ET_o values are estimated using the Penman–Monteith equation (Hess, 1998):

$$ET_o = \frac{0.408\Delta\left(R_n - G\right) + \gamma\dfrac{900}{T+273}U_2\left(e_a - e_d\right)}{\Delta + \gamma\left(1 + 0.34U_2\right)}, \qquad (6.2)$$

where ET_o is the reference crop evapotranspiration (mm d⁻¹), R_n is the net radiation at crop surface (MJ/m²/d), G is the soil heat flux (MJ/m²/d), U_2 is the wind speed measured at 2 m height (ms⁻¹), T is the average temperature (°C), $e_a - e_d$ is the vapour pressure deficit (kPa), Δ is the slope of the vapour pressure curve (kPa⁰C⁻¹), γ is the psychrometric constant (kPa⁰C⁻¹) and 900 is the conversion factor.

The Penman–Monteith method is recommended by FAO for estimating ET_o because it gives reliable results with minimum errors. It also incorporates data for all climate elements and provides values more consistent with actual crop water-use data worldwide (FAO, 2000; Smith et al., 2002). By incorporating humidity as an input in estimating ET_o, it takes into account the effect of aridity on crop water use.

6.7 Case Study: Nyanyadzi Smallholder Irrigation Scheme in Zimbabwe

Nyanyadzi Irrigation Scheme (Figure 6.3) was opened in 1934 and is located in communal lands of Chimanimani District in Manicaland Province in eastern Zimbabwe. The scheme lies within the dry agro-ecological zone of the country on the rain shadow side of Eastern Highlands. It is state-owned and is administered and managed by the Department of Agricultural Rural Extension (AGRITEX) services under the Ministry of Lands, Agriculture, Fisheries, Water and Rural Development. Nyanyadzi Irrigation Scheme is the largest and one most successful state-controlled smallholder scheme in the province and was established as a drought relief project and to:

a. provide food in an area of recurring droughts where peasants were only able to produce a meaningful harvest once in 5 years,
b. reduce government cost in providing famine relief,
c. reduce the peasants' method of shifting cultivation and consequential destruction of natural resources by setting them permanently on good soil where proper agricultural practices would occur, and

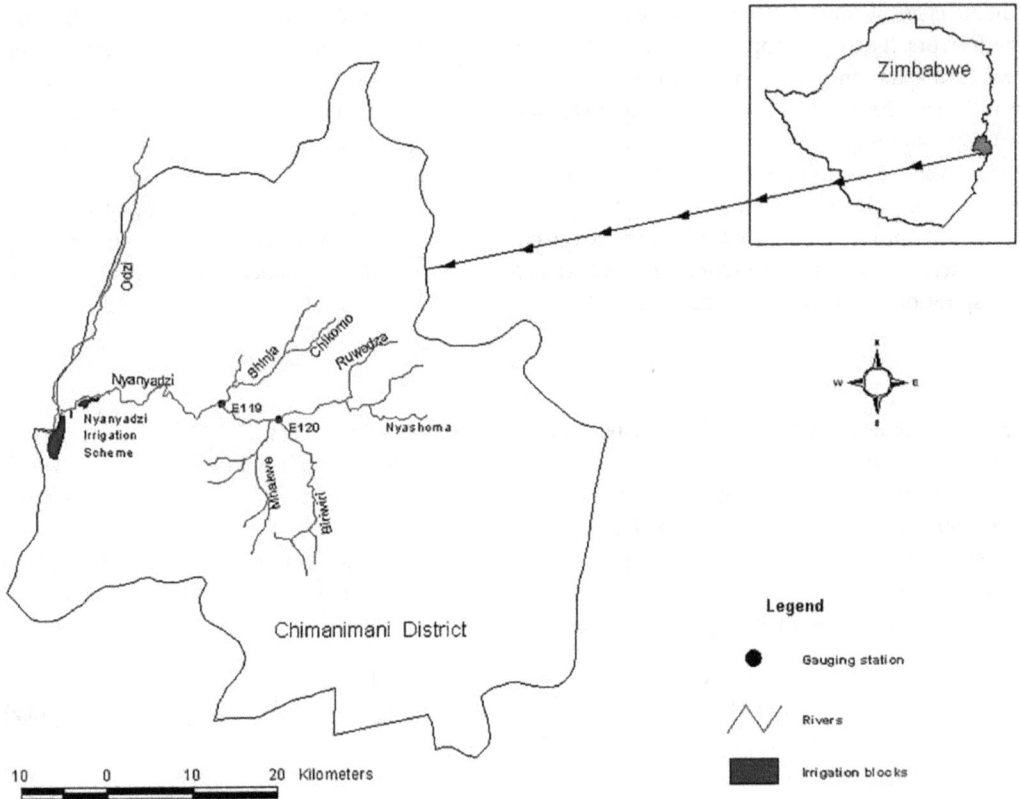

FIGURE 6.3   Location of Nyanyadzi Irrigation Scheme in relation to the river network.

   d. encourage peasant movement from subsistence to a cash economy to practise proper agricultural practices.

6.7.1 Physiography

The scheme lies at an altitude of 530 m in a down-faulted valley of the Save and Odzi Rivers. Considerable north–south and east–west faulting has resulted in a complexity of geological horizons outcropping in the area. The geological formations vary from basement complex granites, limestones and pre-Karoo dolerites intruding into quartzites of the Umkondo system. Soils are of alluvial origin comprising deep, well-draining sand loams and sand clays of high fertility underlain by coarse river sand (Watson, 1969; Ministry of Water Development, 1975).

6.7.2 Climate

River flow gauging station E119 is equipped with automatic recorders. Recording charts were changed regularly. The rating remained fairly stable because little amounts of sediment accumulated upstream of the hydraulic structure. In addition, the downstream gradient is high, thus preventing backwater effect on the water levels. Due to rigorous screening of hydrological and meteorological data before given to the public, the data were assumed to be of high quality.

   The mean monthly temperature is 15°C, and the monthly temperature range is 10°C. October experiences the maximum daily temperature of 31°C, while the lowest temperature of 25°C is experienced in July (Figure 6.4).

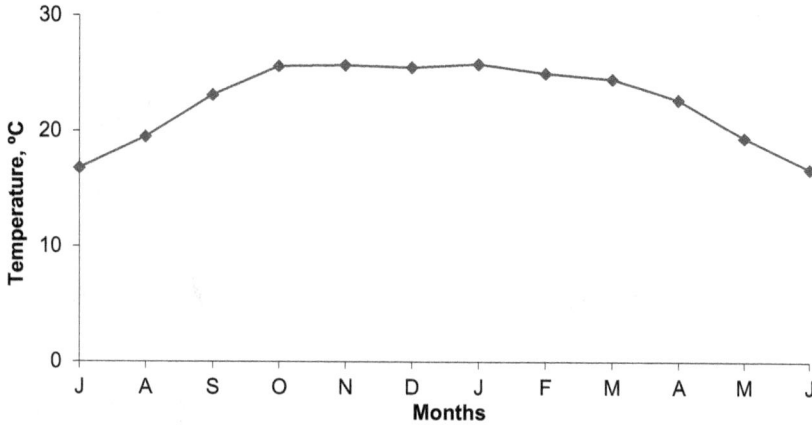

FIGURE 6.4   Mean monthly temperature of Nyanyadzi Irrigation Scheme.

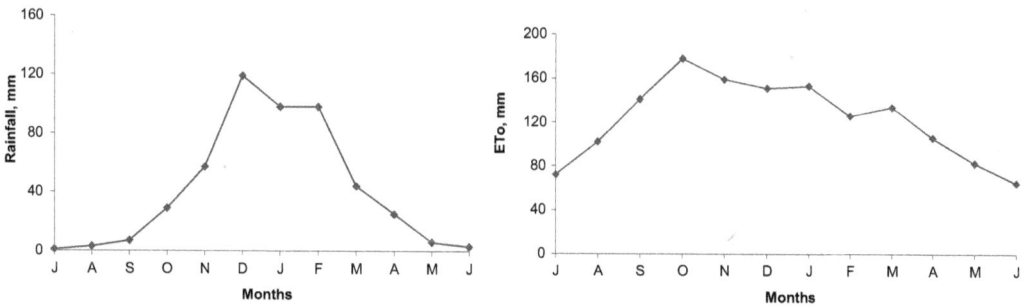

FIGURE 6.5   Mean monthly rainfall and evapotranspiration.

The scheme receives an average of 400 mm of rainfall annually. However, the rainfall is unreliable, erratic and comprises isolated rainstorms. The rain season spans from October to March. However, the onset of the rains may be delayed until late December and terminate beginning of February. In such a dry region of the country, irrigation is crucial to satisfy crop water demand.

The total annual evaporation often rises to 1,900 mm. Annual evapotranspiration rates exceed rainfall even during the rain period in summer season. The mean annual evapotranspiration is 123 mm and the aridity index (mean annual evaporation over mean annual precipitation) is 3.0, implying a semi-arid environmental condition. In such a dry region of the country with annual evapotranspiration rates exceeding rainfall, irrigation is crucial to satisfy crop water demands (Figure 6.5). Thus, the major constraint to crop production in the scheme is water supply.

The 33-year mean annual flow of Nyanyadzi River is $3.5 \times 10^6 \, m^3$. This represents the 'normal' annual flow pattern of the river. Out of the 33 water years (October–September) studied, 15 had mean less than the 33-year average. The annual flows indicate high variability, having a CV of 76.89%. Figure 6.6 shows the temporal variability of Nyanyadzi River annual flows.

In 1992, lowest mean annual flows of $0.11 \times 10^6 \, m^3$ were experienced. This was because of the 1991/92 drought. However, due to improved rainfall in 1997, the highest mean annual flows of $9.65 \times 10^6 \, m^3$ was recorded.

The monthly flow variability depicts a simple regime. The average monthly flow for the 33-year (1975–2001) period is $3.54 \times 10^6 \, m^3$. February has the highest mean monthly flow value of $11.23 \times 10^6 \, m^3$ while in September, the lowest, $0.82 \times 10^6 \, m^3$, was recorded (Figure 6.7).

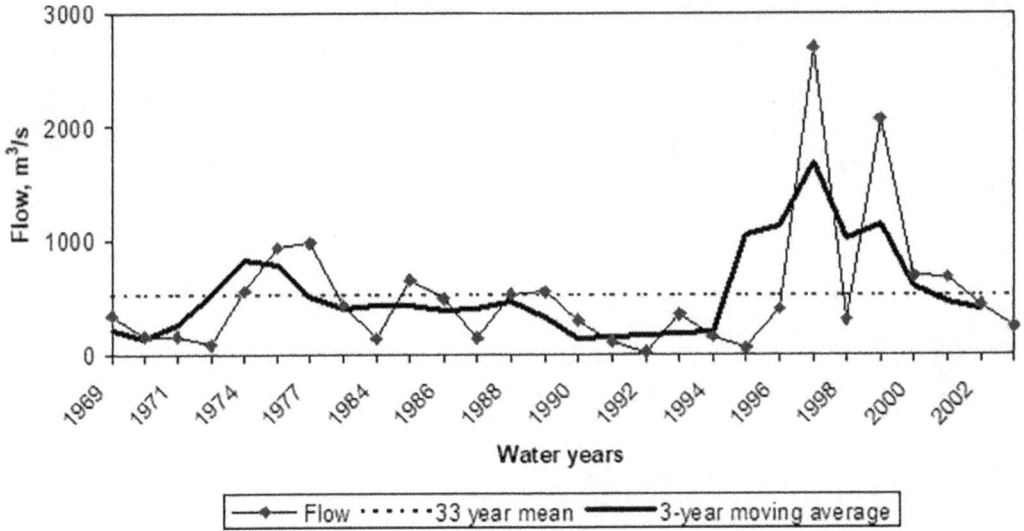

FIGURE 6.6   Mean annual river flow.

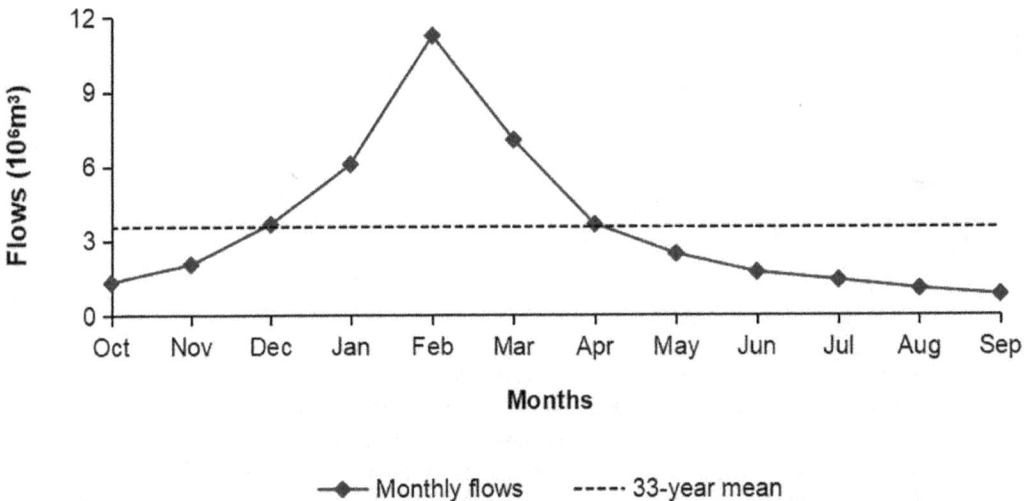

FIGURE 6.7   Mean monthly river flows of Nyanyadzi River.

During the 33 winter seasons from 1969 to 2003, the mean flow is $2.57 \times 10^6 \, m^3$ with a standard deviation of $2.33 \times 10^6 \, m^3$. The 33-year mean summer river flow is $4.87 \times 10^6 \, m^3$ with a standard deviation of $6.8 \times 10^3 \, m^3$.

6.8 Scheme Design Efficiency, Irrigation Water Supply and Management

The scheme has a design irrigation efficiency of 70% below the field gate (Pearce and Armstrong, 1990). This means 70% of water transferred from the field gate reaches the crop root zone. The total losses comprising conveyance, distribution, field application and percolation losses account for 30% of irrigation water supply. Seepage losses along the unlined canal from Nyanyadzi River were estimated to be 25% of flow at the intake (Pearce and Armstrong, 1990). However, the canal was lined in 1996, so seepage losses became negligible since then.

The scheme draws water from Nyanyadzi River and Odzi River (Figure 6.3). From Odzi River, the scheme is entitled to pump 483 L/s. Annual water entitlement from Nyanyadzi River increased from 996 megalitres (ML) in 1937 to 1,022 ML in 1989. An open canal which was lined in 1996 and measuring 2 m wide and 0.5 m deep diverts water from Nyanyadzi River by means of a weir and a gated off-take to Block C and to the night storage dam. Three separate gates along the canal divert water into Block C. Farmers in Blocks A, B and D get additional water from Odzi River which constitutes about 60% of the three blocks' water requirements. Water lifted by pumps from Odzi River is conveyed to the night storage dam through a concrete pipe. From the night storage dam, water is delivered to Blocks A, B and D by means of a lined canal.

Nyanyadzi Irrigation Scheme receives inadequate rainfall for crop production. Farmers depend largely on water supply from rivers. A block system of irrigation is practiced (Figure 6.8). Thus, farmers grow the same type of crop in each block each season. Crops are almost planted and harvested at the same time. Flood irrigation is used in all the four blocks. Unlike Block C which receives irrigation water

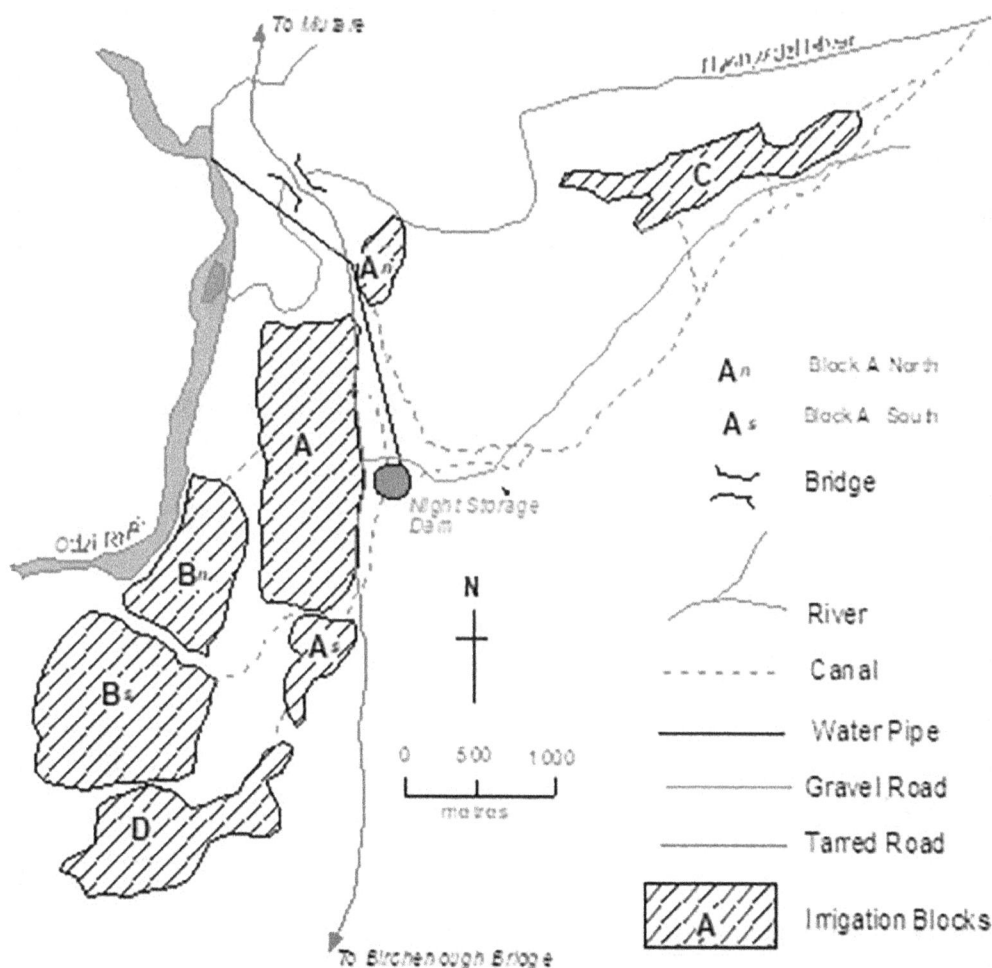

FIGURE 6.8 Nyanyadzi Irrigation Scheme layout with respect to water supply sources.

TABLE 6.1 Scheme Command Areas, Water Sources and Irrigation Turn

Block	Command Area (ha)	Number of Plot Holders	Irrigation Water Source	Irrigation Turn
A	137	161	Night storage dam and Nyanyadzi River	Tuesday, Thursday and Saturday
B	147	269	Night storage dam	Tuesday, Thursday and Saturday
C	65	71	Nyanyadzi River	Monday, Wednesday and Friday
D	69	77	Night storage dam	Tuesday, Thursday and Saturday
Total	418	578		

Source: Nyanyadzi AGRITEX file (2004).

direct from the river, the other three blocks get their supply from the rivers via a night storage dam. Thus, variations in Nyanyadzi River flows have a strong impact on Block C farmers.

Farmers use eight siphons to irrigate a 0.4 ha field at each irrigation turn. Plot sizes range from 1 acre to 4 acres. The blocks receive their water supply from 6 am to 4 pm on rotational basis according to the schedule shown in Table 6.1.

In each block there is an irrigation management committee, agricultural extension worker from AGRITEX, water bailiff and canal supervisors. Besides forcing farmers to follow uniform cropping patterns, AGRITEX personnel monitor whether enough inputs such as water, seed, fertilisers and pesticides applied in the fields by farmers. Water movement along the canal systems to individual plots is controlled by sluice gates operated by water bailiffs with the assistance of field canal supervisors. The irrigation schedule proceeds plot by plot along field canals and each farmer is notified the time of receiving water.

Nyanyadzi Irrigation Scheme has been facing problems of water shortage since its onset (Pearce and Armstrong; 1990; Bolding, 1999). Water shortages were a result of the changing character of river flow and heavy siltation in the river channel, weir, main canal and night storage dam. Inadequate water supply from the river and/or poor water management in the irrigation blocks also resulted in erratic water supply. Siltation is a major problem in the river channel, at the scheme intake weir and in the main canal. Siltation has reduced the storage capacity of the night storage dam by almost 30% (Bolding, 1996). Accumulation of silt reduces the amount of water entering the canal and irrigation water supply, thus negatively affecting crop productivity. Besides causing crop yield reduction, low irrigation water supply often results in clashes between irrigators and other competing upstream water users (Manzungu, 1996). This is pronounced in Block C which depends solely on Nyanyadzi River water supply. Thus, appropriate management solutions need to be implemented to improve crop yields during periods of poor water supply.

6.9 Cropping Patterns

On average, farmers plant three crops in their plots annually. The general cropping pattern is maize and groundnuts during summer (October–March), beans, tomatoes and wheat in winter (April–September). In April, the first bean crop is planted and is harvested in June. Wheat cropping period spans from May to September. Actual planting dates adhered to by farmers vary from season to season and from block to block. The scheme's cropping programmes for the major crops are shown in Table 6.2.

6.10 Block C Irrigation Water Supply and Demand

The study focused on winter season (March–September) when river water supply is the main source of irrigation water. Bean and wheat crops were selected for analysis in this study because they tend to have

TABLE 6.2 Cropping Calendar and Patterns for Winter and Summer Seasons

Crop	Planting Dates	Harvesting Dates	Duration (days)
Maize	15 October–30 November	February–March	130–140
Tomatoes 1	1 February–1 March	April–May	90
Tomatoes 2	1 May	September	90
Beans 1	1 March–30 April	June–July	120
Beans 2	1 June	September	120
Wheat	1 May–25 May	September	120

Source: Field observation (2004).

single planting and harvesting periods. Farmers were thus assumed to give reliable estimates of their winter crops yields during interviews.

Among the scheme's four blocks, Block C was selected for this case study because it obtains its irrigation water supply directly from Nyanyadzi River via a canal which was lined in 1996. The canal is 0.78 m deep and 1.54 m wide. A concrete intake weir 20 m wide with a height of 2 m helps to divert water into the canal. The amount of water supplied to the irrigation block during the crop season was estimated from water abstracted from Nyanyadzi River. Water right No. 1098 of 1934 entitled Block C irrigators 426 ML per annum. Hence, farmers were on average receiving 1,186 m³/day. In 1989, Block C's share of the scheme's annual water commitment increased to 442 ML. This figure translates to 1,228 m³/day.

The farmers with irrigation plots do not have fields for dry land cultivation. Crop production is difficult without irrigation; hence, the resident farmers depend solely on the irrigation scheme for food and income from marketing of maize, beans, wheat, tomatoes and groundnuts. The irrigated fields have been subdivided among family members or for leasing, and hence are getting smaller. By February 2004, farmers' individual plots varied from 0.05 to 4.0 acres

Since the early 1990s, a contract farming scheme of beans and tomatoes between irrigators and private companies was put in place. The companies offer loans and train farmers basic agronomy skills. More than 50% of the cropland is cultivated with tomatoes in winter because the soils are sandy.

The total water supplied to the irrigation block in each cropping season comprises water abstracted from Nyanyadzi River and rainfall. Given that Block C receives its irrigation water supply in 3 days/week, its proportion of the total scheme entitlement was estimated to be three-sevenths. The area under beans was estimated to be 40% of the block area. Irrigation water supply was expressed as an application depth in mm using Eq. (6.3).

$$ID = \frac{V}{A},\qquad(6.3)$$

where *ID* is the irrigation depth in mm, *V* is the volume in m³ and *A* is the irrigated area in km².

6.10.1 Contribution of River Flow towards Block C Water Supply

Irrigation is the main source of water supply because the contribution of rainfall is minimal. Between 1970 and 2003, the average rainfall of 108 mm was received. In 1984, the highest rainfall of 251 mm was received. The lowest rainfall amount of 9 mm was received in 1996. Thus, river flow forms a major contribution towards irrigation water requirements. From 1970 to 2003, the contribution of rainfall to total Block C water supply was minimal as compared to irrigation water from the river because in winter less rainfall is received during the 21 winter seasons studied. On average 84.3 mm of rainfall was received. The highest and lowest winter season rainfall totals were 210.9 and 5.0 mm received in 2003 and 1996, respectively. The contribution of rainfall dry season months was negligible due to dry conditions experienced in the area.

TABLE 6.3   Mean ET_o Values and Effective Rainfall During the Winter Season

Month	ET_o (mm/day)	Effective Rainfall[a] (mm/month)
April	4	22
May	3.3	6
June	2.8	3
July	2.8	1
August	3.7	2
September	4.9	6

[a] Effective rainfall calculated a using the fixed percentage method (Department of Meteorological Services, 1981) which assumes all rainfall to be effective if annual totals were less than 400 mm. Above this threshold value, 66% of the rainfall is lost as runoff.

6.10.2 Block C Irrigation Water Demand

The CROPWAT model was used to estimate bean crop water requirements using hydro-meteorological and hydrological data at Nyanyadzi Irrigation Scheme. The reference crop evapotranspiration ET_o figures presented in Table 6.3 show that dry season cropping is virtually impossible without supplementary irrigation.

The total crop water demand in each cropping season was calculated as a sum of crop irrigation water requirements (ET_c less effective rainfall) and losses. Seepage losses in the field were estimated to be a fixed proportion of 30% of irrigation water supply. The figure was derived from the scheme's design irrigation efficiency of 70% below the field gate (Pearce and Lewis, 1988; Pearce and Armstrong, 1990). Thus, 30% of water transferred from the field gate is lost before reaching the crop root zone. Seepage losses along the unlined diversion canal were estimated to be 25% of flow at Nyanyadzi River intake as observed by Pearce and Armstrong (1990). However, since canal was lined in 1996, its seepage losses were estimated to be negligible since then. The reference crop evapotranspiration figures were determined using CROPWAT model 5.7 based on the Penman–Montieth method.

In order to determine the adequacy of water supply, irrigation water volumes supplied to the irrigation block (depth in mm) were compared with the overall crop water demand according to the RWS ratio:

$$RWS = \frac{S}{D},\tag{6.4}$$

where *RWS* is the relative water supply, *S* is the water supply and *D* is the water demand.

Water supply, *S*, comprises the irrigation depth and effective rainfall, while water demand, *D*, consists of crop water requirements and percolation or seepage losses. Seepage losses along the unlined Nyanyadzi River canal were estimated to be 25% of flow at the intake as observed by Pearce and Armstrong (1990). However, since canal was lined in 1996, its seepage losses were estimated to be negligible.

6.10.3 Bean Crop Yields and Crop Water Requirements

The irrigation of beans, planted on the 1st of April and harvested at the end of June, with a growing season of 120 days is considered in this analysis. The total ET_c for bean crop is 308.5 mm. Table 6.4 shows ET_c figures during different growth stages for the first bean crop planted on 1 April and harvested on 31 July.

The average bean crop yield was 0.91 t/ha from an average cropped area measuring 5.05 ha. Highest yield of 1.56 t/ha was obtained in 1974, while in 1973, 1994 and 1992, no bean crop was harvested (Table 6.5).

TABLE 6.4   Crop Water Requirements for First Bean Crop

Month	Decade (10-Day Period)	Growth Stage	CWR (mm/decade)
April	1	Initial	14.9
	2	Initial	14.0
	3	Development	18.2
May	1	Development	26.5
	2	Development	33.6
	3	Middle	36.0
June	1	Middle	34.1
	2	Middle	32.2
	3	Late	28.6
July	1	Late	31.0
	2	Late	26.2
	3	Late	26.4
		Total	308.5

Source: CROPWAT 5.7.

TABLE 6.5   Bean Crop Yields and Relative Water Supply

Cropping Season	Yields (t/ha)	Yields (t/mm)	RWS (%)
1970	1.23	0.1355	38.17
1971	0.73	0.1371	35.90
1973	0	0	11.69
1974	1.56	0.0741	170.59
1978	0.78	0.0561	146.77
1984	0.85	0.0715	72.90
1985	1.29	0.0397	122.53
1986	0.72	0.0417	127.92
1987	0.48	0.0141	39.59
1988	1.35	0.0165	184.41
1992	0	0	1.70
1994	0	0	8.25
1995	0	0	3.61
1996	0.85	0.0020	99.98
1997	0.91	0.0020	105.26
1998	1.07	0.0027	99.45
1999	1.13	0.0036	85.03
2000	1.06	0.0012	161.58
2001	1.19	0.0018	133.06
2002	1.40	0.0028	115.94
2003	1.36	0.0028	116.70

   In all the 21 winter seasons studied, eight experienced inadequate supply (RWS < 80%) and an average yield of 0.41 t/ha was obtained. The remaining 13 seasons with adequate water supply (RWS exceeding 80%) had average crop yields of 1.07 t/ha. The highest RWS in 1988 yielded 1.08 t/ha. During cropping seasons of zero yields the RWS was less than 4%. The amount of water received (rainfall and irrigation water supply) received in the block was too low to sustain crop productivity. It is important to note that 9 out of the 13 seasons with RWS > 80% were from 1996 to 2003. The increase in irrigation water supply could be attributed to high rainfall and negligible seepage losses since the main canal was lined.

In 1970, the RWS was 38% and yields obtained were 1.23 t/ha. This figure is greater than 1.07 t/ha, the average achieved during seasons of adequate water supply. Thus, high bean yields per unit water received in 1970 can be due to improved water application efficiency since the crop output was 0.14 t/mm of water supplied. The figure is one of the highest during the 21 seasons studied (Figure 6.6).

In the 1978, 1996 and 1997 cropping seasons, supply was adequate (RWS > 100%), but the average yields were 0.65 t/ha. Low yields during the three seasons can be ascribed to water wastage. In 1978, 0.06 t/mm of water supplied was obtained, while 0.02 t/mm was achieved in the 1996 and 1997 seasons (Figure 6.4). Some farmers over-irrigate their fields causing soils to be water logged. In addition, under-irrigation was due to water leakage along the distribution canals. Crop yield data provided by farmers is often open to criticism. Some farmers could have under-estimated their crop yield data fearing not to receive food aid receive from the state. Food aid is often provided to household who harvest virtually nothing from their fields.

6.10.4 Wheat Crop Yields and Crop Water Requirements

The total crop water requirements (CWR) for bean crop is 417 mm. The crop water requirement figures for wheat crop with a growing period of 130 days from 20 May to 1 October are shown in Table 6.6.

The average yield of 1.35 t/ha was obtained from 5.69 ha of cropped area during 21 seasons studied. Highest wheat yields of 1.94 t/ha were obtained in 1974. However, in 1973, 1992, 1994 and 1995 no wheat crop was harvested from the block (Table 6.6). From 1996 to 2003, the average yield was 1.08 t/ha comparably higher than 0.80 t/ha obtained during the 13 seasons before 1996 (Table 6.7).

RWS range from 139% (1986) to 0% (1992 and 1994). The average RWS was 70.68% for the 21 seasons studied. During the four seasons of zero crop yields, water was scarce as RWS was 0.54% on average. In 1973, 18 mm were available for ET_c. No water was available in 1992, 1994 and 1995. Water supply was adequate (RWS > 80%) during 11 seasons and average yields of 1.25 t/ha were harvested. During ten seasons of inadequate water supply, 0.53 t/ha were obtained.

While the RWS was 88% in 1970, the yields harvested were 0.47 t/ha the average yields when water supply was adequate. Thus, water wastage in the fields could have attributed to low yields. With regard to yields per unit water supplied, 0.32 t/mm was obtained during the 1970 season. The figure is less than the average for 21 seasons, which is 0.37 t/mm. However, in 1984, 1.27 t/ha were obtained although the RWS

TABLE 6.6   Crop Water Requirements for the Wheat Crop

Month	Decade (10-Day Period)	Stage	ET_c (mm/decade)
May	3	Initial	15.7
June	1	Initial	14.0
	2	Development	14.0
July	3	Development	17.3
	1	Development	23.8
	2	Middle	30.3
August	3	Middle	37.2
	1	Middle	40.8
	2	Middle	44.4
September	3	Late	49.2
	1	Late	49.5
	2	Late	34.1
	3	Late	36.4
Total			417.5

Source: CROPWAT 5.7.

TABLE 6.7 Wheat Yields and RWS

Cropping Season	Yield (t/ha)	Yields (t/mm)	RWS (%)
1970	0.47	0.32	88.58
1971	0.36	0.26	29.18
1973	0	0	2.16
1974	1.94	0.66	126.50
1978	1.25	0.56	109.35
1984	1.27	0.54	51.69
1985	1.12	0.58	117.39
1986	1.82	0.61	139.12
1987	0.75	0.43	23.98
1988	0.82	0.46	101.02
1992	0	0	0
1994	0	0	0.39
1995	0	0	0
1996	0.82	0.32	61.0
1997	0.81	0.30	60.61
1998	0.87	0.37	80
1999	0.96	0.41	57.51
2000	1.10	0.46	118.68
2001	1.13	0.41	112.20
2002	1.17	0.51	102.72
2003	1.13	0.52	100.59

was 52%. Improved water application efficiency might have caused high yields to be obtained. Wheat yield per unit water was 0.54 t/mm.

6.11 Conclusions

This chapter has highlighted the interlinked concepts of plant or crop water requirements and evapotranspiration. It has been shown that evapotranspiration is consumptive use of water by plants. Specifically, the chapter dovetails on determining adequacy of irrigation water supply during the wheat and bean cropping seasons at Nyanyadzi Irrigation Scheme's Block C in Zimbabwe. It has been shown that low crop yields occurred during seasons of inadequate water supply.

From the research findings, it is necessary to apply short- and long-term intervention strategies to mitigate the effects of low water supply. Such measures include improving water delivery systems in the fields. Significant water savings could be realised by retrofitting the less efficient flood type of irrigation with more efficient irrigation methods. Unlike the flooding method which is less than 75% efficient, conventional sprinkler centre pivot, micro-jet and drip irrigation methods are on average 75%, 80%, 85% and 90% efficient, respectively (DFID, 2003).

Farmers should apply the deficit irrigation in times of water shortage so that irrigation can be applied at crucial crop growth stages. Such practices enable farmers to reduce crop water consumption while at the same time improving water-use efficiency. This can only be realised when information about actual and potential plant water requirement under specified conditions is available. Thus, irrigation scheduling would be determined based on crop transpiration in the field with soil moisture maintained at an optimal condition. Also, farmers can apply precision irrigated agriculture whereby inputs are applied at the right time, in right quantities and at right places. Farmers should plant better crop varieties that produce higher yields, mature earlier and more drought-resistant than the present varieties. This will enable crop water demands to be satisfied with little water supply.

At the intake weir, a concrete structure diverts more water to the scheme. However, downstream streambed irrigators often vandalise the weir to improve water supply to their gardens. This is prevalent in times of water scarcity. During the rains heavy siltation at the weir reduces scheme water supply. Thus, weir should be de-silted on routine basis to keep it free from silt. Furthermore, either a 24-hour security monitoring system at the weir or imposing heavy penalties to those who vandalise it could offer long-term solution to problems of water stealing. This improves water supply to the scheme.

References

Allen, R.G., Pereira, L.S., Raes, D. and Smith, M. (1998) Crop evapotranspiration: Guidelines for computing crop water requirements. FAO Irrigation and Drainage Paper 56, FAO, Rome, Italy.

Bolding, A. (1996) Wielding water in unwilling works: Negotiated management in water scarcity in Nyanyadzi irrigation scheme, winter 1995. In: Manzungu, E. and Van der Zaag, P. (Eds.) *The practice of smallholder irrigation: Case studies from Zimbabwe*. University of Zimbabwe publications, Harare Zimbabwe, pp. 69–101.

Bolding, A. (1999) Caught in the catchment: Past, present and future of Nyanyadzi water management. In: Manzungu, E., Senzanje, A. and Van der Zaag, P. (Eds.) *Water for Agriculture in Zimbabwe: Policy and management options for smallholder sector*. University of Zimbabwe publications, Harare, Zimbabwe, pp. 123–152.

Chancellor, F.M. and Hide, J.M. (1997) Smallholder irrigation: Ways forward. Guidelines for achieving appropriate scheme design. Volume 1. Summary of case studies. Report OD 136, Hydraulics Research, Wallingford, UK.

Department of Meteorological Services (1981) *Climate handbook of Zimbabwe*, Department of Meteorological Services, Harare, Zimbabwe.

DFID (2003) *Handbook for the assessment of catchment water demand and use*. Hydraulics Research, Wallingford, UK.

Doorenbos, J. and Kassam, A.H. (1979) Yield response to water. FAO Irrigation and Drainage paper, No. 33, FAO, Rome, Italy.

Doorenbos, J. and Pruitt, W.O. (1977) Crop water requirements: Guidelines for predicting crop water requirements. FAO Irrigation and Drainage paper, No. 24, FAO, Rome, Italy.

FAO (2000) *Socio-economic Impact of smallholder irrigation in Zimbabwe: Case studies of ten irrigation schemes*, FAO SAFR, Harare, Zimbabwe.

FAO (2003) *Unlocking the water potential of agriculture*, FAO, Rome, Italy.

Kamali, M.I., Nazari, R., Fridhosseini, A., Ansari, H., and Eslamian, S., (2015) The determination of reference evapotranspiration for spatial distribution mapping using geostatistics, *Water Resources Management*, 29(11), 3929–3940.

Lauraya, F.M. and Sala, A.L.R. (1995) *Performance determinants of irrigation associations in national irrigation systems in Bicol, The Philippines: An analysis*. IIMI, the Philippines.

Magadlela, D. (1996) Whose water? Interlocking relations over water in Nyamaropa irrigation scheme. In: Manzungu, E. and Van der Zaag, P (Eds.) *The practice of smallholder irrigation: Case studies from Zimbabwe*, University of Zimbabwe Publications, Harare, Zimbabwe, pp. 102–25.

Makadho, J. (1996) Irrigation timeliness indicators and application in smallholder irrigation in Zimbabwe. *Irrigation and Drainage*, 10, 267–376.

Makadho, J.M. (1994) Water delivery performance. In: Rukuni, M., Svendsen, M., Meinzen-Dick, R., and Makombe, G. (Eds.) *Irrigation performance in Zimbabwe. Irrigation performance project*. University of Zimbabwe, Harare, Zimbabwe, pp. 50–62.

Manzungu, E. (1996) Contradictions and standardisation: The case of block irrigation in smallholder schemes in Zimbabwe. In: Manzungu, E. and Van der Zaag, P. (Eds.) *The practice of smallholder irrigation: Case studies from Zimbabwe*, pp. 47–68.

Manzungu, E. (1999) Rethinking the Concept of water distribution in smallholder irrigation. In: Manzungu, E., Senzanje, A., and Van der Zaag, P. (Eds.) *Water for agriculture in Zimbabwe: Policy and management options for smallholder sector*, University of Zimbabwe Publications, Harare, Zimbabwe, pp. 92–119.

of Zimbabwe Publications, Harare, Zimbabwe, pp. 1–28.

Mate, R. (1996) Juggling with land, labour and cash: Strategies of some resilient smallholder irrigators. In Manzungu, E. and Van der Zaag, P. (Eds.) *The practice of smallholder irrigation: Case studies from Zimbabwe*. University of Zimbabwe Publications, Harare, Zimbabwe, pp. 148–160.

Mati, B. M. (2000) The influence of climate change on maize production in the semi-humid-semi-arid Areas of Kenya. *Journal of Arid Environments*, 46(4), 333–344.

Ministry of Water Development (1975) *Nyanyadzi irrigation scheme: Project Report*. Ministry of Water Development, Harare, Zimbabwe.

Nyamayevu, D., Sun, O. and Chinopfukutwa, G.L. 2018. An assessment of the reliability and adequacy of irrigation water in small holder irrigation schemes. *International Journal of Scientific Engineering Research*, 6, 14–19.

Palmer-Jones, R. (1986) *Why irrigate the north of Nigeria?* Wye college Agrarian development unit, Kent, UK.

Pazvakavambwa, G. T. and Van Der Zaag, P. (2000) The value of irrigation water in Nyanyadzi smallholder irrigation scheme, Zimbabwe. Paper Presented at a WaterNet Symposium: Sustainable Use of Water Resources, Maputo, Mozambique.

Pearce, G R. and Armstrong, A.S.B. (1990) Small irrigation design, Nyanyadzi, Zimbabwe: Summary report of studies on field-level water use and distribution, Report ODI 98, HR Wallingford, UK.

Pearce, G.R. and Lewis, N.S. (1988) Small irrigation design, Nyanyadzi, Zimbabwe: Distribution of water supply, Report ODI 97, Hydraulics Research, Wallingford, UK.

Raes, D., Portilia, J., Kirporir, E., Sahli, A., and Sithole, A. (2003) Regional yield estimates derived from soil water deficits. International Workshop on "Sustainable Management of Marginal Dry Lands, Shiraz, Iran.

Rukuni, M., Svendsen, M., Meinzen-Dick, R., and Makombe, G. (Eds.) (1994) *Irrigation performance in Zimbabwe. Irrigation performance project in Zimbabwe*, University of Zimbabwe, Harare, Zimbabwe.

Sheng-Feng, K., Shih, C.C.C. and Shin-Shen, H. (1999) Implementation of CROPWAT model to evaluate agricultural water demand for ChaiNan Irrigation Association in Taiwan. In: *Proceedings of 99 International Conference on Agricultural Engineering*, Beijing, China, pp. 184–199.

Smith, M. (1993) CLIMWAT for CROPWAT: A climatic database for irrigation planning and management. Irrigation and Drainage Paper 49, FAO, Rome, Italy, p. 114.

Smith, M., Kivumbi, D. and Heng, L.K. (2002) Use of the FAO CROPWAT model in deficit irrigation studies. *Deficit Irrigation Practices*,. 17–27.

Tafesse, M. (2003) Small-scale irrigation for food security in sub-Saharan Africa, CTA Working Document No. 8031, CTA, Ethiopia.

Watson, R.L.A. (1969) The geology of Cashel, Melsetter and Chipinga areas, Zimbabwe. Geological Survey, Bulletin No. 60, Harare, Zimbabwe.

7

Irrigation Practices in Moderately Warm Arid Regions of Sub-Saharan Africa

Nkem J. Nwosu
University of Ibadan

Saurau O. Oshunsanya
University of Ibadan

Saeid Eslamian
Isfahan University of Technology

7.1 Introduction

Irrigation as the term implies includes every operation and process involved in the artificial application of water to the soil for a specific purpose. There is virtually no standard definition for irrigation; however, all definitions point to the fact that water is artificially supplied on land for agricultural use (Ali, 2018). An agricultural engineer views irrigation as a process that involves the conservation and storage of water, water movement from source to application area and subsequent distribution on lands (Newell and Murphy, 1913). Soil physicists/conservationists view irrigation as a practice that involves the artificial supply of water to farming fields such that the soil is not devoid of optimum moisture content for optimum crop/pasture growth (Bressan, 2006).

Irrigation as an art has hitherto existed long before the advent of civilization. Irrigation has been practiced by the early inhabitants of the semi-tropical and moderately arid regions where there is periodic overflow of the desert (Newell and Murphy, 1913). Irrigation has been practiced in the drier parts of

DOI: 10.1201/9781003353928-9

Asia for sustainable grain production. In America, The United States Department of Agriculture noted that owing to the importance of irrigation in crop production, irrigation has been practiced throughout the Midwest, into the South, and in many Eastern states, especially Florida. Urban irrigation now comprises a large component of the irrigation market—for landscaping, golf courses, athletic fields and other recreational uses. In Africa, irrigation became the essential expedient for sustainable crop production during off seasons of rain, regarded as the dry season in West Africa like Nigeria (Adeaga and Eslamian, 2021). Almost 25% of the total cultivated land in the world is irrigated (Eldeiry et al., 2005; Van Vuren and Mastenbrock, 2000).

Irrigation is essentially important in agricultural production due to the inadequacy of natural precipitation during periods of active plant growth and development. This is relative to the environment, as some regions are characterized with high annual precipitation that negates the need for artificial water supply. However, arid areas (with little or spasmodic rainfall characteristics) heavily rely on irrigation for sustainable crop production. Humid regions with high amount and intensity of rainfall possess little sunshine capable of increasing crop production; as such, it is often observed that well-irrigated soils in arid regions tend to give higher crop yields due to the less obstruction of the highly available sunlight on the soil, which aids plant growth.

In America, the Department of Agriculture reported that irrigation made it possible for 16% of U.S. cropland to produce 48% of harvested crop sales in 1997. This corresponds to about 280,000 farms irrigating 55 million acres of crop and grazing lands. In sub-Saharan Africa (SSA), crop production levels have greatly increased due to continuous irrigation of land during all seasons. Irrigation practices in this part of the globe have been motivated by the high population increase and rise in food demands of the teeming population, especially in Nigeria and other parts of North and East Africa, respectively.

Crop production in SSA is basically rain-fed, hence the overreliance on the climatic factors of rainfall in order to supply the needed moisture to satisfy the crop water requirements, to enable it complete its life cycle. However, due to changes in weather patterns all year round, irrigation practices have become a major part of crop production for improved harvest. In light of the increase in population in most SSA countries, crop production has failed to meet the market demands for the populace, thereby compounding the problems of food insecurity. Several strategies have been documented over time to improve the food production levels in many countries of the world, but little attention has been drawn towards the adequate use and application of good irrigation practices, especially in arid and semi-arid regions of Africa. Food production in arid regions has been observed to increase with proper irrigation scheduling. As such, semi-arid regions have lesser moisture than extremely arid regions in order to sustain crop production during off seasons. There has been limited information on the prevailing irrigation practices in the subarid regions of the tropics. In Africa, with the rising population and food demands, information on the irrigation measures for all year round food crop production is essential in tackling the problem of food sufficiency in the tropics. In addition, it is pertinent to access the commonly practiced irrigation systems in moderately arid regions in order to better understand the lapses in the level of crop yields associated with irrigated agriculture in SSA. This chapter therefore seeks to review the concept of irrigation, and the irrigation practices that are mostly conducted in some selected countries in the moderately arid regions of SSA.

7.2 The Concept of Irrigation

Irrigation in simple terms involves the artificial application of water to land for optimum crop growth and development. As a process, irrigation comprises all the activities involved in the transfer of adequate water to soil and crop in order to meet the soil and crop water requirements and establish an optimum soil–water–crop relationship required for optimum crop production. The concept of irrigation involves the holistic process and ways of water application, the need for supplementary moisture and the benefits accruing from such operations.

African agriculture has relied on the benefits of irrigation in order to sustain production levels during periods of rainfall scarcity. Farming in Africa is generally rain-fed, which means farmers completely depend on rainfall to grow crops. This tradition is true, especially in arid and moderately arid environments of the tropics with about 3–4 months of annual rainfall. For crop cultivation to be sustainable in such environments, supplementary moisture is essential for optimum crop production. For instance, in Kano and Niger State in Northern Nigeria, characterized with Sudan Savannah vegetation owing to the low annual rainfall condition, the use of drip and surface irrigation systems have become the bane of crop production in order to sustain the production of hydrophyte crops like grain crops in the region. Ali (2018) reported that agricultural activities in India are majorly dependent on monsoons. Owing to the non-uniform distribution of rainfall in space and time in the region, irrigation becomes an imperative engineering operation to sustain crop production in the region.

The benefits of irrigation in crop production cannot be overemphasized. Irrigation generally increases crop yield by providing adequate moisture that meets the crop water requirements of most cultivated crops in order to complete their life cycle. Irrigation also provides moisture that cools the soil temperature during periods of dry spell (moisture stress). Beneficial soil microorganisms also tend to function well in cool soil environment with adequate water content. Irrigation water also helps in organic matter decomposition as well as ensures that essential soil nutrients are available for plant roots to absorb for optimum growth and development. Soil quality is improved when soil moisture content is optimum for effective soil–plant relationship.

Furthermore, irrigation moisture helps to prevent conditions of famine. Water has been regarded as the most essential requirement for optimum crop production. Hence, during periods of water inadequacies, plants tend to experience moisture stress, which results in loss in turgidity, wilting and ultimate death of crops. These conditions result into poor yields. Irrigation therefore augments the moisture condition of the soils by raising the field moisture capacity of the root zone of crops for optimum use. Fertilization application, especially in most parts of the south western and middle belts in Nigeria, is done with irrigation water during the dry season (period of less rainfall) by the ferti-irrigation method (Newell and Murphy, 1913). This is because moisture is essential for dissolving the granulated fertilizer materials and ensures that they are in absorbable form for plant roots.

In addition, irrigation is also essential for aesthetic purposes (Bressan, 2006). Football fields, lawns and gardens are kept green all year round in areas with limited annual precipitation. The quality of pasture for animal husbandry and nutrition is greatly influenced by the moisture content of the soil. Water is required for the optimum growth of forages and grasses (Bressan, 2006). Animals are generally selective in feeding. Pasture that is extremely dry and devoid of adequate moisture is less consumed in arid and semi-arid environments. However, proper irrigation tends to boost farmer's income and the net income of countries accruing from the sustained export of cash crops (Adelodun and Choi, 2018). Hence, in the wake of the brunt population increase, especially in most tropical regions, irrigation is vital for sustainable food production in order to advance efforts to attain food security.

7.3 Types of Irrigation

Farmers apply water to soil using different techniques with the sole aim of achieving optimum output from the field. Based on the type of crop and nature of the soil, there are various methods of applying water to the soil and/or crop, which culminates in three broad types of irrigation systems. The choice of using any of these irrigation systems greatly depends on the features of the soil, slope and topography of the soil to be irrigated; size and availability of irrigation water source; depth of plant root zone and water table of the soil; and the possible erosion hazards that can accrue from the use of such irrigation systems (Newell and Murphy, 1913; Ali, 2018). All these factors affect the choice of irrigation system employed in crop production either singly or in combination. The three main types of irrigation systems are surface irrigation, sprinkler irrigation and drip or trickle irrigation systems. These are the three predominantly used types of irrigation systems in the tropics, especially in SSA (Adelodun and Choi, 2018). The choice of the system used depends on several factors, which include (but not necessarily limited to) cost, technology, farm size, topography and type of crop to be irrigated (Bacha et al., 2011).

7.4 The Surface Irrigation

This is an irrigation system where water is made to flow and spread harmlessly over the surface of the land. The amount of water allowed on the irrigation field is highly dependent on the time and purpose of application (Newell and Murphy, 1913). As a result, there is a wobbly pattern in water flow under surface irrigation. This sometimes is responsible for the complex hydraulics nature of surface irrigation. Nevertheless, surface irrigation system for any agricultural enterprise can be made suitable and efficient after the following factors (which are also essentially involved in the hydraulics) are considered. They are (but not necessarily limited to):

  i. Surface slope of the field
 ii. Roughness of the field surface
iii. Depth of water to be applied
 iv. Length of run and time required
  v. Size and shape of water-course
 vi. Discharge of the water-course
vii. Field resistance to erosion

Surface irrigation technique is further classified as basin irrigation, border irrigation, furrow irrigation and uncontrolled flooding (FAO, 2019). The mechanism of surface irrigation is quite intriguing. There are four distinct phases in surface irrigation systems (Figure 7.1). They include: the advancement phase, the wetting phase, the reduction phase and the recession phase, respectively (FAO, 2019). Each phase is highly essential and all are significantly interconnected for the success of surface irrigation. The mechanism is as follows: once water has been applied on the field, it flows or "advances" over the soil surface until the applied water covers the whole area. Thereafter, the water applied either flows over as run-off on the field or tends to pond on the surface soil. The wetting phase depicts the interval between the end of the water flow and the point where the inflow water ends. The surface water volume starts to decline

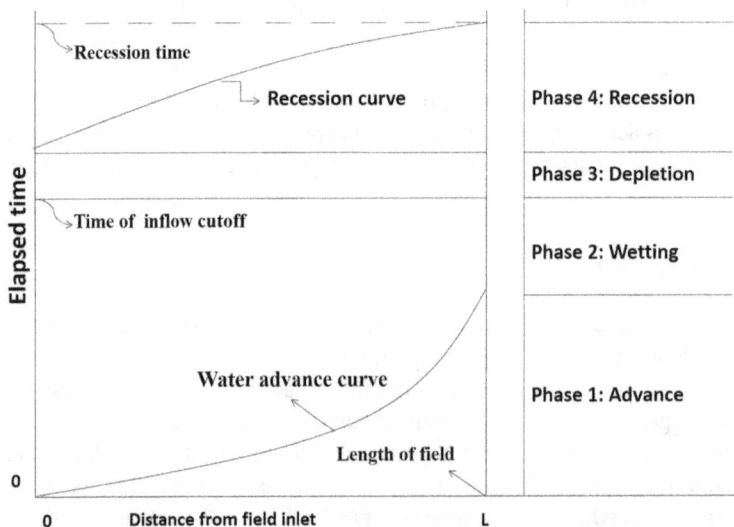

FIGURE 7.1 Hydraulics of surface water irrigation (L indicates final length of field; O indicates initial point of water application).

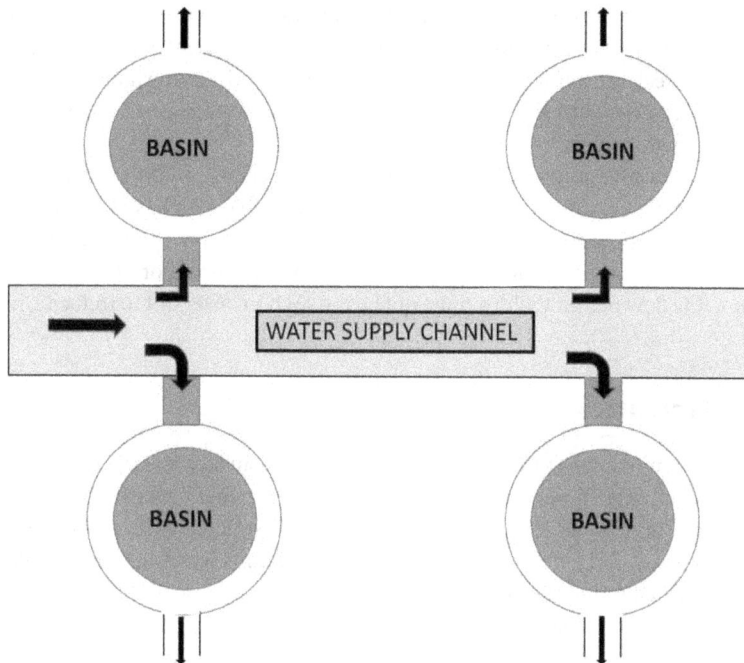

FIGURE 7.2 Basin surface irrigation system.

when there is no more water applied. The depletion phase connotes the interval between the point where the inflow is cut off and the point where there is no ponding water on the soil. The recession commences after the depletion phase and depicts the interval between the bare soil appearances until the surface soil is drained. These phases are used to describe the hydraulics of surface water flow as presented in Figure 7.2.

It is imperative to denote at this point that during surface irrigation, one may not necessarily observe a wetting, depletion or recession phase, respectively. For instance, in basin surface irrigation system, the hydraulics may only involve a depletion phase since water tends to infiltrate vertically down the farming field. Similarly, during the irrigation of paddy, there is sometimes neither an advancement phase nor a recession phase, but the wetting and depletion phases are well observed. This situation arises because water is often added to the already ponding water in the basin.

7.5 Basin Irrigation

Basin irrigation systems are defined as surface-irrigated areas that have no longitudinal slope and complete perimeter dikes to pond water and prevent run-off (North et al., 2010). Basin irrigation is the most common form of surface irrigation, especially in fragmented fields and land holdings. Basins are basically square shaped, although there are some basins with rectangular and irregular configurations. Basins can be furrowed or corrugated, with raised beds for optimal growth of some crops. As long as the inflow of water is undirected and uncontrolled into these field modifications, it is still termed a basin. This form of surface irrigation system is a common practice, used for irrigating most orchards in the tropics. Basin irrigation systems are suitably employed in irrigating flat terrains and soils with low permeability (FAO, 2019). Basin systems are relatively cheap to establish and require a small labour force. When successfully employed, it results in high yields with minimal production risks.

Some crops and/or soils are not compatible with the use of basin irrigation systems; however, basin irrigation is suitable for moderate to low permeable soils that are deep-rooted with closely spaced crops.

Although, in basin irrigation systems, the construction of drainage channels for surface run-off during periods of heavy rainfall makes for a good design, it is not always a necessity. As such, the system can favourably be used to reclaim salt-affected soils. With regards to the pattern of field water application, basins systems tend to serve with less command area and field watercourses than is obtainable using either border and furrow systems, respectively.

Despite the numerous advantages of basin irrigation systems, few limitations still influence their use in tropical soils. For instance, soils with crusting are not compatible with basin irrigation systems. Also, to establish an optimal level of efficiency with the use of basin irrigation, there must be a high-precision land levelling operation for uniform water distribution across the field. For optimal levels of efficiency in basin irrigation, the flow per unit width must not be too high in order not to induce any form of water erosion.

7.6 Furrow Irrigation

Furrow irrigation system is one of the most common types of surface irrigation. This type of surface irrigation is not efficient in arid regions with distinct hydrologic characteristics (like infiltration). The design parameters for this type of surface irrigation are usually selected with little emphasis on the prevailing conditions. Furrow irrigation system is designed to channel water flow along a corrugated path or furrow, thereby ensuring that the surface field is not flooded as observed in Figure 7.3 (FAO, 2019). This system employs the use of basins and/or borders in order to checkmate the influence of variations in the topography of the area to be irrigated as well as the soil crusting. Furthermore, furrow irrigation systems differ from basins and border surface irrigations due to their ability to be independently designed and controlled. Furrow irrigation systems offer more efficient and flexible field water management under varying surface irrigation conditions.

Despite the numerous benefits of furrow irrigation systems, some drawbacks still limit its use, especially in tropical SSA. These include (but are not necessarily limited to) increased salinity between the furrows, increase in the number of required labour for efficient operation, increased erosive potential of the water flow and difficulty in moving farm equipment across the farming fields (FAO, 2019).

7.7 Border Irrigation

Border irrigation is a third form of surface irrigation system and is widely designed to irrigate closely growing crops that are susceptible to mechanical injuries upon lengthy exposure to inundation (Zerihun et al., 2005). Basically, border systems that are well designed and managed tend to uniformly

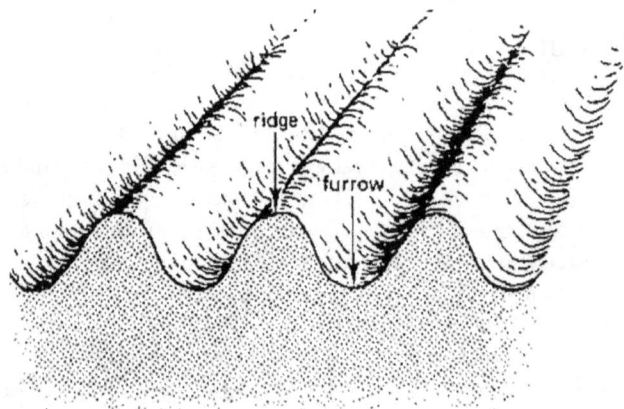

FIGURE 7.3   Schematic diagram of furrow irrigation (FAO, 2019).

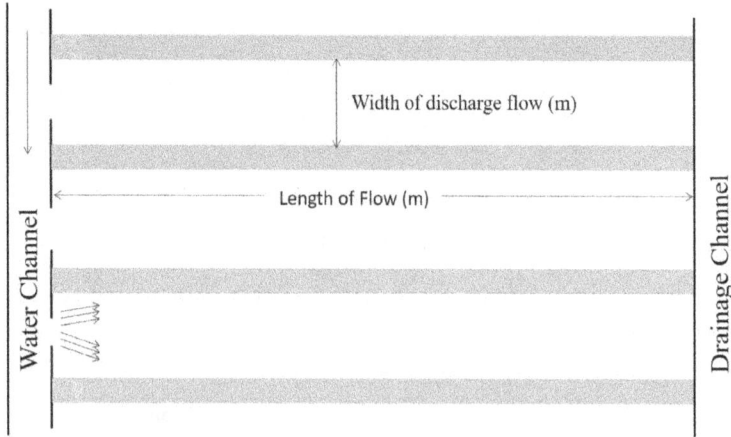

FIGURE 7.4 Border irrigation system.

administer irrigation water at high levels of efficiency and with minimal adverse effects to the environment (Zerihun et al., 2005). The design of border irrigation system aims at maximizing a measure of merit (performance index, which can either be physical or economic) while minimizing some undesirable consequences (Figure 7.4).

7.8 Sprinkler Irrigation

Sprinkler irrigation system supplies water to the field in droplets from the pump through a network of interconnected pipes (Figure 7.5). This type of irrigation is not only designed to supply water to the cropping land, but they are also designed to: cool the soil and crop; provide moisture for seed germination; control wind erosion; apply chemical fertilizers as well as protect frost and increase the development of fruits and buds, respectively.

There are different types of sprinkler irrigation systems. They include:

- Portable Sprinklers: This system of sprinkler irrigation can be moved from place to place due to its portable features (mainlines, submains, laterals and pump) (FAO, 2019).
- Semi-portable Sprinklers: They possess similar characteristics as the portable systems; however, their water source and pumping devices are at fixed locations (FAO, 2019). They cannot be moved from place to place. This system of sprinkler irrigation can be employed for irrigating more than one field in as much as there are extended mainlines.
- Semi-Permanent Sprinklers: This system of sprinkler possesses portable lateral lines with permanently buried mainlines (Newell and Murphy, 1913). The water source and pumping devices are also stationed at fixed locations on the field.
- Permanent Sprinklers: This system of sprinkler is characterized by permanent laterals, mainlines and submains with locations for specific water source and/or pumping devices (Newell and Murphy, 1913). Permanent sprinklers are usually expensive to use and are designed with high technicalities.

Sprinkler systems are employed on agricultural fields that contain sandy soils with uneven slopes. They are basically employed for agricultural, industrial or residential use. Sprinkler irrigation systems are also commonly employed in row cropping and tree crop production. Sprinkler irrigation system has been reported to improve crop yields in some part of the tropics. For instance, in semi-arid regions of India, Indian National Committee on Irrigation and Drainage (1998) reported that sprinkler irrigation saved water and increased yields in different crops (Table 7.1)

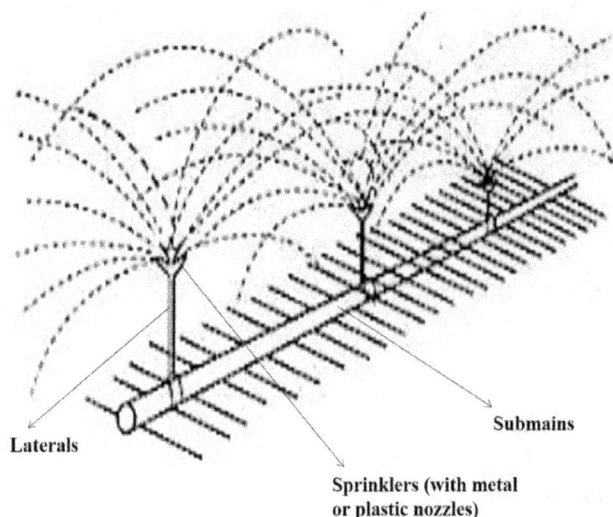

Laterals

Submains

Sprinklers (with metal
or plastic nozzles)

FIGURE 7.5 Sprinkler irrigation system.

TABLE 7.1 Response of Crops to Sprinkler
Irrigation (INCID, 1998)

Crops	Percentage (%) Water Saved	Percentage (%) Yield Increment
Barley	56.0	16.0
Cabbage	40.0	3.0
Cauliflower	35.0	12.0
Cotton	36.0	50.0
Cowpea	19.0	3.0
Garlic	28.0	6.0
Groundnut	20.0	40.0
Maize	41.0	36.0
Onion	33.0	23.0
Potato	46.0	4.0
Sunflower	33.0	20.0
Wheat	35.0	24.0

7.9 Drip Irrigation

Drip irrigation, referred to as trickle irrigation system, is a modern irrigation technique that is slow, steady and ensures precise application of water to crops (Bressan, 2006). In this system, a tank or bucket serves as an irrigation storage system (water source), which is developed by constructing a small bund of earth or stones built across a stream (Figure 7.6). It uses flexible polyethylene tubing with emitters (devices for dripping water) and low-volume sprays. Drip irrigation systems are relatively easy to install. This system does not require trenching, and the only tools needed are pruning shears and a punch. Drip irrigation is efficient for the maintenance of a near-perfect moisture levels within the crop root zone. It safeguards against over-wetness/dryness usually observed in surface irrigation systems. Drip systems can be used to irrigate shrubs and trees, perennial beds, annuals and ground covers (Bressan, 2006). Drip systems have been reported to be widely effective in many crops (Table 7.2).

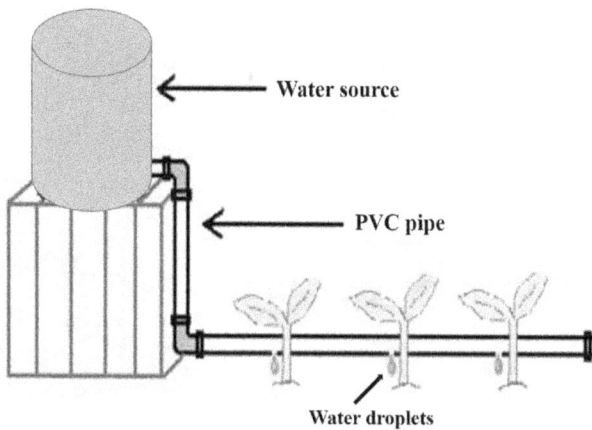

FIGURE 7.6   Drip irrigation system.

TABLE 7.2   List of Crops Suitable for Trickle Irrigation

Crop Types	Examples	Remarks
Vegetables	Pumpkin, Cucumber, Cabbage, Cauliflower, Okra, Onion, Peas, Tomato, etc.	Highly efficient
Flowers	Marigold, Orchids, Rose, Jasmine, etc.	Moderately efficient to highly efficient
Orchard Crops	Oranges, Guava, Pineapple, Banana, Papaya, Mango, Citrus, etc.	Moderately efficient
Plantation Crops	Rubber, Coffee, Coconuts, Tea	Moderately efficient
Forest Crops	Bamboo	Moderately efficient
Spices	Coves, Turmeric	Highly efficient
Oil Seed Crops	Oil Palm. Groundnut, Sunflower, etc.	Moderately efficient

Source: Modified from InDG (2016).

Drip irrigation system is used to cultivate healthy and fast growing crops. It is very effective in maintaining a constant moisture level within the root zone of crops with equally appreciable amount of oxygen. Drip system has the capacity of administering water exactly where it is required. This also helps to reduce water evaporation to its barest minimum. Drip irrigation saves water from wastage and can be used to apply equal amount of water to crops over a wide farm area. Drip irrigation system can be employed on sloppy lands with minimum run-off.

Farming in Moderately Warm Arid Regions of Africa

Agriculture (and food production in particular) has been described as the bedrock of every country owing to its importance in sustainable development and human survival. Farming in the tropics especially in SSA has been associated with increased challenges with regard to sustainability. This is evidently imposed due to the seasonal weather patterns and changing climatic factors of rainfall and temperature over the cropping seasons. Moderately arid regions in Africa are generally characterized with few months of rainfall unevenly distributed annually and consist of countries like Ethiopia, Kenya and some parts of Nigeria (Figure 7.7). This limited rainfall conditions have spear-headed the limitations in the yield potentials of most farming activities and crop production. Agricultural operations in the arid areas are highly dependent on irrigation for sustainable production and all year round crop cultivation. Irrigation is a basic infrastructure and pertinent input necessary for sustainable optimum crop and livestock production in arid regions of the tropics.

The Food and Agricultural Organization of the United Nations in 2013 reported that agriculture in Africa is generally dynamic and diverse in many different structures, organization methods and

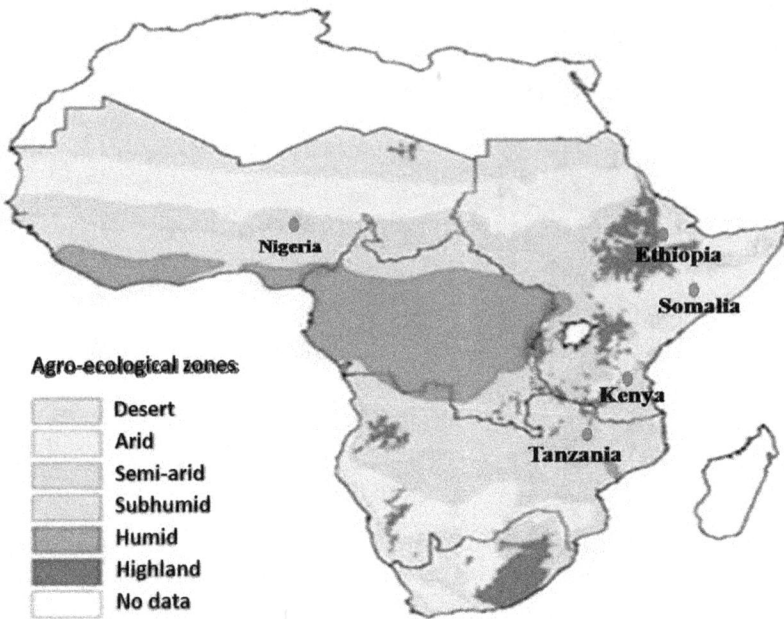

FIGURE 7.7   Agro-ecological zones in sub-Saharan Africa.

farming systems. The diversity in farming operations in the semi-arid regions of SSA is characteristi-
cally observed more within Nigeria in the West and Ethiopia in the East, respectively. Moderately
arid regions also known as semi-arid or semi-desert regions of Africa accounted for almost 25% of
food production in Africa in 2010 (FAO, 2013). The variation in weather factors of temperature and
precipitation especially in semi-arid regions of Africa has resulted in high variation in crop yields,
and the poor water management has limited the productivity in this region to the prevailing climatic
conditions.

Agriculture in the moderately arid regions of SSA is generally being practiced and owned by
small-scale farmers. This buttresses the importance of crop production in this region of the tropics.
Despite the increased level of technology in SSA, farm sizes tend to reduce over a few years owing to
population pressure on land, land degradation and urbanization. Nevertheless, agricultural production in this region of SSA is basically rain-fed, as the weather conditions in the area cannot support
all year round crop and livestock production to meet the needs of the ever-increasing population.
Family labour is still excessively being used in this region as subsistence agriculture dominates this
region with few commercial productions in limited parts of East (Ethiopia and Rwanda) and West
(Nigeria) Africa, respectively. Crop production in moderately arid regions is generally dependent
on irrigation water, soil fertility amendments and good management practices. For instance, in the
upper part of the northern Nigeria, which falls within the semi-arid region of SSA, crops like millet, sorghum, rice, yam, carrot, onions, sugarcane and cabbage are commercially produced using
irrigation water systems in order to support the less than 1,000 mm per annum precipitation, which
is a usual characteristic of the region. The labour for production usually comprises farming families,
and small farm holdings that make up the bulk of agricultural production in this region of West
Africa. Technologies are scarcely adopted by this farming population predominantly made up of the
Fulani tribe of the country. Efforts to boost crop production levels in this region lies on the provision
of accessible and affordable credits facilities and easy access to inputs and market as these regions
comprise the poorly educated people.

7.10 Irrigation Systems in SSA

Irrigation agriculture has helped tropical nations in SSA increase their agricultural production capacities in terms of crops and livestock, respectively. Due to the unreliable rainfall patterns and distribution for all year round crop cultivation, the use of irrigation has become a very necessary technology for continuous crop cultivation and livestock production, especially in arid and semi-arid regions of SSA. FAO (2013) reported that the use of surface irrigation system is observed to be common in all regions of the tropics especially in SSA. This is due to the fact that the small-scale farmers make up the bulk of the farming population in this region of the tropics.

The use of surface irrigation has been reported in farming areas in Mali where the two major rivers in the nation provide the required alternative water source to augment the lapses in water requirement established due to variations in rainfall conditions (Kerga and Dembele, 2018). Large irrigation systems (through dams) have been observed to be very expensive to implement in Mali, while small-scale irrigation (SSI) systems can be implemented at relatively low costs to benefit communities. Few modern farms practice sprinkler irrigation in the nation, but irrigation agriculture in the area is predominantly characterized by traditional surface irrigation using furrows and basins, respectively. In the Eastern region, the use of sprinkler irrigation becomes more pronounced with surface irrigation systems. Werfring (2004) reported that surface irrigation and sprinkler irrigation systems have been used in crop production in Ethiopia. Unlike Mali, Ethiopia has more access to water and these serves as major sources for sprinkler irrigation by commercial farmers in the region. In Kenya, Neubert et al. (2007) reported the use of surface irrigation (furrow and border) while few commercial farms practiced the use of sprinkler irrigation. Kenya has seen increased use of drip irrigation in the nation by small-holder farmers looking to conserve the available water at their disposal for optimum crop production. However, in Nigeria in the Western coast of Africa, surface irrigation is highly predominant and covers a wider area of irrigated lands in the region (Adelodun and Choi, 2018). The use of sprinkler irrigation is commonly observed in modern farms all over the country for commercial crop production. Drip irrigation systems are also observed in the northern region of the country by small-scale farmers producing vegetables and fruits, respectively. This chapter will lay emphasis on the irrigation practices in Ethiopia, Kenya and Nigeria, respectively.

7.11 Irrigation Practices in Ethiopia

Ethiopia is a landlocked East African nation with an approximate land area of 1.14 million km² (Haile and Kasa, 2015). As the second most populous nation in SSA (after Nigeria), the bulk of which are rural dwellers, the populace of Ethiopia is highly dependent on agriculture as their major source of livelihood. Like most countries in SSA, agriculture is the fulcrum of the Ethiopian economy (World Bank, 2006; Makombe et al., 2011). The country is characterized by vast water resources. Awulachew et al. (2007) and Haile and Kasa (2015) reported that Ethiopia has 12 river basins with a combined total run-off volume of 122 billion m³ of water per annum, and an estimated ground water potential (GWP) of 2.6–2.65 billion m³. Despite the vast land area of the nation, about 12 million ha is currently under cultivation (MoA, 2011). Werfring (2004) earlier noted the land area in Ethiopia that is under irrigation ranges between 1.5 and 4.3 million hectares (Mha). However, due to population pressure and poor irrigation schemes and intervention, recent studies have established that the cumulative area currently under irrigation had reduced to a range of 160,000–200,000 ha, thereby making it less than 5% of the country's irrigable land (Haile and Kasa, 2015). Also, MoA (2011) reported that there is a 10%-12% summative irrigation potential of the currently used traditional and modern irrigation schemes in the nation.

Irrigation operations in Ethiopia are considered as a basic strategy for poverty alleviation and food sufficiency. It has been reported to be a very useful technique for the transformation of the rain-fed agricultural system (which depends immensely on weather factors of rainfall and temperature) into the combined rain-fed and irrigation agricultural system. This is geared towards increasing the levels of crop yields and ensuring all year round production of crops in order to advance the effort towards attaining food security and sustainable human development. Historically, irrigation in Ethiopia (like Nigeria) has hitherto been used in ancient times by farmers in the region for crop production (Bacha et al., 2011). Researchers have also highlighted the use of alternative water sources for continuous crop production by small-holder farmers to cater to the needs of their families (Hagos et al., 2009). The use of traditional surface irrigation methods has been recorded especially in the arid and semi-arid northern region of Ethiopia for crop production especially during excessive periods of low precipitations.

Modern irrigation operations commenced in the early 1950s with the bilateral agreement between the Government of Ethiopia and a Dutch company jointly known as HVA-Ethiopia sugar cane plantation (MoA, 2011; Bekele et al., 2012). The bulk of the traditional irrigated fields in Ethiopia are predominantly under surface water sources, while groundwater uses have just been started on a pilot basis in the East Amhara region (MoA, 2011). MoA (2011) noted that pressurized sprinkler irrigation system was once practiced in Fincha State Farm, Eastern Amhara and Southern Tigray, and on some private farms in the Rift Valley. Also, surface irrigation (furrow irrigation and basin irrigation) methods have been used for the production of commercial fruits (like banana), cotton and wheat. Similarly, Awulachew et al. (2007) emphasized that irrigated agriculture was started in Ethiopia in the upper Awash Valley with the aim of large-scale industrial production of sugarcane, cotton and horticultural crops. This laid the foundation for the remarkable emergence of irrigation development and establishment of agro-industrial centres.

Furthermore, there is a need for more improved development in irrigation practices in Ethiopia. This is highly pertinent as it serves a key technique for sustainable and well dependable agricultural development in the country. The agricultural scale and practices of small-holder farmers in Ethiopia can be improved through provision of good and highly efficient irrigation schemes and credits to help facilitate the continuous production of crop all year round. This is due to the fact that in Ethiopia, the bulk of the food crops produced and available in the domestic markets are basically cultivated by small-holder farmers with little or no credit and technology to increase production capacity for the needs of the ever-increasing population. The Ministry of Water Resources (MWR) in Ethiopia through public and private partnerships need to establish modalities that will be geared towards boosting the production capacities of the highly populated small-holder farming families, due to the prevailing weather and climatic conditions, with the aim of ensuring that that crop production can be sustained at the optimum level. For instance, the irrigation scheme in the Wedecha-Belbela Irrigation System that provides essential irrigation services (using furrow irrigation technique) for more than 1,267 households (the largest beneficiary being Goha Weriko with 309 households) is posed with the major problem of seepage losses in primary canals culminating in water shortages during periods of high demand at a particular time. The effect is predominantly observed in the lower part of the irrigation system of the schemes, where farmers are forced to wait for more than 8 hours in order to obtain irrigation water upon its immediate discharge from the source. In addition to this prevailing challenge, there are problems of poor operation of the system, limited availability of control structures, low of interest by the farmers to participate in maintenance and inadequate funds for proper and optimum system operation and maintenance, respectively (Awulachew et al., 2007).

MoA (2011) reported that irrigation development in Ethiopia is still at its infancy state; nevertheless, there have been established plans and programmes by the Ethiopian government to advance effort in reducing poverty and increasing food production in the nation. Haile and Kasa (2015) and Belay and Bewket (2013) maintained the importance of irrigation to poverty alleviation through improvement in crop production in rural areas and for small-holder farming families; this in turn helps improve the food security and rural livelihoods. Irrigation practices by small-holder farmers have now gathered significant focus from local governments to enable farmers to cultivate crops more than once annually. Also, Bacha et al. (2011) in a similar study noted that land productivity, asset ownership, access and use of credits and mean food consumption and expenditure on food and non-food assets were significantly higher for farmers that practice irrigation operations than for farmers who do not.

However, there is a pertinent need for the establishment and improvement of SSI schemes in Ethiopia, in order to more effectively build the resilience of susceptible farming communities and households by ensuring that there is less dependence on the brunt and unpredictable patterns of rainfall in both the semi-arid and arid regions of the country. In addition, the establishment of a good SSI scheme entails detailed attention to management practices at every stage of the implementation in order to achieve high success and efficiency. For instance, in Tigray, SSI schemes have been established to assist the farming households that predominantly comprise widows to improve upon their crop production on their small farms. Farmers in the area have practiced drip irrigation, solar-powered SSI and irrigation systems that use open gravity water canal and pipe conveyance systems for all year round crop cultivation. Also, sprinkler irrigation and surface irrigation (especially furrow and basin irrigation) have also been used by small-holder farmers in this region.

7.12 Irrigation Practices in Kenya

Agriculture has posed significant influence on the economic status in Kenya. Upon the assessment of the impact of the agricultural sector on the economy of Kenya over time, there was a high significant correlation, entailing that agricultural operations have boosted the economy of the nation through the production of food, employment and income, especially for the highly populated small-holder farm families at the primary level. This lays credence to the importance of ensuring crop production all year round in the face of the increase in the number of persons in the nation. With the poor climatic conditions, coupled with the predominantly scarce water resources in this region of East Africa, the use of supplementary water sources is highly imperative for the continuous production of crops at optimum levels. This has been well observed by the symbiotic relationship between the economic well-being of people in such towns (Wanguru, Hola, Bura, Ahero, Marigat, and Nyadorera) and the performance of the irrigation projects around them.

Farming in Kenya, which is characterized by semi-arid and arid ecological zones in Africa, is basically highly dependent on supplementary water to meet the water requirements of the crops produced at every given time. This stresses the importance of irrigation in this region. Farming operation is generally carried out by small-holder farms with their families providing the bulk of the labour force. These farming populations have little capital input and technological know-how to practice the modern sophisticated irrigation systems in order to improve crop production levels on their vast small land holdings. As such, irrigation operation in Kenya is highly characterized by the use of traditional system of water application.

Climatic condition of farm lands in Kenya makes them generally unsuitable for crop production and more unfavourable for optimal livestock production. About 80% of Kenya's land is sandwiched between arid and semi-arid regions of SSA. Artificial water supply to enable crop, livestock and aquatic animal production is very pertinent as a means of securing livelihoods for the population in these areas. Studies have outlined the benefits of good irrigation systems and schemes that tend to multiply crop and livestock production per unit of land by a factor of three to four when compared to rain-fed agriculture. As a result, farmers in the nation can boost their production level to three or four times of what will be conventionally obtained when depending solely on rain-fed farming. In some areas where there is hardly any production because of inadequate soil moisture, this factor could be higher.

Although the use of traditional SSI practices has been predominantly observed in some parts of Kenya for over four centuries, large-scale irrigation schemes have also been used since the colonial era. Since 1966, the large-scale irrigation schemes in Kenya have been solely managed by the National Irrigation Board (NIB), which is a government agency of the Ministry of Water and Irrigation (MWI). It is semi-autonomous and operates relatively independent of the Ministry. Farmers under this scheme have mustered little profits. Hence, the schemes did not reach their full potential, which consequently led to the collapse of all but one of the NIB irrigation schemes (Neubert et al., 2007). In the year 1977, a Small-Scale Irrigation Unit (SSIU) was instituted within the Ministry of Agriculture (MoA) to supplement the NIB. The task of the SSIU (subsequently modified in 1978 to become the Irrigation and Drainage Branch (IDB)) was to provide significant support for the development of small-holder irrigation schemes. Despite the later transfer of this SSIU to the MWI, it still promotes small-holder schemes.

The Regional Development Authorities (RDAs) was established in the 1970s, and charged with the mandate of ensuring the development of their respective regions, including an advanced improvement in irrigated agriculture. These RDAs that were classified as public commercial schemes have been highly involved with most community-based irrigation systems in Kenya today. Some of these community-based schemes are still in existence. The creation of the commercial flower and vegetable farms in the latter part of the 1980s was also based on irrigation systems, and this was predominantly covered by the areas around Naivasha, Eldoret, Nanyuki and the capital, Nairobi. However, three organizational types were identified from the irrigation schemes established by the RDAs. These are:

i. Small-Holder Schemes: These are schemes of varying farm sizes and coordinated by water resource groups or an association of farmers. The produce from these schemes are used to meet subsistence demands as well as for domestic and export markets. This scheme accounts for about 46% of the total area under irrigation in Kenya, covering almost 47, 000 ha of land (Neubert et al., 2007). This scheme is highly characterized by the use of low technological irrigation devices and systems for easy adoption by the local farming populace, which makes up the bulk of the farming population under this system. Under this scheme, farmers predominantly practice drip irrigation and surface irrigation using locally constructed furrows and basins to channel water obtained from nearby streams and rivers and/or reservoir tanks and buckets (for trickle system). The choice of the irrigation system is highly dependent on their low capital inputs and low technology for easy operation of more sophisticated devices procured by the government. Most produce under this system are completely sold at the domestic market with very little or nothing available for export.

ii. Large-Scale Schemes: These schemes are managed wholly by the NIB. The schemes cover a very large area in size and produce from these schemes are solely for the domestic and export markets. The bulk of the Kenya's rice is produced under these NIB schemes. Today there are a total of seven such schemes covering an area of some 13,000 ha. These schemes account for some 12% of Kenya's irrigated land, and about 12% of the farmers active in irrigated agriculture work there (Ngigi, 1999; Government of the Republic of Kenya, 2005; Neubert et al., 2007). With the assistance of technologically sound agronomists, there is proper land and water management under these schemes as modern irrigation techniques for crop and livestock production are utilized. Then, sprinkler irrigation for vegetable, surface irrigation for arable crops using high overhead tanks and high pressure pumps are usually used in this system.

iii. Commercial Flower and Vegetable Farms: These are schemes with higher technicalities and modernized irrigation facilities (having a workforce of approximately 70,000 persons). The produce from this system is more exclusively saddled for export markets. This scheme, which utilizes modern surface irrigation system techniques and devices, with some areas of the western coast practising sprinkler irrigation, with well-sophisticated sprinklers, always records high profit returns. This scheme covers about 43,000 ha, ensuring that about 42% of the land is under irrigation.

Furthermore, some challenges have limited and or impaired the smooth operation of some of these schemes. For instance, the NIB had over 1,000 ha of sugarcane fields under furrow irrigation in Ahero and West Kano Irrigation Schemes in the 70s. But, sugarcane production was terminated in 1980 owing to marketing challenges. The other sugarcane irrigation worth mentioning is the Chemelil Sugar Company, which has 500 ha of irrigated sugarcane. From these, 400 ha are sprinkler-irrigated, while 100 ha are furrow-irrigated. According to NIB irrigation, sugarcane is an essential commodity in humid areas such as the Nyando Region, in which cane yields have to be increased or at least maintained.

Current irrigation practices in Kenya have seen an increase in the number of small-holder farmers employing the services of drip irrigation. For instance, drip irrigation technology has been established in the drier parts of Taita Taveta County in the coast region. This system is readily used by farmers in this region owing to the fact that the system helps the farmers incur fewer expenses while irrigating crops. Nevertheless, Monteiro et al. (2010) reported that there are several irrigation systems currently in use in Kenya agriculture. These include the conventional sprinkler irrigation, sprinkler systems with the aid of a motorized hose-pipe by pinching, low-cost drip irrigation and furrow irrigation. For instance, the Chemelil Sugar Company has over 500 ha of sugarcane—the sprinkler system (about 400 ha) and furrow irrigation (about 100 ha). Also, with the low-cost trickle irrigation kit, farmers in Kenya tend to use buckets or drums as water reservoirs consisting of varying lengths and numbers of drip lines. Monteiro et al. (2010) noted that surface irrigation in Kenya has an irrigated area of about 58,290 ha (occupying about 67% of irrigated area) while drip irrigated area is 2,000 ha (occupying about 2% of irrigated area). In addition, about 26,970 hectares (occupying about 31% of irrigated area) is under sprinkler irrigation system (FAO, 2004). While some of these irrigation systems have been efficient, irrigation systems in Kenya is still governed by the use of traditional furrow and overhead sprinkler systems which utilizes too much of the limited water resources in the country. There is a need to design irrigation systems that are targeted at the poor small-holder farmers and at their small farm sizes in order to improve upon the current production level and consequently advance the nation's effort in food security.

7.13 Irrigation Practices in Nigeria

Agriculture is the bedrock of the Nigerian economy as it tends to provide the bulk of income for up to two-thirds of its population, most of whom are extremely low-income earners (World Bank, 2014). Nigerian agriculture is governed by two distinct seasons—rainy and dry seasons, respectively (Ibe and Nymphas, 2010). The amount of rainfall observed in the country is highly influenced by the location and prevailing ecological zones. This exerts an effect on the nature and properties of soil in the area. Basically, soils in the Southern zone of Nigeria are usually sandy and loose with higher organic matter content compared to soils in the upper north region of the country. These prevailing soil, water and weather conditions in the nation immensely pose a huge influence on the type, frequency and scale of irrigation in farming in different parts of the country.

Adelodun and Choi (2018) reported that irrigation practice across the world is vital to successful green revolution all year round for achieving sustainable development goals in food security, socio-economic and rural development, and Nigeria in particular is yet to fully benefit from the huge investment on irrigation. Farming systems in Nigeria is still highly characterized by subsistence farming where the small-scale farming families produce the bulk of agricultural produce in the food market. Since agriculture in some parts of West Africa is mainly rain-fed, irrigation practices make up less than 20% (accounts for about 1% of the cultivated area) of agricultural operations in the nation (Adelodun and Choi, 2018; FAO-Aquastat, 2017).

Historically, irrigation practices in Nigeria dates back to 700 AD and gained increased attention following the drought of 1970–1975 (Olubode – Awosola and Idowu, 2004). More recently, irrigation practices have been carried out more often in the nation due to the changing climatic conditions and increased pressure for all year round food production to meet the vast population in the nation. Traditional irrigation (consisting of the use of buckets, calabash and earthen pots to obtain water from nearby streams and rivers to irrigate crops) has hitherto existed as the bane of irrigation practices in Nigeria. However, the use of more sophisticated methods of irrigation have become more pronounced, especially in modern commercial farms in the north and south western regions of the country.

In order to ascertain the water resources and potential for irrigation operation in Nigeria, a study was conducted in 1972 and the outcome gave rise to the establishment of the following irrigation schemes: the Bakolori scheme, the Chad Basin scheme and Kano River irrigation scheme (NINCID, 2015). However, owing to the success of the aforementioned schemes in Nigerian agriculture, 11 more schemes under the River Basin Development Authority (RBDA) were established to further improve crop output and ensure food self-sufficiency. The RBDAs include the Niger Basin, Lower Benue Basin, Upper Benue

Basin, Lake Chad Basin, Benin–Owena Basin, Sokoto Rima Basin, Hadejia Jamaare Basin, Cross River Basin, Ogun Osun Basin, Anambra–Imo Basin, and Niger Delta Basin. Nevertheless, the projected efficiency and sustainability of these newly established large-scale public irrigation schemes saddled with the mandate to provide food sufficiency was not met (FAO – Aquastat, 2016). Consequently, most of the new schemes have become obsolete owing to the very high cost of operation and poor maintenance by the schemes' beneficiaries, respectively.

Current irrigation practices in Nigerian agriculture is still highly characterized by surface application of water from nearby streams using buckets, pipes and pumps in most farms in northern Nigeria, which coincides with the moderately arid regions in Africa. Farming activities in this region of the country have been recently boosted by the establishment of the FADAMA irrigation project instituted by the Federal Government to increase and sustain production outputs of agricultural crops. Sprinkler irrigation is predominantly used for vegetable production in farms in the North West, and for orchard farming in the North Central. Furthermore, drip irrigation systems have been used in modern private farms in the upper regions of Kano and Kaduna to checkmate water losses and ensure that the crops are precisely irrigated at minimum costs. In the Southwestern region of Nigeria, surface irrigations are predominantly used as the system of irrigation owing to the low capital inputs and technological know-how of the small-scale farmers that predominantly occupy this region of Nigeria. However, in the International Institute for Tropical Agriculture in Ibadan, sprinkler irrigation is the major system of irrigation. The area has an artificial lake that serves as the source of water for all farming operation (Figure 7.8). The lake as a reservoir is controlled in order to regulate the quantity of water for irrigation and other domestic water uses. This lake is controlled from both ends and is adequately sieved to ensure that the water that is being pumped using a mobile pumping machine (a total of eight) in the area is clean and devoid of particles that will distort the smooth operation of the sprinklers (Figures 7.9 and 7.10).

The pump is used to channel water to the field within an hour of pumping. The pump channels in water to the submains (Figures 7.11 and 7.12), which is then used to sprinkle across the farm area through a network of equal spaced sprinklers (Figure 7.13).The water released is controlled by a nozzle, which regulates the water that enters the submain (Figure 7.14). Drip irrigation systems is also being practiced by some researchers in the area but on a lesser area compared to the land area under sprinkler irrigation.

In addition, unlike the Southern regions where surface irrigation is more commonly practiced, the use of buried pipes and establishment of water sources and basins are commonly used in commercial farms in the north owing to the prevailing weather conditions and limited access to irrigation water. Akinbile et al. (2016) in a study on the assessment of the impact of irrigation systems on food security in

FIGURE 7.8   Artificial Lake of the International Institute of Tropical Agriculture (IITA), Ibadan.

FIGURE 7.9 Connecting pipes to the pump from the reservoir.

FIGURE 7.10 Water pumping device prior to field operation.

FIGURE 7.11 Mobile pumping machine serving water to submains for sprinkler irrigation at IITA.

FIGURE 7.12 Sprinklers with submains for irrigation in the experimental plots at IITA.

FIGURE 7.13 Sprinkler supplying water to the field.

FIGURE 7.14 Nozzle control for regulating water used for sprinkler irrigation in the area.

TABLE 7.3　Irrigation Systems in Selected Countries of sub-Saharan Africa

Countries	Irrigation System	Remark	Reference
Ethiopia	Surface Irrigation		
	• Furrow	Highly Practiced	Werfring (2004)
	• Basin		MOA (2011)
	Sprinkler Irrigation	Moderately Practiced	Bekele et al. (2011)
			MOA (2011)
Kenya	Surface Irrigation		
	• Furrow	Highly Practiced	Neubert et al. (2007)
	Sprinkler Irrigation	Moderately Practiced	FAO (2004), Neubert et al (2007)
			Monteiro et al. (2010)
	Drip Irrigation	Slightly Practiced	Monteiro et al. (2010)
Nigeria	Surface Irrigation		
	• Furrow	Very Highly Practiced	Maurya and Kuzniar (1988), Ogunwale (1989)
	• Border		Akinbile et al. (2016)
	• Basin		
	Sprinkler Irrigation	Moderately Practiced	Ngasoh et al. (2018), Ahaneku (2010), Akinbile et al. (2016)
	Drip Irrigation	Moderately Practiced	Buhari (2018)
			Akinbile et al. (2016)

Southwestern Nigeria observed that 57% of the farmers practiced surface irrigation, while 26% and 13% practiced sprinkler and drip irrigation, respectively.

The advent of improved techniques of irrigation in Nigeria has resulted in increased farming outputs, which are still less than the expected yields to meet the high population demands of the country. As such, there is the need for government intervention and private sector partnership with most commercial and highly efficient subsistent farms in this region of Africa in order to increase crop outputs and increase farmer's income for food security and self-sufficiency.

However, the major irrigation systems practiced in the countries under this study are presented in Table 7.3. Despite the variation in the system of irrigation used for crop and livestock production in semi-arid regions of SSA, the common goal is to improve yield to meet the demands of the ever-increasing population. More systems need to be adopted in regions where irrigation operations are yet to be fully adopted in order to ensure a more sustainable food production and advance our efforts in attaining food sufficiency in SSA.

7.14　Conclusions

Irrigation involves the artificial supply of water to farmland for various agronomic reasons. Moderately arid areas are highly characterized by low rainfall. Agriculture in this region of SSA is basically rain-fed. Moderately arid areas possess limited amount of water resources for irrigation. Irrigation practices in Nigeria, Ethiopia and Kenya were reviewed from their earliest inception to the current status. The study found out that these nations practice diverse irrigation systems but their use is highly dependent on the needs of the small-holder farmers who produce the bulk of food to meet the need of the ever-increasing population. The use of irrigated agriculture in the semi-arid regions of these nations is dependent on government interventions and access to credit facilities and irrigation designs geared for the comfort of the small-scale resource farmers. There are several challenges imposed by some of these irrigation systems and schemes that limit their use in the studied nations. However, for an increase in food production and sufficiency, it is pertinent to integrate the small farming families in the design of appropriate irrigation systems for better livelihood.

Acknowledgements

The authors are grateful for the remarkable contributions and support of Mr. Augustine Akwarandu, Mr. Louis Anajekwu of the International Institute of Tropical Agriculture (IITA – Ibadan), Mr. Cobes Gatarira for the success of this chapter.

References

Adeaga, O. and Eslamian, S. 2021, Assessment of freshwater and conservation in Africa. In: Eslamian, S. and Eslamian, F., eds., *Handbook of Water Harvesting and Conservation, Vol. 2: Case Studies and Application Examples.* John Wiley & Sons, Inc., New Jersey, pp. 121–140.

Adelodun, B. and Choi, K. 2018. A review of the evaluation of irrigation practices in Nigeria: Past, present and future prospects. *African Journal of Agricultural Research*, 13(40): 2087–2097.

Ahaneku, I. E. 2010. Performance evaluation of portable sprinkler irrigation system in Ilorin, Nigeria. *Indian Journal of Science and Technology*, 3(8): 853–857.

Akinbile, C. O., Oyebanjo, O. A., Ajibade, F. O. and Babalola, T. E. 2016. Assessing the impacts of irrigation systems on food security in Southwestern Nigeria. In *27th Annual Conference and Annual General Meeting - Minna 2016*, pp. 344–353.

Ali, A. 2018. Advanced Irrigation Systems for Water Conservation in Arid Regions: Alternative Irrigation Systems for Arid Regions. Retrieved on August 23, 2019 from https://www.slideshare. net/ahmadali476/subsurface-irigation-copy

Awulachew, S. B., Yilma, A. D., Loulseged, M., Loiskandl, W., Ayana, M. and Alamirew, T. 2007. Water resources and irrigation development in Ethiopia. IWMI Working Paper 123. International Water Management Institute, Colombo, Sri Lanka.

Bacha, D., Regasa, N., Ayalneh, B. and Abonesh, T. 2011. Impact of small-scale irrigation on household poverty: Empirical evidence from the Ambo District in Ethiopia. *Irrigation and Drainage*, 60: 1–10.

Bekele, Y., Nata, T. and Bheemalingswara, K. 2012. Preliminary study on the impact of water quality and irrigation practices on soil salinity and crop production, Gergera Watershed, Atsbi-Wonberta, Tigray, Northern Ethiopia. *Momona Ethiopian Journal of Science*, 4(1): 29–46.

Belay, M. and Bewket, W. 2013. Traditional irrigation and water management practices in highland, Ethiopia: Case study in Dangila Woreda. *Irrigation and Drainage*, 62: 435–448.

Bressan, T. 2006. *Drip Irrigation Handbook: The Catalog for Getting Started.* The Urban Farmers Store, p. 29, New York.

Buhari, S. 2018. Why Farmers consider Drip Irrigation: Expert. Publication of the Daily Trust on January 25, 2018. Accessed on September 18, 2019 on https://allafrica.com/stories/201801250595.html

Eldeiry, A. A., Garcia, L. A., El-Zaher, A. S. A. and El-Sherbini, K. M. 2005. Furrow irrigation system design for clay soils in arid zones. *American Society of Agricultural Engineers* 21(3): 411–420.

FAO. 2004. *Socio-Economic Analysis and Policy Implications of the Roles of Agriculture in Developing Countries.* Rome, Italy, p. 12.

FAO-Aquastat. 2016. Food and Agriculture Organization. Aquastat Project. FAO. Rome. Retrieved July 26, 2019, from http://www.fao.org/nr/water/aquastat/main/index.stm.

FAO-Aquastat. 2017. Food and Agriculture Organization. Regional Report-Nigeria. Retrieved on August 4, 2019, from http://www.fao.org/nr/water/aquastat/countries_regions/nga/index.stm

Food and Agricultural Organisation. 2013. Sustaining the Multiple Functions of Agricultural Diversity: Background Paper 1. FAO, Rome. Retrieved in August 23, 2019 from http://www.fao.org/3/x2775e/ X2775E03.htm

Food and Agricultural Organisation. 2019. Furrow Irrigation. FAO, Rome. Retrieved August 23, 2019 from http://www.fao.org/3/s8684e/s8684e04.htm

Government of the Republic of Kenya. 2005. Agriculture Sector Development Strategy.

Hagos, F., Makombe, G., Namara, R. E. and Awulachew, S. B. 2009. Importance of irrigated agriculture to the Ethiopian economy: Capturing the direct net benefits of irrigation, IWMI Research Report 128. International Water Management Institute, Colombo, Sri Lanka, p. 37.

Haile, G. G. and Kasa, A. K. 2015. Irrigation in Ethiopia: A review. *Academia Journal of Agricultural Research*, 3(10): 264–269.

Ibe, O. and Nymphas, E. 2010. Temperature variations and their effects on rainfall in Nigeria. In: Dincer, I., Hepbasli, A., Midilli, A., and Karakoc, T., eds. *Global Warming*. Green Energy and Technology. Springer, Boston, MA. https://doi.org/10.1007/978-1-4419-1017-2_38

INCID. 1998. *Sprinkler Irrigation in India*. Indian National Committee on Irrigation and Drainage, New Delhi, India.

InDG. 2016. Drip Irrigation System. India Development Gateway. Vikespedia. Accessed on September 17, 2019 on http://vikaspedia.in/agriculture/agri-inputs/farm-machinary/drip-irrigation-system

Kergna, A. O. and Dembele, D. 2018. Small – scale irrigation in Mali: Constraints and opportunities. *FARA Research Report*, 2(13): 18.

Makombe, G., Namara, R., Hagos, F., Awulachew, S. B., Ayana, M. and Bossio D. 2011. *A Comparative Analysis of the Technical Efficiency of Rain-Fed and Small-Holder Irrigation in Ethiopia*. International Water Management Institute, Colombo, Sri Lanka, p. 37.

MoA. 2011. *Natural Resources Management Directorates. Small-Scale Irrigation Situation Analysis and Capacity Needs Assessment*, Addis Ababa, Ethiopia.

Monteiro, R. O. C., Kalungu, J. W. and Coelho, R. D. 2010. Irrigation technology in South Africa and Kenya. *Cienca Rural*, 40(10): 2218–2225.

NEPAD. 2013. *African Agriculture, Transformation and Outlook*. Johannesburg, South Africa, p. 72, African Union Development Press.

Neubert, S., Hesse, V., Iltgen, S., Onyando, J. O., Ochoke, W., Peters, V., Seelaff, A. and Taras, D. 2015. *Poverty Oriented Irrigation Policy in Kenya: Empirical Results and Suggestions for Reforms*. DIE, Bonn.

Newell, F. H. and Murphy, D. W. 1913. *Principles of Irrigation Engineering: Arid Lands, Water Supply, Storage Works, Dams, Canals, Water Rights and Products*. McGraw-Hill Book Company, New York, p. 293.

Ngasoh, F. G., Ayadike, C. C., Mbajiorgu, C. C. and Usman, M. N. 2018. Performance Evaluation of Sprinkler Irrigation System at Mambilla Beverage Limited, Kakara – Gembu, Taraba State, Nigeria. *Nigerian Journal of Technology*, 37(1): 268–274.

Ngigi, S. N. 1999. Evaluation of irrigation research and development activities in Kenya: Towards promoting small-scale irrigation technologies. Draft Project proposal for IWMI, Sri lanka.

Nigeria National Committee on Irrigation and Drainage (NINCID). 2015. Country Profile - Nigeria. Federal Ministry of Agriculture & water resources. Abuja, Nigeria. Retrieved July 12, 2019, from www.NINCID.org/cp_nigeria.html

North, S., Griffin, D., Graham, M. and Gillies, M. 2010. Improving the Performance of Basin Irrigation Layouts on the Southern Murray-Darling Basin. Cooperative Research Centres for Irrigation Futures Technical Report 09/10. 66pp. Australia

Olubode-Awosola, O. O. and Idowu, E. O. 2004. Social-Economic Performance of Sepeteri Irrigation Project in Nigeria. Water Resourcess of Arid Areas 287–299. In *Proceedings of International Conference on Water Resources of Arid and Semi-Arid Regions of Africa*, Gaborone, Botswana, 3–6, August 2004.

Van Vuren, G. and Mastenbrock, A. 2000. Management types in irrigation a world-wide inventory per country. Report commissioned by the World Bank. Wageningen University, the Netherlands,

Werfring, A. 2004. Typology of irrigation in Ethiopia. A thesis submitted to the University of Natural Resources and Applied Life Sciences, Vienna. Institute of Hydraulics and Rural Water Management, in partial fulfillment of the degree of Diplomingeieur, Austria.

World Bank. 2006. Ethiopia: Managing water resources to maximize sustainable growth. A World Bank water resources assistance strategy for Ethiopia. The World Bank Agriculture and Rural Development Department. Report No. 36000-ET. Washington, DC, USA.

World Bank. 2014. Transforming Irrigation Management in Nigeria. World Bank indicators. Retrieved July 31, 2019, from http://data.worldbank.org/indicator

Zerihun, D., Sanchez, C. A., Farrell-Poe, K. L. and Yitayew, M. 2005. Analysis and Design of Border Irrigation Systems. *American Society of Agricultural Engineers*, 48(5): 1751–1764.

8

Choosing the Proper Smart Irrigation Technique in Temperate Semi-Arid Zones

Mahmoud
Abdellaoui
ENET'COM

Saeid Eslamian
*Isfahan University
of Technology*

8.1 Introduction

The good posture and the best stature of a stable, seine, powerful, robust, balanced society is revealed above, all on the self-sufficiency of the nobility of every member of society; that is why we find that every strategic development plan in every country of the world is based on one of many pilots, which is self-sufficiency of food and the good quality of food for the people of the country in question. A prosperous one in general case, and especially in convenient and prosperous life of a people is generally manifested by the abundance of goods and the ease of good living conditions and is of interest and concern to all sectors of life and especially the agricultural sector. The evolution and progress of a country involves the prosperity of agriculture. This prosperity of agriculture is concretized by the implementation of several suitable techniques of irrigation in the country. In this context and for this purpose, we have contributed to the development of two different irrigation systems suitable for semi-arid temperate zones, which is the case of the Sidi Bouzid region in Tunisia.

DOI: 10.1201/9781003353928-10

In this book chapter, two technical solutions of irrigation appropriate to the semi-arid zones will be presented and these two systems are applied in the region of Sidi Bouzid, Tunisia. With this technical recipe, which is a support and makes it a complete-finished demo project and set up by any interested person wanting to carry out his/her irrigation project. It is expected that the WSN will be used commonly in applications in intelligent agriculture, personal healthcare, home automation, industrial control and monitoring, asset and inventory tracking, home security and so on (ZigBee, WLAN link, Radio link-RF link, AIS, GPRS, LoRa and wireless personal area networks). Most of the developments and experimental deployments of WSNs are intended to provide for citizens in towns various applications like smart city-smart home. However, there're some researches to share the technology with farmers in irrigation of agricultural parcels. Jones [1] reported the result of deployment from an agricultural WSN point of view and precisely in greenhouses with tomatoes, melon, and potatoes. In this work [2], considering placing the sensor nodes with fairly high precision, the authors used a planned network rather than a network with ad-hoc routing. For reliability, they provided route diversity and multiple transmission. The work results demonstrated some of the value that a wireless sensors network might deliver to an agricultural setting. In addition, to demonstrate that the total cost of ownership of wireless sensors network is lower than a wired network, they could also show an example, that is, frost prevention where a live network is more valuable than a standard data logger. The works [3,4] presented Lofar Agro project that concentrated on monitoring micro-climates in a crop field; on the other hand, it presented the automated agriculture based on WSN [5]. The pilot project concerned the protection of potato crop against phytophthora, a fungal disease that can spread easily among plants and destroy a complete harvest within a large region. The authors described a precision agriculture architecture based on WSN. They deployed a small-scale sensor network (100 nodes) and acquired valuable experiences. These papers are various references for WSN applications in the agriculture field. However, they have weak points. They can gather the environment of the field. To take draw from the work [6], presenting the Vessel Finder Free-Applications Android's version tracking application, providing real-time data on the position and movements, utilizing a large network of the terrestrial AIS receivers. Vessel Finder Free aims to reveal the comprehensive picture of the global AIS coverage that Vessel Finder Pro will provide. As the same manner (well advised), and in the present project, the development is the ecosystem stages of Internet of Things (IoT) and cloud computing of things for remote control and supervision of the agricultural parcels irrigation and their identification-supervision through a distanced interface via the Internet of Things (IoT) and Cloud Computing of Things, the new Wireless Smart Sensors Network (WSSN) based on a miniaturized agricultural smart sensors nodes, the wireless sensors network (WSN), PHP on chip, Raspberry Pi, LoRa system communication and Android supervision interface. There're two proposed systems. The contributions of our book chapter are summarized as follows: Section 8.1 describes a few introductory words allowed to introduce the project and its context. Section 8.2 presents architecture for automated irrigation biological greenhouses agriculture system I using the WSN. It's studied and elaborated from sensor/actuator node hardware in the bottom to management sub-system in the top and is evaluated in the real deployment. The way to control the environment using the feedback from the gathered information is described in this first part. This part gives the methods-simulation results and experiences of real deployment in Sidi Bouzid area with many parcels of irrigation using system I and the ideas to improve the automated system of each parcel. Section 8.3 describes the new irrigation management system architecture (System II) using the new WSSN based on a miniaturized agricultural smart sensor node. A brief comparison between our project and other projects manifesting itself in a comprehensive description of the basic formulas of irrigation and its effective management; and having the determinants factors of irrigation optimization applied in semi-arid and arid areas [2]. Then, it enumerates the deployment, identification and supervision with IoT platform of advanced irrigation agriculture. Section 8.4 gives discussion of the results. Finally, the chapter book concludes with a summary of our work and a statement of future work.

8.2 Automated Irrigation Biological Greenhouses Agriculture (System I)

8.2.1 General

The desire to connect all electronic computing devices together has increased. In addition, it is more convenient and effective to use wireless links when we consider the large number of pervasive devices in the environment. In the coming years, it is expected that the WSN will be used commonly in applications especially in intelligent agriculture because the WSN might deliver some of the added value to an agricultural setting. In this first part and in the following paragraph, a precision irrigation architecture based on WSN is described and a large-scale sensors network is deployed.

8.2.2 System I Architecture

The large agricultural area of Sidi Bouzid is divided into many automated irrigation parcels. Each automated irrigation parcel has fully developed WSN, PHPoC/Raspberry Pi gateways, a management sub-system, 18 sensors nodes, an actuator node and one sink node. These equipment are deployed in each parcel according to a planned position, which had been decided by agriculturists considering the cultivation; in fact, for remote control of advanced irrigation and operated during the whole of the year. Four sensors called: A-node, B-node, C-node and D-node are placed on the four corners of each agricultural parcel. These four sensors are engaged to calculate and to determine the automated irrigation biological greenhouses agriculture parcel area (widthwise/breadthwise-overall length) based on the packet propagation time or on the signal strength. By various localization technics such as Time of Arrival (ToA), Time Difference of Arrival (TDoA), Direction-Angle of Arrival (AoA), and when the nodes (A-sensor node, B-sensor node, C-sensor node, D-sensor node) are synchronized, one-way packet is used (if the radio environment is not too disturbed) to determine the delay and then to estimate the distance between the different smart sensor nodes (between A-B, or B-C, or C-D and finally between D and A) [7,8]. In the system I called Automated Irrigation Biological Greenhouses Agriculture based on WSN (see Figure 8.1) and their associated equipment, two industrial PC-based gateways are installed to transform data over RS-232 from sink node to data over TCP/IP to servers. WLAN APs with directional antennas provide the long-range wireless link between WSN and the management sub-system which is about 1 km away from the irrigation parcel. The management sub-system with a Data Base-server and a Web-server manages WSN and provides easy interface to farmers with hand-help devices such as a PDA (Personal Digital Assistant). Among the 18 sensors nodes deployed in each automated irrigation parcel, the CH-node is an agriculture sensor node developed to be deployed in an irrigation agriculture parcel and to sense the environment of the agriculture parcel (see Figure 8.1).

8.2.3 Irrigation WSN Deployment and Management

The node is embedding all of 16 bits MCU, IEEE 802.15.4 compatible with 2.4 GHz band transceiver, a CPLD (Sleep timer to wake up the MCU from power down mode) and sensors (temperature and humidity sensor, water level and insect powder in soil, diseases of plants, …) in one PCB to reduce possibilities of defects. A sensor node is equipped with a battery of the lithium-ion rechargeable cells and monitors the voltage level of battery for maintenance purpose. To protect CH-node and all other equipment sensors included from the pelting rain, gust of wind, hailstorm, snowstorm, fog and humidity in irrigation agriculture parcel, protection paint is used based on meteorological weather chart. So, it's a protected area. The software of CH-node is based on the fourth version of ANTS-EOS improvement of the initial version [9] by our research team. EOS is a light-weight C-based multi-threaded operating system, which is developed to support multiple sensors network platforms. The network protocols of EOS are customized for this project to a light-weight CSMA-based MAC [10] and a robust multi-hop wireless

FIGURE 8.1 Different Configurations of the greenhouses with tomatoes, potatoes, carrots, onions, ...

sensors network routing protocol that are implemented as follows [7]: When CH-node has a packet to send, it checks CCA. If the channel is idle, then the packet is transmitted. If there is no acknowledgement from the recipient of the packet, MAC layer retransmits the packet up to three times. To simplify the route discovery, a tree-level which is the same as hop count from the sink node is preprogrammed in PHPoC/Raspberry Pi (EEPROM) with a network-wide unique 16 bits' address. The level of the sink node (see Figure 8.2) is "0", level "1" is assigned to the children of the sink node, level "2" is assigned to the irrigation agriculture parcel of the sink node, and so on. When CH-node initializes its parent–child relationship, it begins the joining process by broadcasting the search parent packet including a cost function, in this case, its tree-level. Whenever a sensor node receives the search parent packet, it checks the cost function and unicasts the invite packet to the joining node if the received tree-level is one-level

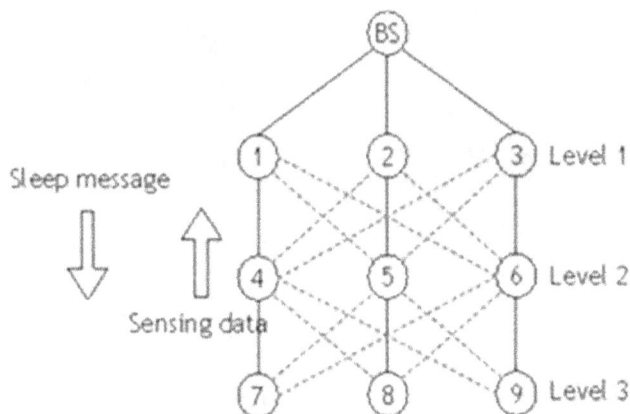

FIGURE 8.2 Runtime cycle of the sink node levels.

larger than its own tree-level [7]. For the specified time, the joining node waits for the invite packets from the prospective parent nodes and stores RSSI values with the invite packets [11]. After the specified time, the joining node chooses its parent according to the cost function; in this case, the largest RSSI value and sends the joint packet to its parent. The sensor node, which receives the joint packet, adds the joining node to its children list. Finally, the joining process is finished and the parent–child relationship is established. Whenever the parent of CH-node is dead, the CH-node initiates the joining process by trying to search a new parent with immediate lower tree-level as quickly as possible.

The process carries on; in the second step and after setting up, from the network topology, CH-node runs its application software [7].

The application software begins its active period by turning on its sensors and sensing the environment and the characteristics of the irrigation architecture parcel. The application software reads temperature, humidity, water level in the soil, ingredient ratio, fruitfulness level of soil, fertilizing... of an irrigation agriculture parcel from sensors and reports the result to the sink via its parents. If it receives any packet from its children during this active time, it relays the packets to its parent. After the transmission of its sensing data, CH-node waits for its working schedule such as sensing period in sleep order message from the sink. As it receives its sensing period, the application software turns off its attached sensors and puts the transceiver to power down mode [7]. Finally, its sets up the internal sleep timer, goes to its sleep period and waits for the expiration of the timer. After the expiration of the timer, the application restarts its next active periods by turning on the transceiver and the sensors and continues to sense the environment and the characteristics of the irrigation architecture parcel [12]. To map out an advanced Irrigation Greenhouses Agriculture Sidi Bouzid large area, we used a video-image smart sensor called ID-node (IDentifier-node) that is designed to identify, to specify, to state and to determine the type of truck farming (potatoes, carrots, tomatoes, onions, melon, cabbage, ...) of the irrigation agriculture parcel (see Figure 8.3) [7]. It has an additional node to CH-node in order to search/to select/to capture/to compare/to decide/to execute and to choose from the table list of the type of plants or the various agriculture gardening and market garden produce used in the database server. The application software of ID-node waits for the command from the application server via the sink when any user wants to say the characteristics of any parcel from Sidi Bouzid large area irrigation agriculture [7,13–17]. In the Automated Irrigation Biological Agriculture WSN System, a main sensor in the system called AC-node (ACtuator-node) is under tacked to control the state of the sounding rod of pump. It has an additional relay board to CH-node and ID-node to control the switch ON/OFF of the water pump in the irrigation agriculture parcel. The application software of AC-node waits for the command from the sink based on the information issuing from the CH-node, the ID-node and the AC-node. Whenever it receives the command from the application server via the sink, it controls its relay to turn ON or OFF

FIGURE 8.3   Remote control and exchange data inside and outside of the greenhouses.

the pump-sounding rod. The opening duration of the pump-sounding rod versus discharge/delivery of pump and a function of necessary water quantity to cultivate the soil [7]. The sink node is developed to gather the sensing information from CH-nodes & ID-nodes and transmits commands to CH-nodes, ID-nodes and AC-nodes. As the core component of the gateway, the sink node has an additional interface board to CH-nodes to provide RS-232 serial link feeling in PHP on chip/Raspberry Pi to the gateway.

The main MCU module has the same hardware specification with CH-node except that the sink node is not equipped with any smart sensors. The sink node keeps the sensing schedule from the application server and schedules the operation of the CH-node by sending sleep order message every sensing period. The order-based sleep scheme to conserve power consumption using the sleep order message is used as follows [7]: Whenever the setting schedule (sensing period) is set or ordered by the application server, the sink node keeps the schedule and it spreads the sleep order message over its network every sensing period. Whenever CH-node receives the sleep order message, it sets the expiration time of its sleep timer to the value of the duration field included in the message. When the timer is expired, CH-node senses its irrigation agriculture parcel parameters and the voltage level of its battery, sends the data to the sink and waits for the next sleep order message. To use this power saving scheme, it is required that the parent node of a node awakes earlier than the node in order to relay the sensing data of the child node. Since the sleep order message is spread out from the sink level by level, the requirement is met [7].

The initial sensing period is set to the minimum, 20 seconds to test operation of CH-nodes, ID-nodes and AC-nodes quickly and then the sensing period is set to 5 minutes. After running from the gateways to the application server, CH-nodes and AC-nodes are placed one by one from the vicinity of the sink. So, after that, the gateway transforms data over RS-232 from a sink to data over TCP/IP to server. The gateway is connected to AP via a WLAN link. The AP is connected to the management sub-system via WLAN link. The management sub-system consists of a database server, an application server and a web-server. The application server receives data from WSN and stores them in the database server and provides the way to configure the WSNs. The schedule of sensing is configured by the application server. The whole sensing data are stored in the database server and can be accessed by users using a PC or a PDA via the web-server.

8.2.4 Evaluation Performance

The automated Irrigation Biological Greenhouses Agriculture System was operated continuously during 2 months [Just anything else, the supervisor (farm manager) repeats the same process, reliances the process execution (death warrant) and the work in progress], that's based on the sleep scheme having a strong point that it doesn't require any complex "time synchronization" schemes, the parent node should be awoken earlier than the child node to relay the sensing data of the child node because the sleep order message is spread out from sink to nodes level by level, the parent node goes to sleep mode earlier than the child node. Into account of many interfering sources and obstacles, the communications range of CH-nodes, ID-nodes and AC-nodes was up to 100 m. Considering the stable RF link and the sensing range, the separation is limited up to 70 m so that it won't be a bursting of link.

Originally, CH-nodes, AC-nodes and ID-nodes were planned to be woken up from the sleep mode by an external timer in on-board CPLD to reduce the current drawing. However, the power consumption of the CPLD was measured relatively high and was useless. We lost some of the sensing data because the battery exhaustions of some smart sensors' nodes. It is recommended to use a real-time clock using an internal timer and an external low frequency crystal in order to minimize the power consumption.

Although, the order-based sleep scheme has a strong point that it doesn't require any complex time synchronization schemes, the parent node should be awakening earlier than the child node to relay the sensing data of the child node. According to that, because the sleep order message is spread out from sink to nodes level by level, the parent node goes to sleep mode earlier than the child node, the parent node wakes up earlier than the child node. Although, we used very accurate sensors, they showed the same output levels in the same place. By calibrating the smart sensors, the sensing result is expected to be more accurate. In system I, to optimize the line of sight communication range to be more than 100 m and to reduce the effects of the many interfering sources existing in greenhouses, it is recommended for the sensors to be isolated from the other interfering components in the PCB and the enclosure.

8.2.5 Results

In this first part of the book chapter, the study is reported, the design and the performance results of the real deployment of system I which is designed and implemented to realize an Automated Irrigation Biological Greenhouses Agriculture (see Figure 8.3). From the sensor node hardware to the management system, the whole system architecture is explained. For low power consumption, the order-based sleep scheme is used. In the previous part, studying on more another smart irrigation system is used based on WSSN and developing more miniaturized agricultural smart sensor node.

8.3 Smart Irrigation Biological Plantation Agriculture System (System II)

8.3.1 General

In order to reach the best quality, it is considered that it is essential for farmers to develop a good irrigation strategy to attain marketable products and to reduce product losses. To have a reliable advanced intelligent agriculture, we are deploying cutting-edge technologies in plantations of high-quality standards and to ease farmer daily works. There is an investigation on more another smart irrigation system based on wireless smart sensors network based on a miniaturized agricultural smart sensors node.

8.3.2 Irrigation WSSN Architecture and Deployment

At the beginning, Sidi Bouzid agriculture large area is divided into many smart irrigation parcels. Each parcel was studied and analysed in order to know their specific needs. In addition, each smart irrigation parcel has a fully developed WSSN, PHPoC/Raspberry Pi gateways [18], a management sub-system, 18 smart sensors nodes, an actuator node and one sink node. These equipment are deployed in each parcel according to a planned position, which had been decided by agriculturists considering the cultivation; in fact, for remote control of advanced irrigation and operated during the whole of the year. Four smart sensors called: A-smart node, B-smart node, C-smart node and D-smart node are dropped down on the four corners of each smart irrigation biological plantation agriculture parcel. These four smart sensors are engaged to calculate and to determine the smart irrigation biological plantation agriculture parcel area (widthwise/breadthwise-overall length) based on the packet propagation time or on the signal strength. By various localization techniques such as: ToA, TDoA, AoA, … and when the smart sensors nodes (A-smart sensor node, B-smart sensor node, C-smart sensor node, D-smart sensor node) are synchronized, one-way packet is used (if the radio environment is not too disturbed) to determine the delay and then to estimate the distance between the different smart sensor nodes (between A-B, or B-C, or C-D and finally between D and A) [7,8]. About irrigation architecture conception and based on the following idea: in traditional forms of industry, like agriculture, where technology can often be a foreign concept, the simplicity to effectively deploy such devices is greatly appreciated. So, the proposed smart irrigation system is controlled and monitored thanks to different soil moisture smart sensors that measure many parameters and characteristics like humidity and water flow in strategic points of the area. To notify, that the precision agriculture object aims to optimize the production by taking account of local soil and climatic variation. Our Advanced Intelligent Agriculture Research (AIAR) is a research team that works on technology-based support for agriculture, offering solutions for crop monitoring, remote control, supervision and management. We have developed WSSN, based on Libelium technology, Waspmote Plug & Sense–Smart sensor nodes, Meshlium gateways, probes data logger, communication protocols, such as GPRS, LoRa, RF, Sigfox and Smart Agriculture IoT–Cloud Computing, etc., to develop accurate irrigation strategies for farmers [19,24,66]. In the agricultural WSSN system architecture and to map out an advanced Irrigation Plantation Agriculture in Sidi Bouzid large area, we used a video-image smart sensor called ID-node (Identifier-node) that is designed to identify, to specify, to state and to determine the type of plants(red-yellow peach, apricot, olive, vine, pomegranate, …).

To reduce the farmer task by including models to help farmers in decision and alerts, we introduce the IoT approach and cloud computing in the platform. The platform allows to collect information from the smart sensor nodes and other information such as geo-referenced camera pictures. With this platform, farmers have been able to monitor without interruptions soil water status to have irrigation always under control. We noticed that the Smart sensor technology can play a key role especially in Smart agriculture project from mobility to energy efficiency to environmental sustainability. The Waspmote smart sensors devices, used in this project, were praised for their installation ability and the ease at which they could be configured. For Smart agriculture, Waspmote Plug & Sense-smart sensors nodes can detect, monitor and inform on a wide number of agricultural issues, such as soil moisture, temperature, leaf wetness, atmospheric pressure, solar radiation, wind speed, humidity, rain full and many other relevant purposes. The apricot fruit, the peach fruit, the pomegranate fruit and the olive fruit are the most sensitive fruits in terms of quality which are given by size, sweetness and dry manner. Beyond the apricot–peach–pomegranate and olive yards, smart agriculture has a host of applications from regulated irrigation to monitoring and controlling climate conditions, Libelium is addressing every possibility. We have chosen Libelium Waspmote Plug & Sense-smart sensors platform for its wide range of smart sensors and the easy development of software for data acquisition and transmission. Waspmote users—such as agriculture consultancies—can add value layers to the platform by using the open source API and programming environment. The equipment included three Meshlium Wireless gateways, 18 Waspmote nodes equipped with smart sensors to measure temperature, humidity and soil moisture in real time. Meshlium's job is to gather all the data from the smart sensor nodes and send it to the cloud computing through an IoT [17,20,24,66].

Besides, the Meshlium gateways were installed, defining smart sensor zones within the apricot yard, peach yard and olive yard. This was followed by the deployment and calibration of the Waspmote smart sensor nodes, starting with the mote located closest to a Meshlium and finishing on the point farthest away. In the new irrigation management system, one of the many contributions to the service is by soil moisture monitoring with in situ probes. The deployed probes data logger is based on Waspmote smart sensor platform with Smart sensor probes. It enables connection with a wide range of smart sensor probes that does not have to be from the same manufacturer. In smart Irrigation, soil moisture probes are located underground together with Waspmote smart sensor platform, which are put inside water-proof boxes that ensure high durability. Besides, these devices are powered by a long-life battery with 1-year autonomy.

To broaden the study, complete the information and data, explore the horizon in this area and put this project in its context of applicability and especially in its climatic and meteorological conditions while having at first a constructive comparison between the temperate semi-arid climate of our project [2] and the temperate arid climate of A.M. Al-Omran's work [21]. So, our project [2], concerning irrigation in temperate semi-arid climates, is compared to the Al-Omran's work [21] for temperate arid regions and winked/peeked out while showing brightness.

To broaden the study and to explain the good functioning of this new intelligent irrigation system and to carry out a good management of the Smart Irrigation Plantation System, we have formed tree clusters be it peach or apricot or olive. Each cluster is composed of ten trees of the same nature (peach or apricot or olive or pomegranate), having a Cluster Head (CH) which is a smart sensor equipped with an antenna and a radio transmission system between this CH and the Sink itself. It is found at the level of the electric pump of the sounding. Each tree is equipped with three Waspmotes sensors located at three different depths below the tree and a Libelium-technology-type transceiver located at the top of the tree. These 30 Waspmotes sensors of the ten trees forming the cluster encompass all members of the cluster. Each Waspmote from each tree detects and captures soil moisture under and around the tree at different depths, and that depending on the degree of salinity found in the water and in the soil. These two main parameters, whether humidity/drought or salinity/softening (becoming pleasantly soft), are determined from the following formulas and expressions presented in the Section 8.3.3.

In our project [2], the expressions of the formulas of these parameters are programmed and saved in the software of each Waspmote, as well as each smart sensor and each transceiver. The three Waspmotes of each tree detect and capture the different values of the humidity of the considered tree and the transceiver of the type Libelium technology sending them to the smart sensor. This smart sensor takes care of collecting, processing, analysing and defining the average value of the humidity for each tree according to the formulas recorded, the algorithm and the programme saved in its database and in its management software. The smart sensor, based on its references and data stored in its software, provides for the strength and direction of the wind and especially the hot south Saharan wind and especially the level of heat ardeur-farveur in the summer months while determining the maximum threshold of drought and its period. This smart sensor also determines the temperature in the tree's environments whether during the day (light period) or during the night (dark period). Based on all the parameters mentioned above, the sink calculates the exact amount of water needed and useful for the needs of each tree in the orchard. The sink locates and focuses on the tree to irrigate and sets the amount of water needed for irrigation and the time allocated for the irrigation of such a tree. From that moment on, the sink recalls the itineraries to follow to reach the indicated tree and is interested in its irrigation while automatically opening all the solenoid valves of the drip irrigation channels. The sink orders using radio signals, the intelligent irrigation system to perform and complete the irrigation of such a tree while beginning with the automatic opening of the electric pump borehole. With this new intelligent irrigation system, we have progressed tremendously and we have evolved considerably in the right direction in this area while overcoming the limitations and errors of traditional and conventional techniques and the fatal losses of the means proposed before while trying to appear and root the following advantages and opportunities:

- Realize good water management during the irrigation without loading it and without bearing the burden of the loss of water without requirement, without necessity and without a valid cause. "The exact amount of water is allocated for the time and the time it takes, depending on the irrigation time allocated and the period allocated to this irrigation: no surplus or insufficiency."

- Perform and certify the quality–quantity–price ratio. It is necessary to manage and balance the energy consumption in electricity (reduce the amount of the electricity bill).
- Select only the tree that needs to be irrigated without excess water or insufficient water. This intelligent irrigation system can irrigate a single tree among hundreds and more trees in the orchard. This intelligent irrigation system performs the irrigation operation only for the tree or trees that need to be irrigated according to its information and data analysed by the pilot (sink).
- The opening and the closing of the electric pump of the sounding are carried out automatically. The opening/closing of the solenoid valves of the main and secondary drip irrigation channels is automatically activated.

The new irrigation management system architecture deployment, illustrated by Figure 8.4, is based on smart sensors' technology and consists in allowing remote control of the irrigation system to facilitate the management of the water network (see Figure 8.5). Then, new irrigation management system allows an automatic control of the electronic valves that close or open the water flow. The system optimizes water consumption because it irrigates with the proper amount according to weather conditions and the

FIGURE 8.4 Main steps of the smart irrigation plantation agriculture system.

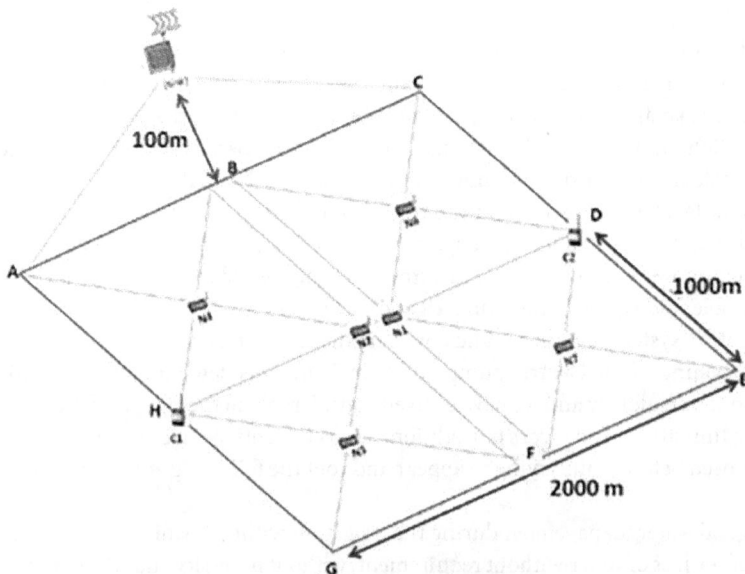

FIGURE 8.5 Smart network topology of application IoT intelligent irrigation.

plant's needs. It is saving resources such as water with Internet of Things technology and contributing to enhance the environment too. To provide and prospect the need of water for the parcel irrigation with a good estimate and precise accurate quantity of water [22]. Some apricot–peach–pomegranate and olive roots are too dry while others are waterlogged. Three soil moisture smart sensors are simultaneously placed at different depths, under each tree (apricot/peach/olive/pomegranate), to assess the local water retention in the soil. By measuring evapo-transpiration, it is possible to work out how much irrigation water is being actually absorbed by the plants [23]. Using smart sensor data to automatically adjust irrigation to match local conditions help to conserve water, and it is equally applicable to match local conditions conserve water and to apricot–peach–pomegranate and olive yards. Avoiding overwatering also helps prevent certain crop diseases including rot, fungi and bacteria, which thrive in wet conditions [24].

8.3.3 Smart Irrigation and Other Irrigation Systems: Analytical Foundations—Management and Optimization

Since about 30 years, several works, many studies as well as some systems and methods of optimization of agricultural irrigation in different regions of the world characterized and differentiated it mainly by its climatic and meteorological conditions (see Figure 8.6). In this section of this book chapter, we are interested in works, studies and irrigation systems intended for use in semi-arid and arid temperate regions since these two types of climate have some common properties. Among these works and proposed irrigation systems, we will make a comparison between our project subject of this book chapter [2] and the project of A.M. Al-Omran [21].

First, there is a start with a reminder of the history of the work done in the field of irrigation from the nineties of the last century until now. The ecosystem of arid and semi-arid regions is impoverished by scarcity of water resources and is dominated by sandy soil. Sandy soils are particularly critical for water management due to their low water-holding capacity, high infiltration rate, high evaporation, low fertility levels and deep percolation, all factors that lead to lower water use efficiency (WUE), thereby decreasing crop productivity [25]. The water shortage and increasing demand for water in agriculture and other sectors force adoption of irrigation strategies in these regions. This may allow saving irrigation water for agricultural sector [26,27]. An approach to attain the objective of saving water and increasing WUE is through using deficit irrigation programme (DI) and degree of deficit irrigation through the whole growth stage or at certain stages of the growth [28,29]. Deficit irrigation has been extensively studied on several crops [28,30] and was recommended for arid and semi-arid regions [31]. DI is studied on tomato and found that the dry mass yield did not decrease under DI compared with full irrigation. Moreover, DI can save up to 50% of irrigation water and increase WUE by 200%, with satisfactory yield.

The adoption of deficit irrigation requires the knowledge of crop evapotranspiration (ET_c), crop response to water deficit, critical stages of growth under water deficit and economic impacts of yield reduction [32,33] and concluded that seasonal crop ET values were greater during reproduction growth stage of the crop. Reference [34] concluded that cucumber yield significantly decreased in a linear relationship with increasing water deficit. However, no significant change was observed when water was applied above 100% ET_c. Reference [35] studied the effect of deficit irrigation on yield and water use of grown cucumber in China and reported that the WUE decreased when increasing the irrigation water applied from stem fruiting to the end of the growth stages. However, the WUE increased with the increase of irrigation water from cucumber fruit setting to first fruit repining. There is increasing pressure on the agricultural sector to create ways to improve water use efficiency by taking full advantage of available water. Drip irrigation using a deficit and/or the PRD irrigation strategy and the use of plant residues or natural and industrial amendments are ways to improve the chemical and physical properties of soils to increase WUE.

Deficit irrigation (DI) is a strategy that is aimed to decrease water applied during insensitivity stage without a significant decrease in the production of crops [32]. The expectation is that any yield reduction will be insignificant compared to the benefits gained from the saving of water. The goal of deficit irrigation is to increase crop WUE by reducing the amount of water applied during the act of watering or by reducing the number of irrigation events. DI involves the use of appropriate irrigation schedules, which mostly derive from field trials, which are necessary because some crops are sensitive to water deficit during growing season changes [36]. Subsequent work concluded that the deficit irrigation on squash has no significant differences in yield between 100% ET_c and 85% ET_c treatments.

The work on the yield response factor (K_y) to water for many crops has been documented in the literature where crops have a value of K_y lower than one can tolerate the water deficit. On the contrary, crops showing a K_y greater than 1 show a yield decrease more than proportional to the applied ET decrease, which means that the crop might not tolerate any irrigation deficit.

Reference [37] reported that K_y value for cucumber grown in Turkey ranged between (0.196 and 1.31) depending on the water stress growth stage, while it concluded that these values ranged between 0.71 and 0.85 in field experiment in Egypt. The value of K_y for green bean was 1.23, while the values for Safflower and eggplant were 0.97 and 1.37, respectively [38]. In agricultural policy, the most important goal for improving WUE is to produce more food and economical yield using less water. Reference [39] concluded that the main reasons to enhance WUE and are: to meet poverty and human food demands; to evaluate water resources for agriculture: to cope with drought stress and water losses because of evaporation rates. Reference [40] proved that partial root-zone drying is a good technique for deficit irrigation in agronomic and horticultural farms. In that technique of Partial Root-Sone Drying, the irrigation water can be conserved and saved till 50% without significant reductions in yield. References [41 and 42] summarized the advantages of PRD like: (i) WUE and nutrient use efficiency improved; (ii) reduction in irrigation water and (iii) fruit quality can be improved. Thereby, the partial root-zone drying irrigation is favourable and recommendable in arid and semi-arid regions.

Reference [43] reported on irrigated pepper that irrigation use efficiency was 12%, 20.1% and 17.1% for conventional irrigation, regulated deficit irrigation and PRD, respectively. They reported that PRD saved irrigation water by 50%. Reference [44] performed a deficit irrigation study on cucumber crop at 40%, 60%, 80% and 100% crop evapotranspiration (ET_C). They reported that the highest value of water productivity was 61.9 kg/m³ at 40% ET_c and (DI) strategy increased water productivity and soil salinity of the field. Reference [45] conducted an experimental study for evaluating the impact of deficit irrigation (DI) at (50%, 75% and 100% ET_c) on tomato using saline and non-saline water (EC 3.6 and 0.9 dS/m) under greenhouse; they reported that using saline water resulted in about 22% and 24% reduction in yield for first and second seasons, respectively. They recommended to use 75% ET_c for irrigating tomato crop as a good policy for (DI) under greenhouse. In a field experimental work using surface and subsurface drip irrigation conducted by Reference [46], the results showed that increasing the amendment rate of clay deposits from 1% to 2% increased the yield from 24.9 to 26.9 ton/ha and increased WUE from 2.26 to 2.76 kg/m³. The subsurface drip irrigation improved WUE to 2.89 kg/m³ compared to 2.43 kg/m³ of surface drip irrigation. In another study, under greenhouse conditions using the full irrigation (FI), deficit irrigation (DI) and partial root-zone drying (PRD) irrigation for tomato, Reference [47] reported that biochar applied to soil at rate 0% and 5% by weight resulted in an increasing of soil moisture content and yield production. The results showed a significant increase in the water use efficiency (WUE) by 35% and 15% for PRD and DI compared to FI and the fruit yield of tomato increased by 20% and 13% for FI and PRD with biochar addition compared with the untreated soil. A study conducted [48] on the effects of varied levels of water application on WUE on bell pepper; they reported that WUE was 39.2, 30.7, 6.31 and 1.45 kg/m³ for 120%, 100%, 80% and 40% ET_c, respectively. Total yield was 26,953, 26,880, 13,605 and 5,107 kg/ha for 120%, 100%, 80% and 40% ET_c, respectively. The results showed that the volume of irrigation water and yield relationship was linear. In addition, there were no significant differences between 120% ET_c and 100% ET_c. Reference [49] investigated the effects of three irrigation intervals based on cumulative pan evaporation (18–22) mm, (38–42) mm and (58–62) mm on WUE and yield of bell pepper. They reported that when plant-pan coefficient=0.5, WUE was 7.6, 6.1 and 5.7 kg/m³ for 18–22, 38–42 and 58–62 mm, respectively, the yield values were 28.1, 22.3 and 21.6 kg/ha for 18–22, 38–42 and 58–62 mm, respectively.

The deficit irrigation strategy has received very little attention in agricultural sector in semi-arid/arid region; therefore, the objectives of this chapter were: (i) introducing deficit (DI) and partial root-zone drying irrigation system (PRD) to arid regions and (ii) studying the effect of DI or/and PRD at different stages of cucumber growth on yield and WUE.

The history and scan of the evolution of irrigation allowed us to carefully examine the various studies, contributions and proposals concerning agricultural irrigation, which allowed us to deduce that all expressions and the majority of formulas are used throughout the world; hence, there is unanimous agreement on the basics of basic study of the irrigation sector. As far as we are concerned, all the formulas, which we will describe in the following, are programmed and implemented in the main elements of our intelligent irrigation system, be it at the PC control and control level, or level actuators, that level smart sensors, or others... Since there is a total agreement between the basic formulas of our approach [2] and the other approaches especially concerning Al-Omran's work [21], the only major difference is that our project [2] is to realize a completely computerized and automated irrigation system (Hardware and Software) and that all the formulas are programmed, implemented and inserted in the software of this system while taking into account the temperate semi-arid climates in order to have complete, efficient and effective irrigation management and optimization. For this, we will roughly describe and recall the expressions of the formulas used. Smart irrigation can be achieved by different methods such as empirical equations, lysimeters and field methods. One of the most used equations is Penman-Monteith [51]. Using the Penman-Monteith equation, based on climate data on the farm in both greenhouse and open field, to estimate the water needs and then calculate the total irrigation water requirements based on the quality of irrigation water and soil salinity, taking into account the values of crop coefficient K_c for each month and irrigation efficiency. The combined FAO [50] Penman-Monteith [51] method was used to calculate ET_o or using pan evaporation methods through the following equation:

$$ET_O = \frac{0.408\,\Delta\left(Rn-G\right)+\gamma\left(\dfrac{900}{T+273}\right)U_2\left(e_s-e_a\right)}{\Delta+\gamma\left(1+0.34U_2\right)} \tag{8.1}$$

where:

ET_O = Reference evapotranspiration (mm/day)
Rn = Net radiation at the crop surface (MJ/m² per day)
G = Soil heat flux density (MJ/m² per day)
T = Mean daily air temperature at 2 m height (°C)
U_2 = Wind speed at 2 m height (m/sec)
e_s = Saturation vapour pressure (kPa)
e_a = Actual vapour pressure (kPa)
$e_s - e_a$ = Saturation vapour pressure deficit (kPa)
Δ = Slope of saturation vapour pressure curve at temperature T (kPa/°C)
γ = Psychrometric constant (kPa/°C)

Leaching requirements:

$$LR = \left(EC_{iw}\right)/\left(2Max\ EC_e\right)\times\left(1/LE\right) \tag{8.2}$$

where: LR is the leaching requirements, $ECiw$ is the salinity of irrigation water (dS/m), max ECe is the maximum tolerable salinity of soil for pepper crop (dS/m) (max ECe=8.6) and LE is leaching efficiency (LE=90%) [52].

Uniformity Distribution:

$$UD = \left(Q\tfrac{1}{4}\right)/\left(Q_{mean}\right)\times100 \tag{8.3}$$

where: *UD* is the uniformity distribution, $Q\frac{1}{4}$ is the mean of the lowest quarter of the observed discharge values of emitter and Qmean is the average discharge of all the emitters [53].

Storage efficiency was estimated as (K_s=0.91) according to [53]. Then, irrigation efficiency was calculated by the following equation:

$$Effirri = EU \times K_s \tag{8.4}$$

In order to apply the deficit irrigation, one should determine crop water requirements for each crop, which can be achieved through crop evapotranspiration empirical equations (*ET_c*).

Pan Evaporation: The calculation of evapotranspiration was based on pan evaporation method according to Reference [53] as follows:

$$ET_c = E_o \times K_p \times K_c \tag{8.5}$$

where *ETc* is the maximum daily ET in (mm), *Eo* is the evaporation from class A pan in (mm), *Kp* is the pan coefficient and *Kc* is the crop coefficient of pepper. *Kc* of bell pepper crop was recorded as *Kc*-ini=0.6, *Kc*-mid=1.15 and *Kc*-end=0.9 from FAO standard tables [54].

The third main element of the irrigation strategy is the Lysimeter Method. Twelve non-weighing reinforced concrete lysimeters were used to grow the crops at the farm in an open field and greenhouse. A graded gravel layer of 0.10 m was spread at the bottom of each lysimeter to facilitate drainage. Experiments were conducted using alfalfa as a reference crop, and potato, tomato, carrot, onion and cucumber as the main crops. The irrigation water was controlled and the amount of drainage water was recorded weekly. Each Lysimeter was irrigated in such a way that about 10% of water contributed to drainage. Water balance method by difference of soil moisture content between two irrigations by measuring the changes in moisture content after and before irrigation at the root zone using a device to measure moisture (Terra Sen Dacom) at depths of 10–120 cm all year, after verifying the accuracy of moisture sensitive, calibrated sensors with direct method (gravimetric laboratory method) with data from the sensors. The total amount of irrigation is calculated by the following equation:

$$ET = P + I - Dr \pm \Delta S \tag{8.6}$$

where

ET=Consumptive use (mm)
P=Precipitation (mm)
I=Irrigation added (mm)
Dr=Drainage (mm)
ΔS=change in soil water content (mm)

The daily crop water requirements (mm/day) were estimated by the following equation:

$$GWR = (ET_c) / \left((1 - LR) \, x \left(Eff_{irr} \right) \right) \tag{8.7}$$

For (olive or pomegranate or date palm) water requirements can be determined according to our work [2] and Al-Omran et al.'s work [21] by the following equation:

Calculating the gross water requirements (*GWR*)

$$GWR = \frac{ET_c \times S_e}{(1 - LR) \times Effir} \tag{8.8}$$

GWR=Gross water requirement (m³/ha).
ET_c=Crop evapotranspiration (m³/ha).

Effir.=Efficiency (%), 90%.
LR=Leaching requirements.
S_e=Percentage of evapotranspiration area.

$$S_e = \frac{\text{Shaded area per tree}}{\text{Actual area}} \times 100 = \frac{\pi R^2}{10\text{ m} \times 10\text{ m}} \times 100 \tag{8.9}$$

where

S_e=Percentage of evapotranspiration area.
R=Radius of tree (m).
Shaded area=Area of the shade of one tree measured at noon.

Another important factor to take into account in the smart irrigation strategy is the *Crop Water Production Function*. The relationship between crop yield and water application is called crop water production function (CWPF). The typical relationship between the applied water and yield for the most crop is concave "curvilinear" as more of applied water goes to drainage or loss. The CWPF can be divided approximately into four sections [39,55]. The CWPF clearly showed a linear relationship between applied water and the yield. Then, as the level of water increases, the increase in the yield diminishes. Yield reaches a theoretical value. Beyond it, any further increase of applied water will not increase the yield and in some cases the yield will decrease as applied water increases beyond the maximum value. The shape of CWPF has very important implication for irrigation water management such as introducing the deficit irrigation programme [56,57]. A useful way to express the water production function is on a relative basis, where actual yield (Y_a) is divided by maximum yield (Y_m) and actual evapotranspiration (ET_a) is divided by crop evapotranspiration (ET_c). The relationship between evapotranspiration deficit ($1-(ET_a/ET)$) and yield depression ($1-(Y_a/Y_m)$) is always linear [58] and the slope is called yield response factor of the crop (K_y). This relationship is expressed by the following equation:

$$\left(1-(Y_a/Y_m)\right) = K_y \left(1-(ET_a/ET_m)\right) \tag{8.10}$$

The crop water production function reflects the benefit of applied water in production of dry matter or yield. It presents the relationship between the quantity of applied water and the yield production. The quadratic polynomial function of Helweg (1991) was expressed as follows:

$$Y_a = b_0 + b_1 W + b_2 W^2 \tag{8.11}$$

where:

Y_a: Crop production or yield, ton/ha
W: Applied irrigation water, m³/ha
b_0, b_1 and b_2 are the fitting coefficients

When yield approaches its maximum value, the slope of the water productivity function against water applied goes to zero; therefore, the maximum applied water (W_{max}) was calculated by differentiating the WPF (Eq. 8.11) and equalized with zero. Then, the maximum predicted yield (Y_{max}) can be calculated by substituting the W_{max} in Eq. (8.11):

$$\partial Y / \partial W = +b_1 + 2b_2 W = 0 \tag{8.12}$$

$$W_{max} = -b_1 / 2b_2 \tag{8.13}$$

$$Y_{max} = b_0 + b_1 W_{max} + b_2 W_{max}^2 \tag{8.14}$$

Another main parameter to take into account in the smart irrigation strategy is the *Water Use Efficiency* and the *Water Productivity*. According to Al-Omran's work [21] and the results of our project [2], the results of crop evapotranspiration (ET_c) for each treatment and applied water (AW) are presented because our results agree with these previous findings (these results are similar). The irrigation treatments were started by measuring evaporation from class A pan. The maximum amount of applied water to the cucumber crop was 727 mm on treatment T_{1-100}, while the minimum applied water was 267 mm for T_{12} treatment. The applied water of traditional practice by the farmers in the region was 1562 mm. The calculated ET_c using pan evaporation method ranged between 223 and 617 mm for the different treatment. Water use efficiency (WUE) and crop water productivity (CWP) values increased when water amount decreased with exception of the traditional irrigation. The highest value of CWP was at the highest stressed treatment (T_{12}), which recorded 12.7 kg/m³. Moreover, decreasing irrigation water to the level of 80% of ET_c did not affect the growth and yield. An attempt was made to establish a relationship between water consumed and yield. According to the mathematical analysis of the crop water production function (WPF), the predicted maximum yields were 7.58 and 8.96 kg/m² and the corresponding predicated applied water was 1,290 and 980 mm for summer and fall, respectively. A comparison of the water productivity functions of the partial root-zone drying irrigation system (PRD) and the conventional methods is achieved. These results were in agreement with those reported by References [26,59]. Reference [35] reported a polynomial relationship between ET and yield. The study also concluded that the treatment T_{1-100} had the highest yield; however, treatments $T_{2,3,4,5,6-80}$, and also T_{12-40} gave fairly good marketable yield while economically saving water, fertilizers and pesticide. The result indicated that the water productivity (WP) increased with decreasing the amount of applied water; the increased values were from 9.9 to 12.7 kg/m³ for T_{1-100} and T_{12-40}, respectively. On the other hand, the WP of the traditional irrigation treatment recorded lowest value (3.7 kg/m³). It was evident that over irrigation as the traditional method leads to the lower water productivity, however lack of irrigation as of treatments T_{12-40} lead to very high-water productivity but yield quantity and quality decreased to be unacceptable. Similar results were reported by Reference [60]. There are many explanations for increasing WP with DI; some of them are that the DI can increase the ratio of yield over crop water consumption (evapotranspiration) by the following: (i) reducing the water loss by unproductive evaporation, (ii) increasing the proportion of marketable yield to the totally produced biomass (harvest index) and (iii) adequate fertilizer application and avoiding bad agronomic conditions during crop growth such as water logging in the root zone, pests, diseases, etc. [32,39,61].

With regard to the optimization of irrigation, the purpose of which is to have good and better agricultural production, several factors come into play and must be taken into account in any optimization operation. These factors include:

- Effect of water quality
- Water requirement
- Water use Eefficiency
- Effect of soil amendments
- Deficit irrigation

Semi-arid area is impoverished by water resources and the irrigation water management is considered as the most important point to investigate. There is an urgent need for the methods and practices that reduce the excessive amount of water applied in irrigation without decline in productivity. and ton optimize the irrigation system, the water quality-requirement, water use efficiency, soil amendments and deficit irrigation are the strategies that involves the need that express optimizing the irrigation method with very good irrigation manner and solution. The main aim of this paragraph is to investigate the impact and the effect of water quality, of WUE, of soil amendments and of deficit irrigation on different crops.

In the semi-arid temperate region of Sidi Bouzid-Tunisia (see Figure 8.6), salinity ranges from 0.1g/L (fresh water) to 6g/L and more (saline water). So, the effect of water quality, on the optimization of

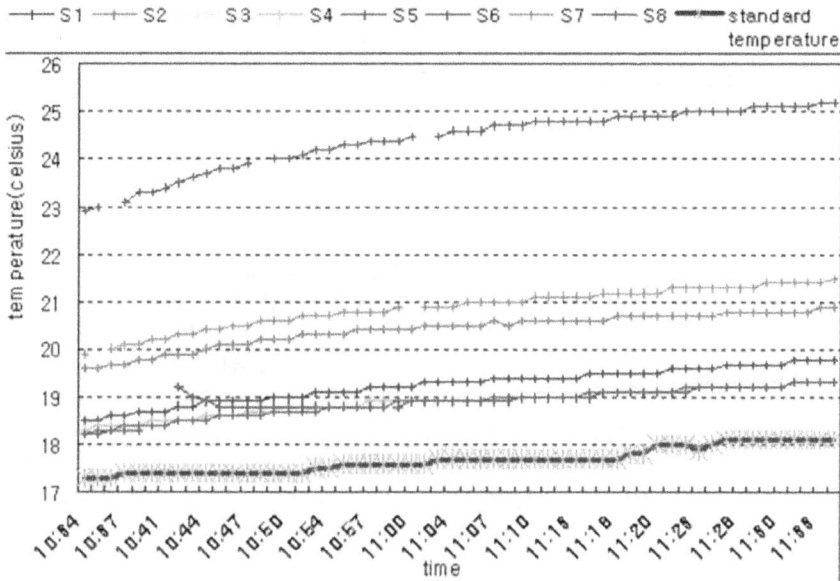

FIGURE 8.6 Climatic conditions and variations of measured temperature in the Sidi Bouzid area.

smart irrigation and especially on production and productivity, is decisive. The interaction between water quality (fresh and saline) and deficit irrigation (*DI*) had significant effects on total yield, WUE and irrigation water use efficiency (IWUE) during both growing seasons. The following shows a comparison between our results [2] and the results of Al-Omran's work [21]. The highest average yield was (173/171.83) ton/ha under freshwater irrigation at 60% of ET_c during the first season, whereas the highest average yield with saline water irrigation was (124/128. 91) ton/ha at 100% of ET_c. The lowest yields under freshwater irrigation were (166/164.41) and (154/150.83) ton/ha, produced during the first season at 100% and 80% of ET_c, respectively. The lowest yields under saline water irrigation were (8.0/10.35) and (6.1/8.98) ton/ha under 80% and 60% of ET_c, respectively. Meanwhile, the values of WUE and IWUE at 60% of ET_c tended to increase (compared to 100% ET_c) when the amount of water irrigation decreased, with the highest values of (36.2/34.90) and (28.5/27.70) kg/m³, respectively, under freshwater irrigation; and (12.2/18.25) and (11.3/14.48) kg/m³, respectively, under saline water. The lowest values of WUE and IWUE were (24.01/20.04) and (172/15.90) kg/m³, respectively, under freshwater irrigation; and (14.5/15.71) and (10.7/12.47) kg/m³, respectively, under saline water irrigation. On the other hand, WUE and IWUE values under 80% of ET_c were (24.3/22.98) and (21.0/18.23) kg/m³ under freshwater irrigation; and (13.2/15.77) and (10.7/12.51) kg/m³ under saline water irrigation.

The final average yields of tomato irrigation during the second season were affected by different levels of water irrigation, i.e., the amount of water applied was positively related to the final yield. The highest yield was (303/303.20) ton/ha, produced at 100% of ET_c, while the lowest yield was (253/252.71) ton/ha, produced under freshwater irrigation at 60% of ET_c. The same trend was observed with yields under saline water irrigation, which were (210/234.05), (186/200.01) and (150/172.41) ton/ha at 100%, 80% and 60% of ET_c, respectively. Higher values of WUE and IWUE were obtained when amount of irrigation water decreased. Under freshwater irrigation at 60%, 80% and 100% of ET_c, WUE decreased in the following order: (74/72.26), (60.2/58.43) and (54/52.01) kg/m³, respectively, while IWUE decreased in the following order: (57/57.35), (46/46.37) and (40/40.81) kg/m³, respectively. Under saline water irrigation at 60%, 80% and 100% of ET_c, WUE decreased in the following order: (50/49.29), (43/42.89) and (42/40.15) kg/m³, respectively, while IWUE decreased in the following order: (39/39.12), (34/34.04) and (31/31.50) kg/m³, respectively. These results agree with previously published reports [62].

Climatic and meteorological conditions have a decisive impact on irrigation and the agricultural season. In this paragraph, we will give an overview on the effect of PRD on crop response factor (ky) and yield at different treatment of irrigation systems. Agro-meteorological conditions, air temperate, humidity and wind speed, affect the evaporation and applied water during the four growth stages (initial, crop development, mid-season and late stage). According to A.M. Al-Omran's work [21] and the results of our project [2], the total evaporation amount was (60/87.1), (89/113.9), (216/486.2) and (120/176.9) mm for first, second, third and fourth growth stage, respectively. Irrigation water requirement was calculated according two irrigation levels 75% ET_c and 50% ET_c (PRD75%ET_c and PRD50%ET_c). Applied water for PRD75%ET_c irrigation level was (19/30.4), (32/55.9), (191/298.3) and (78.6/103.6) mm for four growth stages. While the applied water for PRD50%ET_c level was 20.2, 37.3, 198.9 and 69.1 mm. Reference [58] reported that total irrigation water needs are (600–900) mm per season for pepper. Consequently, our findings explain that the irrigation water amounts declined/were saved by 35% and 57% for PRD75%ET_c and PRD50%ET_c), respectively. These findings agree with the concept of PRD that irrigated the pepper plants by half of the amounts of water applied.

The rapid and huge expansion of environmental pollution around the world, whether in the sea (oceans, seas, oued, gulf,...), or in the air (space free, atmospheric layers, layer of ozone,...), or in the earth (in the different layers of soil or subsoil,...), has a negative, dramatic and disastrous impact on agriculture in general and on irrigation and especially on the quality of the harvest if we do not take the necessary precautions. In this context, we are going to make a substantial review of the WUE and the effects of irrigation and deficit irrigation while making a qualitative and quantitative comparison of our results [2] and those of Al-Omran's work [21]. Water use efficiency can be improved by either increasing yield or decreasing the amount of irrigation water applied, and growers usually aim to decrease the water use of crops while saving yield and quality [63]. The tomato crop grown under deficit irrigation levels of 80% and 60% of ET_c showed higher WUE and IWUE than those grown under 100% of ET_c. The WUE and IWUE under deficit irrigation levels of 100%, 80% and 60% of ET_c and different soil amendments and types of irrigation water during first season (2017) was determined from our results [2] and from the results of A.M. Al-Omran [21]. The WUE values increased significantly by average values of (15.3/14.67) % and (75.2/74.18) % at 80% and 60% of ET_c, respectively, as compared to 100% of ET_c. The application of soil amendments (biochar, polymer and mixture) improved WUE, the highest WUE value of (20.4/22.95) kg/m³ was observed under biochar treatment at 100% of ET_c, while the polymer treatment at an irrigation level of 80% of ET_c resulted in a WUE value of (22.9/24.63) kg/m³, but the mixture treatment at 60% of ET_c resulted in a WUE value of (40.5/44.70) kg/m³, compared with other soil amendments. This result may be related to the effect of the soil amendment to increase the specific surface area of soil and to keep water moving while increasing the water-holding capacity. WUE tended to increase more under freshwater irrigation compared to saline water. The percentage of increase in WUE values at 100% of ET_c were (45/55.00) %, (11/16.49) %, (19.5/21.10)% and (21/21.07) %, respectively, at 80% of ET_c, the corresponding values were 38.82%, 34.32%, 61.13% and 49.66%, respectively; and at 60% of ET_c, the corresponding values were 71.43%, 130.95%, 74.54% and 103.39%, for control, biochar, polymer and mixture treatments, respectively.

The effect of deficit irrigation and soil amendments on WUE and IWUE during the second season in 2018 was presented according to our project [2] and to Al-Omran's work [21]. Average WUE values increased by (10.5/12.99) % and (30/39.73) % at 80% and 60% of ET_c, respectively, as compared with 100% of ET_c. On the other hand, WUE increased when the amount of water irrigation decreased. The highest WUE values of (70.5/72.71) and (80.5/81.52) kg/m³ were for polymer and mixture treatments, respectively, under 60% of ET_c. The lowest WUE values of (45.5/47.25) and (47.5/49.09) kg/m³ were for the control and mixture treatments, respectively, under 100% of ET_c. However, the WUE was higher under freshwater irrigation compared with saline water, and the values of percentage increase were (24/24.86%), (49.5/50.71%), (8/9.93)% and (35/38.43)% for control, biochar, polymer, and mixture, respectively, under 100% of ET_c; the corresponding values under 80% ET_c were (18.5/19.30)%, (60.1/65.83)%, (38.5/42.95)% and (16.7/19.25)%, respectively, while under the high stress level of irrigation at 60% of

ET_c, the corresponding values were $(42.5/46.35)$%, $(21.6/26.69)$%, $(48.7/51.25)$% and$(60.1/63.66)$%, respectively. The results are essentially due to the accumulation of salt in the root zone, thereby increasing the osmotic pressure, which leads to reduced absorption of available water by plants. Improved WUE is attributed to the application of soil amendments that improve the soil microenvironment for crop growth, thereby significantly enhancing the crop yield and WUE. Generally, the application of biochar and polymer improved yield and WUE; these results are similar to previously reported findings [64,65].

The best conclusion said: "Reduced water demand can be achieved by adopting improved farm, advanced irrigation systems and performed deficit irrigation." In this study, many appropriate procedures were developed and presented showing that full irrigation with 80% of ET_c was the best treatment in terms of water productivity and final yield; decreasing irrigation water to 40% ET_c caused very high-water productivity while decreasing the final yield. In addition, the crop water productivity values increased when water amount decreased. So, under semi-arid conditions/arid conditions, WUE and water productivity values increased when the amount of applied water decreased. The yields, for different irrigation levels of fresh and saline waters during the first season, are summarized and presented as follows:

* The total yield under fresh water at 100% of ET_c was significantly higher with biochar compared to the polymer treatment, i.e., biochar and polymer yields were $(1.90/188.33)$ ton/ha and $(1.30/124.33)$ ton/ha, respectively.

* Under saline water for irrigation, the yield was significantly lower by $(1.0/14.44)$ % in the polymer treatment, in contrast to the biochar and mixture treatments. This could be due to the interaction between water and soil amendments. Moreover, irrigation using saline water led to increasing salt accumulation in the soil, which affected soil productivity adversely.

8.4 Advanced IoT for Smart Irrigation

Remember the goal of this study is to promote agricultural research, to deliberate on important issues of agricultural research and to select the better system among the existing or the proposed versus nature of plants (apricot or peach or olive or pomegranate). For this, there is some instalations at the same time and to be placed on the different footings: one connected at apricot–peach orchard whereas the other at olive orchard and pomegranate orchard, two different wireless smart sensors systems to monitor soil water status to plan irrigation in an olive orchard-in pomegranate orchard and in peach-apricot orchard [66]. Data have been recorded with the same system but information has been transmitted to the platform by two different wireless connections: GPRS and Sigfox (see Figure 8.7). Two Waspmote Plug & Sense-smart agriculture have been deployed with Watermark smart sensors in different depths (see Figure 8.8) to control soil moisture with fruit diameter smart sensor to measure the size of the fruit (see Figure 8.9), and temperature and humidity smart sensors to monitor environmental conditions. One of the smart sensor platforms is connected to a GPRS Shield and the other to Sigfox. The first one represents the classical widely used data communication network and the second one the rapidly diffusing LPWAN technologies. The information collected by the smart sensors has been sent to the platform that includes both GPRS and Sigfox technologies. To manage GPRS stations, however, a server had to be configured. A Meshlium IoT Gateway has been used for embedding Meshlium management system making data handling easier. Farmers can get valuable information to schedule irrigation timing to avoid stress conditions, which is fundamental on apricot–peach and olive plants [19,24,66]. Agriculture smart sensor networks using Waspmote send data using LoRa communication system. Alarms can also be sent to the mobile phone network using Waspmote's LoRa board. Data gathered by Waspmote smart sensor platform can be sent to a gateway or directly to the cloud. The information collected in the Meshlium Gateway can be visualized in a platform, which concentrates and allows knowing the state in each parcel (see Figure 8.7). We carry out the application, which can be controlled with computers, smart phones and also tablets. Multi-protocol router is used to gather all the data from the smart sensor nodes and leaving them in the cloud computing. The new smart agriculture IoT kit is factory programmed and enables monitoring of environmental parameters in agriculture [17,19,66]. The IoT kit includes a visualization plugging in Meshlium where you can check data in real-time, display a graphic with every measured parameter between two time periods or geo-locate the nodes via GPS and compare different parameters in the same mode (see Figure 8.7) [66].

8.5 Results

Table 8.1 describes the smart irrigation management (system II) in contrast of soil water over the growing season at the specified depth. This sophisticated monitoring brings extreme precision to crop growing in apricot yards/peach yards or olive yards/pomegranate yards by enabling irrigation and climate control to be matched to local conditions (air temperature, air humidity, soil temperature, soil moisture, leaf wetness, atmospheric pressure, solar radiation, wind speed, rainfall….) According to Table 8.1, we deduce that the smart sensor node was able to successfully monitor soil water as measured by the soil moisture smart sensors/Waspmote Plug& sense-smart sensor nodes or Watermark sensors. The irrigation scheduling by system II using a WSSN in conjunction with Meshlium gateways-Probes data logger-GPRS-Sigfox-LoRa communication protocols

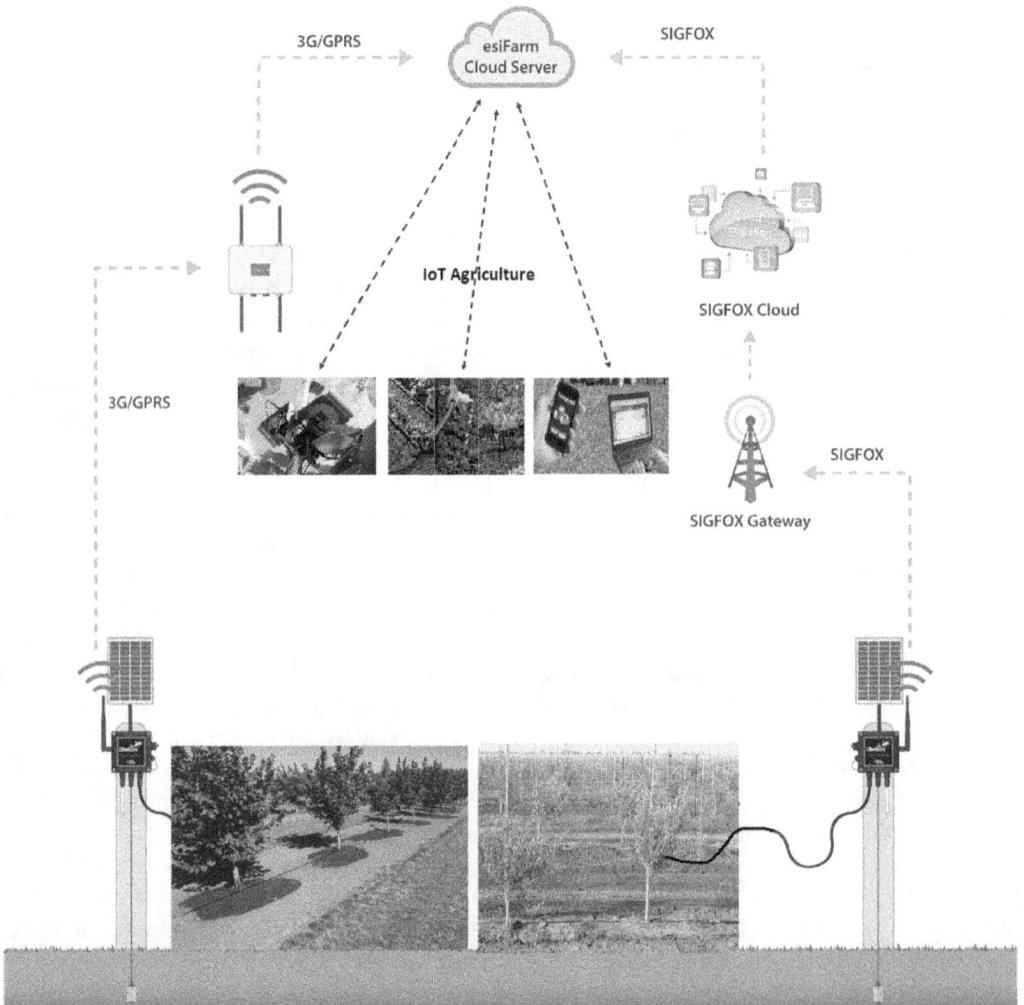

FIGURE 8.7 Functioning diagram of the new irrigation management (system II) deployment.

TABLE 8.1   Irrigation Scheduling by System II

Irrigation per day (mm)	10	20	40	12	31	28	30	32	27	29	36	39	43	48	53	61	55	53	50	45	30	38	40	30	28
Cumulative rain-watermarks at 0.4 m (mm)	20	5	0	15	0	2	0	0	2	0	0	0	0	0	0	0	0	0	0	0	30	2	0	2	3
	2	0	5	2	10	7	20	11	5	2	1	2	0	0	0	1	11	22	10	10	30	2	10	2	3
Soil temperature at 0.2 m (°C)	15	15.7	16	16.3	16.5	16.8	17	18	23.5	28.2	30	32	34	36.5	40	45	43	41	40	35	33	31	30	25	23
2016 growing season date	10	22	2	22	12	20	2	22	2	12	2	12	2	22	1	11	22	31	10	20	30	10	20	21	31
	November		December		January		February		March		April		May		June		July		August		September		October		
	Autumn		Winter						Spring						Summer						Autumn				

and smart agriculture IoT–Cloud Computing can aid in determining when to begin watering based on the driest zone in a field (see Figure 8.7). It can also be used to determine the optimum amount of water to apply across the field. However, to optimize irrigation applications, the smart irrigation management is best used in conjunction with Waspmote Plug sense-smart sensor nodes. That is sufficient for installing the smart sensor nodes to be easy as sticking the smart sensor probes in the underground of tree to cover the soil moisture and soil temperature smart sensors and turning the Waspmote on. A control system that enables a centre pivot irrigation system to supply water at rates relative to the needs of each parcel versus nature of plants was developed. We notice that system II allows to improve and to have a good production. The new smart irrigation management is a variable rate irrigation system interface to provide a fully automated, closed-loop irrigation system, a dynamic system capable of addressing varying water needs in smart agriculture with diverse irrigation management zones. Thanks to this smart irrigation management system, the water pump has been cut down near a 60% in the irrigation parcel. Moreover, this reduction is not just about money, the water usage has been reduced too. With Internet of Things technology, the water resources are being saved and it contributes to enhance the environment too. Controlling the irrigation system and detecting any incidents that may have occurred can nowadays be checked in real-time (see Figure 8.7).

Figure 8.8 shows that any irrigation is not uniformly effective. This event of outstanding importance should be local variations in soil, drainage and evaporation. For this, three soil moisture sensors/three probes or three Waspmotes are simultaneously placed at different depths in the underground tree. The local water retention in the soil is assessed. By this, it is possible to work out how much irrigation water is being actually absorbed by the plants.

Figure 8.9 shows the importance of production quality of apricot fruit/peach fruit or olive fruit or pomegranate fruit. Accurate dendrometers, capable of measuring changes in diameter of few micrometres, allow the measurement of water intake of individual apricot/peach or olive/pomegranate from irrigation.

8.6 Discussions

In the second part of this book chapter, the new smart irrigation plantation agriculture system is established. The system II was able to successfully monitor soil water status and soil temperature for the entire

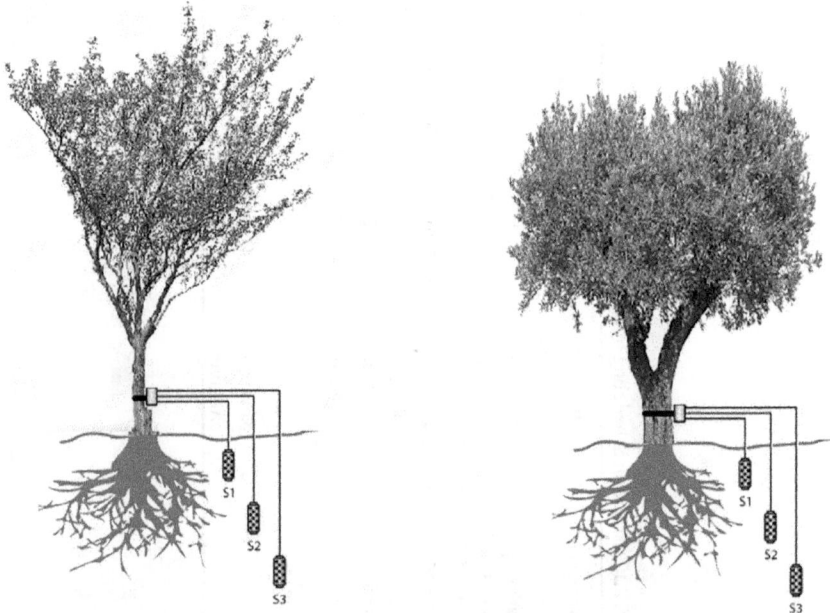

FIGURE 8.8 New Waspmotes sensors boards platform enable extreme smart agriculture.

FIGURE 8.9 Apricot fruit diameter dendrometer smart sensors.

2016 growing season. We report the results of real deployment of the new smart irrigation management system (system II), which are designed and implemented to realize automated smart agriculture.

The smart sensor nodes reliably recorded and transmitted the readings of the watermark/waspmote/probe sensors and allowed us to successfully implement our irrigation scheduling protocol. The stability of WSSN is taken into account in software and hardware design.

The smart irrigation control system was put in place and shows good and continuous improvements to meet various irrigation demands and requirements.

8.7 Conclusions

This book chapter has been presented to those interested in this field a complete, targeted, efficient, advanced study of a technical synopsis of two irrigation systems intended for temperate and semi-arid zones. In addition, this book chapter has offered to the farmers, agricultural designers, etc through advanced technology, two intelligent irrigation systems that enable control, supervision, and monitoring, remoted by the technology of the Internet of Things and based on two successful networks the WSN and the WSSN.

With this, anyone interested in the field can choose the appropriate irrigation system and highlight it without difficulty while respecting the geographical and atmospheric conditions.

Acknowledgements

This work has been accomplished at WIMCS-Research Team, ENET'COM, Sfax-University, Tunisia.

Part of this work has been supported by APIA-Tunisia Agriculture Ministry and MESRSTIC Scientific Research Group-Tunisia.

References

1. R.B.D. Teibel, P.B. Jones, "Report from the field: Results from an agricultural wireless sensor network", in *Proceedings of 29th IEEE LCN'04*, Tampa, Florida, USA, November 15–17, 2004.
2. Madani, S., Weber, D., Mahlknecht, S., "A hierarchical position based routing scheme for data centric wireless sensor networks," in INDIN 2008. *Industrial Informatics*, 6th IEEE International Conference, DOI: 10.1109/INDIN.2008.4618245.

3. A. Baggio, "Wireless sensor networks in precision agriculture", in *Proceedings of real WSN'05*, Stockholm, Sweden, June 20–21, 2005.

4. K. Langendoen, A. Baggio, O. Visser, "Murphy loves potatoes: Experiences from a pilot sensor network deployment in precision agriculture", in *14th International Workshop on Parallel and Distributed Real-Time Systems (WPDRTS)*, Rhodes, Greece, April, 2006.

5. Yoo, S. et al., "A²S based on WSN", Report, Research Gate.

6. VFF. Vessel Finder Free-Applications Android/Google Play. Accessed 2, 22, 2023. https://play.google.com/store/apps/details?id=com.astrapaging.vff&hl=en&gl=US

7. Abdellaoui, M. *Multitasks-Generic-Intelligent-Efficiency-Secure WSNs and Their Applications.* LAMBERT Academic Publishing (LAP), Geneva, Switzerland, 2017, pp. 186–323.

8. O. Mezghani M. Abdellaoui, "An efficient localization method for mobile nodes tracking in wireless sensors networks", Proposed at ICC'17 Conference, 21–25 May 2017, Paris, France.

9. D. Kim, T. S. Lopez, S. Yoo, J.Sung, J. Kim, Y.Kim Y. Doh, "ANTS: An evolvable network of tiny sensors", *Lecture Notes in Computer Science*, Vol. 3824, 2005, pp. 142–151.

10. I. Bouabidi, M. Abdellaoui, "Energy efficient cross-layer MAC protocol and secure authentication via an implementation of data confidentiality and integrity in WSN", 2016, FTC, San Francisco, USA, USA, December 6–7.

11. O. Mezghani, M. Abdellaoui, "Mobi-Sim: An emulation and prototyping platform for protocols validation of mobile WSN", *Advances in Science, Technology and Engineering Systems*, Vol. 2, No. 1, 2017, pp. 108–120.

12. M. Mezghani, M. Abdellaoui, "WSN efficient data management and intelligent communication for load balancing based on Khalimsky Topology and mobile Agents", in *Intelligent Systems and Applications*. Springer, Cham, Switzerland, 2016, pp. 385–404. DOI: 10.1007/978-3-319-33386-1/19

13. I. Bouabidi, M. Abdellaoui, "Cross layers security approach via an implementation of data privacy and by authentication mechanism for mobile WSNs", *ASTES Journal*, Vol. 2, No. 1, 2017, pp. 97–107.

14. H. Masayuki, et al., "Advanced sensor network with field monitoring servers and Metbroker", in *CIGR, International Conference*, Beijing, China, 2004, No. 30–125 A.

15. T. Fukatsu, et al., "Field monitoring using sensor nodes with a web server", *Journal of Robotics and Mechatronics*, Vol. 17, No. 2, 2005, pp. 164–172.

16. D. He et al., "A crop field remote monitoring system based on web-server embedded technology and CDMA Service", in *International Symposium on Applications and the Internet Workshops (SAINTW'07)*, Hiroshima, Japan, 2007.

17. M. Abdellaoui, "Dynamic spectrum access of virtualized-operated networks over MIMO-OFDA dedicated to 5G cognitive WSSNs for IoT intelligent agriculture", Chapter Book-Springer: Advances in Information Security and Communication. 2019.

18. PHPoC IoT Gateway, http://www.phpoc.com/phpoc_iot_gateway.php; Building a Raspberry Pi Gateway, https://www.mysensors.org/build/raspberry

19. F. Gargouri, M. Abdellaoui, "Smart sensors & internet of things platform for remote control and identification of advanced irrigation agriculture project", Advanced Materials World Congress 2017, American Sensors and Actuators Summit, December 03–08, 2017, Miami, USA.

20. J. Gubbi, R. Buyya, S. Marusic, M. Palaniswami, "Internet of Things (IoT): A vision architectural elements and future direction", *Future Generation Computer Systems*, Vol. 29, 2013, pp. 1645–1660.

21. A. M. Al-Omran, A. Al-Khasha, S. Eslamian, "Irrigation water conservation in Saudi Arabia", in *Handbook of Water Harvesting and Conservation, Vol. 2: Case Studies and Application Examples*, Ed. By Eslamian, S. & Eslamian, F., John Wiley & Sons, Inc., New Jersey, 2021, pp. 373–384.

22. X. Kenui, X. Deqin, L. Xiwen, "Smart water-saving irrigation system in precision agriculture based on wireless sensor network", *Transaction of the CSAE*, Vol. 26, No. 11, Nov. 2010, pp. 170–175.

23. G. Veldis, M. Tucker, G. Perry, G. Kiven, C. Bednarz, "A real-time wireless smart sensor array for scheduling irrigation", *Computers and Electronics in Agriculture*, Vol. 61, 2008, pp. 44–50.

24. M. Abdellaoui, "Smart Irrigation Agriculture to improve production", in *European Advanced Materials Congress*, August 22–24, 2017, Stockholm, Sweden.
25. A. M. Al-Omran et al., "Effect of drip irrigation on squash (*Cucurbita pepo*) yields and water use efficiency in sandy calcareous soils amended with clay deposits", *Agricultural Water Management*, Vol. 73, No. 1, 2005, pp. 43–55.
26. A. R. Al-Harbi et al., "Effect of drip irrigation levels and emitters depth on growth", *Journal of Applied Science*, Vol. 8, 2008, pp. 2764–2769.
27. A. M. Al-Omran et al., "Impact of irrigation water quality, irrigation systems, irrigation rates and soil amendments on tomato production in sandy calcareous soil", *Turkish Journal of Agriculture and Forestry*, Vol. 34, 2010, pp. 59–73.
28. C. Kirda et al., "Yield response of greenhouse grown tomato to partial root drying and conventional deficit irrigation", *Agricultural Water Management*, Vol. 69, No. 3, 2004, pp. 191–201.
29. C. Patanè et al., "Effects of soil water deficit on yield and quality of processing tomato under a Mediterranean climate", *Agricultural Water Management*, Vol. 97, No. 1, 2009, pp. 131–138.
30. A.R. Sepaskhah et al., "Deficit irrigation planning under variable seasonal rainfall", *Biosystems Engineering*, Vol. 92, 2005, pp. 97–106.
31. J.A. Zegbe-Dominguez et al., "Deficit irrigation and partial root-zone drying maintain fruit dry mass and enhance fruit quality in Pet pride processing tomato (*Lycopersicone Sculentum* Mill.)", *Scientia Horticultural*, Vol. 98, 2003, pp. 505–510.
32. L.S. Pereira et al., "Irrigation management under water scarcity", *Agricultural Water Management*, Vol. 57, No. 3, 2002, pp. 175–206.
33. S.O. Agele et al., "Evapotranspiration, water use efficiency and yield of rainfed and irrigated tomato", *International Journal of Agriculture & Biology*, Vol.13, 2011, pp. 469–476.
34. K.H. Agele et al., "Effect of deficit irrigation and fertilization on cucumber", *Agronomy Journal*, Vol. 101, 2009, pp. 1556–1564.
35. X. Mao et al., "Effects of deficit irrigation on yield and water use greenhouse grown cucumber in the north china plain", *Agricultural Water Management*, Vol. 61, No. 3, 2003, pp. 219–228.
36. A. Istanbulluoglu, "Effects of deficit irrigation regimes on yield and water productivity of safflower (*Carthamus tinctorius* L.) under Mediterranean climatic conditions", *Agricultural Water Management*, Vol. 96, No. 12, 2009, pp. 1792–1798.
37. S. Ayas et al., "A Deficit irrigation effects on cucumber (*Cucumis sativus* L. Mara ton) yield in unheated greenhouse condition", *International Journal of Food, Agriculture & Environment*, Vol. 7, 2009, pp. 645–649.
38. S. Lovelli et al., "Yield response factor to water (Ky) and water use efficiency of *Carthamus tinctorius* L. and *Solaum melongeua* L.", *Agricultural Water Management*, Vol. 92, 2007, pp. 73–80.
39. S. Geerts et al., "Deficit irrigation as an on-farm strategy to maximize crop water productivity in dry areas", *Agricultural Water Management.*, Vol. 96, 2009, pp. 1275–1284.
40. A.R. Sepaskhah et al., "A review on partial root-zone drying irrigation", *International Journal of Plant Production.*, Vol. 4, No. 4, 2010, pp. 241–258.
41. K. Liepaja et al., "Influence of irrigation management as partial root-zone drying on Raspberry canes", *Bulgarian Journal of Agricultural Sciences*, Vol. 24, No. 4, 2018, pp. 648–653.
42. M.M. Giuliaui et al., "Deficit irrigation and partial root-zone drying techniques in processing tomato cultivated under a Mediterranean climate conditions", *Sustainability*, 2017. DOI: 10.3390/SU9122197
43. K. Dorji et al., "Water relations, growth, yield and fruit quality of hot pepper deficit irrigation and partial root-zone drying", *Scientia Horticultural*, Vol. 104, 2005, pp. 137–149.
44. A.R. Al-Harbi et al., "Salinity and deficit irrigation influence tomato growth, yield and water use efficiency at different developmental stages", *International Journal of Agriculture & Biology*, Vol. 17, No. 2, 2015, pp. 241–250.

45. A.M. Al-Omran et al., "Impact of deficit irrigation on soil salinity and cucumber yield under greenhouse condition in an arid environment", *Journal of Agricultural Science and Technology*, Vol. 15, No. 6, 2013, pp. 1247–1259.

46. A.M. Al-Omran et al., "Management of irrigation water salinity in greenhouse tomato production under calcareous sandy soil and drip irrigation", *Journal of Agricultural Science and Technology*, Vol. 14, No. 4, 2012, pp. 939–950.

47. S.S. Alhtar et al., "Biochar enhances yield and quality of tomato under reduced irrigation", *Agricultural Water Management.*, Vol. 13, 2002, pp. 37–44.

48. O. Aladenola et al., "Response of greenhouse grown bell pepper (*Capsicum annuum* L.) to variable irrigation", *Canadian Journal of Plant Science*, Vol. 94, No. 2, 2013, pp. 303–310.

49. S.M. Sezen et al., "Effects of drip irrigation regimes on yield and quality of field grown bell pepper", *Agricultural Water Management.*, Vol. 81, No. 1, 2006, pp. 115–131.

50. FAO, "Guidelines for soil profile description: Food and Agriculture Organization of the United Nations, Land and Water Development – Rome 1977".

51. P. Monteith, "Yield Response to field Crops-to Deficit Irrigation & in irrigation practices", Nielsen editions(eds) and Water Report #22 FAO, Rome, 2000.

52. E.V. Maas et al., "Crop salt tolerance—current assessment", *Journal of the Irrigation and Drainage Division*, Vol. 103, No. 2, 2002, pp. 115–134.

53. D. Karmeli et al., "Trickle irrigation design", Rain Bird Sprinkler Manufact. Corp. Glendora, CA", 1973.

54. R.G. Allen et al., "Crop Evapotranspiration: Guidelines for computing crop water requirements", FAO Irrigation and Drainage paper No. 56 Report #7(300pp), FAO, Rome, Italy, 1998.

55. T. Foster et al., "Simulating crop water production functions using crop growth models to support water policy assessments", *Ecological Economics*, Vol. 152, 2018, pp. 9–21.

56. M. English, "Deficit irrigation I: Analytical framework", *Journal of Irrigation and Drainage Engineering*, Vol. 116, 1990, pp. 399–412.

57. M. English et al., "Perspectives on deficit irrigation", *Agricultural Water Management*, Vol. 32, No. 1, 1996, pp. 01–14.

58. J. Dorenbos et al., "Yield response to water", FAO Irrigation and Drainage paper No. 33 Report #10(193pp), FAO, Rome, Italy, 1986.

59. Y. Zhang et al., "Water yield relations and optimal irrigation scheduling of wheat in the Mediterranean region", *Agricultural Water Management.*, Vol. 38, 1999, pp. 195–211.

60. Y. Zhang et al., "Effect of soil water deficit on evapotranspiration, crop yield and water use efficiency in the North China Plain", *Agricultural Water Management*, Vol. 64, 2004, pp. 107–122.

61. P. Steduto et al., "Resource use efficiency of field grown sunflower, sorghum, wheat and chickpea & II: Water use efficiency and comparison with radiation use efficiency", *Agricultural and Forest Meteorology*, Vol. 130, 2018, pp. 269–281.

62. R. Chandra et al., "Comparative effects of deficit irrigation and partial root-zone drying (PDR) on growth, yield and water use efficiency of Rabi Maize", *International Journal of Current Microbiology and Applied Science*, Vol. 7, No. 2, 2018, pp. 1073–1080.

63. H. Kirnak et al., "Effects of Deficit irrigation on growth, yield and fruit quality of eggplant under semi-arid conditions", *Australian Journal of Agricultural Research*, Vol. 53, No. 12, 2002, pp. 1367–1373.

64. K.C. Uzoma et al., "Effect of cow manure biochar on maize productivity under sandy soil condition", *Soil Use Management*, Vol. 27, No. 2, 2002, pp. 205–212.

65. S. Wan et al., "Effect of drip irrigation with saline water on tomato (*Lycopersicom esculentum* Mill) yield and water use in semi-humid area", *Agricultural Water Management*, Vol. 90, No. 1–2, 2007, pp. 63–74.

66. M. Abdellaoui, Printed Book (420 pages, 2Parts), "IoT system & Multi-standards system: Design, Performance and Applications", 2019, CAMBRIDGE Scholars Publishing (CSP).

III

Chinese
Irrigation History

9

History of Irrigation in China: Schedule and Method Development

Wenzhi Zeng,
Chang Ao, and
Guoqing Lei
Wuhan University

9.1 Introduction

The history of irrigation in China is as long as that of civilization, which is mainly determined by its unique natural environment. China has a vast territory with widely varying topography and climatic conditions across the country, which have a direct impact on irrigation projects (Fan et al., 2021). For example, irrigation projects without dams are mainly distributed along the banks of rivers, while irrigation projects dominated by ponds and weirs are mainly distributed in mountainous and hilly areas (Tan, 2005). In addition, the development of irrigation is obviously promoted or restricted by social systems. For example, during the Warring States Period of the Qin and Han Dynasties in Chinese history (600 BC–220 BC), due to the change of production relations and the release of social productivity, a rapid development of science and technology emerged, which led to the construction of Zhengguo canal, Dujiangyan and other irrigation-based water conservancy projects. Otherwise, the development of social civilization also promotes the development of irrigation. During the Tang and Song Dynasties (600 AC–1100 AC), ancient Chinese civilization reached its peak, which led to the advancement of engineering construction technology and the popularization of irrigation projects in more than ten centuries. This chapter will mainly explain its relationship with nature, society and civilization during the

DOI: 10.1201/9781003353928-12

construction and technological progress of irrigation projects in the historical period of China, and provide background information for the following chapters.

9.1.1 Irrigation and Nature

9.1.1.1 Precipitation

China has a typical East Asian monsoon climate with the majority of precipitation occurring in the summer season, making it suitable for crop growth. However, the temporal and spatial distribution of precipitation in China is very different. Specifically, from the southeast coast to the northwest inland, the average annual precipitation can drop from 1,600 mm to less than 200 mm. Moreover, 60%–80% of the annual precipitation in the eastern region tends to be concentrated in June to September, which often makes torrential floods occur in the eastern region. In addition, droughts often hit parts of the country in almost every year, resulting in crop irrigation requirements not being met. The differences in climatic conditions, especially in precipitation, make irrigation projects developed in different ways. For instance, underground irrigation projects, such as Kariz, were developed in extremely arid regions such as Northwest China, while irrigation canal systems are mostly used in semi-arid regions such as the Ningxia and Inner Mongolia provinces and Northeast China. Paddy fields are mostly distributed in humid areas such as South China.

9.1.1.2 Topography

China's overall terrain is high in the northwest and low in the southeast. More exactly, mountains account for about 33%, plateaus account for about 26%, hills account for about 10% and plains and basins account for about 31% of the total areas. Under the complex terrain, the labourers of the past dynasties created a variety of irrigation engineering methods (Table 9.1).

9.1.2 Irrigation and Society

As early as the Spring and Autumn Period of China (approximately 645 BC–221 BC), irrigation had been one of the most important things of the government. More exactly, the first thing for the ruler of the kingdom to do was to promote the water conservancy and eliminate the water hazards. This ruling concept is closely linked with the social activities of ancient China through the Emperor Yu Tames the Flood. Take irrigation management as an example. As early as 600 BC, the vassal states of China established the position of Sikong, which was dedicated to the construction and management of irrigation projects. This position is similar to the Minister of Agriculture or the Minister of Water Resources nowadays. In the Han Dynasty (202 BC–220 AC), irrigation affairs were managed by local officials, and the central government dispatched imperial censors to inspect local irrigation projects. At the beginning of the Sui Dynasty (581 AC–618 AC), six ministries were established, and irrigation was under

TABLE 9.1   Irrigation Engineering Type in Areas with Different Topographies

Topography	Source of Water	Irrigation Engineering Type
Piedmont or alluvial fan plain	Rivers	Diversion water without dam, Regulating sluice or dam
	Snow water	Well or canal
Plains in the middle and lower reaches of the river	Rivers in coastal areas	Polder
	Estuary or delta	Reject salty dam, sluice or canal
Mountains and hills	Mountain stream	Low-lying and narrow pond for water storage (Beitang)
	Spring water or groundwater	Canal system
Other	Well water	Pumping irrigation

the management of the Ministry of Industry. In the Ming and Qing Dynasties, the Governor of Rivers was established, holding the power to check and balance the chief executives of several provinces and mobilize the army to carry out irrigation and flood control work. In addition, the "Shuibu Style" was formulated in the Tang Dynasty, which was the earliest existing irrigation-related management regulations applied across the whole country. On the other hand, in addition to government actions, the civil organizations also often built irrigation and drainage projects. Therefore, families or individuals who were able to manage these irrigation canals and drainage ditches often had a great impact on local public affairs. Once floods and droughts occurred, whether the irrigation and drainage projects or other water conservancy projects could effectively play a role that would have a huge impact on the stability of the society and politics at that time. Meanwhile, the local and overall conflicts of interest that may lead it also often directly influence the policy formulation and implementation.

9.1.3 Irrigation and Civilization

Manufacturing tools are believed to be indicative of the origin of human civilization, and the development of irrigation projects also depends on the tools of production (Li, 2020). The earliest well that could be used for irrigation in China was built about 5,600 years old. At the end of the primitive commune, because of the use of stone and wooden tools, large-scale irrigation and flood control like the Emperor Yu Tames the Flood appeared. In the Spring and Autumn and Warring States Periods, the use of iron tools made it possible to reclaim wasteland and engage in agricultural production on a large scale, and it also provided important conditions for the construction of large-scale irrigation projects. Therefore, the development of irrigation was based on the improvement of scientific and technological advancement. Meanwhile, the construction of large-scale irrigation projects has promoted the development of social civilization and scientific progress, vice versa. For example, for the design of artesian irrigation and drainage, topographic surveys were required, which promoted advances in mathematics. In the Warring States Period, the "Guanzi·Dudi" recorded the calculation methods of the slope of the open canal in the irrigation project, the energy dissipation of the water jump and the design method of the pressure pipe. In addition, there are many solar terms related to crop growth and crop water demand in the traditional Chinese 24 solar terms.

9.2 Irrigation Project

After the Qin state unified China, the establishment of a centralized dynasty created conditions for the construction of large-scale irrigation projects. After the development for thousands of years, China has built a large number of irrigation projects. This chapter only selects the most representative ones for a brief introduction.

9.2.1 Dujiangyan

Dujiangyan is a water conservancy project built by the Qin state to unify the whole China (Zou, 2015), which was built in the last year of King Zhaoxiang of Qin state (approximately 256 BC–251 BC). Located in the northwest of Chengdu on the upstream of the Minjiang River, it is the starting point for the Minjiang River to enter the Chengdu Plain. It has the good condition of topography for artesian water diversion. The water inlet of Dujiangyan was excavated using the remaining range of the Minshan Mountains, known as the Precious-bottle-neck, which was the first key project to be constructed in Dujiangyan irrigation system. According to the previous researches, Dujiangyan had its current scale in the Tang Dynasty at the latest, and it was formed by water diversion project (Fish-mouth), diversion project (Baizhang dike and Herringbone dike), regulation project (Feisha Spillway) and water inlet (Precious-bottle-neck) composed of the hinge of the irrigation project. The Fish-mouth is built on the top of the central island of the Minjiang River, where the river water is divided into two parts. The west part is called Waijiang, where the rains move downward along the Minjiang River. The east part is called Neijiang, where the water flows through the comprehensive regulation of the Baizhang dike, the Herringbone dike, the Feisha Spillway and the Precious-bottle-neck, and the remaining water flows through the Feisha Spillway to enter the Minjiang River. As the Neijiang is narrow and deep while the Waijiang is wide and shallow, about 60% amount of the river water flows into the Neijiang in the dry season, which guarantees the production and domestic water use of the Chengdu Plain. Furthermore, most of the river water could be drained away from the Waijiang during the flood seasons due to the wider surface area of the Waijiang when compared with Neijiang.

Therefore, the scientific layout and reasonable elevation control make Dujiangyan suitable for both irrigation and drainage. It has been in operation for more than 2,000 years and still plays an important role in the local agricultural development. In 2000, it was included in the UNESCO World Cultural Heritage List. Moreover, it was also included in the World Heritage List of Irrigation Engineering by the 69th International Executive Council of the International Irrigation and Drainage Commission in 2018.

9.2.2 Zhengguo Canal

Zhengguo canal, which was built in 246 BC with an irrigation area of 2.8 million acres, lies in the central part of Guanzhong, Shaanxi (Ren, 2019). It was the largest irrigation channel in ancient China, enabling Qin to complete the preparation for the war to unify China economically. Although the Qin Dynasty only maintained about 15 years after the unification of the six states to establish a unified China, the continued development of the Yinjing Irrigation District based on Zhengguo canal has become the guarantee for the economic and social development of the Guanzhong area. After the Qin Dynasty, Zhengguo canal was further developed and gradually became an irrigation engineering system. Specifically, a new main canal for water supply named Baiqu was built in the south of the original Zhengguoqu in the Han Dynasty. In Tang Dynasty, three main canals named Taibai canal, Zhongbai canal and Nanbai canal were further expanded. During the Northern Song Dynasty, large-scale reconstruction of Zhengguo canal was carried out, and Fengli canal was built after 40 years. The outfall of the Zhengguo canal continued to move up in the Yuan Dynasty and was called as the "Wangyushi canal". Moreover, the head of the Zhengguo canal moved up to the exit of the Jing River gorge and was called as the "Guanghui canal" in the Ming Dynasty. In the Qing Dynasty, as the head of the Zhengguo canal moved up to the limit, it was much more difficult to divert water. Therefore, the local government had to guide the mountain spring water into the canal, also known as the "Longdong canal". During the period of the Republic of China, a 68 m long and 9.2 m high dam was built on the Jing River. The old canal system was still used and the irrigation area reached 33,000 ha. After the founding of the People's Republic of China, the related canal systems have undergone many upgrades and reconstructions to better benefit the people in Guanzhong area.

Currently, in the Fuping country of Shaanxi Province, the head works, the ancient canals and the barrage of Zhengguo canal are still preserved. Moreover, the ruins about the reestablishments or reconstructions of the Zhengguo canal are also preserved nearby, which indicate the important role that Zhengguo canal played for the agricultural irrigation in different dynasties. In 2016, Zhengguo canal was also included in the World Heritage List of Irrigation Engineering by the 67th International Executive Council of the International Irrigation and Drainage Commission.

9.2.3 Water-Saving Irrigation

Today, China still attaches great importance to the development of agricultural irrigation. Since the end of the 1990s, China has carried out the rehabilitation and water-saving reform of large-sized irrigation districts, and vigorously promoted water-saving irrigation (Li, 2010; Yang et al., 2015). The effective irrigation area of farmland has increased from 50 million ha in 1995 to 68 million ha in 2018. Furthermore, the water-saving irrigation area expanded from 15.2 million ha in 1998 to 36.1 million ha in 2018. In the water-saving irrigation regions, the high-efficiency water-saving irrigation area has increased from 5.9 million ha in 2000 to 21.9 million hectares in 2018. Meanwhile, the total amount of agricultural water use decreased from about 400 billion m³ in 1995 to about 370 billion m³ in 2018. Moreover, the proportion of agricultural water used in the total water consumption of China dropped from about 83% in 1995 to about 61.5% in 2018. Therefore, the water-saving irrigation has achieved remarkable success in today's China (Zhuang et al., 2019). In the future, during the 14th Five-Year Plan period, China will accelerate the modernization of irrigation districts to ensure national food security.

9.3 Irrigation Science and Technology Inventions

9.3.1 Development of Irrigation Science in China

Irrigation originates and develops along with farming agriculture. The theory and technology of irrigation are derived from human's understanding of natural phenomena and the practices of agricultural production. In ancient Chinese myths and legends, Emperor Yu found the principle that water flowed downward. He took the method of dredging to effectively control the flood. He also advocated the excavation of field ditches for irrigation and drainage. Emperor Yu's success in flood control was benefited from his deep observation of natural phenomena and rational utilization of natural laws (Tan, 2005; Wu et al., 2010).

About 725–645 BC, Guan Zhong proposed that people who were good at governing the country must pay attention to the water disasters. It was suggested that water diversion irrigation should adopt corresponding engineering measures by obeying the principle that water flows downward. For example, in order to irrigate high-altitude farmland, water-blocking structures such as weirs and dams were required to be built in the upstream of water flow. It is also necessary to select the reasonable slope of the channel to create conditions for water diversion. When the channel passes through the unavoidable road, river or valley, a variety of buildings, such as inverted siphon and water drop, should be built.

"Shangshu·Yugong", written in the first century AD, was the first geographical monograph in the existing Chinese history. In this book, China was divided into nine states. The soil of each state was classified for the first time, and the soil fertility was overall judged. It was considered that the best land in China at that time was distributed in Yongzhou (now Shaanxi and Shanxi Province), Jizhou (now most of Hebei Province), Xuzhou (now most of Shandong Province) and other places near rivers. At that time, these places had a dense population and a relatively better irrigation condition.

Huainanzi, a book written in the second century AD, analysed the water quality of the Yellow River and its main tributaries. In this book, suitable crop varieties for different water qualities were proposed. For example, the water quality of Yellow River was moderately turbid, which was suitable for planting millet; Fenshui, a tributary of the Yellow River, was suitable for planting hemp and other similar crops.

In the book "Nongshu" (written around 1300 in the Yuan Dynasty), the famous agronomist Wang Zhen systematically summarized the situation and specific experience of farmland water conservancy construction in the south of the Yangtze River, and a lot of irrigation technologies have been recorded. He emphasized the importance of irrigation and believed that different irrigation methods should be adopted for different water sources and terrain conditions. If the height of water source was higher than the arable land, irrigation could be applied directly, while if the height of water source is lower, irrigation must be conducted by mechanical water extraction. For different situations, Wang Zhen also summarized the corresponding supporting engineering measures.

Xu Guangqi (1562–1633), a famous scientist in the late Ming Dynasty, believed that water and soil were important resources of the country. A country must maximize the soil fertility and build water conservancy projects to become rich and strong. He emphasized the importance of irrigation for agricultural production. In the book "Nongzheng Quanshu", he systematically summarized Chinese water conservancy science and technology related to agricultural production of the 17th century. In 1630, he wrote the book "Water for Dry Land", which introduced detailed theories and methods of water resources and water use. He believed that the construction of water conservancy projects was not only necessary for controlling drought and eliminating waterlogging, but could also regulate the regional climate. According to the different conditions of water resources, five technical measures for dryland irrigation such as water storage, diversion, water transfer, water conservation and water lifting were summarized. There was also a special chapter systematically introducing western water science and technology, which was the earliest introduction of western modern water conservancy science and technology in China.

Due to the diversification of Chinese topography and climate, ancient Chinese people also created various characteristic engineering forms in accordance with local conditions, such as "Beitang" in hilly areas, irrigation canals in plain areas, and polders in lakeside and riverside areas. The formation, development and continuous improvement of these technical categories and projects contributed to the birth of the Chinese traditional irrigation science.

9.3.2 Invention and Creation of Irrigation Technology

9.3.2.1 Irrigation Technology in Ancient China

Many irrigation and drainage machines were invented in ancient China, and only two representative machines were introduced briefly here.

1. "Fanche" is a water-lifting machine that is turned by wheels (Figure 9.1). It was first recorded 1,700 years ago in the Donghan Dynasty. This machine is suitable for short-distance water lifting in plain areas. The water-lifting height is about 1–2 m. The machine can be used directly to lift water from the water canal to the farmland serving as an auxiliary facility for irrigation projects.

   The vertical "Fanche" is used to lift water in the well, and the transmission device of the water wagon consists of flat and vertical wheels to switch the power direction. After the Tang and Song Dynasties, "Houfanche" was a relatively common irrigation water-lifting machine, which had a high water-lifting efficiency. Wang Zhen, a scientist in Yuan Dynasty, described the structure and operational method in detail in his book "Nongshu". The structure of "Fanche" that used different powers, including feet, hands and animals were drawn in the book. After the Yuan Dynasty, there began to appear "Fanche" powered by water. During the Ming and Qing Dynasties, "Fanche" powered by wind was developed.

FIGURE 9.1 Photos of "Fanche", which is also known as keel waterwheel.

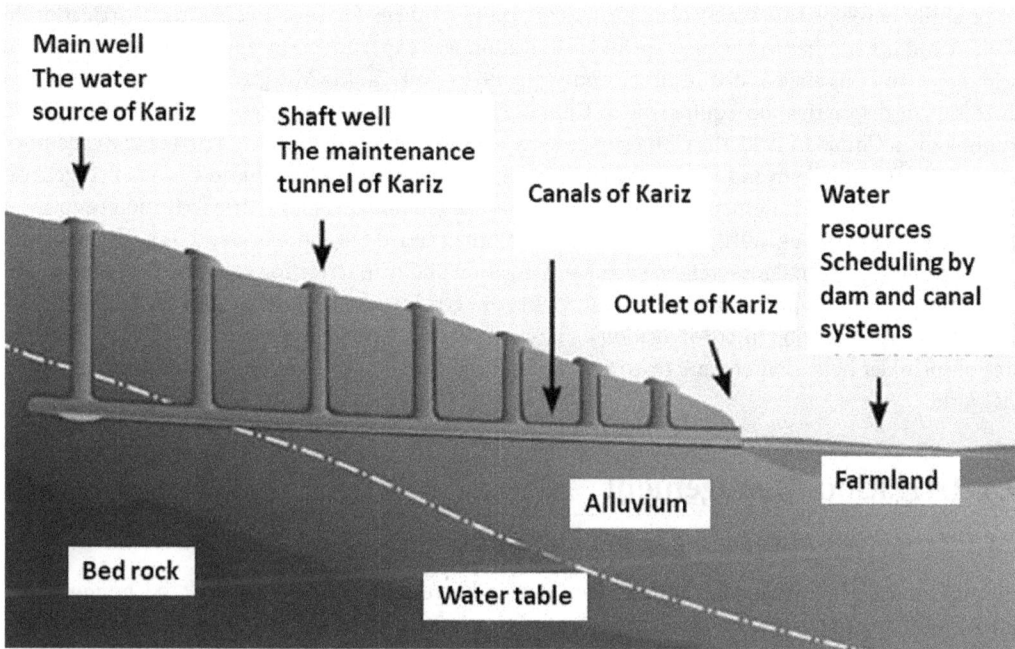

FIGURE 9.2   Schematic diagram of the Kariz (Cui et al., 2012).

2. Kariz is a kind of water conservancy project that applies underground culverts to gather ground-water (Figure 9.2) (Cui et al., 2012). The gathered groundwater is then flowed out of the ground to irrigate the land. Nowadays, Kariz is mainly distributed in Turpan and Hami. The Kariz is composed of shafts and underground channels. The shafts are arranged from high to low along the terrain, and the depth of the well changes with the change of the ground slope. After digging the shaft, a tunnel (underdrain) is excavated at the bottom of the well to connect the wells so that the water can be led out of the underground to store in the pond, and finally transported to the field through the channel. The shape of cross-section of the well is an upright oval with a height of 1–1.5 m. The length of the tunnel is generally about 30 km. Xinjiang has an arid climate, leading to the little surface runoff and large evaporation. Therefore, the use of Kariz can effectively collect groundwater to avoid the large amount of evaporation that occurs in general water delivery projects by delivering water in underground channels, which could fully utilize valuable water resources.

9.3.2.2 Development of New Irrigation and Drainage Technology

Among the new water conservancy projects in the Republic of China, the motor drainage station and the pumping irrigation station were the most eye-catching engineering facilities at that time. The pumps rented from the Shanghai Concession were first used to drain water in Furongwei, Wujin County, Jiangsu Province, in the late Qing Dynasty. In 1915, the water pump produced by Jiangsu Changzhou Heavy Iron Factory (China's self-produced water pump) was used in Furongwei (Tan, 2005). It took only 9 years from the first use of the leased water pump to the independent production of drainage machinery in China. In the 1930s, the first generation of electromechanical irrigation and drainage projects appeared in the relatively affluent and flood-prone areas in the middle and lower reaches of the Yangtze River, the Pearl River Delta and the southeast coast.

Sprinkler irrigation and micro-irrigation started relatively late in China (Li et al., 2012). In the 1950s, China used refraction sprinklers for vegetable sprinkler irrigation for the first time. In the 1960s, China successfully developed turbine-worm sprinklers and tried them for vegetable sprinkler irrigation in Wuhan and other cities. After that, sprinkler irrigation was experimentally applied to cash crops and field crops, and this stage lasted until the end of the 1970s. In 1974, the Mexican government presented three sets of drip irrigation equipment to China, which initiated the development of drip irrigation technology in China. In 1980, the Chinese Academy of Water Sciences, Liaoning Provincial Academy of Water Sciences and Shenyang Plastic 7th Factory jointly developed and produced China's first generation of drip irrigation equipment. This equipment fills the gap that there is no drip irrigation equipment in China (Yao and Wang, 2011). From 1981 to 1986, China carried out more in-depth improvement and application experimental research on sprinkler irrigation and drip irrigation equipment products, and expanded the scope of promotion. After that, China directly introduced foreign advanced manufacturing technology to develop micro-irrigation equipment products from a high starting point. By now, the area of sprinkler irrigation and micro-irrigation in China has reached 10 million ha, ranking second in the world.

9.4 Irrigation Management

9.4.1 Irrigation Management Regulations

The appearance of Chinese water conservancy laws and regulations could trace back to Chunqiu (770–476 BC). During that time, the representatives from each country would meet in a fixed time to make regulations on the establishment of institutions and personnel and the management of irrigation water (Gu and Chen, 2008; Zhou, 1997). In 111 BC, the "Guanzhong Liuqu" canal developed a "water law" to effectively manage channels and irrigation water, which was the first irrigation management system in China (Xin, 1986). In Tang Dynasty, "Shuibuling" was formulated and promulgated by the Ministry of Works, which was the first existing national water conservancy law in China. The basic facilities, operation, organization and personnel allocation of the irrigation area were specified in detail. Some concepts reflecting the principle of modern water right were put forward. Among them, the provisions on rotation irrigation are still widely used today. In Northern Song Dynasty, one of the most important contents of Wang Anshi Reform was the enactment and promulgation of "the restriction of farmland and water conservancy" in 1069, which was an administrative regulation to encourage and regulate large-scale farmland and water conservancy construction (Hu, 2012). In the following dynasties, the water conservancy provisions of the national law and special water conservancy regulations had been gradually improved. The relevant institutional regulations for irrigation management had become more and more detailed, which had played an important role in regulating and guaranteeing the development of Chinese irrigation industry (Tan, 2005).

In modern China, water conservancy legislation began in the 1930s. In 1931, the National Conference of the Interior passed a proposal for compiling water conservancy regulations, determining water rights and developing water conservancy. Two years later, the Ministry of Interior developed a "Draft of Water Conservancy Law". In July 1942, the first Chinese "Water Conservancy Law" with modern water conservancy characteristics was promulgated.

9.4.2 Irrigation Management Agency

The government of ancient China also very valued the management of irrigation projects (Gu and Chen, 2008; Liu, 2002). Judging from the establishment of institutions in the past dynasties, the central and local government agencies generally have dedicated water conservancy agencies to be responsible for the construction and management of water conservancy projects. During the Qin and Han dynasties, Zhengguo canal was directly managed by a special agency established by the state, but it was changed to be managed by the local government after the Tang Dynasty (Wang, 2006). Until the Tang Dynasty, the irrigation project management organizations were relatively fully equipped. Irrigation projects organized and constructed by the state are generally managed by the central or local governments. More exactly, the governments usually authorize and dispatch special officers to station at the location of the project to drive the management authorities (Liu, 2002). Some classical examples about the above management style are Zhengguo canal and Dujiangyan irrigation project.

Differently, the irrigation projects built by the private sectors, such as field canals, ditches and small water source projects, are mostly privately managed. No matter the irrigation projects are managed by the governments or private sectors, manpower is very important. Taking irrigation projects of Guanzhong plain in the Tang Dynasty as examples (Xin, 1986), where staff were distributed to different irrigation canals and water delivery gates to ensure that there are at least one person to be responsible for one irrigation project. Moreover, there were also specific staff to guard the water outlet. In addition, the staff also often inspected these irrigation projects to find and solve the potential issues timely. After the Song Dynasty, irrigation projects were mainly privately run or government-supervised privately run (Wang, 2006). There are two types to determine the management officer for the irrigation projects. One is selection system and the other is rotation system. The selection system is to select managers from the water users in the irrigation area based on their seniority, and most of them have high prestige among the water users. In the rotation system, the water users who own a certain industry take turns in rotation, and they rotate regularly. Many irrigation projects have exerted good benefits for a long time due to their emphasis on irrigation management. After the Qing Dynasty was overthrown in 1911, the government of the Republic of China began to imitate the experience of European and American countries and make major changes to the establishment of state institutions (Zheng, 2005). The earliest establishment of water conservancy committees was a basin committee, which tried to establish a water administrative system with basins as a unit. At the beginning of the Republic of China, irrigation was under the jurisdiction of the Ministry of Industry and Commerce. In 1927, the government was reorganized, and the irrigation affairs was assigned to the Ministry of Industry.

In 1943, the Water Conservancy Commission was established as the competent authority for water conservancy in all provinces. In 1947, the Water Conservancy Commission was renamed to the Ministry of Water Resources.

Today, in the People's Republic of China, the irrigation and drainage affairs are also mainly managed by the Ministry of Water Resources (Feng, 2009). However, it should be noted that the management responsibilities of farmland construction projects have been integrated into the Ministry of Agriculture and Rural Affairs.

9.5 Irrigation Schedule and Method Development

The construction climax of ancient irrigation projects was reached in the Qin and Han dynasties, and the formulation of irrigation schedule was also sprouted during this period. Shengzhi Si, an outstanding agronomist in the Western Han Dynasty, had already applied the concept of irrigation schedule in his famous agronomy book "Book of Shengzhi Si" (Wan, 1957). In the same period, works such as "Huainanzi" and "Rites of the Zhou Dynasty" also had begun to pay attention to the close relationships between irrigation schedule and varying irrigated water quality, soil and climate (Gu and Chen, 2008). Subsequently, with the developments of irrigation projects and agricultural equipment, from the Three Kingdoms to the Tang Dynasty, the ancient Chinese irrigation area continually expanded, and the irrigation schedule became more and more mature. Ancient agricultural irrigation experts gradually realized that the irrigation schedule should be adjusted according to crop types. Sixie Jia, an outstanding agronomist of the Northern Wei Dynasty, written a famous book, "Qi Min Yao Shu" (Li, 2003), that systematically summarized the agricultural and pastoral production experiences of working people in the middle and lower reaches of the Yellow River before the sixth century, and who introduced the detailed relationships between crop water demand and different kinds of climates and soils (Liu, 2002). The roasting technique at the end of the tillering stage of rice is mentioned in this book. Meanwhile, the books "Agricultural book", written by Song Dynasty agricultural expert Fu Chen, and "Shen's

Agriculture Book", written at the end of the Ming Dynasty, also considered the roasting technique that not only can promote the development of rice roots but can also increase soil temperature and improve soil permeability (Li, 2003; Zhou, 2006). Yingxing Song, a famous agronomist in Ming Dynasty, scientifically demonstrated that the wheat is resistant to flooding in the seedling stage, but the irrigation amount must be strictly controlled when it is close to maturity, excessive irrigation will easily cause the wheat stalks to fall and reduce the yield. In the Qing Dynasty, Shaanxi agronomist Can Yang also talked about the importance of timely irrigation when discussing millet agronomy. Additionally, the ancients also conducted research studies on the regulation of soil temperature by irrigation (Zhou, 1997). The earliest record can be found in the Western Han Dynasty's "Book of Shengzhi Si", and Guangqi Xu's "Agricultural Policy Book" written in the later Ming Dynasty, which aimed at solving the problem about the low water temperature irrigated by well water and spring water, the method of increasing the water temperature by extending the water flow pathway and increasing the sunlight was introduced. This method was also recorded in the "Book of Fengyu Zhuang" written by Zengyi Pan in the eighth year of Daoguang in the Qing Dynasty (Liu, 2002). After the establishment of the People's Republic of China, on one hand, with the rapid upgrading of machinery and equipment, the technologies of drip irrigation, sprinkler irrigation and other micro-irrigation have gradually become popular. Tingwu Lei, Yuehu Kang, etc., the contemporary agricultural water conservancy agronomists, have implemented the irrigation schedules for new irrigation technologies under the different crops-water-fertilizer management situations. On the other hand, as the understanding of the crop water consumption law gradually deepened, to alleviate the shortage of agricultural water, deficit irrigation became a research hotspot. Shaozhong Kang, Taisheng Du and other experts put forward the split-root alternate irrigation technology based on physiological information about crop water demand; practice proved this irrigation method achieved good water-saving and quality-preserving effects (Du et al., 2005; Hu et al., 2005).

9.6　Irrigation Education

China is a great agricultural country with a magnificent irrigation management history, but the education histories of water conservancy and irrigation are not long. According to legend, Anshi Wang (1021–1086 AD, a famous politician, and officer in Northern Song Dynasty) introduced the knowledge about water conservancy engineering in Taixue (the highest national institution in ancient China), but which was suspended with the failure of the Wang Anshi Reform (Lu, 2001). Then until the 35th year of Kangxi (1696 AD), Yan Xiqi set up a private "Zhangnan Academy" to teach hydrology (irrigation) and fire science in Feixiang County, Hebei Province. But this education attempt about irrigation remain could not continue because of a fire in the academy. Officially, the educations of water conservancy and irrigation emerged in the late Qing Dynasty during the period of educational system reform (Yuan, 2014). In 1895 AD, Xuanhuai Sheng founded Beiyang Western Learning School in Tianjin, offering five specialties for learners, including engineering and electricity, etc. Especially, the subject of "Water Conservancy Machinery" was involved in engineering education. In the same year, Franya (a British missionary) had written the "Regulations of Gezhi College's Teaching of Western Studies" for Shanghai Gezhi College. The regulations stated that the academy opened a hydrology subject "Water Stress", which was divided into hydrostatics and hydrodynamics. In 1908 AD, under the proposal of Peifen Lu and others, the River Engineering Research Institute was established and began to focus on cultivating professional river workers. During the Republic of China, with the gradual improvement of the educational system, eight national universities and seven provincial universities began to offer related water conservancy and irrigation majors (in 1936 AD). After the establishment of the People's Republic of China, school diploma education, on-the-job continuing education and social education related to water conservancy and irrigation have developed rapidly. Tsinghua University, Wuhan University, Hohai University and other well-known universities all opened irrigation-relevant majors (Wang, 2006). Nowadays, modern irrigation education is carried out by the second-level disciplines of water conservancy and hydropower engineering and agricultural water and soil engineering subordinate to the first-level disciplines of water conservancy engineering and agricultural engineering. The discipline content involves the construction and management of irrigation engineering, the relationship between soil–plant–atmosphere, irrigation and drainage, farmland improvement, survey and conservation of water-soil resources, etc. (Wei, 1987; Xiong and Kang, 1996).

9.7 Conclusions

The closest relationships between irrigation and water directly affect human activities. In the thousands of years of China's history, due to the specific geographical environment and the agricultural bases of society and economy, a unique irrigation science and irrigation engineering system has been formed. It is worth noting that many irrigation projects not only serve agricultural irrigation but also consider domestic water, flood control, waterway transportation and even food processing, which have a huge impact on the natural environment and human life. For example, Dujiangyan created the Chengdu Plain 2,260 years ago. While most of the ancient irrigation projects belonged to empirical design, construction and management, some projects or measurements might cause damages to the natural environment to meet special needs. For example, to make the Beijing-Hangzhou Canal unblocked, weirs were built at the intersection of the Huai River, the Yellow River and the Beijing-Hangzhou Canal in 1128 AD, forcing the Huai River to merge with the Yellow River due to the formation of Hongze Lake. As a result, the sediment of the mainstream of the Yellow River was deposited on the Huaibei Plain, causing the flood discharge difficulty in the Huai River.

After the establishment of the People's Republic of China, as the concept of sustainable development was widely accepted, irrigation science and irrigation engineering technology have been greatly developed. However, the development of science and technology is inherited, and science itself has no historical limitations. Modern irrigation science and irrigation engineering should take the essence of ancient irrigation engineering, and use new technologies to reveal the scientific value of ancient irrigation engineering, thereby opening up a new path for the development of irrigation science and irrigation engineering.

References

Cui, F., Wang, S.M., and Zhao, Y. (2012). The Values of Agro - Cultural Heritage of Xinjiang Karezes and Its Protection and Utilization. *Journal of Arid Land Resources and Environment*, 26: 47–55.

Du, T.S., Kang, S.Z., Hu, X.T., Fucang Zhang F.C. and Gong, D.Z. (2005). Research Progress on Alternate Irrigation of Fruit Tree Root System. *Journal of Agricultural Engineering*, 172–178.

Fan, J., Galoie, M., Motamedi, A., Liao, Y., and Eslamian, S., (2021). Rainwater Conservation Practices in China, in *Handbook of Water Harvesting and Conservation, Vol. 2: Case Studies and Application Examples*, Ed. By Eslamian, S. & Eslamian, F., John Wiley & Sons, Inc., New Jersey, 363–282.

Feng, G.Z. (2009). Reviewing the 60-Year History and Understanding the Law of Farmland Water Conservancy Development. *China Water Resources*, 7–9.

Gu, H. and Chen, M.S. (2008). Irrigation Civilization in Ancient China. *China Rural Water Conservancy and Hydropower*, 1–8,14.

Hu, X. (2012). Reference Significance of Wang Anshi Farmland Water Conservancy Law to Water Conservancy Reform. Times Finance, 309–309.

Hu, X.T., Kang S.Z., Zhang J.H., Zhang F.C., Li, Z.J. and Zhou, L.C. (2005). Laboratory Experiment and Water-Saving Mechanism of Alternate Control Drip Irrigation for Tomato Vertical Root Zone. *Journal of Agricultural Engineering*, 1–5. www.ChinaAgriSci.com

Li, G.P. (2003). "Book of Agricultural Book" and "Three Talents" Theory--Compared with "Qi Min Yao Shu". *Journal of South China Agricultural University (Social Science Edition)*, 2: 101–108. doi: 10.3969/j.issn.1672-0202.2003.02.017

Li, H.B. (1992). "Shui Bu Shi" and Water Conservancy Management in Tang Dynasty. *China Water Resources*, 35–36.

Li, Y. (2020). Characteristics of Chinese Traditional Irrigation Techniques from the Perspective of Irrigation Engineering Heritage. *Study on Natural and Cultural Heritage*, 4: 94–100.

Li, Y.H. (2010). Water Saving Irrigation in China. *Irrigation and Drainage*, 55: 327–336.

Li, Z.L., Zhao, W., Sun, W. and Fan, Y. (2012). Application Prospect of Sprinkler Irrigation Technology in Water-Short Areas of Northern China. *Transactions of the Chinese Society of Agricultural Engineering*, 28: 1–6.

Liu, Y.W. (2002). Ancient Chinese Irrigation Technology and Water Management. *Journal of Shanxi Agricultural University (Social Science Edition)*, 1: 243–246. doi: 10.3969/j.issn.1671-816X.2002.03.016

Lu, H.S. (2001). The Rise and Early Development of Modern Water Conservancy Higher Education. *Journal of Shanxi University (Philosophy and Social Sciences Edition)*, 104–106.

Ren, Y.J. (2019). Research on Protection and Development of Ancient Chinese Water Conservancy Project Heritage-Taking Zhengguo Canal as an Example. *Cultural Heritage*, 159: 146–147.

Tan, X. 2005. *History of Irrigation and Flood Control in China*. China Water and Power Press, Beijing.

Wan, G.D. (1957). *Compilation and Explanation of the Shengzhi Si's Book*. Zhonghua Book Company. China.

Wang, K. (2006). *Research on Optimization of Hierarchical Structure of Water Conservancy Higher Education in My Country*, Hohai University, Nanjing, China.

Wei, N. (1987). Introduction to Agricultural Engineering. *Journal of Agricultural Engineering*, 3: 112–114.

Wei, S. and Chen, J. (2010). Turnover Chain, Drum Wheels—Analysis of the Origin and Development of Ancient Chinese Waterwheels. *China Water Resources*, 1: 59–61.

Wu, P.T., Jin, J.M. and Zhao, X.N. (2010). Impact of Climate Change and Irrigation Technology Advancement on Agricultural Water Use in China. *Climatic Change*, 100: 797–805.

Xin, Y. (1986). Water Conservancy in Guanzhong during the Qin and Han Dynasties. *History Monthly*, 112–114.

Xiong, Y.Z. and Kang, S.Z. (1996). Discipline of Agricultural Water and Soil Engineering in China and Its Development Forecast. *Journal of Agricultural Engineering*, 1(1): 10–13.

Yang, S.H., Peng, S.Z., Xu, J.Z., He, Y.P. and Wang, Y.J. (2015). Effects of Water Saving Irrigation and Controlled Release Nitrogen Fertilizer Managements on Nitrogen Losses from Paddy Fields. *Paddy and Water Environment*, 13: 71–80.

Yao, Z.X. and Wang, S.J. (2011). The Drip Irrigation Development Progress and Suggestion of China. *Agricultural Engineering*, 1(2): 54–58.

Yuan, B. (2014). *Historical Investigation of Modern Chinese Water Culture*, Shandong Normal University, China.

Zheng, Q.D. (2005). The Development of Farmland Water Conservancy in the Period of the Republic of China Government. *Research on Chinese Economic History*, 20–29.

Zhou, G.X. (2006). Research on Rice Fertilization Techniques in "Shen's Agricultural Book". *Journal of Nanjing Agricultural University (Social Science Edition)*, 6: 69–72. doi: 10.3969/j.issn.1671-7465.2006.01.013

Zhou, K.Y. (1997). Ancient Chinese Farmland Irrigation and Drainage Technology. *Ancient and Modern Agriculture*, General Administration of Press and Publication of the People's Republic of China. 1–12.

Zhuang, Y.H., Zhang, L., Li, S.S., Liu, H.B., Zhai, L.M., Zhou, F., Ye, Y.S., Ruan, S.H. and Wen, W.J. (2019). Effects and Potential of Water-Saving Irrigation for Rice Production in China. *Agricultural Water Management*, 217: 374–382.

Zou, L.H. (2015). A Successful Example of Water-Control: Dujiangyan Irrigation Project. *Journal of Xihua University (Philosophy and Social Sciences)*, 6: 31–33.

10

History of Irrigation in China: Legislations and Farmer's Incentives

Muhammad Touseef
and Lihua Chen
Guangxi University

Saeid Eslamian
*Isfahan University
of Technology*

10.1 Introduction

The People's Republic of China is located in the southeast Asia surrounded by Mongolia and Russian Federation to the north, the Democratic People's Republic of Korea and the Pacific Ocean to the east, Vietnam, the Lao People's Democratic Republic, Myanmar, Bhutan, Nepal and India to the South and Islamic Republic of Pakistan, Afghanistan, Tajikistan, Kazakhstan and Kyrgyzstan to the west. The average altitude in China ranges from 4,000 m in the west to less than 100 m in east. The total area is 9.6 million km². Land is composed of mountains (33%), plateaus (26%), valleys (19%), plains (12%) and hills (10%). China has 23 provinces, three municipalities, five autonomous regions and special administrative regions of Hong Kong and Macau. Total cultivated area was 124.30 million ha in 2009. Out of the total cultivated land arable land was estimated 110 and 14.30 million ha under permanent crops.

The Climate of China varies from region to region. The country's northeast where Beijing sits is hot in summer and freezing cold in winters. In the southeast there is plenty of rainfall with semi-tropical summers and cold winters. South China is affected by the East Asia monsoon climate. North China is generally under the influence of dry cold air masses from Siberia. Average annual rainfall is 645 mm, while in some regions in southwest and in the coastal areas of southeast, the mean annual precipitation exceeds 2,000 mm. In northeast and northwest regions, the mean annual precipitation is between 400 and 900 mm.

China is conventionally classified into four main agroclimatic zones. Arid zone is located mainly in the inland river basins in the west and northwest. This zone is suitable for irrigated cotton, grains, vegetables and fruits. Livestock is the predominant land-use in this zone. Semi-arid zone covers the upper and central regions of Yellow River Basin in central China. Wheat, maize and cotton are the main crops. Semi-humid zone subjected to both floods and droughts. Songhua-Liao River basin in northeast of the country is in this zone. Potentially fertile land with short growing season but western part is suffering from waterlogging and alkaline soils. The humid zone lies in the south and southwest. Flooding season lasts from July to September. Drought can limit crop yield in early or late flooding seasons.

The average river run-off generated within the country is 2,711.5 km³. Precipitation contributes 98% to the total run-off while 2% come from melting glaciers. China can be divided into nine main river basins. The basins in north are Song-Liao or Heilong-Songhua, the Huai, the Huang (Yellow), the Hai-Luan and the interior or endorheic river basins. The total average annual interval renewable surface-water resources (ISRWR) in these five main river basins are 535 km³, which is 20% of total country's ISRWR. In South China there are four river basin groups, Chang (Yangtze), Zhu (Pearl River Basin), the southeast and southwest river basin groups. These four basin groups contribute 80% of the country's IRSWR, estimated to be 2,176.2 km³.

There are more than 50,000 rivers with a basin area of over 100 km², while 1,500 rivers have basin area exceeding 1,000 km². The total drainage area of rivers flowing to the sea covers about 65% of the territory. The volume of water flowing to nine neighbouring countries is estimated to be almost 719 km³/year. There are 12 main rivers that enter China from six neighbouring countries. The average annual volume of water entering the country is just over 17 km³.

The total area of glaciers in China is about 58,651 km². Country's glacier storage is around 5,100 km³ in total. The amount of mean annual glacier melt water is about 56 km³. The average annual groundwater resources for the whole country are an estimated 828.8 km³. The part which reaches the river as baseflow, or comes from river seepage, called "overlap", is an estimated 727.9 km³. Around 70% water resources are in southern China and 30% in northern China.

There are about 2,300 natural lakes (excluding seasonal ones) with a total storage of 708.8 km³, of which the freshwater portion is 31.9%. At the end of 2005, the total number of reservoirs and dams were 85,108 with a total capacity of 562 km³. Large reservoirs (>1,000 million m³), 2,934, medium reservoirs (10–100 million m³) and 81,704 were small reservoirs (0.1–10 million m³).

China has a long history of irrigation. The first canals to divert and wells to lift water for irrigation were constructed 4,000 years ago. Since the founding of People's Republic China in 1949, irrigation has experienced a period of vigorous development. The area equipped for irrigation increased from 16,000,000 ha in 1949 to 62,938,226 ha in 2006 (Figure 10.1).

After 1949, irrigation using groundwater was developed rapidly to promote agricultural production. In north China, insufficient surface-water resources have meant that since 1950 the Government has had to rely on groundwater for the development of irrigation projects. In 1985, an area of 11.1 million ha was irrigated using tube wells. In 2006, groundwater irrigation area using around 4.8 million tube wells was an estimated 19 million ha; 31% of the total area equipped for irrigation was 63 million ha. In addition, 17 million ha is power-irrigated using surface water. This means that 57% of the total area equipped for

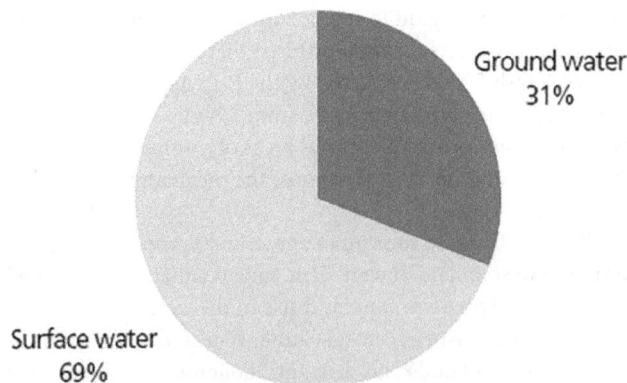

FIGURE 10.1 Source of Irrigation water on area equipped for full control irrigation (total 62,938,226 ha in 2006).

irrigation, or 36 million ha, used power irrigation. In 1995, the power irrigation area was 29 million ha and the total installed capacity of water-lifting machines for irrigation and drainage was 68,240 MW.

China can be divided into three main irrigation zones:

- The zone of perennial irrigation, where annual precipitation is less than 400 mm and irrigation are necessary for agriculture. It covers mainly the northwest regions and part of the middle reaches of the Huang River.
- The zone where annual precipitation ranges from 400 to 1,000 mm, strongly influenced by the monsoon, with a consequently uneven precipitation distribution. Irrigation here is necessary to secure production. This zone includes the Hang Huai Hai plain and northeast China.
- The zone of supplementary irrigation, where annual precipitation exceeds 1,000 mm. Irrigation is still necessary for rice cultivation to improve cropping intensity, and supplementary irrigation is sometimes required for upland crops. This zone covers the middle and lower reaches of the Chang, Zhu, Min and some parts of southwest China.

The irrigation potential is roughly 70 million ha. The maximum possible area that could be equipped for irrigation by 2050 is about 66 million ha, of which 63 million ha for annual or food crops. China uses the expression effective irrigation to indicate the area of food crops, not to be confounded with the area irrigated. In 1996, the total area equipped for irrigation, including farmland, forests, orchards and pastures was 52.90 million ha.

Surface irrigation area, mainly for cereals, vegetables and cotton, is estimated as 59.3 million ha, which is 94.3% of the total irrigation practiced in 2006. Sprinkled irrigation was introduced in China in early 1950s. The first sprinkle irrigation was installed in Shanghai in 1954. Sprinkler and localized irrigation were considerably developed in the late 1970s. In 1976, the area of sprinkler irrigation was about 67,000 ha. It increased until 1980, but then large areas were abandoned owing to the poor quality of equipment and poor management. Then, in 2006, the area expanded to about 2.8 million ha, which is 4.5% of the total area equipped for irrigation. Localized irrigation was practiced on about 0.8 million ha or 1.2%.

This chapter will focus on the long history of irrigation system in China, the old and existing and agriculture policies and the incentives offered to farmers.

10.2 Irrigation Development: Current and Past

Four ancient Chinese irrigation systems were recognized as Heritage Irrigation Structures (HIS) by International Commission on Irrigation and Drainage. China has the world's oldest irrigation system known as "Dujiangyan". The only non-dam irrigation system exists till date. This marvellous irrigation system was built over 2,200 years ago. This system is still used to irrigate over 668,700 ha of farmland and drain floodwater, and it provides water resources to more than 50 cities in the Sichuan province. Dujiangyan system arose to confront the frequent flooding of the Minjiang River, a tributary of Yangtze River. Li Bing, the founder of this system, suggested an artificial levee which could move some of the water to another area and then make channel in Mount Yuli to send the excess water to the dry Chengdu plain. This project was funded by King Zhao of Qin dynasty. They built the levee by creating long sausage like baskets of bamboo filled with stones, known as Zhulong. These were held in place with wooden tripods called Macha. Natural topographic and hydrologic features aid the system in irrigation, draining sediment, controlling flooding and controlling water flow. The most amazing features of the engineering project were the creation of a channel through Mount Yuli. It is worth nothing that workers did this before gunpowder and explosives were invented. It took 8 years to create 20 m wide channel through the mountain. The completion of the system brought an end to flooding in the area and helped make Sichuan the most productive agricultural region in China. Li Bing's ambitious project is known recognized as "Treasure of Sichuan".

Another structure inducted is Lingqu in southern China's Guangxi province. The canal was built around 214 BCE, making it the first of its kind to connect two river valleys. It also allowed boats to travel the 2,000 km from Beijing to Hongkong for the first time. Jiangxiyan in Zhejiang province has also been added to HIS. Construction of the facility spanned the Yuan and Sun dynasties and took place between 1,220 and 1,333. The final addition is Changqu in Central China's Hubei province. Changqu was built during the 770–476 BCE. The objectives of recognition include tracing the history and understanding the evolution of irrigation in civilizations across the world, as well as preserving these historical irrigation structures.

10.2.1 Chinese Grand Canal

The Grand Canal and the Great Wall in China are the landmark projects of the Chinese Civilization. From the warring states to the early 20th century, in the historical period up to 2,500 years, the Grand Canal has been continuing construction and applications for water transportation across the major rivers to the final destinations of the capital. The project remains deep imprints of the political, economic, technological and cultural patterns during the various historical periods and demonstrates the unyielding national spirit and great creativity of the Chinese. In the process of Chinese Civilization, countless canals were dug during the various historical periods. However, independent engineering projects such as the Grand Canal were built through continuous construction (Tan et al., 2019). The first Grand Canal consists of Tongji Canal (Bian Canal), Yongji Canal (Weihe River), Huaiyang Canal, Jiangnan Canal and Zhedong Canal during the Sui, Tang and Song Dynasties, with Louyang or Kaifeng as the Center. The construction of the Grand Canal started with the canals between the Yangtze River and Huaihe River as well as between Yellow River and Huaihe River and lasted for nearly 1,000 years. Another Grand Canal started from the Yuan, Ming and Qing Dynasties, consists of Tonghui Canal, North Canal, South Canal, Huitong Canal, Huaiyang Canal, Jiangnan Canal and Zhedong Canal from Beijing to Hangzhou and to Ningbo, with a total length of 2,000 km (Liu et al., 2013). The canal was continuously used for 600 years. All the canals together are called the Chinese Grand Canal (Jindong and Jing, 2019).

The Chinese Grand Canal was started to construct in late spring and autumn periods and was the product of territory expansion by the vassal states. With the political and military demands, artificial waterways were built and continuously improved, which were the infant of the Grand Canal. Chinese Grand Canal is continuing waterway more than 1,000 km with relatively independent engineering system. Chinese Grand Canal retained or still used generally refers to the Beijing-Hangzhou Grand Canal and Sui-Tang Grand Canal. Since the Qin-Han dynasties unified the China's political system, the canal became the transportation lifeline to maintain the political and economic connection to the central government, and a large-scale water conservancy project under the national management. The political, economic, cultural and technological development in different historical periods gave the specific context of the canal cultural value.

Tonghui Canal, one of the sections of Grand Canal above Shichahai, was an outstanding example of the canal planning. The project provided water source guarantee for the Tonghui Canal and was also with the municipal and environmental functions. The rivers and lakes derived from the project the beautiful landscape environment of the ancient Beijing. In Tonghui Canal, 24 regulating sluices were used to control water and navigation depths. Currently the only remaining sluices include Guangyuan, Gaoliang, Chengqing and Qingfeng.

The current North Canal section of the Grand Canal was mainly formed in Jin Dynasty. After Tonghui Canal was constructed in Yuan Dynasty, North Canal became a part of Beijing-Hangzhou Grand Canal and was included in the Grand Canal management system. In 1912, Chaobai River changed course into Ji Canal. Due to water source shortage, North Canal water transport was terminated. North Canal became weak since the end of Qing Dynasty. Currently, the North Canal is mainly used for flood discharge, irrigation and water transmission. On the other side, South Canal, called Weihe River on the upper reaches of Shandong Linqing, originated from the east foothill of Taihang Mountains within the territory of Shanxi, and merged mainly with Zhanghe River, Tuhai River and Ziya River along the way. The South Canal was the mainstream of southern system of Haihe River. The major water transport problems included great variations of water amount during the dry and wet season, and frequent embankment breakage in the flood season. Therefore, in the Yuan Dynasty, two types of engineering measures were adopted. The first type was to construct flood protection channels. The second type was to utilize the irrigation water of Weihe River upstream to guarantee the canal water source in winter and spring. These two measures were used till the Qing Dynasty.

North and South Canals have minimum engineering facilities, the curvature design provided the better navigation conditions and security for the canal water transport. The curvature design made floods slowdown in the canal curve segments to enhance the navigation safety. To solve drainage problem in the flood season and the water supply problem in the dry season of the South and North Canals, water drainage sluices were constructed at the east bank of the canal and Jianhe was excavated to drain flood to the east into the sea.

The Huitong Canal is also known as the Shandong Canal. It is the most important reach of the Grand Canal, connecting the Haihe River, the Yellow River and the Huaihe River. In 1128, the Yellow River burst its banks at Kaifeng of Henan province and changed its course, which started a history of 700 years of flowing south. After the Yellow River changed its course, the reach of the canal between the Yellow River and Huaihe River was occupied by the new channel of the Yellow River. As a result, the waterway connecting the North and South was interrupted. After the capital of Yuan Dynasty was established in Beijing in 1172, it became a matter of urgency to restore the waterway system connecting the North and South to ensure the food supply for the capital city of Beijing. As a result, the construction of the Grand Canal with Beijing as the destination of transportation started. In the Yuan Dynasty, two new canals were constructed and connected to the previous canal that passed through the northern and southern areas in the east of China. The two new canals were the Tonghui Canal and the Huitong Canal built in 13th century (Zhang and Zhou, 2020).

The Huaiyang Canal originating from the Han Ditch excavated by King Fu Chai of the Wu Kingdom in 486 BC. Huaiyang Canal connects the Yangtze River and Huaihe River and is the oldest canal with the definite record of excavation time. The Canal runs from Huaian to Yangzhou across numerous lakes. As a result, it was called "Lake Channel". After the Yellow River changed its course so that the lower reach of the Huaihe River became part of the Yellow River in the Southern Song Dynasty, Qingkou in Huaian or the North mouth of Huaiyang Canal was the confluence of Yellow River, the Huaihe River and Huaiyang Canal. The Huaiyang Canal runs through Huaian and Yangzhou, two prefecture-level cities in Jiangsu province, with a total length of 160 km. It is in low-lying area with a dense network of rivers, lakes and relatively rich water resources. Originating from the Han Ditch excavated in the spring and autumn period, the Huaiyang Canal has experienced 2,500 years of development and is now still playing an important role in navigation. The Huaiyang region has a subtropical monsoon climate with an average temperature of 14°C–15°C. Average precipitation ranges from 800 to 1,100 mm in North to South order. The maximum monthly rainfall generally occurs during the plum rain season in June or during the typhoon season from August to September.

The Jiangnan (South of the Yangtze River) Canal is one of the reaches of the Grand Canal, which were first constructed with the longest continuous operation time and the best natural conditions. Its excavation started during the spring and autumn period (770–476 BC), its entire channel was realigned during the Sui Dynasty, and its southern reach became gradually straight after the Yuan Dynasty. The region that Jiangnan Canal runs through is abundant in rainfall with dense river network and relatively rich water resources. The history of the Jiangnan Canal reflects the whole process of the construction and operation of the Grand Canal in China. As one of the areas with the most developed economy in China, the region where the Jiangnan Canal is located boasts important cities and towns, all of which are closely related to the canal and embellish it like bright particles of pearls. The Jiangnan Canal starts in Zhenjiang City of Jiangsu province in the North and ends in Hangzhou City of Zhejiang province in the South with a total length of 340 km. It connects the Yangtze River, the Taihu Lake and Qiantang River and is one of the reaches of the Grand Canal that were first constructed with the longest continuous operation time and best natural conditions. The whole channel of the Jiangnan Canal now remains navigable.

The Zhedong Canal refers to the southmost part of the Grand Canal, connecting the rivers around it and the sea. It is not only one of the important starting points of the ancient silk road at sea. The construction of the Zhedong canal started with the excavation of the Shanyin Canal of the Yue Kingdom in the spring and autumn period, but it was not until Song Dynasty that the Zhedong Canal became a continuous waterway with a perfect system of control projects. The Zhedong Canal consist of artificial channels and natural rivers, including the artificial reach from Xiaoshan to Shangyu and the natural reach from Shangyu to Ningbo using such natural rivers as the Yao River and Yong River for navigation. Affecting by the region-specific hydrological conditions such as physical geography, typhoons and the tide. In the process of the formation of the Zhedong Canal, water projects were used to integrate the regional river network and form a canal system with comprehensive water system such as navigation, flood control, drainage, water diversion and irrigation. The canal system played a crucial rule in the regional water structure.

10.3 Irrigation Water Management and Legislations

China is the foremost populated nation within the world. Over the last six decades, the population of China has expanded from 0.5 to 1.3 billion, the total irrigated area has increased almost monotonically from 15.9 to 61.7 million ha and grain output has increased from 113.2 billion kg (249.6 billion lb) to 571.2 billion kg (1,259.5 billion lb) (Shen, 2015).

Irrigation stabilizes crop production, improves crop quality, reduces rural poverty and allows for diversification in farm production. Approximately half of the national cropland is irrigated and produces 75% of the nation's food, 80% of its cotton and oil-bearing crops and 90% of its vegetables and fruits. The crop production from irrigated areas is much higher than that from non-irrigated lands, especially in northwest China where the production of paddy crop in irrigated land is about three times higher than that in non-irrigated land. Advances in agricultural science and technology (including agricultural irrigation) have allowed the total production of grain crops to increase over fivefold from 1949 (113.2 billion kg [249.6 billion lb]) to 2011 (571.2 billion kg [1,259.5 billion lb]), with an annual growth rate of 6.42%. The average per capita grain production in 2011 was 425 kg (937 lb) compared to 208.9 kg (460.6 lb) in 1949, representing a twofold increase during that period (Yu et al., 2015).

The role that irrigation can play in ensuring future food security is unclear. On one hand, agricultural irrigation is expected to continue to contribute to the stabilization of food prices, increases in farmers' income and increases food supply for over 1.3 billion Chinese. Simultaneously, the potential to develop additional water resources and infrastructure in the future may be limited. Water security and water controversy among different sectors are becoming increasingly drastic. In addition, climate change is projected to worsen the situation. The warming trend is expected to continue, which is likely to accelerate evaporation and soil dryness. Climate change also modifies precipitation geographically and temporally, placing more stress on the water and food security of China (Wang et al., 2018).

The distribution of water resources is spatially and seasonally uneven. About 81% of water resources are found in the south of China, which accounts for 36.5% of the total land area. Northern China accounts for 63.5% of the total land area and has 40% of the Chinese population but owns only 19% of the total water in China. For example, the Huang-Huai-Hai region is estimated to have the greatest potential for food production in China but is limited by its access to less than 10% of the total national water resources. The amount of water per hectare of cultivated land in southern China is 28,695 m^3 (6.3 million gal), compared to only 9,645 m^3 (2.1 million gal) in northern China. The average supply of groundwater is about four times greater in the south than in the north. In addition, the distribution of water resources is seasonally uneven (Deng et al., 2006). Precipitation occurs mainly in spring and summer. In most areas of China, the accumulated rainfall over those four consecutive months accounts for over 70% of annual rainfall (Figure 10.2).

Over the past 50 years, China has constructed a vast and complex bureaucracy to manage its water resources (Eslamian, and Zamani, 2021). To understand the functioning of this system, it is important

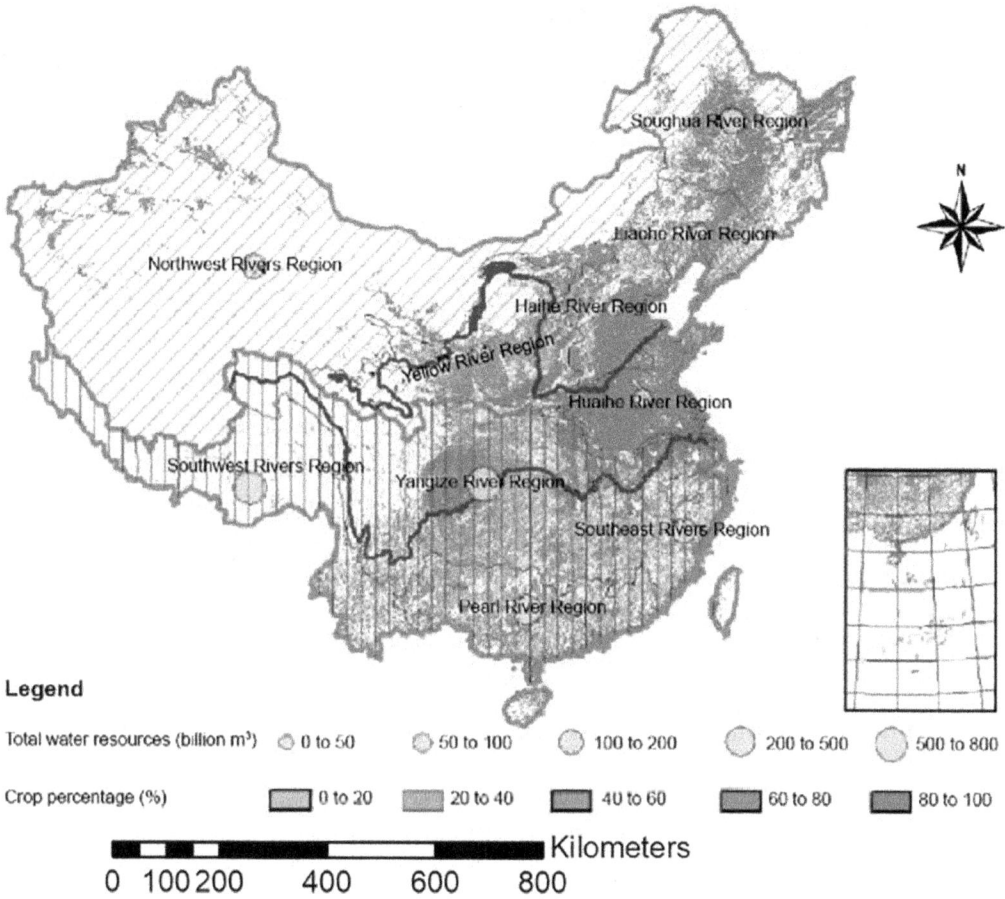

FIGURE 10.2 Distribution of cropland and total amount of water resources in China. There are six water resource regions in northern China and four water resource regions in southern China (Zhu et al., 2013).

to first understand that, until recently, water saving has never been a major concern to policymakers. Instead, the system was designed to construct and manage systems to prevent floods that have historically devastated the areas surrounding the major rivers, and effectively divert and exploit water resources for agricultural and industrial development. Indeed, China's success in accomplishing this latter goal is largely why the nation faces water-shortage problems today. Water policy is created and executed primarily by the Ministry of Water Resources (MWR) (Ministry of Water Resources, 2009). The MWR has run most aspects of water management since China's first comprehensive Water Law was enacted in 1988, taking over the duties from its predecessor, the Ministry of Water Resources and Electrical Power. The policy role of the MWR is to create and implement national price and allocation policy and oversee water conservancy investments by providing technical guidance and issuing laws and regulations to the subnational agencies (Figure 10.3) (Lohmar et al., 2003).

The national government invests in developing the water resources from all large rivers and lakes and projects that cover more than one province. Local governments oversee projects that are within their administrative districts. Historically, investment from national funding sources has been heavily biased toward new investments, while local governments have been responsible for maintenance funds. Under the 1988 Water Law, the MWR is not solely responsible for all water-related policies; other ministries in China also influence water policy for both rural and urban areas. The diverse uses of water and diverse objectives and interests of water management agencies often result in conflicts and inefficient

**The vertical and horizontal structure of the
Ministry of Water Resources**

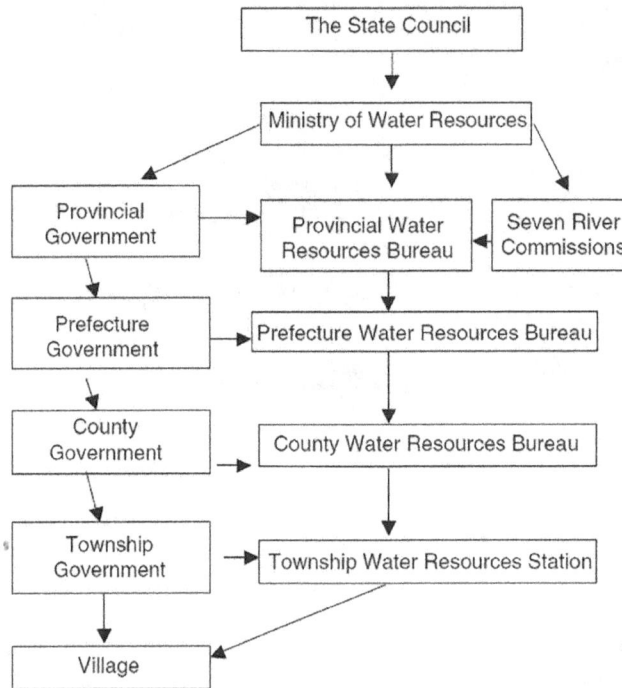

FIGURE 10.3   Ministry of Water Resources Structure (Lohmar et al., 2003).

water use. In the use of agricultural water, the MWR shares its duties with the Ministry of Agriculture, particularly in developing local delivery plans and extending water-saving technology. In urban areas, Urban Construction Commissions (or Bureaus) are charged with managing the delivery of water to urban industrial and domestic users. Urban Construction Commissions also have taken responsibility for managing groundwater resources that lie beneath municipalities' land area. Groundwater levels, both urban and rural, are monitored jointly by the Ministry of Geology and Mining (MGM) and its local associates. In theory, the MGM's information about the groundwater level is used when deciding whether to grant groundwater pumping permits, though local water bureaus do not always use the information. China's State Environmental Protection Agency has the responsibility for managing industrial wastewater and municipal sewage treatment. Last, in the area of price-setting, the MWR, in conjunction with the State Price Bureau and acting with the approval of the State Council sets comprehensive assessment of the profits and costs, especially the potential ecological costs of such projects. Meanwhile, adjusting planting and harvest dates based on temporal climate change effects, such as precipitation seasons, might be a useful adaptation to variations in temporal water distribution (Yao et al., 2017).

10.4 Farmer's Incentives

Despite the improving water management environment in China, the fact remains that in many parts of northern China, ground- and surface-water sources are being depleted and current water-use levels are not sustainable with the current water supply system. As noted earlier, agricultural users will not be given priority for any additional sources of water that become available. Indeed, while it is the stated goal of China's leaders to increase irrigated area, they also explicitly acknowledge that this expansion will occur without any additional water allocations to agricultural users. Thus, using water more efficiently is the only method to increase irrigated area and effectiveness without increasing total agricultural water demand in northern China. Even with what seems to be an impending water crisis, farmers have hardly begun to adopt water-saving technologies or practices. The reasons for this are found in the nature of the incentives faced by China's farming community (and those in other sectors). Until the 1970s, water was considered abundant in most parts of China and was not even priced for agricultural users so there was no incentive for users to save water. Collectives had de facto rights over the water in their communities—whether that water was underground or in nearby lakes, rivers or canals. Facing low or free water prices, farmers naturally used as much water as they wanted. Even today, most farmers "save" water only when their deliveries are curtailed, not because the price is too high or because they are given other incentives. Shortly after the agricultural reforms that began in 1978, the central government encouraged the adoption of a system of volumetric surface-water pricing. While the prices were set by the Price Bureau in Beijing and modified by provincial price bureaus, adoption of prices did not begin all at once in all locations, but instead could diffuse gradually as experience was gathered. Hence, the current price structure exhibits substantial variation across the country and considers both scarcity and the ability to pay. Typically, for a specific end-use (agriculture, industry, domestic) in a specific province, prices are uniform, although there is flexibility for local exceptions. In terms of ability to pay, agricultural users pay lower prices than domestic users, who in turn pay less than industrial users. For example, in Hubei Province, the price for agricultural users is 0.04 RMB per cubic metre, while domestic and industrial users pay 0.08 and 0.12 RMB per cubic metre, respectively. In terms of scarcity, different prices prevail in different provinces, with prices increasing substantially as water scarcity becomes more severe (generally, as one moves from south to north). For example, in the late 1990s, agricultural surface water was priced at about 0.01 RMB per cubic metre in the southern province of Guangdong, 0.04 RMB per cubic metre in the central provinces of Hubei and Henan, and 0.075–0.10 RMB per cubic metre in the northern province of Hebei, where water shortages are most acute. Despite increasing water prices, current pricing policies do not effectively encourage water saving and in fact contribute to China's water problems in other ways. Since China's farmers have provided farms and each farm have several small plots, charging each farmer according to how much water they use (volumetric pricing) is very costly and difficult to monitor. Some observers argue that water prices are so low that demand is relatively inelastic; thus raising water prices would only raise revenues and not decrease the demand for water by a significant amount. Raising revenues, however, would still be good since the low prices fixed by the provincial price bureaus are often insufficient for irrigation districts to cover their operating costs.

10.5 Prospects for Irrigation Water

While reforms that unify water management authority have helped to allocate water more rationally among users, the formal extension of water rights may provide for even more effective water allocation. A workable system of water rights, however, also requires sound legal institutions to enforce contracts and resolve conflicts. Presently, the transfer of water licences or water-use rights is technically prohibited in China because all water is state-owned property (although water transfers do happen under certain circumstances, as indicated in the examples above). With rising water shortages and the need to allocate water more rationally, the MWR is considering modifications to the law that will formally permit transfers of water rights. Following through with reforms that establish more secure rights, and making these rights tradable, will further increase the flexibility and rationality of water allocation in China, and may even increase rural incomes and hasten the development of rural areas. The efficacy of water markets and a system of water rights, however, will depend heavily on establishing a transparent and independent legal system to enforce contracts and resolve disputes. In addition, maintaining effective infrastructure, via significant and well-targeted investment, will also improve the functioning of water markets (Huang et al., 2006).

The objectives of future water pricing reform are to promote high efficiency of water use and to realize water saving and proper allocation of water resources, to guarantee steady and sustainable development of water supply projects. The specific measures are as follows (Webber et al., 2008):

1. Strict execution of principles for determination of water tariff, i.e., pricing according to supply cost for grain crops and supply cost plus minor profit for cash crops, thus ensuring financing of operation, maintenance and rehabilitation of irrigation projects.
2. Adopting basic pricing and water volume pricing. Basic pricing is to ensure normal operation and maintenance in any operation conditions, water volume pricing is to set tariff based on the water supplied, which is aiming at covering the depreciation and overhaul cost.
3. Set water quota and supply water according to it, higher pricing for any water using exceeding the quota.
4. Adopting different tariffs for different seasons or fluctuating pricing based on relationship of supply and demand. Higher pricing when supply is smaller than demand, and lower pricing when supply is larger than demand.
5. Properly decentralizing pricing approval authority. Pricing authority stipulated in Water Tariff Method is too concentrated and adjustment procedure is too complicated. The power for water pricing and adjustment should be decentralized to County or above Pricing Department according to jurisdiction of projects.

10.6 Conclusions

China has successfully harnessed its limited water resources to achieve remarkable gains in agricultural and industrial production, but in important agricultural areas of northern China, the exploitation of existing water resources has gone beyond sustainable levels. Policymakers in China, however, are responding to this situation to avert a more serious water crisis in the future. At all levels of the water management system, policies and institutions to encourage better water management and water conservation are being established. These trends are encouraging, yet it is still unclear whether China can adapt to a world where water is relatively scarce, while maintaining levels of agricultural production and increasing industrial production. More thorough and rigorous research is needed to answer some of the salient questions regarding these policy changes, the potential they hold for inducing water conservation and the effects they will have on China's agricultural production. Improving the storage and delivery capacity of irrigation systems will improve the performance of these systems and could affect agricultural production in several ways. As part of a national campaign to increase infrastructure investment overall, China has dramatically increased national investment in water conservancy in the past few years and plans to continue such levels of investment. To ascertain how these improvements will affect agriculture, however, depends on several unknown relationships. Among the most important is to better understand how effective the increased investment dollars have been at improving surface-water storage and conveyance infrastructure and the extent to which these investments improve the reliability of surface-water deliveries, especially at the ends of the water delivery systems. Researchable questions include: How are these investments allocated? Do they go to the most water-stressed or least efficient systems? Another important relationship to understand is how more reliable surface-water systems will help reduce farmers' reliance on ground water and decrease their groundwater withdrawals, and also the extent to which increased reliability encourages adoption of water-saving irrigation practices or other water conservation efforts. In addition, understanding how these changes in upstream irrigation districts affect downstream users will also be critical to understanding the overall effect these investments will have on the hydrological system and China's economy. China has also established a variety of institutional responses intended to solve the problem of conflicts between users. Generally, these responses seek to increase the power of a higher level of the bureaucracy, to "internalize" the conflict. Most of these responses, however, are new, experimental and difficult to implement, and therefore still have much to prove before offering solutions to water problems in China. A better understanding of how such responses are adopted and implemented, and how both the losers and the winners of these changes are affected, will further our capacity to determine their ultimate success and how they will affect economic activity. A system of water rights along river basins is also being considered by water policymakers in China to resolve conflicts between users. Understanding what preconditions are needed to implement a system of water rights, and how a system of water rights will affect water allocation and agricultural production, it will assist policymakers in their decisions that whether and how to establish a system of water rights in China. The incentives faced by farmers and local water managers to conserve water and how they go about adopting water conservation practices will be a fruitful area for further research. A wide variety of institutional responses have been established to encourage farmers and local leaders to adopt water-saving practices including reforming irrigation management, raising water prices and reforming water fee collection and investing in water-saving irrigation technology. Understanding how these institutions work, which type are more effective, what the determinants of adopting such measures are, and how they affect agricultural production are important questions that call for more rigorous research. The role of water prices, the adoption of water-saving irrigation practices and how these affect crop choice and yields will also play an important role.

References

Deng, X.-P., et al. (2006). "Improving agricultural water use efficiency in arid and semiarid areas of China." *Agricultural Water Management* **80**(1–3): 23–40.

Eslamian, S., and Zamani, N., (2021). Examples of rainwater harvesting and utilization in China, in *Handbook of Water Harvesting and Conservation, Vol. 2: Case Studies and Application Examples*, Ed. By Eslamian, S. & Eslamian, F., John Wiley & Sons, Inc., New Jersey, 283–290.

Huang, Q., Rozelle, S., Howitt, R., Wang, J., and Huang, J. (2006). *Irrigation Water Pricing Policy in China*, Chinese Academy of Sciences.

Jindong, C. and Jing, P.E.N.G. (2019). "Introduction of Beijing-Hangzhou Grand Canal and analysis of its heritage values." *Journal of Hydro-Environment Research* **26**: 2–7.

Liu, J., et al. (2013). "Water conservancy projects in China: Achievements, challenges and way forward." *Global Environmental Change* **23**(3): 633–643.

Lohmar, B., Wang, J., Rozelle, S., Huang, J., and Dawe, D. (2003). *China's Agricultural Water Policy Reforms: Increasing Investment, Resolving Conflicts, and Revising Incentives*. Market and Trade Economics Division, Economic Research Service, U.S. Department of Agriculture. Agriculture Information Bulletin Number 782.

Ministry of Water Resources (2009). *China Water Statistical Yearbook 2009*. China Statistics Press, Beijing.

Shen, D. J. W. P. (2015). "Groundwater management in China." *Water Policy* **17**(1): 61–82.

Tan, X., et al. (2019). *The Technical History of China's Grand Canal*, World Scientific, China Water and Wastewater Press, Beijing.

Wang, X.-J., et al. (2018). "The new concept of water resources management in China: ensuring water security in changing environment." *Environment, Development and Sustainability* **20**(2): 897–909.

Webber, M., et al. (2008). "Pricing China's irrigation water." *Global Environmental Change* **18**(4): 617–625.

Yao, L., et al. (2017). "China's water-saving irrigation management system: Policy, implementation, and challenge." *Sustainability* **9**(12): 2339.

Yu, X., et al. (2015). "A review of China's rural water management." *Sustainability* **7**(5): 5773–5792.

Zhang, W. and K. Zhou (2020). *Canal and Navigation Lock. Thirty Great Inventions of China*. Springer, Singapore, 213–269.

Zhu, X., et al. (2013). "Agricultural irrigation in China." *Journal of Soil and Water Conservation* **68**(6): 147A–154A.

IV

American
and European
Irrigation
Developments

IV

11

Irrigation Water Use in the United States

R. Deepa
Florida Agricultural and Mechanical University

S. Suchithra
New York University
Abu Dhabi

11.1 Introduction

Agriculture constitutes about 80% of the US consumptive water use from ground and surface water as a whole, while, in the western states, it is 90% (United States Department of Agriculture-Economic Research Service, USDA-ERS, 2020). Irrigated agriculture satisfies the world food demand but disturbs the natural hydrological cycle by increasing the crop water demand (Zohaib and Choi, 2020). Irrigation water use includes water that helps in regulating the crop physiological phases in agriculture and horticulture (USGS, 2020). Irrigated water also consists of pre-irrigation, frost protection, fertilizer application, pesticide application, and nutrient leaching from the root zone. The irrigation water withdrawals include those consumed as evapotranspiration from plants and the ground and the recharges from aquifers. Irrigation is an anthropogenic climate constraint that directs the surface water and energy balance that linearly relates to increase in soil moisture (Sacks et al. 2010; Zaussinger et al., 2019). The land surface gets cooled due to irrigation on local and regional scales through evapotranspiration increase. Climate has a significant bearing on the amount of irrigation water applied and the demand for irrigation over a region (Schlenker et al., 2005, 2007). Due to the rising world population and higher living standards, lack of water will significantly constrain agriculture in the coming decades. This is because natural water availability will get affected by the mean temperature and changing precipitation patterns (Kummu et al., 2016). Moreover, future predictions warn of the intensification of hydrological cycles that increase the frequency and intensity of floods and droughts, which further disrupt the agricultural water availability (Allan and Soden, 2008; Döll and Siebert, 2002). It is estimated that more than a billion people in the United States have no access to safe, affordable drinking water.

In this context, sustainable irrigated agriculture, i.e., producing more crop yield per drop, would be imposed by the environmental and water policymakers to provide more water for environmental uses and urban areas (Gleick, 2002; Rockström et al., 2007; Jury and Waux, 2005). Therefore, in the chapter,

DOI: 10.1201/9781003353928-15

an overview of irrigated water use in the conterminous United States and future water stress is explored with respect to future socio-economic pathways. The topics covered in this chapter include (i) groundwater use for irrigation, (ii) surface water use for irrigation, (iii) irrigation water use and climate change, (iv) Water Footprint (WFP) and irrigation water use, and (v) smart irrigation water use, and the final section concludes the chapter.

11.2 Irrigated Acreage and Water Use in the United States

The use of irrigated water in agriculture fluctuates across the United States (Figure 11.1). The western states show more significant irrigation withdrawals compared to the east. The consumptive use (water incorporated into crops, livestock, and human consumption) of withdrawn irrigation water is high in the state of California and Idaho. The water withdrawals for thermoelectric power are more in the eastern states Illinois, Alabama, Michigan, Florida, North Carolina, and New York. Industrial water use was higher in Indiana and Louisiana in 2015 as depicted in Figure 11.1. A significant increase in the irrigated area is observed in the whole United States (Figure 11.2, right panel) from 2012 to 2017. The change in irrigated acreage depends on the coupled ocean-atmospheric interannual variability (Elias et al., 2016), such as El Nino Southern Oscillation. According to 2018 Irrigation and Water Management Survey from US National Agricultural and Statistical Services (NASS), the total amount of irrigated water used in 2018 was 83.4 million acre-feet, a decrease in 5.8% compared to 2013. Grain, oil seed crops, nursery, greenhouse, and hay crops comprise a significant portion of irrigated farmland acres in the United States. The mode of irrigation in 2018 was through the sprinkler systems rather than gravity irrigation. The five states namely California, Nebraska, Arkansas, Texas, and Idaho constituted approximately 50% of the irrigated acres in 2018, and 56% of the total irrigation.

The irrigation system shifted according to the shift in water resources over time from gravity irrigation (35%–42%) to more efficient pressurized sprinkler and drip irrigation systems (58%–65%). In sprinkler and drip irrigation systems, applicators such as center-pivot, surface rip, sliding roll or wheels, and micro-sprinkler are used. Furrow, regulated, and uncontrolled floods are all examples of gravity

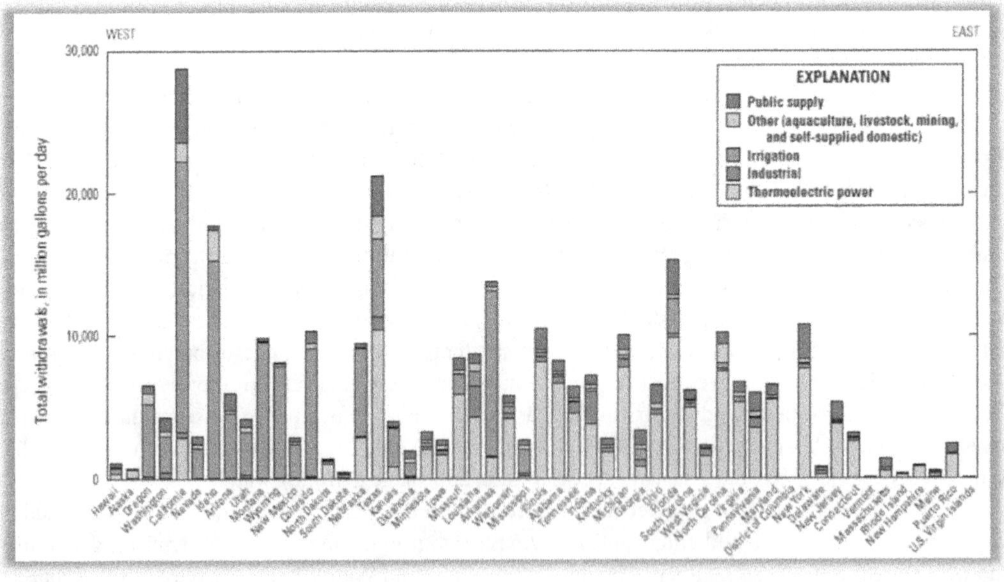

FIGURE 11.1   Bar chart representing the state-wise total water withdrawals from west to east in the United States in 2015 (Dieter et al., 2018, Maupin et al. 2014).

Top States — Irrigated Acreage and Water Use, 2018			
Irrigated Acres		**Water Applied (acre-feet)**	
	million		million / avg per acre
California	8.4	California	24.5 / 2.9
Nebraska	7.7	Idaho	6.6 / 1.9
Arkansas	4.2	Texas	5.3 / 1.3
Texas	4.1	Arkansas	5.1 / 1.2
Idaho	3.4	Nebraska	4.9 / 0.6
Colorado	2.5	Arizona	4.4 / 4.7
Kansas	2.4	Washington	4.1 / 2.2
Montana	2.1	Colorado	3.8 / 1.6
Washington	1.9	Oregon	2.7 / 1.7
Mississippi	1.7	Montana	2.5 / 1.2
U.S. Total	**55.9**	**U.S. Total**	**83.4 / 1.5**

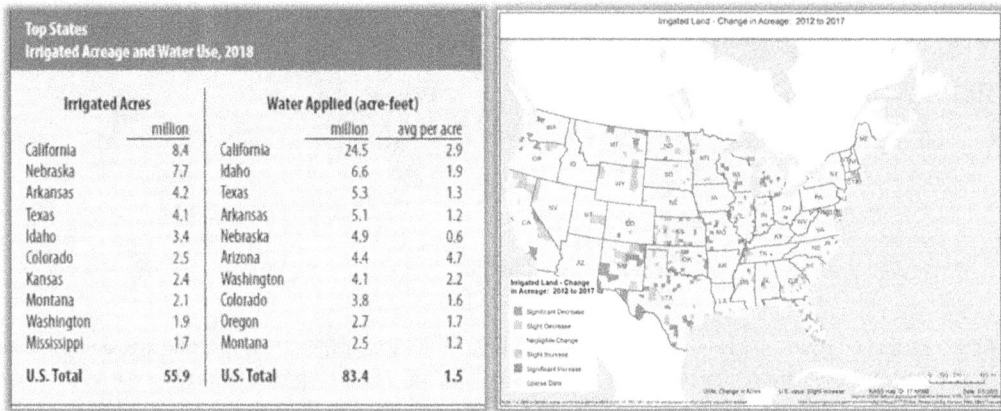

FIGURE 11.2 Irrigated acreage and water use (left panel) in the United States in 2018 and change in irrigated land acreage (right panel) (USDA, NASS).

irrigation methods (Stubbs, 2016). Artificial intelligence is becoming more prevalent and efficient in irrigation system scheduling using drones, sensor networks, and data analytics.

11.3 Groundwater Use for Irrigation

Groundwater is a definitive source of fresh water for irrigation in major agricultural production areas. In the United States, the primary use of groundwater is for irrigation (65%). The states Kansas (80%), Arkansas (60%), Mississippi (68%), Florida (64%), and Hawaii (63%) are most dependent on groundwater in terms of percentage of total freshwater withdrawals (Dieter et al. 2018). At the same time, California (16%), Arkansas (10%), Texas (10%), and Idaho (5%) are the most significant groundwater users with respect to all groundwater withdrawals (Mandler, 2017). Aquifers, the major source of groundwater, act as the primary buffers against drought (Siebert and Doll, 2007, Siebert et al., 2010) and crop production. The ease in groundwater availability makes it preferable over surface water for irrigation purposes. The utilization of groundwater for agriculture is a significant driver for declining freshwater ecosystems (Perkin et al., 2017). The groundwater withdrawal varies across the United States, including the aquifer recharges. The aquifer's recharge rate depends on the intensity and duration of precipitation, geology of the land surface, and land-cover/land-use patterns. In the semi-arid and arid regions, the groundwater withdrawals (intensive agriculture) can exceed aquifers' recharge, leading to depletion. The Ogallala aquifer in the United States is one of those aquifers, the single most significant source of groundwater that supports $35 billion in US market value of agricultural products (Scanlon et al., 2012). Even though the aquifer replenishes through rainfall, streams, and springs, the overuse of aquifer for irrigation pumping makes it challenging to recharge at a fast rate. Thus, to identify and classify the sources of water used for irrigation, databases are required to collect, analyze, and distribute on a global scale. Subsequently, an estimate of irrigated area with groundwater was provided by Shah et al. (2007) as 83–576 million ha. This estimate has certain disparities due to the differences in the definition of irrigated land (Siebert et al., 2010). Later, in 2010, Thenkabail et al. developed a global irrigation map based on land use, crop type, and water resources based on remote sensing. Some of the uncertainties for this method were coarse resolution of satellite data and errors in pixel classification. The methods in quantifying the differences between ground and surface water were not clearly said. A global information system on agriculture and water management by Food and Agriculture Organization (FAO), known as AQUASTAT (Siebert et al., 2013), provides a quantitative estimate of the global irrigated area (89 million ha) with groundwater (Burke, 2002). AQUASTAT disseminates information that pertains to water

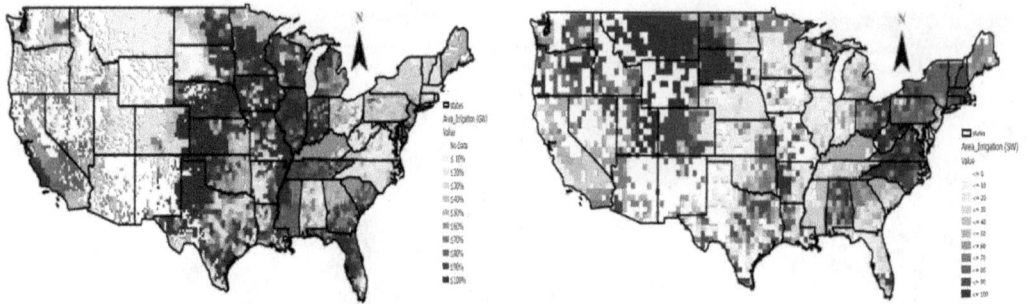

FIGURE 11.3 Left panel: Area equipped for irrigation with groundwater (percentage of total area equipped for irrigation); right panel: Area equipped for irrigation with surface water (figures prepared using FAO's AQUASTAT database).

resources, water use, and agricultural management, thereby setting the United Nations Sustainable Development Goal 6.4 on water stress and water use efficiency. Figure 11.3 represents the percentage of total area equipped for groundwater and surface water irrigation. This AQUASTAT database used in this work is a geospatial raster data set with a resolution of 5 minutes. The Ogallala aquifer region and the mid-west United States, California, Florida, South Georgia, and South Carolina, are heavily dependent on groundwater irrigation (Figure 11.3, left panel). Three types of water withdrawal are identified in AQUASTAT, namely agricultural (69%), municipal (12%), and industrial water withdrawal (19%).

11.4 Surface Water Use for Irrigation

Seventy percent of freshwater use in United States is dependent on surface water sources (USGS, 2015). Surface water resources include water in rivers, streams, lakes, reservoirs, and creeks essential to our everyday lives. Surface water uses include public uses, irrigation, thermoelectric power industry, etc. The major agricultural producing states, such as California, and the southern Great Plains, extract more surface water for crop irrigation and livestock production. Climate and anthropogenic activities affect the spatial and temporal distribution of surface water. Surface water bodies exhibit interannual variabilities across the contiguous United States (Zhou et al., 2018). The southwestern United States is a hotspot of water deficiencies identified from hydrological model assessments and projections (Gaupp et al., 2015). Extensive water withdrawals for agriculture (Caldwell et al., 2012) and thermoelectric power plants (Melillo et al., 2014) aggravate the water deficiencies in these areas. The area of surface water bodies in the southeastern United States is approximately 1 ha/km² higher than the western half of the United States. Trends in water withdrawals in the United States during 1950–2015 have indicated that surface water withdrawals for irrigation were higher in 2015 compared to 2010 by 2%. Figure 11.3 (right panel) shows the percentage of area equipped for surface water irrigation in the United States. The Northwest and Northeast regions of the United States are the zones where surface water is used for irrigation on a considerable scale. The surface water bodies influence the land water storage (Proulx et al., 2013) and groundwater dynamics (Brunner, 2009). The interannual variability of surface water body area forces water users to extract groundwater frequently, decreasing land water storage.

11.5 Irrigation Water Use and Climate Change

Anthropogenic emissions of greenhouse gases alter average temperature and precipitation patterns (IPCC, 2007). There is concern about future agricultural irrigation requirements in some regions of the United States and the direct impacts of climate change on crop yield (Adams, 1990; Fischer et al., 2005). The changes in temperature and precipitation will favor irrigated crop production. Most climate change models predict an increase in irrigated acreage in the major irrigated regions. Another primary concern for greenhouse gas emission is the substantial increase in future irrigation water demand (Wada et al., 2013). Multimodel and multiclimate projections show an irrigation water demand increase of >20% by 2100 over North and Central America. A measure of water supply for irrigation, water yield is estimated to increase in the western United States in 2030. A decrease is predicted in the central and southeast regions with 1961–1990 as the base period. Also, it is projected to decrease in the Lower Mississippi and Texas Gulf basins which are driven by temperature and precipitation patterns (Rosenberg and Rosenberg, 2003). The 2095 scenario predicts a water yield of 39% in the Ohio basin, 57% in the western United States, and 76% in the Upper and Lower Colorado. For instance, a recent study by Duan et al. (2019) suggested that future changes in irrigation water use will aggravate the water stress in the east and lessen in the west. Local flow, upstream flow, and upstream water consumption also contribute to surface watersheds' water stress. The sectorial demand for water and supply gets affected by climate change. According to the National Climate Assessment report 2013, the southwest, southeast, and Great Plains in the United States are highly vulnerable to water shortages. There will be a likelihood of increasing trends in these areas. The projected water withdrawals under the A1B scenario from 2005 to 2060 from 12% to 41% and for the period 2005–2090, it is from 35% to 52% (Brown et al., 2013). For the A1 scenario, the projected withdrawals are from 67% to 103%, and for the B2 scenario, the rise is from 9% to 15%. These projections are based on the different socio-economic conditions and the baseline being 2005 withdrawal, 480 km³.

The need for a database on climate change sensitivity to water withdrawals and irrigation involving multiple water-use sectors, its efficiencies, associated drivers, and consumption use paved the way for developing Aqueduct Water Risk Atlas by World Resources Institute (Luck et al., 2015). It is made possible by an 'Aqueduct Alliance', a coalition of public and private sector undertakings at the cutting edge of water stewardship (Albaji et al., 2020). They used the Coupled Model Intercomparison Project Phase 5 and General Circulation Models to create decadal water stress forecasts of supply and demand. Under three combinations of climatic and socio-economic situations, Aqueduct contains indications of water availability, order, stress, and seasonal fluctuation. For the years 2020, 2030, and 2040, RCP 4.5/Shared Socio-Economic Pathways (SSP)2, RCP6.0/SSP2, and RCP8.5/SSP3 have been developed (Figure 11.4). SSPs are designed to create new scenarios for exploring future societal and economic circumstances with adaptation and mitigation challenges. It represents the allocation of mitigation and adaptation challenges in two-dimensional space (Fujimori et al., 2017; Vuuren et al., 2011). These projections are helpful to policy makers for decadal-scale planning, adaptation, and investment.

Land use is considered as one of the core drivers in developing the SSPs in the impact, adaptation, and vulnerability scenarios. The baseline scenario for SSP3 is a high deforestation rate and

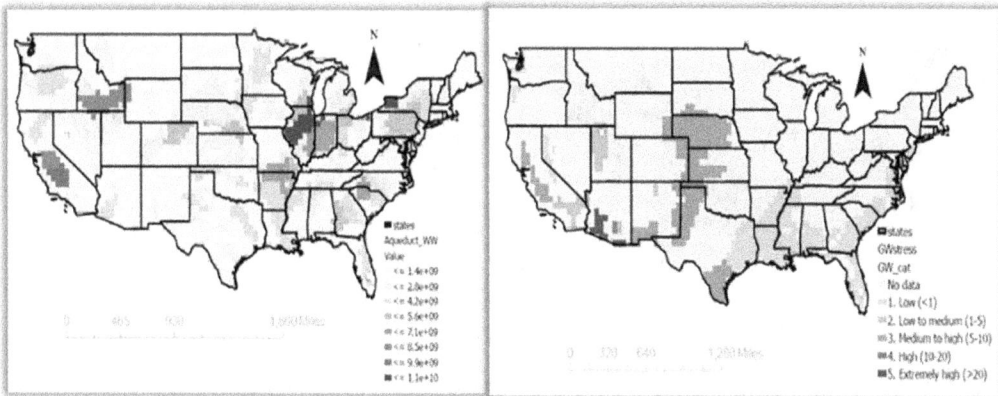

FIGURE 11.4 Left panel: Water withdrawal in the United States and right panel: groundwater stress from World Risk Atlas (WRI) Aqueduct database.

areal increase in cropland and pastureland compared to the SSP2 scenario. The corresponding water stress for the three time periods, 2020, 2030, and 2040 for the SSP2 and SSP3 pathways are shown in Figure 11.5. The dark regions represent a corresponding decrease in water stress, while the light areas decrease from the baseline. The SSP2 (RCP 4.5) scenario in Figure 11.5a and b has near-normal values of pressure, except the south Texas region. North Dakota, South Dakota, and Montana (Figure 11.5b) show a decrease lesser than usual for the 2020 timeline. The water stress projections for the 2030 timeline for the SSP2 depict an increase in water stress in the south Texas and southwest region of United States. The 2040 projections say the water stress will become still worse in the western part of the United States (Figure 11.5e and f). An exciting feature noticed from the projections is that the southeast remains near normal except Georgia. The increase in cropland during SSP2 and SSP3 creates more demand for irrigation and an increase in water stress. The future temporal variation of

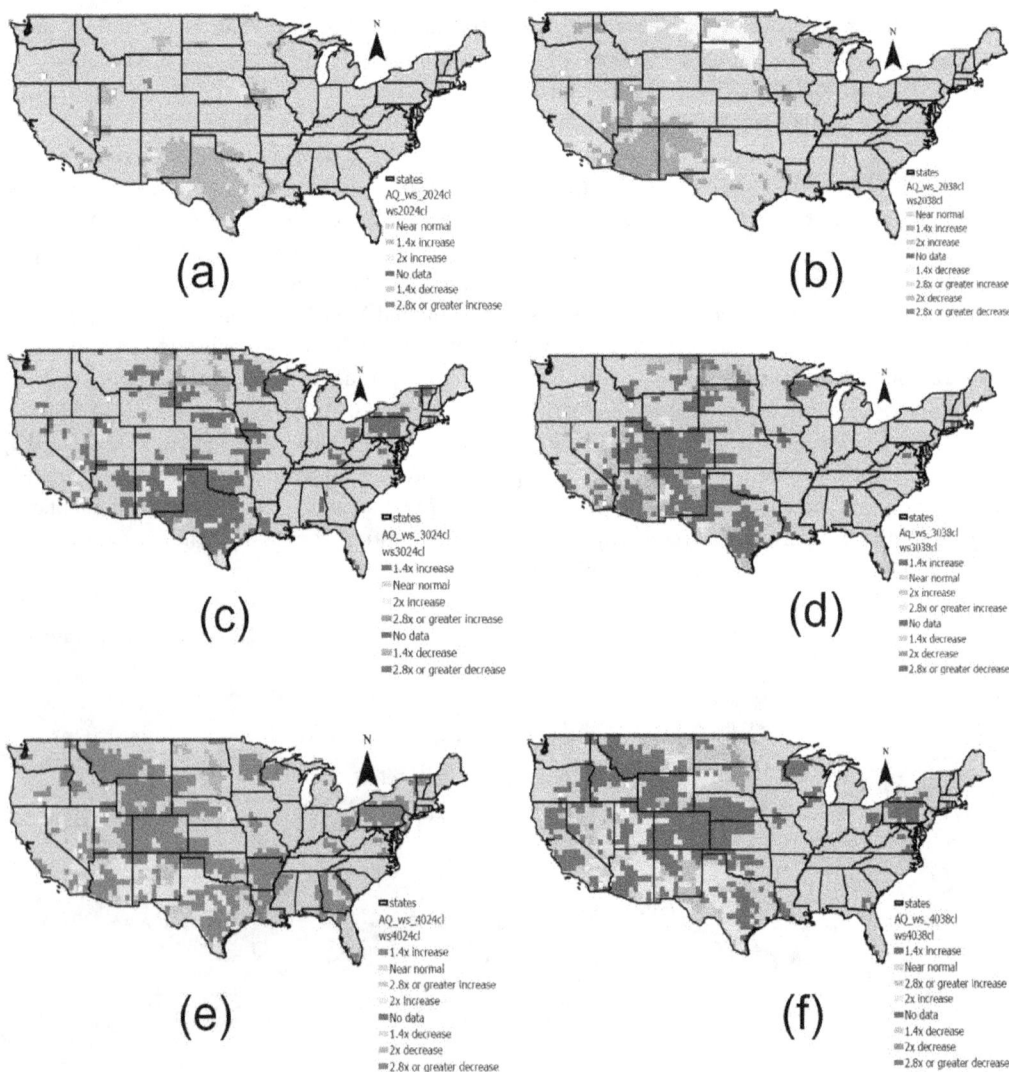

FIGURE 11.5 Projections of water stress during 2020, 2030, and 2040 for diverse societal and economic scenarios: (a) 2020 SSP2 RCP 4.5, (b) 2020 SSP3 RCP 8.5, (c) 2030 SSP2 RCP 4.5, (d) 2030 SSP3 RCP 8.5, (e) 2040 SSP2 RCP 4.5, and (f) 2040 SSP3 RCP 8.5 scenarios.

water stress projections indicates a northward shift in water stress as time progresses in the Ogallala aquifer region.

In 2030, the Texas region of the aquifer and New Mexico region depict an increase in water stress (Figure 11.5c) and slowly migrates north and westward in 2040. This will affect the agriculture and ecosystem services in this region.

11.6 Water Footprint and Irrigation Water Use

Globally, majority of water use is in the agricultural sector (Morillo et al., 2015), specifically irrigated agriculture. Arid regions receive irrigation more frequently, but at the same time supplemental irrigation is also required for humid regions to act as a buffer for rainfall availability (Rey et al., 2016; Sweet et al., 2017). It is diagnosed as a water-intensive sector, and therefore, sustainable use of water resources and management is essential. The concept of WFP was introduced by Hoekstra (2003) to devise such an approach. Lately, it is incorporated within the framework of Life Cycle Assessment (Bayart et al., 2010; Ferrandis, 2015). WFP is used as an indicator to measure the consumption of water resources by an individual, product, nation or country and its impacts on water quality. This process is referred to as WFP assessment (Hoekstra et al., 2011). The WFP has three components: green (rainwater), blue (irrigated water), and gray (polluted water). WFP accounting deals with four steps, namely defining the goal and scope of study, inventory/accounting, sustainability assessment, and response formulation. Alternately, WFP is used as a water-use efficiency indicator. Recent studies have shown that WFP accounting in conjunction with irrigation indicators serves as a tool for categorizing irrigation hotspots. Relative Irrigation Supply (RIS) is defined as the ratio of total annual volume of water extracted to the total theoretical irrigation requirement of the crop per irrigation season (Levine, 1982). The universal acceptance of RIS as a performance indicator (Malano and Burton, 2001) made it a powerful tool on farm scale and irrigation district scales (Rodríguez-Díaz et al., 2008) and a regional sustainable development base (Geng et al., 2014). The variability in climate, soil and water resources availability creates challenges for implementing irrigation technologies.

11.7 Smart Irrigation Water Use

The major components of agricultural production are irrigation and the nutrient application. Data science has transformed the smart farming method, paving the way for present-day agriculture by assisting farmers in making better decisions. The use of artificial intelligence and machine learning techniques in data-driven agriculture paved the way for long-term water resource management (Jha et al., 2019). The site-specific or precision irrigation water application within the agricultural field allows water-use efficiency and reduces water usage. Precision agriculture deals with splitting the field into small plots for augmented production (Mckinion et al., 2001). It began in the early 1980s and depends on spatial information, information technology, and better decision tools (Neupane and Guo, 2019). Geographic Information Systems (GIS), remote sensing, Global Navigation System Satellites, and variable-rate technology are some of the major technological advancement tools in precision agriculture. Of these, variable-rate irrigation technology helps to apply irrigation water in a timely and accurate manner considering the temporal variations in soil moisture and demand at different phenological phases of plant growth (Al-Kufaishi et al., 2006). Recent decades have witnessed the terms for data-driven agriculture such as digital farming and smart farming, through which young farmers can transform the existing land to sophisticated farms to meet the global demands of the agri-food chain (Nierenberg et al., 2019). For example, across 1,200 million ha and 250,000 farms in the United States, 10%–5% of farmers use Internet of things (IoT). IoT uses sensors and other instruments that translate the farming into data, and also increases agricultural productivity by 70% in the future (Sarni et al., 2016; Alpha Brown, 2018). In addition to IoT, big data analytics, robotics, and artificial intelligence also become popular. The beginning of robotics in agriculture increased the productivity manifold and also led to the reduction in farm costs (Reddy, 2016). Solving problems in agriculture using robotics and machine learning require higher costs for implementation stopping farmers from adopting those methods. Irrigation control optimization methods based on machine learning are in practice that train a model which control sprinklers to save wastage of water and pollution of water resources (Murthy et al., 2019). Machine learning is used for water management, plant growth, and soil management from an irrigation perspective. Two types of learning are supervised and unsupervised. Supervised learning utilizes labeled data as input while unsupervised learning uses unlabeled data as input (Janani and Jebakumar, 2019). Using Artificial Neural Network models that use different algorithms such as Levenberg–Marquardt and Black Propagation algorithm using MATLAB as platform for water and energy savings, recurrent Neural Network Algorithms, and Multivariate Vector Machine (Torres et al., 2011), Patil and Deka (2016) developed a smart irrigation decision support system based on partial least squares regression

TABLE 11.1　Software Applications Used in Crop Data Management and their Main Features in the United States

Software	Company	Headquarters	Relevant Features
ADAPT	AgGatekeeper	Washington DC, USA	Farm Management Information System (FMIS). Open-source
AGERmetrix	AGERpoint	Florida, USA	Platform for analyzing crop data (mobile and Light Detection and Ranging (LiDAR) mapping).
AgHub	GiSC	Texas, USA	Data collection and storage in a secure manner. IBM's Weather Operations, Main Street Data Validation, and Market Vision integration.
AgVerdict	AgVerdict (Wilbur-Ellis)	California, USA	Data integrity, decision-making, Variable Application Rate (VRA), soil sampling, and crop recommendations.
APEX™ JDLink	John Deere	Illinois, USA	Access to agricultural, equipment, and agronomic data is available through an online platform.
CASE IH AFS software	CASE IH	Wisconsin, USA	To produce yield or VR1 prescription maps, view, update, organize, analyze, and use precision farming data. Communicate maps in different formats.
Connected Farm	Trimble Agriculture	California, USA	Real-time input, access, and sharing of records (pictures, reports). Crop scouting, grid sampling, fleet management, and contracts are all integrated into one system. Farm Core platform for farm's operation.
Cropio	New Science Technologies	New York, USA	System for managing productivity. Land surveillance from afar. Field and crop conditions are updated in real time. Smartphone app and a web-based service.
ESE™ Agri solution	Source Trace	Massachusetts, USA	A unified and up-to-date database of farmers. Photographs, notes, activities, and the location of field visits should all be recorded. Farm to fork tracking. Farmer's unique ID.
Field View™	The Climate Corporation	California, USA	Data visualization and connection, crop performance analysis, and field health imaging.
Granular	DowDuPont	California, USA	Various software are available depending on the requirements. Models for decision-making should be created. Advisory and training services are available. 230 crop subspecies are supported. Cloud-based.
SMS	AgLeader	Iowa, USA	Soil samples, grids, and areas are all things to consider. Choosing better yield potential based on historical reports. Mobile app.

Source: Adapted from Saiz-Rubio and Rovira-Más (2020).

and adaptive neural fuzzy inference algorithms. GIS in irrigation management makes the data analysis and manipulation much easier and efficient.

11.8 Conclusions

The urge for more food production and the subsequent use of irrigated water for agriculture demand more sustainable options in the agri-food sector. This can be achieved through integrating technology, scientific knowledge, and agronomy by a thorough understanding of the environmental conditions required for crop growth and development. The dependency and accuracy of factors such as soil moisture and climate variables need to be considered for site-specific irrigation. GIS, along with remote sensing techniques, helps in mapping the seasonal changes in the phenological phases of plant growth. Spatial analytics and data science, combined with machine learning and artificial intelligence, provide future avenues for variable-rate irrigation applications. Sustainable irrigation and agriculture call for best management practices, decision support tools identifying the drivers of water stress, and corresponding adaptation strategies at the local, national, and international levels. Delineation of agricultural regions into water stress zones using water scarcity indicators and hotspot identification will be helpful to reduce the irrigation WFP. The future water stress scenarios project a near-normal condition in the Southeast United States and a decrease in the northern United States. The states Texas, New Mexico, Arizona, Utah, and Colorado will be competing for water in the SSP2 and SSP3 scenarios. Therefore, the additional irrigation requirement for agriculture in those states will also be higher. Adaptation strategies need to be implemented to cope with future scenarios. Switching to crop varieties that require less water will be a good option in future water stress scenarios and to meet the considerable demand to feed the millions.

References

Adams, W. M. (1990). How beautiful is small? Scale, control and success in Kenyan irrigation. *World Development*, 18(10), 1309–1323.

Al-Kufaishi, S. A., Blackmore, B. S., & Sourell, H. (2006). The feasibility of using variable rate water application under a central pivot irrigation system. *Irrigation and Drainage Systems*, 20(2–3), 317–327.

Allan, R. P., & Soden, B. J. (2008). Atmospheric warming and the amplification of precipitation extremes. *Science*, 321(5895), 1481–1484.

Alpha Brown (2018). https://www.alphabrown.com/single-post/2018/04/02/How-the-Internet-of-things-in-Agriculture-impact-on-the-African-economies. Accessed on 8 May 2020.

Bayart, J. B., Bulle, C., Deschênes, L., Margni, M., Pfister, S., Vince, F., & Koehler, A. (2010). A framework for assessing off-stream freshwater use in LCA. *The International Journal of Life Cycle Assessment*, 15(5), 439–453.

Brown, T. C., Foti, R., & Ramirez, J. A. (2013). Projected freshwater withdrawals in the United States under a changing climate. *Water Resources Research*, 49(3), 1259–1276.

Brunner, G. (2009). Near and supercritical water. Part II: Oxidative processes. *The Journal of Supercritical Fluids*, 47(3), 382–390.

Burke, J. J. (2002). Groundwater for irrigation: Productivity gains and the need to manage hydro-environmental risk. Intensive Use of Groundwater Challenges and Opportunities, 478, United Nations Publication ST/ESA/205.

Caldwell, P. V., Sun, G., McNulty, S. G., Cohen, E. C., & Myers, J. M. (2012). Impacts of impervious cover, water withdrawals, and climate change on river flows in the conterminous US. *Hydrology and Earth System Sciences*, 16, 2839–2857.

Dieter, C. A., Maupin, M. A., Caldwell, R. R., Harris, M. A., Ivahnenko, T. I., Lovelace, J. K., Barber, N. L., & Linsey, K. S. (2018). Estimated use of water in the United States in 2015, 1441, US Geological Survey, USA.

Döll, P., & Siebert, S. (2002). Global modeling of irrigation water requirements. *Water Resources Research*, 38(4), 8–1.

Duan, K., Caldwell, P. V., Sun, G., McNulty, S. G., Zhang, Y., Shuster, E., Liu, B., & Bolstad, P. V. (2019). Understanding the role of regional water connectivity in mitigating climate change impacts on surface water supply stress in the United States. *Journal of Hydrology*, 570, 80–95.

Elias, E., Rango, A., Smith, R., Maxwell, C., Steele, C., & Havstad, K. (2016). Climate change, agriculture and water resources in the Southwestern United States. *Journal of Contemporary Water Research and Education*, 158(1), 46–61.

Albaji, M., Eslamian, S., Naseri, A. & F. Eslamian, 2020, Handbook of Irrigation System Selection for Semi-Arid Regions, Taylor and Francis, CRC Group, USA, 317 Pages.

Ferrandis, C. (2015) Environmental Management. Water Footprint. Principles, Requirements and Guidelines, ISO 14046.

Fischer, G., Shah, M., Tubiello, F. N., & Van Velhuizen, H. (2005). Socio-economic and climate change impacts on agriculture: An integrated assessment, 1990–2080. *Philosophical Transactions of the Royal Society B: Biological Sciences*, 360(1463), 2067–2083.

Fujimori, S., Hasegawa, T., Masui, T., Takahashi, K., Herran, D. S., Dai, H.,… & Kainuma, M. (2017). SSP3: AIM implementation of shared socioeconomic pathways. *Global Environmental Change*, 42, 268–283.

Gaupp, F., Hall, J., & Dadson, S. (2015). The role of storage capacity in coping with intra-and inter-annual water variability in large river basins. *Environmental Research Letters*, 10(12), 125001.

Geng, Q., Wu, P., Zhao, X., & Wang, Y. (2014). A framework of indicator system for zoning of agricultural water and land resources utilization: A case study of Bayan Nur, Inner Mongolia. *Ecological Indicators*, 40, 43–50.

Gleick, P. H. (2002). Water management: Soft water paths. *Nature*, 418(6896), 373–373.

Hoekstra, A. Y. (2003). Virtual water trade: A quantification of virtual water flows between nations in relation to international crop trade. In *Proceedings of the International Expert Meeting on Virtual Water Trade 12, Delft, the Netherlands* (pp. 25–47).

Hoekstra, A. Y., Chapagain, A. K., Mekonnen, M. M., & Aldaya, M. M. (2011). *The Water Footprint Assessment Manual: Setting the global Standard*. Routledge Earthscan, London (UK).

IPCC. (2007). *The Physical Science Basis. Contribution of Working Group I to the Fourth Assessment report of the Intergovernmental Panel on Climate Change* (p. 996). Cambridge University Press, Cambridge, United Kingdom and New York.

Janani, M., & Jebakumar, R. (2019) A study on smart irrigation using machine learning. *Cell & Cellular Life Sciences Journal*, 4(2), 1–8.

Jha, R. N., Ansari, M. S., & Thakur, M. (2019). Efficiency evaluation of different agricultural machinery in rice cultivation at Agricultural Machinery Testing and Research Centre (AMTRC), Nawalpur, Sarlahi, Nepal. *International Journal of Bio-resource and Stress Management*, 10(6), 621–627.

Jury, W. A., & Vaux, H. (2005). The role of science in solving the world's emerging water problems. *Proceedings of the National Academy of Sciences*, 102(44), 15715–15720.

Kummu, M., Guillaume, J. H., de Moel, H., Eisner, S., Flörke, M., Porkka, M., Siebert, S., Veldkamp, T. I. E., & Ward, P. J. (2016). The world's road to water scarcity: Shortage and stress in the 20th century and pathways towards sustainability. *Scientific Reports*, 6, 38495.

Levine, G. (1982). Relative water supply: An explanatory variable for irrigation systems. Technical Report No. 6. Cornell University, USA.

Luck, M., Landis, M., & Gassert, F. (2015). *Aqueduct Water Stress Projections: Decadal Projections of Water Supply and Demand Using CMIP5 GCMs*. World Resources Institute, Washington DC.

Malano, H. M., & Burton, M. (2001). Guidelines for benchmarking performance in the irrigation and drainage sector (No. 5). Food and Agriculture Organization, Italy.

Mandler, Ben. 2017. *Groundwater Use in the United States*. American Geosciences Institute. March 9, 2017. https://www.americangeosciences.org/geoscience-currents/groundwater-use-united-states.

Maupin, M.A., Kenny, J.F., Hutson, S.S., Lovelace, J.K., Barber, N.L., & Linsey, K.S., (2014). Estimated use of water in the United States in 2010: U.S. Geological Survey Circular 1405, 56 p., http://dx.doi.org/10.3133/cir1405.

McKinion, J. M., Jenkins, J. N., Akins, D., Turner, S. B., Willers, J. L., Jallas, E., & Whisler, F. D. (2001). Analysis of a precision agriculture approach to cotton production. *Computers and Electronics in Agriculture*, 32(3), 213–228.

Melillo, J. M., Richmond, T. T., & Yohe, G. (2014). *Climate Change Impacts in the United States: Third National Climate Assessment*. US Global Change Research Program, Washington, DC.

Morillo, J. G., Díaz, J. A. R., Camacho, E., & Montesinos, P. (2015). Linking water footprint accounting with irrigation management in high value crops. *Journal of Cleaner Production*, 87, 594–602.

Murthy, A., Green, C., Stoleru, R., Bhunia, S., Swanson, C., & Chaspari, T. (2019). Machine Learning-based Irrigation Control Optimization. In *Proceedings of the 6th ACM International Conference on Systems for Energy-Efficient Buildings, Cities, and Transportation*, pp. 213–222.

Neupane, J., & Guo, W. (2019). Agronomic basis and strategies for precision water management: A review. *Agronomy*, 9(2), 87.

Nierenberg, D., Powers, A., & Papazoglakis, S. 2019. Data-driven nutrition in the digital age. *Sight and Life*, 33(1), 2019.

Patil, A. P., & Deka, P. C. (2016). An extreme learning machine approach for modeling evapotranspiration using extrinsic inputs. *Computers and Electronics in Agriculture*, 121, 385–392.

Perkin, J. S., Gido, K. B., Falke, J. A., Fausch, K. D., Crockett, H., Johnson, E. R., & Sanderson, J. (2017). Groundwater declines are linked to changes in Great Plains stream fish assemblages. *Proceedings of the National Academy of Sciences*, 114(28), 7373–7378.

Proulx, R. A., Knudson, M. D., Kirilenko, A., VanLooy, J. A., & Zhang, X. (2013). Significance of surface water in the terrestrial water budget: A case study in the Prairie Coteau using GRACE, GLDAS, Landsat, and groundwater well data. *Water Resources Research*, 49(9), 5756–5764.

Reddy, B. S. (2016). Soil fertility management in semiarid regions: The sociocultural, economic and livelihood dimensions of farmers' practices—A case of Andhra Pradesh. In Ghosh, N., Mukhopadhyay, P., Shah, A., Panda, M. (eds) *Nature, Economy and Society*. (pp. 195–223). Springer, New Delhi. https://doi.org/10.1007/978-81-322-2404-4_10.

Rey, D., Holman, I. P., Daccache, A., Morris, J., Weatherhead, E. K., & Knox, J. W. (2016). Modelling and mapping the economic value of supplemental irrigation in a humid climate. *Agricultural Water Management*, 173, 13–22.

Rockström, J., Lannerstad, M., & Falkenmark, M. (2007). Assessing the water challenge of a new green revolution in developing countries. *Proceedings of the National Academy of Sciences*, 104(15), 6253–6260.

Rodríguez-Díaz, J. A., Camacho-Poyato, E., Lopez-Luque, R., & Pérez-Urrestarazu, L. (2008). Benchmarking and multivariate data analysis techniques for improving the efficiency of irrigation districts: An application in Spain. *Agricultural Systems*, 96(1–3), 250–259.

Rosenberg, G., & Rosenberg, A. (2013). U.S. Patent No. 8,496,193. Washington, DC, USA: U.S. Patent and Trademark Office.

Sacks, W. J., Deryng, D., Foley, J. A., & Ramankutty, N. (2010). Crop planting dates: An analysis of global patterns. *Global Ecology and Biogeography*, 19(5), 607–620.

Saiz-Rubio, V., & Rovira-Más, F. (2020). From smart farming towards agriculture 5.0: A review on crop data management. *Agronomy*, 10(2), 207.

Sarni, W., Mariani, J., & Kaji, J. (2016). From dirt to data the second green revolution and the internet of things. *Deloitte Review*, 18, 4–19.

Scanlon, B. R., Faunt, C. C., Longuevergne, L., Reedy, R. C., Alley, W. M., McGuire, V. L., & McMahon, P. B. (2012). Groundwater depletion and sustainability of irrigation in the US High Plains and Central Valley. *Proceedings of the National Academy of Sciences*, 109(24), 9320–9325.

Schlenker, W., Hanemann, W. M., & Fisher, A. C. (2005). Will US agriculture really benefit from global warming? Accounting for irrigation in the hedonic approach. *American Economic Review*, 95(1), 395–406.

Schlenker, W., Hanemann, W. M., & Fisher, A. C. (2007). Water availability, degree days, and the potential impact of climate change on irrigated agriculture in California. *Climatic Change*, 81(1), 19–38.

Siebert, S., & Döll, P. (2007). 2.4 Irrigation water use–A global perspective. *Central Asia*, 14, 10–2.

Siebert, S., Burke, J., Faures, J. M., Frenken, K., Hoogeveen, J., Döll, P., & Portmann, F. T. (2010). Groundwater use for irrigation—a global inventory. *Hydrology and Earth System Sciences*, 14(10), 1863–1880.

Siebert, S., Henrich, V., Frenken, K., & Burke, J. (2013). *Global Map of Irrigation Areas Version 5*. Rheinische Friedrich-Wilhelms-University/Food and Agriculture Organization of the United Nations, Bonn, Germany/Rome, Italy, 2, 1299–1327.

Stubbs, M. (2016). *Irrigation in US Agriculture: On-Farm Technologies and Best Management Practices*. Congressional Research Service, Washington, DC.

Sweet, S. K., Wolfe, D. W., DeGaetano, A., & Benner, R. (2017). Anatomy of the 2016 drought in the Northeastern United States: Implications for agriculture and water resources in humid climates. *Agricultural and Forest Meteorology*, 247, 571–581.

Thenkabail, P. S., Hanjra, M. A., Dheeravath, V., & Gumma, M. (2010). A holistic view of global croplands and their water use for ensuring global food security in the 21st century through advanced remote sensing and non-remote sensing approaches. *Remote Sensing*, 2(1), 211–261.

Torres, A. F., Walker, W. R., & McKee, M. (2011). Forecasting daily potential evapotranspiration using machine learning and limited climatic data. *Agricultural Water Management*, 98(4), 553–562.

USDA, ERS (2020). https://www.ers.usda.gov/. Accessed on 8 May 2020.

USGS, (2020). https://www.usgs.gov/mission-areas/water-resources/science/irrigation-water-use?qt-science_center_objects=0#qt-science_center_objects. Accessed on 8 May 2020.

Van Vuuren, D. P., Edmonds, J., Kainuma, M., Riahi, K., Thomson, A., Hibbard, K., Hurt, G. C., Kram, T., Kreym V., Lamarque, F. J., Masui, T., Meinshausen, M., Nakicenovic, N., Smith, J. S., & Rose, K. S. (2011). The representative concentration pathways: An overview. *Climatic Change*, 109(1–2), 5.

Wada, Y., Wisser, D., Eisner, S., Flörke, M., Gerten, D., Haddeland, I., Hanasaki, N., Masaki, Y., Portmann, F. T., Stacke, T., Tessler, Z., & Schewe, J. (2013). Multimodel projections and uncertainties of irrigation water demand under climate change. *Geophysical Research Letters*, 40(17), 4626–4632.

Zaussinger, F., Dorigo, W., Gruber, A., Tarpanelli, A., Filippucci, P., & Brocca, L. (2019). Estimating irrigation water use over the contiguous United States by combining satellite and reanalysis soil moisture data. *Hydrology and Earth System Sciences*, 23(2), 897–923.

Zhou, L., He, J., Qi, Z., Dyck, M., Zou, Y., Zhang, T., & Feng, H. (2018). Effects of lateral spacing for drip irrigation and mulching on the distributions of soil water and nitrate, maize yield, and water use efficiency. *Agricultural Water Management*, 199, 190–200.

Zohaib, M., & Choi, M. (2020). Satellite-based global-scale irrigation water use and its contemporary trends. *Science of the Total Environment*, 714, 136719.

12

Irrigation Developments in Brazil

Timóteo Herculino
da Silva Barros,
Cassio Hamilton
Abreu-Junior,
Rubens Duarte
Coelho, and
Flávia Rosana
Barros da Silva
University of São Paulo

Jonathan Vasquez
Lizcano
*Corporación Colombiana de
Investigación Agropecuaria*

12.1 Introduction

The aim of this chapter is to present the past, present and future of the Brazilian irrigation areas based on historical events and on sustainability concepts and efficient use of water by all the actors involved in the agricultural business, as well as the use of remote sensing technologies.

Irrigation and agriculture in Brazil have developed in parallel throughout history, as both of them are interdependent and their economic effect can be considered as a synergy. This development would not have happened without the support of the federal government. An example of this is the long-term structured plan in the 1970s that encouraged the creation of Embrapa and other state research agencies, as well as some subsidies for rural credit and infrastructure projects to allow the transport of agricultural products, in addition to a national strategy for the integration of the Amazon. This milestone allowed Brazil to move from a food commodity importer to become an outstanding global agricultural exporter player.

At the beginning of this millennium, the commercialization of flex vehicles in Brazil helped the ethanol industry to boost ethanol waste water application through sprinkler irrigation equipment (travelers) in sugarcane areas. After that came the intensification of sustainable agriculture aiming to reduce greenhouse gas emissions and increasing carbon sequestration, and at the same time the focus on water use efficiency (WUE) (water productivity) in agriculture.

With irrigation equipment, the agricultural frontier in Brazil increased. The development that the country has experienced was possible due to the active participation of federal government, producers and agricultural researchers; this experience can be replicated by other countries that present a production system similar to Brazil.

DOI: 10.1201/9781003353928-16

In this chapter, a brief history of irrigation in Brazil will be presented. A description of the total potential irrigated area in Brazil based on future climate change scenarios is presented. Finally, advances on irrigation management based on private companies are discussed.

12.2 Agriculture in Brazil

Brazilian agriculture has undergone a revolution in recent decades, elevating it as the largest producer of food in tropical areas in the world. In the 1940s, the country was a food importer and an exporter of a few agricultural commodities (Nehring, 2016). At that time, Brazilian agriculture had rudimentary aspects regarding production (Mueller and Mueller, 2016). The soybean was a novelty for most farmers in Brazil, with no supply for the national market, and still far below international trade (Mueller and Mueller, 2016). Between the 1950s and 1960s, manual labor prevailed in agricultural production; in numerical terms, less than 2% of rural properties had agricultural machinery (Embrapa, 2020). Today, Brazil is one of the world's largest producers and exporters of a variety of agricultural products, such as soybeans, corn, cotton, orange, coffee and beef (Hogeboom et al., 2018; Nehring, 2016).

In the 1970s, the country was an importer of food commodities and had an average international trade deficit of $1.8 billion/year (Garcia et al., 2017). To solve this imbalance, the federal government implemented a structured long-term plan, including subsidies for rural credit; investments in agricultural research through the creation of Embrapa and other state research agencies; a national plan for the integration of the Amazon; and infrastructure projects to allow the transport of agricultural products (Chaddad, 2015). Fifty years later, its agricultural sector accounts for more than 20% of the country's GDP (Stabile et al., 2019). As a result of impressive productivity gains, Brazil was able to achieve food security, real food prices decreased, households began to spend a decreasing share of their income on food and Brazil became one of the main agricultural producers and exporters in the world (Nehring, 2016; Stabile et al., 2019).

Environmentalists who led global efforts to transition to a low-carbon economy often note Brazil's long and unique experience in large-scale production of fuel ethanol as an alternative to gasoline (Santos et al., 2018).

Vehicles powered by ethanol in Brazil were another interesting option for Brazilian consumers at the end of 2003. Especially, due to the introduction of "flex" vehicles: those vehicles, capable of running on any combination of gasoline and hydrated ethanol (Anfavea, 2020). In 2014, bi-fuel light vehicles, which represented more than 50% of the national vehicle fleet, were responsible for more than 90% of passenger car sales, with ethanol being offered practically all around the nation (Costa and Henkin, 2016).

In addition, the use of ethanol derived from sugarcane as an alternative to gasoline can contribute to an approximate 13% reduction in GHG from the entire energy sector, when gases reabsorbed by carbon sequestration by sugarcane harvest are taken into consideration (Garrett et al., 2018; Santos et al., 2018). Studies have found that the use of ethanol instead of gasoline (or mixed fuel) has the potential to mitigate the release of toxic lead, carbon monoxide and sulfur pollutants (Cançado et al., 2006; Goldemberg et al., 2008).

The promotion of sustainable agricultural intensification in Brazil has the potential to generate benefits for mitigating the GEE emissions and increasing carbon sequestration. In fact, in recent decades, national agricultural development policies have associated increased productivity with increased sustainability, along with income generation (Sorrensen, 2009). In this sense, participatory programs, such as the Sustainable Rural Development Programs in Brazil, have been successful in driving the transition from degraded rural areas to sustainable production systems, providing knowledge, technical assistance and incentives for small farmers to plan, implement and monitor sustainable interventions (Finger and El Benni, 2013).

Soil plays a key role in climate regulation, as it can be a source or a sink for atmospheric carbon dioxide, depending on the practices to which the soil is subjected. (Severo Santos and Naval, 2020). So, the soil acts as an important compartment of temporary carbon storage, depending on its quality and capacity to support biomass production (Minasny et al., 2017). Other sources of reduction of the negative impact of carbon are related to appropriate management practices in crops of greater expression in the country, such as sugarcane, soybeans, corn, and cotton (Coelho et al., 2019; Perin et al., 2019; Ribeiro et al., 2017).

Biofuel from 1 ha of sugarcane plantation has the potential to avoid the emission of about 14 Mg CO_2 eq/year in relation to the use of fossil fuels (Betts, 2011). When compared to other vegetable raw materials of ethanol such as corn, sugarcane is the most diligent in mitigating GHG emissions (Bordonal et al., 2018; Severo Santos and Naval, 2020). As a result, global demand for sugarcane ethanol is increasing year after year (Minasny et al., 2017). In Brazil, which is considered the world's largest producer of sugarcane ethanol, the area cultivated with sugarcane is expanding and the most widespread scenario of land-use change is the conversion of pastures into cane fields (Bordonal et al., 2018).

The production estimate for total ethanol, which includes the sum of ethanol made from sugarcane and also from corn, is expected to reach 35.5 billion L and only the derivative of cane is 33.8 billion L, showing an increase of 7.2% in relation to the last harvest 2018/2019 (Conab, 2020). The production of ethanol from sugarcane and corn in the Center-South Region registered a new record of growth in the 2019/20 season, surpassing by 7.3% the 30.99 billion liters produced in the last harvest, months before the official end of the current cycle, which will end in March (Conab, 2020). The great advantage of integrating ethanol production in these corn-producing areas is the possibility of decreasing the supply of corn on the market in high-yield crops, which ensures greater financial gain for rural producers, with the possibility of storing ethanol.

With the world in full growth, it is estimated that by 2030 the world population will exceed 8.5 billion people and that most of this population growth will occur in countries like China, India and Indonesia (Ndraha et al., 2017). In general terms, the increase in the purchasing power of the world society together with the changes in food preferences and patterns has caused Brazil to position itself as one of the main suppliers and providers of food in the world (Bonan and Levis, 2006; Huang et al., 2017).

12.3 Irrigation History in Brazil

Irrigation corresponds to agricultural practice that supplies water to plants in a full or partial way (ANA, 2017a). Irrigation is part of our daily lives, whether on the lawns of soccer fields and residential condominiums or when we consume rice, beans, vegetables, fruits, biofuels and vegetables— foods produced largely under irrigation (ANA, 2017b; Testezlaf, 2017).

According to Wisser et al. (2010) and Garcia et al. (2017), irrigation is a way of supplying water artificially to crops, in order to maintain their evapotranspiration (ET) rate and thus meet the global demand for food, providing better growing conditions and higher average crop yields (Alcamo et al., 2003; Zohaib et al., 2019). Exercised since the remote civilizations that developed in arid regions, large-scale irrigation is, however, a recent practice in humid regions, that is, where the amount and the average spatial and temporal distribution of the rains are able to supply crop water needs (ANA, 2017a).

In Brazil, irrigation started between the end of the 19th century and the beginning of the 20th century in the rice fields of Rio Grande do Sul, having been consolidated as an important irrigation hub since then. The start of the operation of the Cadro reservoir in 1903, whose construction started in 1881(ANA, 2017a), was an important milestone in this process. It also registered the occurrence of specific irrigation initiatives in the Semi-Arid areas in this initial phase, in particular with the construction of public warehouses for multiple uses (ANA, 2017a).

The National Water Agency reported the main historical milestones for the development of irrigated agriculture in Brazil, in which the following stand out: 1903, beginning of the operation of the Cadro reservoir for rice irrigation in the Rio Grande do Sul; 1909, creation of the Inspectorate of Works Against

Drought (IOCS), called the Federal Inspectorate of Works Against Drought (IFOCS) in 1919, transformed into DNOCS in 1945; and 1926, creation of the Rio Grande do Sul Rice Union. It gave rise to the IRGA in 1940; already in 1934, approval of the Water Code (Federal Decree no. 24.643/1934), which outlines guidelines that allow the public authorities to control and encourage the industrial use of water. The first version of this code appeared in 1907; 1948, creation of the São Francisco Valley Commission, called the São Francisco Valley Superintendence in 1967, transformed into CODEVASF in 1975.

In 1970, the National Integration Program (PIN) was created, a government program instituted by Decree-Law No. 1,106, of June 16, 1970, during the government of General Emílio Garrastazu Médici. It aimed to implement economic and social infrastructure works in the North and Northeast of the country (FGV, 2020). Five years later, the São Francisco Valley Development Company—CODEVASP was created, which started to operate in irrigation programs, in support of production and commercialization, in the reinforcement of the socio-economic infrastructure.

At the end of the 1970s, the National Irrigation Policy was approved for the first time (Federal Law No. 6,662/1979), Art 1- The National Irrigation Policy aims at the rational use of water and soil resources for the implantation and development of irrigated agriculture, with the following basic assumptions: (i) preeminence of the social function and public utility of the use of water and irrigable soils; (ii) encouragement and greater security for agricultural activities, primarily in regions subject to adverse climatic conditions; and (iii) promotion of conditions that can increase agricultural production and productivity, later in 2013 the policy was revoked by Law No. 12,787 of 2013).

In 1981, the *National Program for the Rational Use of Irrigable Lowlands* (PROVÁRZEAS) was created, with the aim of promoting the rational and gradual use of national lowland areas at the level of rural property, taking a considerable step toward irrigated agricultural activities in the country; in 1982, the Irrigation Equipment Financing Program (PROFIR) was presented; in 1986, the National Irrigation Program (PRONI) and the Northeast Irrigation Program (PROINE) were created; in 1988, when the Constitution of the Federative Republic of Brazil was announced, it dealt with some articles on the use of water resources and irrigation.

In the late 1990s, a new Water Law was imposed (Federal Law No. 9,433/1997); establishing the *National Water Resources Policy* in 1997, creating instruments for the management of water resources in the federal domain (those that cross more than one state or border) and the National Water Resources Management System (SINGREH); and in 2000, the National Water Agency (ANA) was created—Federal Law No. 9,984/2000; federal entity implementing the National Water Resources Policy and coordinating the National Water Resources Management System.

In 2001, CONAMA Resolution 284, dated 08/30/01, which provides for the environmental licensing of irrigation projects, was approved. In this resolution, irrigation projects were classified into several categories, according to the effective size of the irrigated area, by property, individual and the irrigation method used, such as sprinkling, localized and superficial; in the year 2008, the Permanent Forum for the Development of Irrigated Agriculture was created by Decree 1,869/2008, by the Minister of State for National Integration as an instance of exchange, articulation and diffusion of knowledge, experiences and cooperation for the discovery of solutions, constituted by a network of specialists and Brazilian institutions and aimed to contribute to the development of the political-institutional, technical and managerial capacity of its members, also operating as an instrument of national integration for the management of knowledge related to irrigated agriculture and agricultural drainage.

At the global level, irrigation is used as an essential tool to meet food security, since in the coming decades, population growth will occur particularly in populous, emerging and less developed countries (Batjes, 1996; Romano et al., 2011; Sato et al., 2007). This implies that these countries will be faced with the need to increase their food supply by increasing production in their own territory, perhaps in combination with the increase in imports, as is the case of several countries (Alexandratos, 1996; Chen, 2011, 2008; Huang et al., 2017; Ndraha et al., 2017).

In this way, irrigation has been supporting 40% of agricultural productivity worldwide using only 20% of the total arable land (Kumar et al., 2015; Puma and Cook, 2010). However, irrigation uses almost

70% of the freshwater available and is a critical anthropogenic intervention in the natural processes of the earth's surface (Wisser et al., 2010). These concerns are reported by several researchers (Ozdogan et al., 2010). For now, the main alternative solution to this concern in improving irrigation efficiency of water use worldwide (Finger and El Benni, 2013).

For many regions, the rapid expansion of irrigation during the 20th century significantly altered the hydrological cycle and the energy balance on the earth's surface, motivating research on the possible impacts that modern irrigation rates have on regional and global climate. These studies, and others, support the notion that current irrigation significantly alters the climate in some areas (Bonan and Levis, 2006; Collins et al., 2006; Lobell et al., 2009; Qiu et al., 2009), with the magnitude of the climate response depending on the spatial extent of irrigation and the degree to which a region's climate regime is linked to its land surface processes (Kumar et al., 2015; Wisser et al., 2010).

12.4 WUE in Brazilian Agriculture

Water productivity is defined by the relationship between organic production (dry matter produced) and the water consumed and not recoverable for this production; the process is usually evaluated in terms of ET, the sum of transpiration by culture (T) and soil evaporation (E) (Radhouane, 2008; Santos et al., 2019). The author assesses the production of biomass in relation to transpiration. Only transpired water is considered useful water because the evaporation of the soil does not result in the exchange of assimilated carbon. The productivity of water for biomass is defined as the dry matter of the aerial part produced (g or kg) per unit of transpired water (mm or m^{-3}). When using drip irrigation, it can be considered that the plant's ET is approximately equal to its transpiration, as evaporation losses are minimized (Barbosa et al., 2014; Liu et al., 2016; Mahdavi et al., 2017).

Water productivity is defined by plant physiologists as WUE; this term is used in infrared gas analyzer equipment (IRGA), which presents the output files based on this name. WUE is a concept introduced 100 years ago, showing a relationship between plant productivity and the use of water used. To facilitate understanding, WUA is defined as the amount of carbon assimilated as biomass or grain produced per unit of water used by a given crop of agricultural interest or not. One of the main questions asked is how plants will respond to climate change with changes in temperature, precipitation and CO_2 that affect the process (Hatfield and Dold, 2019). The concept is easy because the stomata on the leaf surfaces that absorb CO_2 for photosynthesis are the same pathways that allow water vapor to escape (Taiz and Zeiger, 2015).

The term efficiency of water use has no existing theoretical limits as a reference, as it should be for efficiency in the sense of engineering (maximum of 100%). On the other hand, the term efficiency is widely used in economics without referring to a maximum theoretical value (Kothari et al., 2019; Vu and Allen, 2009). However, the excessive use of the European Union (EU) has already caused some confusion and found objections in the scientific community, mainly because the meaning of the term depends on the specificities of the numerator and the denominator that define it. Therefore, both the term water productivity and WUE in this text are similar.

12.5 Potential for Expansion of Irrigated Agriculture

The analysis of the expansion potential of irrigated agriculture in Brazil brings together explanatory variables to point out areas that can be expanded. They tend to focus on physical and environmental aspects and need the application of robust economic models, as well as field research in order to provide a better understanding of this expansion. However, they provide perspectives and direction for both the private sector and public policies (ANA, 2017a).

Recently, the *Ministry of National Integration* published the study Territorial Analysis for the Development of Irrigated Agriculture in Brazil, which evaluated the irrigable area of the country using basins and microbasins as a territorial unit of analysis. The procedure for calculating the additional

TABLE 12.1 Indicators of Additional Irrigable Area by Region

Brazil/Region	Expansion Potential by Soil-Relief Suitability Class (×1.000 Ha)					Effective Potential (×1.000 Ha)	Effective Potential (%)
	High	Medium	Low	Total	Total (%)		
North	4.818	9.043	8.895	22.755	29.9%	679	6.1%
Northeast	1.881	3.481	3.551	8.912	11.7%	1.277	11.4%
Southeast	3.353	3.926	7.25	14.528	19.1%	3.318	29.6%
South	2.361	2.484	4.556	9.401	12.3%	2.313	20.7%
Midwest	9.388	6.93	4.279	20.597	27.0%	3.611	32.2%
Total	21.8	25.863	28.531	76.195	100.0%	11.198	100.0%

Data Base: ANA.

irrigable area was similar to that used in the design of irrigation projects in the field, taking into account: the water demand of the reference crops (in this case, corn and beans); the quantitative balance between water uses and surface water availability; and the area available for agricultural activities. In addition to the definitions of territorial classes, other aspects were also analyzed, such as land dynamics, logistical quality and environmental importance (ANA, 2017a).

The study used data from irrigated areas in 2012, provided by ANA, which basically consisted of projections from the 2006 Agricultural Census. Based on the current irrigated area estimate, an update of the irrigable potential was made, totaling 76.19 million hectares (Mha) as can be seen in Table 12.1. Considering the soil-relief aptitude classes, this value is distributed in: with high aptitude; 25.86 Mha with medium aptitude; and 28.53 Mha with low fitness. Or, still, 47.66 Mha with high-medium aptitude, which represent areas of greater potential for effective expansion of irrigation and implementation of new centers. The midwest stands out for the concentration of 43.1% of high fitness areas and 34.2% of high-medium areas (ANA, 2017a).

The effective potential considers only areas with high or medium soil suitability; high relief fitness; high logistical quality (existencwe of production and electric energy outlets); exclusion of other areas of environmental protection; and territorial classes that indicate expansion of irrigation, that is, combinations in which there is both the potential for further expansion and irrigated agriculture already established (referring to the presence of infrastructure, support services, technology, technical assistance, etc.) (ANA, 2017a).

The *National Water Agency*—ANA together with the Brazilian Institute of Geography and Statistics highlighted that the irrigated area in the country has been growing at average rates above 4% per year since the 1960s. The first data were recorded with 462 thousand ha in 1960, in the 1990s; that number later reached 3 million irrigated hectares. In 2015, this number jumped to 6.95 million irrigated hectares (Figure 12.1), thus placing Brazil among the countries with the largest irrigated territories in the world (ANA, 2017b).

Irrigation information is essential for several applications in planning, expansion strategies, development of key crops and hydrological issues (Gourdji et al., 2013; Zohaib et al., 2019), as resource management for living with drought and understanding of water and energy cycles, biosphere–atmosphere interactions and climate dynamics (Alcamo et al., 2003; Ambika et al., 2016; Boucher et al., 2004; Zhang and Lin, 2016; Zohaib et al., 2019).

Figure 12.1a shows the evolution of the irrigated area since 1960, with an area of just under 0.5 million irrigated hectares; in 2015, a considerable number of 6.95 million hectares was reached. The average incorporation of the area is estimated at around 200 thousand ha/year, so in 2030 the country will have approximately 10.09 million irrigated hectares. This increase corresponds to an increase of 45% (ANA, 2017b).

Figure 12.1b represents the additional irrigable area based on physical and occupation criteria (soils, relief, area available for agriculture, conservation units, surface water availability, water requirements for reference crops, other uses of water). This potential is presented by soil-relief aptitude class (high,

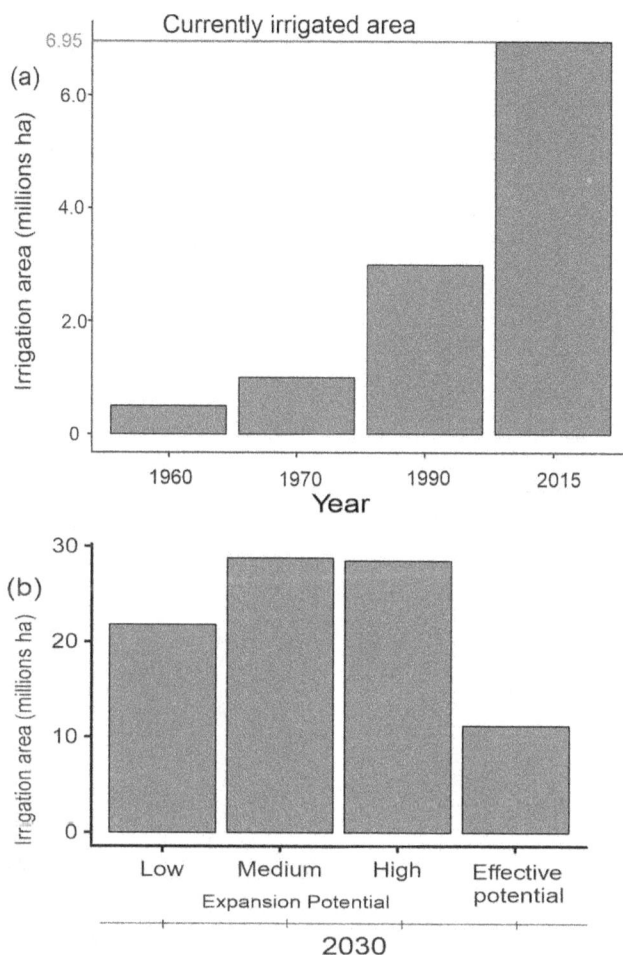

FIGURE 12.1 History of irrigated area in Brazil (a) and additional irrigable area based on physical and occupation criteria (b).

medium and low), whose sum can reach 76.2 Mha. The area with effective potential represents additional irrigable land considering areas with greater soil and relief capacity; good logistical quality (existence of production and electric energy outlets); exclusion of other areas of environmental protection; and the presence of irrigated agriculture (established or under development) which, under the conditions explained, may reach 11.2 Mha (ANA, 2017b).

12.6 Distribution of Irrigation Methods

According to Figure 12.2, the Sectorial Chamber of Irrigation Equipment—CSEI, referring to the Brazilian Association of Machinery and Equipment Industry—ABIMAQ, brought the estimate of the irrigated area from 2000 to 2018 grouped by the type of the system. In the period from 2000 to 2018, the total irrigated area was 6,023,087 against 5,822,337 ha in 2017, which represents an increase of 3.45% (ABIMAQ, 2020).

CSEI-ABIMAQ estimates reiterate the growth trend of the area irrigated by the localized irrigation systems and central pivots in the country. In the last 6 years, the pivots have performed better than the historical record, with more than 40% in relation to the total increase in mechanized irrigated area.

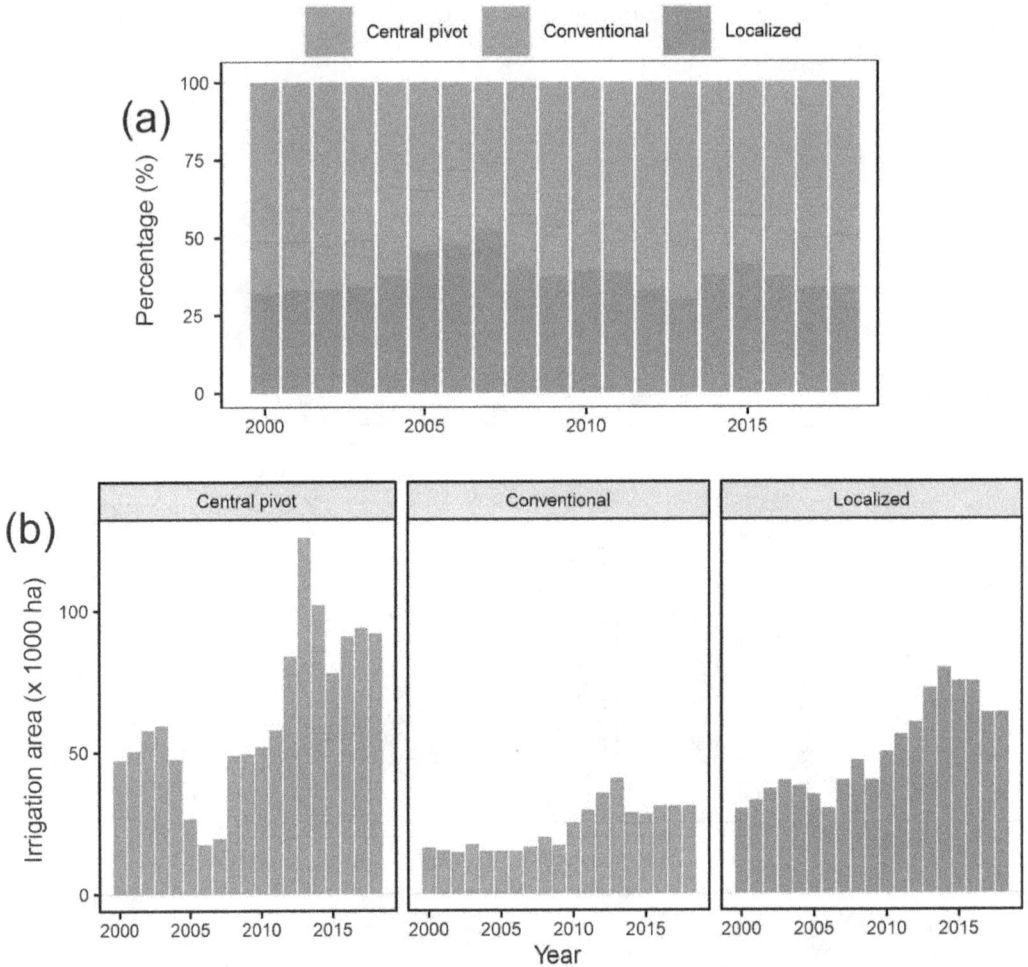

FIGURE 12.2 Percentage of irrigated area in Brazil from 2000 to 2018 by the central pivot, located (sprinkler, cannon) and located (a), the evolution of the irrigated area in the same period in each method (b).

Figure 12.3 highlights the evolution of the irrigated area accumulated by each irrigation method, in which the area of pivots with almost 1.2 million irrigated hectares can be highlighted. It is worth mentioning that since 2008, the average growth rate has been 12.6% (ABIMAQ, 2020).

Central pivot irrigation systems are among the most popular for irrigating field crops and are used on more than half of the sprinkler-irrigated land in the United States, Brazil, Argentina and other countries (Moreno et al., 2012). The first central pivot was installed in Brazil in 1979 in the Tietê river basin in the municipality of Brotas, São Paulo, SP. The equipment, capable of irrigating an area of 76 ha, was implanted on the banks of the Minhoca stream, a tributary of the Gouveia stream and a tributary of the Jacaré Pepira river, one of the least polluted watercourses in the State of São Paulo (MAPA, 2019).

In 2017, Brazil had 23,181 pivot points with 1,476,101 ha equipped for irrigation by central pivots, which corresponds to about 20% of the total irrigated area and 30% of the mechanized irrigated area. Figure 12.4 shows the distribution of pivots in the country; currently, it is the fastest-growing system in the country: in the last 7 years (2012–2018), the additional average equipped area was 94 thousand ha/year—a trend that should be maintained or intensified by 2030. Among the Brazilian states, six states account for 91.8% of the area equipped by pivots, namely, Minas Gerais 30.6%, Goiás 18.4%, Bahia 14.7%,

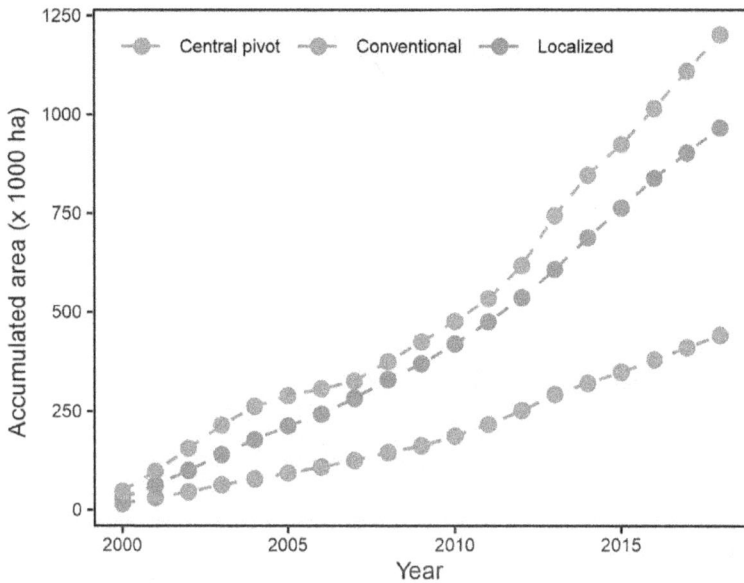

FIGURE 12.3 Evolution of the irrigated area accumulated by each irrigation method.

São Paulo 12.9%, Mato Grosso 7.7% and the Rio Grande do Sul 7.5%. The other states account for 8.2% of the total area (MAPA, 2019).

12.7 Advances in the Management of Irrigated Agriculture

Agricultural production strategies have changed dramatically over the past decade (Takatsuka et al., 2018). Many of these changes were driven by economic decisions to reduce inputs, water and maximize profits through environmental guidelines that demand more efficient and safe use of resources used in agriculture. (Luiz Lourenzani et al., 2016). They are selecting cultivars and adjusting planting dates to accommodate anticipated weather patterns, for example, El Niño or La Niña events (Jones, 1999; Melandri et al., 2019). They also depend on biotechnological innovations for pest suppression, for example, crops protected against insects and diseases (Prasara-A, 2016).

Water stress on crops induces stoma closure, which limits perspiration and reduces evaporative cooling, resulting in high temperature in the leaves or tops of plants (Jones et al., 2009). Leaf temperature can provide information to estimate stomatal conductance or water vapor in leaves (Jones, 2004) and has been widely used to develop tools to improve the management of irrigated plants (Al-Kayssi and Mustafa, 2016; Eksteen et al., 2014; Evett et al., 2012; Jackson et al., 1981; Melandri et al., 2019; Roberts et al., 1990; Veysi et al., 2017).

The research has been developing several vegetation ant thermal indices to improve the understanding of water demand in crops. As an example, we can mention the Culture Water Stress Index, from the English Crop Water Stress Index (CWSI) (Ray, 1982), which is a theoretical method based on the relationship between actual and potential ET, and the water available in the soil. In this case, they are estimated values for generalized cultures, in which the culture coefficient K_c is not used.

As the leaf temperature is influenced by environmental conditions, the CWSI relates the observed temperature to the minimum (non-stressed) temperature and the maximum (non-breathable) temperature of a reference culture under similar environmental conditions (Gonzalez-Dugo et al., 2014; Idso et al., 1980; Jackson et al., 1988, 1981). Thus, when the plants are under water stress, there is an increase in stomatal resistance, which in turn restricts the loss of latent heat of vaporization, leading to an increase

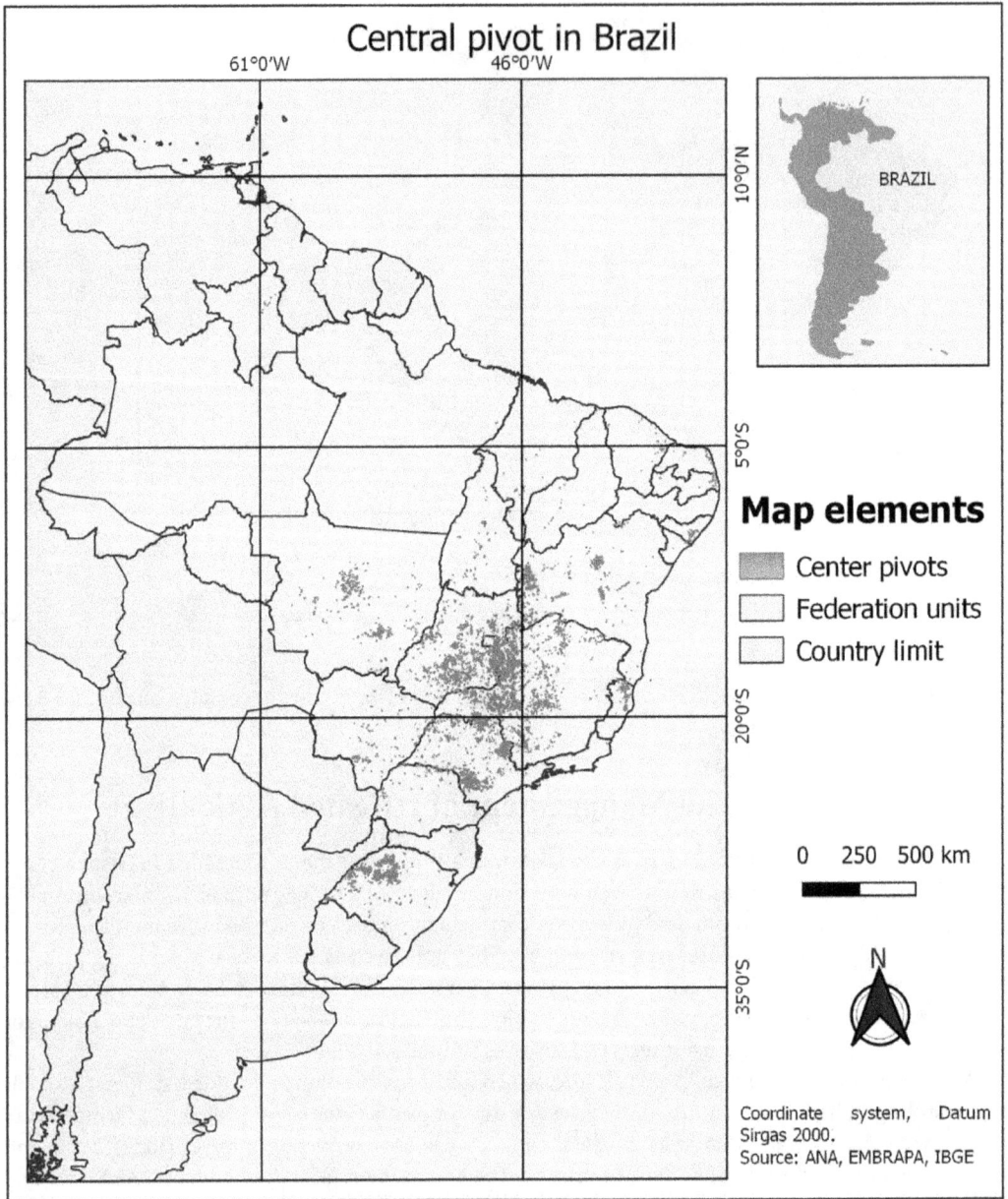

FIGURE 12.4   Distribution of central pivots in Brazil.

in leaf temperature in relation to the environment (Çolak et al., 2015; Garcia-Tejero et al., 2017; Jackson, 1982).

One of the products generated by (EOS—Earth Observing System) is the Moderate Resolution Imaging Spectroradiometer (MODIS) sensor for Vegetation Indices (VIs). It performs spectral transformations of two or more bands and allows the visualization of vegetation properties and temporal intercomparisons of terrestrial photosynthetic activity and structural variations of the crop canopy. VIs, such as the Vegetation Index for Normalized Differences (NDVI), and the Enhanced Vegetation Index (EVI) allow for the monitoring of seasonal, interannual and long-term variations in the structure of vegetation, phenological and biophysical parameters of plants of agricultural interest (Khorsandi et al., 2018).

Among the difficulties encountered to increase the accuracy of the sensing systems are the costs of acquiring equipment and making data collection operational in short periods of time or the low spatio-temporal resolution of low-cost methods (Jones and Sirault, 2014; Gonzalez-Dugo et al., 2013). Thus, the farmer must select the most cost-effective method taking into account the resolution required for effective management in the field.

Recently, there is a lot of discussion about the use of UAVs/drones for precision agriculture, considered as an agricultural management strategy that uses information and communication technologies to integrate real-time information about the field in assisting decision-making, aiming at optimizing the use of inputs and maximizing profitability and taking into account the spatio-temporal variability of information collected at specific points in the field (Makanza et al., 2018; Srivastava et al., 2019; Xia et al., 2018).

12.8 UAVs, Drones and Satellites in Irrigation

Today, most agricultural fieldwork is carried out with human-powered machines in modern agriculture (Natarajan et al., 2016). Due to the maximization of mechanized agriculture, farmers have no practical experience in detecting abnormal characteristics in field conditions (Bu et al., 2017). Some remote sensing processes have been proposed for the advancement of precision agriculture and precision irrigation, mainly for established crops over relatively large areas (Croft et al., 2014). Another important factor is the wide range of satellite and temporal and spatial images available from American and European agencies (Malveaux et al., 2014).

The use of drones is growing rapidly and the use is quite diverse; one of the areas that have evolved the most in the use of this equipment is agriculture. The numbers are growing rapidly; 48% of the drones are used for agricultural purposes, with growth forecast for 80% of the market and US $ 82 billion in annual sales (Watkins et al., 2020). It is worth mentioning that drones equipped with cameras are able to collect images with higher resolution than satellites and can fly at reduced heights, eliminating the risk of cloud obstruction, a factor that brings many problems in the interpretation of images, especially in regions with high incidence of clouds such as in the region. In addition, they do not present safety risks and the high costs characteristic of manned flights (Malveaux et al. 2014).

The popularly called "Drones" are a category of UAV called a multicopter (multi-propeller helicopter) that has become very popular for various uses, such as professional or personal filming and photos, military and agricultural applications, etc.(Gago et al., 2015), which advantageously resulted in the reduction of its cost, size and weight over time (Srivastava et al., 2019). Other advantages of this vehicle are greater stability, lower speed and low flight altitude. In contrast, their carrying capacity and flight time are limited (Gago et al., 2015) and the costs of software for image processing exceed and are much higher than the cost of purchasing the equipment (Simelli and Tsagaris, 2015).

The term "Unmanned Aerial Vehicle" (UAV) is recognized worldwide and includes several aircraft that are autonomous, semi-autonomous or remotely operated. According to ABA (Brazilian Aeromodelling Association), the definition for UAV is: "a vehicle capable of flying in the atmosphere, outside the ground effect, which was designed or modified to not receive a pilot human and that is operated by remote control or autonomous." The Unmanned Aerial System means the set of unmanned aerial vehicles, their flight controls and their operating system, that is, the union of all activities that are interconnected in the flight plan. UAVs have different classifications regarding altitude and range. According to altitude, there are models ranging from 600 to 15,200 m since in relation to range it is quite relative and can be from 2 to 200 km away; however, this is useful for most UAVs, and there is equipment with a much greater range (Malveaux et al., 2014).

Several studies have shown that canopy temperature is related to physiological processes (Castro-Nava et al., 2016; Dejonge et al., 2015; Nelson et al., 2018; Ray, 1982) or even used to detect nutritional deficiencies (Angus et al., 1998).

The use of thermal images can be considered as a substitute for the direct measurement of water loss by the leaves, based on the principle that the temperature reduction is proportional to the rate of water loss due to the evaporative cooling process (Brown and Escombe, 1905; Lapidot et al., 2019). Interest in the use of thermal images to assess the plant's water status has increased as the price of thermal cameras decreases and the quality is rapidly improving (Petrie et al., 2019). This includes the use of land-based systems and unmanned aerial vehicles (drones) and orbital satellites to assess the water status of agricultural crops, which opens up the possibility of automating this process (Costa et al., 2019; Egea et al., 2017; Khanal et al., 2017; Klem et al., 2018; Petrie et al., 2019), becoming a valuable tool in the management of precision irrigation.

12.9 Conclusions

1. The Federal Government of Brazil has been the main engine for the development of Brazilian irrigation through the creation of lines of financing, subsidies and research centers since the 1970s.
2. Brazil has become a rudimentary agricultural producer and one of the main players in the world food market. In parallel, Brazil, driven by the commercialization of flex vehicles, developed to a high degree the culture of sugarcane and its associated products; this agribusiness is still evolving.
3. The use and inclusion of the concept of WUE in Brazilian agriculture meant that it focused on agricultural sustainability, thus contributing to the reduction of greenhouse gas emissions and the development of precision irrigation practices.
4. Climate change has led Brazil to explore the use of aeronautical and remote sensing technology in agriculture for better water management.

References

ABIMAQ, 2020. Associação Brasileira da Indústria de Máquinas e Equipamentos [WWW Document]. Câmara Setorial Equipamentos Irrig. (CSEI), da Assoc. Bras. da Indústria Máquinas e Equipamentos. URL http://www.abimaq.org.br/site.aspx/detalhes-imprensa-ultimos-releases?codNoticia=56p7vjbdz2I= (accessed 2.22.20).

Alcamo, J., Döll, P., Henrichs, T., Kaspar, F., Lehner, B., Rösch, T., Siebert, S., 2003. Global estimates of water withdrawals and availability under current and future "business-as-usual" conditions. *Hydrol. Sci. J.* 48, 339–348. https://doi.org/10.1623/hysj.48.3.339.45278

Alexandratos, N., 1996. China's projected cereals deficits in a world context. *Agric. Econ.* 15, 1–16. https://doi.org/10.1016/S0169-5150(96)01191-7

Al-Kayssi, A.W., Mustafa, S.H., 2016. Impact of elevated carbon dioxide on soil heat storage and heat flux under unheated low-tunnels conditions. *J. Environ. Manage.* 182, 176–186. https://doi.org/10.1016/j.jenvman.2016.07.048

Ambika, A.K., Wardlow, B., Mishra, V., 2016. Remotely sensed high resolution irrigated area mapping in India for 2000 to 2015. *Sci. Data* 3. https://doi.org/10.1038/sdata.2016.118

ANA, 2017a. *Atlas Irrigação Uso da Água na Agricultura Irrigada.* Agência Na. ed. Brasília.

ANA, 2017b. *Atlas Irrigação - Uso da Água na Agricultura Irrigada.* Agência Nacional de Águas, Brasília.

Anfavea, 2020. Associação Nacional dos Fabricantes de Veículos Automotores [WWW Document]. URL http://www.anfavea.com.br/estatisticas.html (accessed 1.8.20).

Angus, J.F., Van Herwaarden, A.F., Fischer, R.A., Howe, G.N., Heenan, D.P., 1998. The source of mineral nitrogen for cereals in south-eastern Australia. *Aust. J. Agric. Res.* 49, 511–522. https://doi.org/10.1071/A97125

Barbosa, F.D.S., Coelho, R.D., Maschio, R., Lima, C.J.G.S., Silva, E.M.D., 2014. Drought resistance of sugar-cane crop for different levels of water availability in the soil. *Eng. Agríc., Jaboticabal* 34, 203–210. https://doi.org/10.1590/S0100-69162014000200002

Batjes, N.H., 1996. Development of a world data set of soil water retention properties using pedotransfer rules. *Geoderma* 71, 31–52. https://doi.org/10.1016/0016-7061(95)00089-5

Bonan, G.B., Levis, S., 2006. Evaluating aspects of the community land and atmosphere models (CLM3 and CAM3) using a dynamic global vegetation model. *J. Clim.* 19, 2290–2301. https://doi.org/10.1175/JCLI3741.1

Bordonal, R.D.O., Carvalho, J.L.N., Lal, R., de Figueiredo, E.B., de Oliveira, B.G., La Scala, N., 2018. Sustainability of sugarcane production in Brazil. A review. *Agron. Sustain. Dev.* https://doi.org/10.1007/s13593-018-0490-x

Boucher, O., Myhre, G., Myhre, A., 2004. Direct human influence of irrigation on atmospheric water vapour and climate. *Clim. Dyn.* 22, 597–603. https://doi.org/10.1007/s00382-004-0402-4

Brown, H.T., Escombe, F., 1905. Researches on some of the physiological processes of green leaves, with special reference to the interchange of energy between the leaf and its surroundings. *Proc. R. Soc. B Biol. Sci.* 76, 29–111. https://doi.org/10.1098/rspb.1905.0002

Bu, H., Sharma, L.K., Denton, A., Franzen, D.W., 2017. Comparison of satellite imagery and ground-based active optical sensors as yield predictors in sugar beet, spring wheat, corn, and sunflower. *Agron. J.* 109, 299–308. https://doi.org/10.2134/agronj2016.03.0150

Cançado, J.E.D., Saldiva, P.H.N., Pereira, L.A.A., Lara, L.B.L.S., Artaxo, P., Martinelli, L.A., Arbex, M.A., Zanobetti, A., Braga, A.L.F., 2006. The impact of sugar cane-burning emissions on the respiratory system of children and the elderly. *Environ. Health Perspect.* 114, 725–729. https://doi.org/10.1289/ehp.8485

Castro-Nava, S., Huerta, A.J., Plácido-de la Cruz, J.M., Mireles-Rodríguez, E., 2016. Leaf growth and canopy development of three sugarcane genotypes under high temperature rainfed conditions in Northeastern Mexico. *Int. J. Agron.* 2016, 2561026. https://doi.org/10.1155/2016/2561026

Chaddad, F., 2015. *The Economics and Organization of Brazilian Agriculture: Recent Evolution and Productivity Gains, The Economics and Organization of Brazilian Agriculture: Recent Evolution and Productivity Gains.* Elsevier Inc. https://doi.org/10.1016/C2014-0-00991-4

Chen, M.F., 2008. Consumer trust in food safety - A multidisciplinary approach and empirical evidence from Taiwan. *Risk Anal.* 28, 1553–1569. https://doi.org/10.1111/j.1539-6924.2008.01115.x

Chen, M.F., 2011. Consumer's trust-in-food-safety typology in Taiwan: Food-related lifestyle matters. *Heal. Risk Soc.* 13, 503–526. https://doi.org/10.1080/13698575.2011.615825

Coelho, R.D., Lizcano, J.V., da Silva Barros, T.H., da Silva Barbosa, F., Leal, D.P.V., da Costa Santos, L., Ribeiro, N.L., Júnior, E.F.F., Martin, D.L., 2019. Effect of water stress on renewable energy from sugarcane biomass. *Renew. Sustain. Energy Rev.* 103, 399–407. https://doi.org/10.1016/J.RSER.2018.12.025

Çolak, Y.B., Yazar, A., Çolak, İ., Akça, H., Duraktekin, G., 2015. Evaluation of crop water stress index (CWSI) for eggplant under varying irrigation regimes using surface and subsurface drip systems. *Agric. Agric. Sci. Procedia* 4, 372–382. https://doi.org/10.1016/j.aaspro.2015.03.042

Collins, W.D., Rasch, P.J., Boville, B.A., Hack, J.J., McCaa, J.R., Williamson, D.L., Briegleb, B.P., Bitz, C.M., Lin, S.J., Zhang, M., 2006. The formulation and atmospheric simulation of the Community Atmosphere Model version 3 (CAM3). *J. Clim.* 19, 2144–2161. https://doi.org/10.1175/JCLI3760.1

Conab, 2020. Conab - Boletim da Safra de Cana-de-açúcar [WWW Document]. Cia. Nac. Abast. - CONAB. URL https://www.conab.gov.br/info-agro/safras/cana/boletim-da-safra-de-cana-de-acucar (accessed 1.9.20).

Costa, J.M., Egipto, R., Sánchez-Virosta, A., Lopes, C.M., Chaves, M.M., 2019. Canopy and soil thermal patterns to support water and heat stress management in vineyards. *Agric. Water Manag.* 216, 484–496. https://doi.org/10.1016/J.AGWAT.2018.06.001

Costa, R.M.D., Henkin, H., 2016. Estratégias competitivas e desempenho da indústria automobilística no Brasil. *Econ. e Soc.* 25, 457–487. https://doi.org/10.1590/1982-3533.2016v25n2art7

Croft, H., Chen, J.M., Zhang, Y., 2014. The applicability of empirical vegetation indices for determining leaf chlorophyll content over different leaf and canopy structures. *Ecol. Complex.* 17, 119–130. https://doi.org/10.1016/j.ecocom.2013.11.005

Dejonge, K.C., Taghvaeian, S., Trout, T.J., Thomas, L.H., 2015. Comparison of canopy temperature-based water stress indices formaize. *Agric. Water Manag.* 156, 51–62.

Egea, G., Padilla-Díaz, C.M., Martinez-Guanter, J., Fernández, J.E., Pérez-Ruiz, M., 2017. Assessing a crop water stress index derived from aerial thermal imaging and infrared thermometry in super-high density olive orchards. *Agric. Water Manag.* 187, 210–221. https://doi.org/10.1016/J.AGWAT.2017.03.030

Eksteen, A., Singels, A., Ngxaliwe, S., 2014. Water relations of two contrasting sugarcane genotypes. *F. Crop. Res.* 168, 86–100. https://doi.org/10.1016/j.fcr.2014.08.008

Embrapa, 2020. Trajetória da agricultura brasileira - Portal Embrapa [WWW Document]. URL https://www.embrapa.br/visao/trajetoria-da-agricultura-brasileira (accessed 1.7.20).

Evett, S.R., Agam, N., Kustas, W.P., Colaizzi, P.D., Schwartz, R.C., 2012. Soil profile method for soil thermal diffusivity, conductivity and heat flux: Comparison to soil heat flux plates. *Adv. Water Resour.* 50, 41–54. https://doi.org/10.1016/j.advwatres.2012.04.012

FGV, 2020. Programa De Integracao Nacional (PIN) | CPDOC - Centro de Pesquisa e Documentação de História Contemporânea do Brasil [WWW Document]. URL http://www.fgv.br/cpdoc/acervo/dicionarios/verbete-tematico/programa-de-integracao-nacional-pin (accessed 1.16.20).

Finger, R., El Benni, N., 2013. Farmers' adoption of extensive wheat production - Determinants and implications. *Land Use Policy* 30, 206–213. https://doi.org/10.1016/j.landusepol.2012.03.014

Gago, J., Douthe, C., Coopman, R.E., Gallego, P.P., Ribas-Carbo, M., Flexas, J., Escalona, J., Medrano, H., 2015. UAVs challenge to assess water stress for sustainable agriculture. *Agric. Water Manag.* 153, 9–19. https://doi.org/10.1016/j.agwat.2015.01.020

Garcia, E., Ramos Filho, F., Mallmann, G., Fonseca, F., 2017. Costs, benefits and challenges of sustainable livestock intensification in a major deforestation frontier in the Brazilian Amazon. *Sustainability* 9, 158. https://doi.org/10.3390/su9010158

Garcia-Tejero, I.F., Hernández, A., Padilla-Díaz, C.M., Diaz-Espejo, A., Fernández, J.E., 2017. Assessing plant water status in a hedgerow olive orchard from thermography at plant level. *Agric. Water Manag.* 188, 50–60. https://doi.org/10.1016/j.agwat.2017.04.004

Garrett, R.D., Koh, I., Lambin, E.F., le Polain de Waroux, Y., Kastens, J.H., Brown, J.C., 2018. Intensification in agriculture-forest frontiers: Land use responses to development and conservation policies in Brazil. *Glob. Environ. Chang.* 53, 233–243. https://doi.org/10.1016/j.gloenvcha.2018.09.011

Goldemberg, J., Coelho, S.T., Guardabassi, P., 2008. The sustainability of ethanol production from sugarcane. *Energy Policy* 36, 2086–2097. https://doi.org/10.1016/j.enpol.2008.02.028

Gonzalez-Dugo, V., Zarco-Tejada, P., Nicolás, E., Nortes, P.A., Alarcón, J.J., Intrigliolo, D.S., Fereres, E., 2013. Using high resolution UAV thermal imagery to assess the variability in the water status of five fruit tree species within a commercial orchard. *Precis. Agric.* 14, 660–678. https://doi.org/10.1007/s11119-013-9322-9

Gonzalez-Dugo, V., Zarco-Tejada, P.J., Fereres, E., 2014. Applicability and limitations of using the crop water stress index as an indicator of water deficits in citrus orchards. *Agric. For. Meteorol.* 198, 94–104. https://doi.org/10.1016/J.AGRFORMET.2014.08.003

Hatfield, J.L., Dold, C., 2019. Water-use efficiency: Advances and challenges in a changing climate. *Front. Plant Sci.* https://doi.org/10.3389/fpls.2019.00103

Hogeboom, R.J., Knook, L., Hoekstra, A.Y., 2018. The blue water footprint of the world's artificial reservoirs for hydroelectricity, irrigation, residential and industrial water supply, flood protection, fishing and recreation. *Adv. Water Resour.* 113, 285–294. https://doi.org/10.1016/j.advwatres.2018.01.028

Huang, J.K., Wei, W., Cui, Q., Xie, W., 2017. The prospects for China's food security and imports: Will China starve the world via imports? *J. Integr. Agric.* https://doi.org/10.1016/S2095-3119(17)61756-8

Idso, S.B., Reginato, R.J., Hatfield, J.L., Walker, G.K., Jackson, R.D., Pinter, P.J., 1980. A generalization of the stress-degree-day concept of yield prediction to accommodate a diversity of crops. *Agric. Meteorol.* 21, 205–211. https://doi.org/10.1016/0002-1571(80)90053-9

Jackson, R.D., Idso, S.B., Reginato, R.J., Pinter, P.J., 1981. Canopy temperature as a crop water stress indicator. *Water Resour. Res.* 17, 1133–1138. https://doi.org/10.1029/WR017i004p01133

Jackson, R.D., Kustas, W.P., Choudhury, B.J., 1988. A reexamination of the crop water stress index. *Irrig. Sci.* 9, 309–317. https://doi.org/10.1007/BF00296705

Jones, H., Sirault, X., 2014. Scaling of thermal images at different spatial resolution: The mixed pixel problem. *Agronomy* 4, 380–396. https://doi.org/10.3390/agronomy4030380

Jones, H.G., 1999. Use of thermography for quantitative studies of spatial and temporal variation of stomatal conductance over leaf surfaces. *Plant, Cell Environ.* 22, 1043–1055. https://doi.org/10.1046/j.1365-3040.1999.00468.x

Jones, H.G., 2004. Application of thermal imaging and infrared sensing in plant physiology and ecophysiology, in: *Advances in Botanical Research.* Academic Press, Vol. 41, pp. 107–163. https://doi.org/10.1016/S0065-2296(04)41003-9

Jones, H.G., Serraj, R., Loveys, B.R., Xiong, L., Wheaton, A., Price, A.H., 2009. Thermal infrared imaging of crop canopies for the remote diagnosis and quantification of plant responses to water stress in the field. *Funct. Plant Biol.* 36, 978–989. https://doi.org/10.1071/FP09123

Khanal, S., Fulton, J., Shearer, S., 2017. An overview of current and potential applications of thermal remote sensing in precision agriculture. *Comput. Electron. Agric.* 139, 22–32. https://doi.org/10.1016/J.COMPAG.2017.05.001

Khorsandi, A., Hemmat, A., Mireei, S.A., Amirfattahi, R., Ehsanzadeh, P., 2018. Plant temperature-based indices using infrared thermography for detecting water status in sesame under greenhouse conditions. *Agric. Water Manag.* 204, 222–233. https://doi.org/10.1016/j.agwat.2018.04.012

Klem, K., Záhora, J., Zemek, F., Trunda, P., Tůma, I., Novotná, K., Hodaňová, P., Rapantová, B., Hanuš, J., Vavříková, J., Holub, P., 2018. Interactive effects of water deficit and nitrogen nutrition on winter wheat. Remote sensing methods for their detection. *Agric. Water Manag.* 210, 171–184. https://doi.org/10.1016/J.AGWAT.2018.08.004

Kothari, K., Ale, S., Bordovsky, J.P., Thorp, K.R., Porter, D.O., Munster, C.L., 2019. Simulation of efficient irrigation management strategies for grain sorghum production over different climate variabllity classes. *Agric. Syst.* 170, 49–62. https://doi.org/10.1016/J.AGSY.2018.12.011

Kumar, S. V., Peters-Lidard, C.D., Santanello, J.A., Reichle, R.H., Draper, C.S., Koster, R.D., Nearing, G., Jasinski, M.F., 2015. Evaluating the utility of satellite soil moisture retrievals over irrigated areas and the ability of land data assimilation methods to correct for unmodeled processes. *Hydrol. Earth Syst. Sci.* 19, 4463–4478. https://doi.org/10.5194/hess-19-4463-2015

Lapidot, O., Ignat, T., Rud, R., Rog, I., Alchanatis, V., Klein, T., 2019. Use of thermal imaging to detect evaporative cooling in coniferous and broadleaved tree species of the Mediterranean maquis. *Agric. For. Meteorol.* 271, 285–294. https://doi.org/10.1016/J.AGRFORMET.2019.02.014

Liu, J., Basnayake, J., Jackson, P.A., Chen, X., Zhao, J., Zhao, P., Yang, L., Bai, Y., Xia, H., Zan, F., Qin, W., Yang, K., Yao, L., Zhao, L., Zhu, J., Lakshmanan, P., Zhao, X., Fan, Y., 2016. Growth and yield of sugarcane genotypes are strongly correlated across irrigated and rainfed environments. *F. Crop. Res.* 196, 418–425. https://doi.org/10.1016/j.fcr.2016.07.022

Lobell, D., Bala, G., Mirin, A., Phillips, T., Maxwell, R., Rotman, D., 2009. Regional differences in the influence of irrigation on climate. *J. Clim.* 22, 2248–2255. https://doi.org/10.1175/2008JCLI2703.1

Luiz Lourenzani, W., Bernardo, R., Marques Caldas, M., 2016. Produção de biocombustível e alteração da composição agropecuária no Centro-Oeste do Brasil Biofuel production and changing of agricultural composition in Midwest region of Brazil. *INTERAÇÕES* 17, 561–575. https://doi.org/10.20435/1984-042X-2016-v.17-n.4(02)

Mahdavi, S.M., Neyshabouri, M.R., Fujimaki, H., Majnooni Heris, A., 2017. Coupled heat and moisture transfer and evaporation in mulched soils. *Catena* 151, 34–48. https://doi.org/10.1016/j.catena.2016.12.010

Makanza, R., Zaman-Allah, M., Cairns, J., Magorokosho, C., Tarekegne, A., Olsen, M., Prasanna, B., 2018. High-throughput phenotyping of canopy cover and senescence in maize field trials using aerial digital canopy imaging. *Remote Sens.* 10, 330. https://doi.org/10.3390/rs10020330

Malveaux, C., Hall, S., Price, R.R., 2014. Using drones in agriculture: Unmanned aerial systems for agricultural remote sensing applications. *Am. Soc. Agric. Biol. Eng. Annu. Int. Meet.* 6, 4075–4079. https://doi.org/10.13031/aim.20141911016

MAPA, 2019. *Ministério da Agricultura, Pecuária e Abastecimento (MAPA) República Federativa do Brasil Ministério do Desenvolvimento Regional*, 1st ed. Ministério da Agricultura, Pecuária e Abastecimento, Brasilia.

Melandri, G., Prashar, A., Mccouch, S.R., Van Der Linden, G., Jones, H.G., Kadam, N., Jagadish, K., Bouwmeester, H., Ruyter-Spira, C., 2019. Association mapping and genetic dissection of drought-induced canopy temperature differences in rice. *J. Exp. Bot.* https://doi.org/10.1093/jxb/erz527

Minasny, B., Malone, B.P., McBratney, A.B., Angers, D.A., Arrouays, D., Chambers, A., Chaplot, V., Chen, Z.S., Cheng, K., Das, B.S., Field, D.J., Gimona, A., Hedley, C.B., Hong, S.Y., Mandal, B., Marchant, B.P., Martin, M., McConkey, B.G., Mulder, V.L., O'Rourke, S., Richer-de-Forges, A.C., Odeh, I., Padarian, J., Paustian, K., Pan, G., Poggio, L., Savin, I., Stolbovoy, V., Stockmann, U., Sulaeman, Y., Tsui, C.C., Vågen, T.G., van Wesemael, B., Winowiecki, L., 2017. Soil carbon 4 per mille. *Geoderma*. https://doi.org/10.1016/j.geoderma.2017.01.002

Moreno, M.A., Medina, D., Ortega, J.F., Tarjuelo, J.M., 2012. Optimal design of center pivot systems with water supplied from wells. *Agric. Water Manag.* 107, 112–121. https://doi.org/10.1016/j.agwat.2012.01.016

Mueller, B., Mueller, C., 2016. The political economy of the Brazilian model of agricultural development: Institutions versus sectoral policy. *Q. Rev. Econ. Financ.* 62, 12–20. https://doi.org/10.1016/j.qref.2016.07.012

Natarajan, R., Subramanian, J., Papageorgiou, E.I., 2016. Hybrid learning of fuzzy cognitive maps for sugarcane yield classification. *Comput. Electron. Agric.* 127, 147–157. https://doi.org/10.1016/j.compag.2016.05.016

Ndraha, N., Hsiao, H.I., Chih Wang, W.C., 2017. Comparative study of imported food control systems of Taiwan, Japan, the United States, and the European Union. *Food Control*. https://doi.org/10.1016/j.foodcont.2017.02.051

Nehring, R., 2016. Yield of dreams: Marching west and the politics of scientific knowledge in the Brazilian Agricultural Research Corporation (Embrapa). *Geoforum* 77, 206–217. https://doi.org/10.1016/j.geoforum.2016.11.006

Nelson, W.C.D., Hoffmann, M.P., Vadez, V., Roetter, R.P., Whitbread, A.M., 2018. Testing pearl millet and cowpea intercropping systems under high temperatures. *F. Crop. Res.* 217, 150–166. https://doi.org/10.1016/J.FCR.2017.12.014

Nepstad, D., McGrath, D., Stickler, C., Alencar, A., Azevedo, A., Swette, B., Bezerra, T., DiGiano, M., Shimada, J., Da Motta, R.S., Armijo, E., Castello, L., Brando, P., Hansen, M.C., McGrath-Horn, M., Carvalho, O., Hess, L., 2014. Slowing Amazon deforestation through public policy and interventions in beef and soy supply chains. *Science* 344(6188), 1118–1123. https://doi.org/10.1126/science.1248525

Ozdogan, M., Yang, Y., Allez, G., Cervantes, C., 2010. Remote Sensing of Irrigated Agriculture: Opportunities and Challenges. *Remote Sens.* 2, 2274–2304. https://doi.org/10.3390/rs2092274

Perin, V., Sentelhas, P.C., Dias, H.B., Santos, E.A., 2019. Sugarcane irrigation potential in Northwestern São Paulo, Brazil, by integrating Agrometeorological and GIS tools. *Agric. Water Manag.* 220, 50–58. https://doi.org/10.1016/J.AGWAT.2019.04.012

Petrie, P.R., Wang, Y., Liu, S., Lam, S., Whitty, M.A., Skewes, M.A., 2019. The accuracy and utility of a low cost thermal camera and smartphone-based system to assess grapevine water status. *Biosyst. Eng.* 179, 126–139. https://doi.org/10.1016/J.BIOSYSTEMSENG.2019.01.002

Prasara-A, J., 2016. Sustainability of sugarcane cultivation: Case study of selected sites in north-eastern Thailand. *J. Clean. Prod.* 134, 613–622. https://doi.org/10.1016/j.jclepro.2015.09.029

Puma, M.J., Cook, B.I., 2010. Effects of irrigation on global climate during the 20th century. *J. Geophys. Res.* 115, D16120. https://doi.org/10.1029/2010JD014122

Qiu, Y., Cai, W., Guo, X., Pan, A., 2009. Dynamics of late spring rainfall reduction in recent decades over Southeastern China. *J. Clim.* 22, 2240–2247. https://doi.org/10.1175/2008JCLI2809.1

Radhouane, L., 2008. Caract??ristiques hydriques du mil (*Pennisetum glaucum* (L.) R. Br.) en pr??sence de contraintes hydriques. *Comptes Rendus - Biol.* 331, 206–214. https://doi.org/10.1016/j.crvi.2007.12.004

Ray, D.J., 1982. Canopy temperature and crop water stress. *Adv. Irrig.* 1, 43–85. https://doi.org/10.1016/B978-0-12-024301-3.50009-5

Ribeiro, R.V., Machado, E.C., Magalhães Filho, J.R., Lobo, A.K.M., Martins, M.O., Silveira, J.A.G., Yin, X., Struik, P.C., 2017. Increased sink strength offsets the inhibitory effect of sucrose on sugarcane photosynthesis. *J. Plant Physiol.* 208, 61–69. https://doi.org/10.1016/J.JPLPH.2016.11.005

Roberts, J., Nayamuth, R.A., Batchelor, C.H., Soopramanien, G.C., 1990. Plant-water relations of sugarcane (Saccharum officinarum L.) under a range of irrigated treatments. *Agric. Water Manag.* 17, 95–115. https://doi.org/10.1016/0378-3774(90)90058-7

Romano, G., Zia, S., Spreer, W., Sanchez, C., Cairns, J., Araus, J.L., Müller, J., 2011. Use of thermography for high throughput phenotyping of tropical maize adaptation in water stress. *Comput. Electron. Agric.* 79, 67–74. https://doi.org/10.1016/j.compag.2011.08.011

Santos, A.S., Gilio, L., Halmenschlager, V., Diniz, T.B., Almeida, A.N., 2018. Flexible-fuel automobiles and CO_2 emissions in Brazil: Parametric and semiparametric analysis using panel data. *Habitat Int.* 71, 147–155. https://doi.org/10.1016/j.habitatint.2017.11.014

Santos, L.C., Coelho, R.D., Barbosa, F.S., Leal, D.P.V., Fraga Júnior, E.F., Barros, T.H.S., Lizcano, J. V., Ribeiro, N.L., 2019. Influence of deficit irrigation on accumulation and partitioning of sugarcane biomass under drip irrigation in commercial varieties. *Agric. Water Manag.* 221, 322–333. https://doi.org/10.1016/J.AGWAT.2019.05.013

Sato, F.A., Da Silva, A.M., Coelho, G., Da Silva, A.C., De Carvalho, L.G., 2007. Coeficiente De Cultura (Kc) Do Cafeeiro (Coffea arabica L.) No Período De Outono-Inverno Na Região De Lavras –MG. *Engenharia Agrícola* 27, 383–391.

Severo Santos, J.F., Naval, L.P., 2020. Spatial and temporal dynamics of water footprint for soybean production in areas of recent agricultural expansion of the Brazilian savannah (Cerrado). *J. Clean. Prod.* 251. https://doi.org/10.1016/j.jclepro.2019.119482

Simelli, I., Tsagaris, A., 2015. The Use of Unmanned Aerial Systems (UAS) in Agriculture, in: *Proceedings of the 7th International Conference on Information and Communication Technologies in Agriculture, Food and Environment.* HAICTA, Kavala, Grecce, pp. 730–736.

Sorrensen, C., 2009. Potential hazards of land policy: Conservation, rural development and fire use in the Brazilian Amazon. *Land Use Policy* 26, 782–791. https://doi.org/10.1016/j.landusepol.2008.10.007

Srivastava, K., Bhutoria, A.J., Sharma, J.K., Sinha, A., Pandey, P.C., 2019. UAVs technology for the development of GUI based application for precision agriculture and environmental research. *Remote Sens. Appl. Soc. Environ.* 16, 100258. https://doi.org/10.1016/j.rsase.2019.100258

Stabile, M.C.C., Guimarães, A.L., Silva, D.S., Ribeiro, V., Macedo, M.N., Coe, M.T., Pinto, E., Moutinho, P., Alencar, A., 2019. Solving Brazil's land use puzzle: Increasing production and slowing Amazon deforestation. *Land Use Policy* 91, 104362. https://doi.org/10.1016/j.landusepol.2019.104362

Taiz, L., Zeiger, E., 2015. *Plant Physiology*, 6th ed. Sinauer Associates.

Takatsuka, Y., Niekus, M.R., Harrington, J., Feng, S., Watkins, D., Mirchi, A., Nguyen, H., Sukop, M.C., 2018. Value of irrigation water usage in South Florida agriculture. *Sci. Total Environ.* 626, 486–496. https://doi.org/10.1016/J.SCITOTENV.2017.12.240

Testezlaf, R., 2017. Irrigação: Métodos, sistemas e aplicações, 1st ed. Unicamp - FEAGRI, Campinas, SP, Brasília. 56.p.

Veysi, S., Naseri, A.A., Hamzeh, S., Bartholomeus, H., 2017. A satellite based crop water stress index for irrigation scheduling in sugarcane fields. *Agric. Water Manag.* 189, 70–86. https://doi.org/10.1016/J.AGWAT.2017.04.016

Vu, J.C.V, Allen, L.H., 2009. Growth at elevated CO_2 delays the adverse effects of drought stress on leaf photosynthesis of the C4 sugarcane. *J. Plant Physiol.* 166, 107–116. https://doi.org/10.1016/j.jplph.2008.02.009

Watkins, S., Burry, J., Mohamed, A., Marino, M., Prudden, S., Fisher, A., Kloet, N., Jakobi, T., Clothier, R., 2020. Ten questions concerning the use of drones in urban environments. *Build. Environ.* 167, 106458. https://doi.org/10.1016/j.buildenv.2019.106458

Wisser, D., Fekete, B.M., Vörösmarty, C.J., Schumann, A.H., 2010. Reconstructing 20th century global hydrography: A contribution to the Global Terrestrial Network- Hydrology (GTN-H). *Hydrol. Earth Syst. Sci.* 14, 1–24. https://doi.org/10.5194/HESS-14-1-2010

Zhang, T., Lin, X., 2016. Assessing future drought impacts on yields based on historical irrigation reaction to drought for four major crops in Kansas. *Sci. Total Environ.* 550, 851–860. https://doi.org/10.1016/j.scitotenv.2016.01.181

Zohaib, M., Kim, H., Choi, M., 2019. Detecting global irrigated areas by using satellite and reanalysis products. *Sci. Total Environ.* 677, 679–691. https://doi.org/10.1016/j.scitotenv.2019.04.365

13

Irrigation Management in Romania

Rares Hălbac-
Cotoară-Zamfir
*Politehnica University
of Timișoara*

Luca Salvati
University of Macerata

Saeid Eslamian
*Isfahan University
of Technology*

13.1 Introduction

Romania has about 15 million ha of agricultural land, of which 9.5 million ha are cultivated. Agriculture was always an important sector in the Romanian economy until 1990, accounting for about 24% of the total employed workforce. Until 1989, the trade balance of exports of agricultural products was generally positive, but it became negative after sociopolitical changes.

IPCC studies (Parry et al., 2007; Metz et al., 2007; IPCC, 2012) indicate that climate change will have a negative impact on agricultural productivity by reducing it, through a negative evolution of productivity and income stability in agriculture, and also indicating the existence of areas that already have a significant level of food insecurity. Developing agriculture adapted to climate change is crucial to achieving food security and achieving the goals of mitigating the effects of climate change.

Romania presents a considerable risk to climate change, with their effects being clearly reflected by the changes in the temperature and precipitation regime. Since the second half of the 20th century, the most affected areas according to the relevant international reports and analyses of the climatological data for the period of 1901–2010 are in the south, southeastern and eastern parts of the country.

A significant part of Romania's agricultural area is experiencing the negative effects of drought, insufficient water reserves and poorly functional irrigation facilities. The absence or high degree of degradation of the irrigation infrastructure led to approximately 48% of the total agricultural area (7.1 million ha in 2006) being affected by these phenomena (the most affected areas were the Romanian Plain, southern Moldavia and Dobrogea). It has been estimated that in the last 35 years, Romania suffered average annual weather losses to the amount of over 8 million European Union (EU) (representing 0.26% of GDP), of which 34% were drought related. The main risks that Romania faces in the short and medium term consist of a significant increase in the average annual temperature, a decrease in precipitation and a general occurrence of extreme climatic events.

Extremely severe and severe drought in southern and eastern areas of Romania, coupled with high water consumption during July–August, led to the soil water reserve often being under the wilting point on large agricultural areas. In these areas, complex agricultural drought is a climatic hazard phenomenon that causes the worst consequences that have ever been recorded in agriculture.

Droughts can last from a few days to several months, resulting in high agricultural production variability, especially in regions with a currently high vulnerability and low adaptation potential, affecting a full year of agricultural production and having negative consequences on food supply and the national economy.

Under the current pedo-climatic conditions in Romania, irrigation has a complementary character to precipitation and has an important role in obtaining high and relatively stable crops from year to year, in ensuring food safety and in achieving a surplus for export and in environmental protection. Because the precipitation recorded relatively low values in southern and southeastern Romania (Danube plains, South Moldova, Dobrogea), irrigation systems are highly needed here to ensure large and stable agricultural yields. However, given the variability and irregularity of precipitation pattern and volumes, there are also periods when an additional water support from irrigation is not necessary. Most of Romanian irrigation facilities are old and generate high water and energy consumptions, having a negative impact on Romania's water reserves, a country classified as one with low water reserves (the average water quantity available per capita is 2,660 m³ water/inhabitant/year, including the Danube, just over half the European average 4,230 m³ water/inhabitant/year). A significant percentage of irrigation facilities are also at an advanced stage of degradation, almost 75% of irrigation arrangements being not functional and/or inefficient in terms of water and energy consumption and expensive for farmers.

This chapter aims to familiarize the international scientific community with aspects of the evolution of irrigation works in Romania, how they have been managed throughout history and the perspectives of these systems in the national, European and international context taking into account the current challenges of society.

13.2 Beginning of Irrigations in Romania from Ancient to Modern Times

The southern and southeastern areas of Romania were and still are the most arid parts of this country, especially the southeast where there is a hydrographic network of low-intensity. In ancient times there were higher and more constant flows in comparison to the current situation (around 0 AC, this area was covered by forest, which contributed to these more consistent flows) (Botzan, 1980; Cazacu et al., 1989; Blidaru et al., 1997).

The aridity of the Dobrogea step is also spoken about in the poet Ovidiu's "Pontic" epistles, with a few mentions about the irrigation of the Tomis fortress gardens, this being a common agricultural process.

Thus, meeting the needs of the population with vegetables, medicinal and aromatic plants could only be obtained with the help of irrigated agriculture, through small earthworks and perishable materials, their traces being difficult to identify after more than 2,000 years. The conquest of South-East Romania (the Dobrogea of today) by the Roman Empire introduced to this area the technique of water use for the most diverse purposes as well as brought about an economic boom in an arid and relatively poor area. In the northern half of Dobrogea, with richer hydrographic resources and a dense vegetation relief, the water courses identified in that area could have been used for irrigation, using for this purpose the wheel with cups. In the area of the former Histria fortress, it was possible to practice gravity irrigation as well as the one with wheels with cups for crops that were not close to the water source. In the southern part of Dobrogea, the water resources were more reduced, the water deficit in irrigation being complemented by the use of meso-thermal sulfur springs (Canarache, 1951; Bleahu, 1963; Mindru, 1969; Alexandrescu, 1970; Aricescu, 1977; Botzan, 1980; Avram and Bounegru, 1986; Blidaru et al., 1997).

In the southern part of Romania (today's Oltenia), the Roman conquest brought a new organization of agricultural land, which was divided among veterans and poor Romanian citizens. The colonists, part of them coming from the arid areas of the empire and familiar with the technique of water harvesting and lifting it in the areas where it was needed, successfully introduced the irrigation technique. Numerous hydraulic milling installations have been identified in this area, installations which were also used for irrigation of vegetable gardens. The rivers in this area were fed by many streams with high flows, which provided excellent conditions for both gravity irrigation and traditional hydraulic mechanisms (Bodor, 1957; Branga, 1980; Ammianus, 1982; Botzan, 1984; Barbulescu, 1984).

In antiquity, the gravitational irrigation of some vegetable gardens could be done without a pond for water accumulation. The necessity of the ponds for this purpose came as the population increased, in order to ensure a higher summer flow, to cater to the higher demands due to a reduction in the natural flow rate of the watercourses. This theory does not exclude the setting up of ponds for fishery, especially when a milling plant was operating along the water. The technique of domestic irrigation was improved with the practical elements brought by the soldiers of the Roman Empire coming from the southern areas of the empire, where agriculture was conditioned by irrigation. Here are historical evidences about the presence of a Syrian unit stationed in southern of Romania (on the northern bank of Danube River) for more than 150 years (Cantor, 1967; Petrescu-Dambovita, 1978; Branga, 1980; Cazacu et al., 1989; Blidaru et al., 1997).

We can observe a distinction between irrigation in plain areas and those in the piedmont areas, especially if we are referring to the type of irrigated crops. In the plain areas there are preferred irrigation of vegetables and cereals, while in the piedmont area are mainly irrigated vegetables and meadows (some of them with fruit trees).

To the north of the Carpathian Mountains, in the present-day Transylvania, the technique used for the gravitational irrigation of the meadows for their fertilization, in order to avoid stagnation of excess water through an adequate approach of the natural slopes, seems to have originated in the Roman period. During the Roman administration in the Dacia province, the waters, pastures and meadows were the property of the emperor who leased them. Those who leased had the right and ability to make arrangements for using water for milling and irrigation (Conea, 1935; Giurescu, 1973; Botzan and Albota, 1980; Botzan, 1984; Barbulescu, 1984).

It should be mentioned that the area with the most extensive traditional irrigation does not coincide with the most arid area of the country, but it is where the necessity of soil irrigation has met favorable conditions of application. Gravitational irrigation, applied to meadows, orchards and vegetable gardens, from derivations that powered simultaneously and hydraulic installations, does not have a higher density, as in the southwest of Transylvania.

Starting with the 12th century, water mills are becoming more and more numerous in Europe. They also enter the space between the Danube and the Carpathian Mountains. Thus, in 1247, a water mill is first documented north of Danube River. In the eastern part of today's Romania, the Moldavian region, the water mills are mentioned since the 14th century (Hoffmann, 1981; Giurescu, 1973; Botzan, 1979; Botzan, 1984; Botzan, 1989).

Over the next 200 years, water mill documents are considerably multiplying. These covered the entire area between the Danube and the Carpathian Mountains, being documented water mills powered by ponds (in the plains) as well as water mills fed by gravitational debris in the piedmont areas. These mills had mainly economic purposes other than irrigation but, in the next decades, they were used also for irrigating gardens, meadows and pastures (Cazacu et al., 1989; Botzan, 1984; Botzan, 1989).

In Moldova (eastern part of the actual Romanian territory), in the XIV century, according to the historical documents, a significant number of ponds could be identified. These ponds had a double role—irrigation and pisciculture, with their number increasing to over 1,500 in XVII century (Dumitriu-Snagov, 1979; Semenescu, 1983).

At the beginning of the 18th century, the conscription records from Transylvania brought numerous evidences of the existence of irrigated gardens and meadows along the streams passing through the villages north of the Carpathian Mountains (the geographical area known as Tara Fagarasului) (Irimie, 1968; Botzan and Albota, 1980; Blidaru et al., 1997).

It is to be noticed that in all the villages where the conscription took place during the period 1721–1722, there are mentioned fountains along the water courses, a custom which is still preserved today, with these areas being identified by significant surfaces of irrigated meadows. The practice of surface irrigation (by overflow) is more clearly mentioned a hundred years later, and the advantages of this practice are also stated. According to a study conducted at the middle of the 20th century, at the beginning of the 19th century, in Tara Fagarasului, over 6,000 ha were irrigated in 20 villages (Mateescu, 1975; Nicolau et al., 1970; Nedelea, 1971; Semenescu, 1983; Cazacu et al., 1989; Blidaru et al., 1997).

The Romanian peasants from Tara Fagarasului practiced a technique of irrigation of the meadows and orchards perfectly adapted to the hydrography and the shape of the land, with very simple means and procedures. With rocks from the river and bundles of twigs, they built a spur into the unstable riverbed that diverted the water into an adduction channel (sometimes branched) that turns back through a loose shape. This channel branches into the irrigable land through gutters from which the watering furrows split off being oriented on the level curves. The water flows out of these furrows in the meadows, wetting the gap between the two furrows and feeding with meadow the excess water (Irimie, 1968; Telegut, 1971; Nedelea, 1971; Botzan, 1984).

In the 18th and 19th centuries, a series of irrigation channels were executed to the south and east of the Carpathian Mountains, which are still functional today. In 1865, Ion Ionescu de la Brad proposed the irrigation of several meadows and sandy areas (Mindru, 1969). Seven years later, P.S. Aurelian recommended irrigation as a general measure for increasing agricultural production. In 1912, A. Davidescu presented an ante-project for irrigation of over 1 million ha in Baragan of which, approx. 70% to be irrigated from inland waterways and the difference with Danube water (Telegut, 1971; Semenescu, 1983; Blidaru et al., 1997).

13.3 Irrigation in Modern Romanian State (Since 1918 Until Nowadays)

There is not much information about irrigation systems' development until the second half of the 19th century. Even in Romania there were favorable conditions for extending the irrigation systems. Until 1944, the total irrigated surface was about 15,000 ha mostly covering vegetables and rice crops. The design, implementation and exploitation of these arrangements were made in total inadequate conditions.

Starting in 1953, the expansion of irrigated surfaces has seen a higher rate in 1965 reaching a total irrigated area of about 230,000 ha (Nicolau et al., 1970). Five years later, the irrigated area increased by almost 800,000 ha. At the end of 1990, 3.2 million ha were irrigated in Romania, most of them in large and very large irrigation systems (over 200.000 ha) (Blidaru et al, 1997) (Figure 13.1).

Until 1990, the irrigation sector benefitted from important governmental incentives. Much of the irrigated land was located on high terraces above the level of water supply, which, in some cases, assumed pumping heights of up to 75 m. The specific costs of irrigation through networks were kept secret by the communist authorities (especially regarding the cost of energy consumed to pump water). At that time (1985–1990), the state was already in a situation where it could no longer provide the necessary funds for the essential maintenance works and exploitation of the irrigation infrastructure. As a consequence, it was a matter of time before irrigation systems started to experience a continuous degradation.

After the political changes that occurred at the end of 1989, the new government instituted reforms that had a serious impact on agricultural productivity. The most important reform took place in 1991, by applying Law No. 181/1991. Thus, the land was returned to the former owners (those confiscated by the communists immediately after the end of Second World War). Agricultural Production Cooperatives (CAPs), communist-type structures for agricultural land management, were divided into many agricultural units, each of which was split into plots.

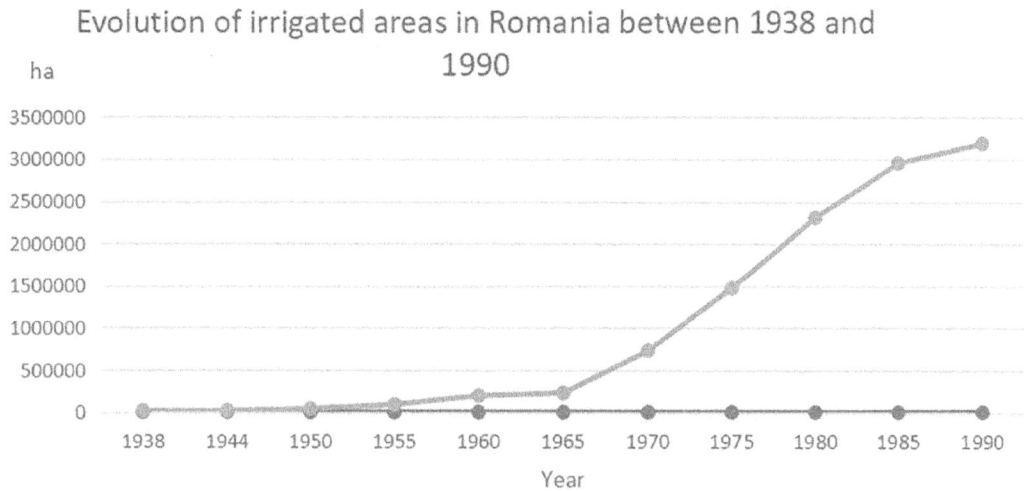

FIGURE 13.1 Evolution of irrigated areas in Romania between 1938 and 1990.

Restoration or establishment of land ownership rights should have been made within the limit of at least 0.5 ha per person and up to a maximum of 10 ha per family, all areas being calculated in cultivation equivalent. As the main result, this land reform has led to an enormous fragmentation of land.

The existing and functional national irrigation system before 1989 became vulnerable after this reform, the main risk factors being:

- designed and equipped for very large farms, irrigation facilities have become inoperable/nonviable in the conditions of "breaking up" agricultural land and in the absence of an effective governmental policy to stimulate landowners' association/land consolidation;
- the implementation of market economy principles and the criteria of economic efficiency have made irrigation facilities unviable in the absence of coherent and effective governmental policies aimed at their energy-saving nature (high costs of pumping electricity), the lack of investment in reconversion of pumping water supply to gravity, the wear of irrigation infrastructure and the irrigation facilities and equipment.

At the beginning of the last decade of the 20th century, a study developed by the World Bank emphasized that only about 50% of the over 3.1 million ha of lands arranged with irrigation infrastructure could be considered economically viable. The study was based on an analysis of several factors like the complementary nature of irrigation, the high pumping height of many irrigation systems and the low degree of irrigation use. As a result, Romanian government developed a new strategy on irrigation aiming to encourage economic irrigation and optimal use of available resources, seeking to maximize the areas where irrigation was economically viable, given the amount of subsidies provided at that time. However, due to government changes, conservative policies and consequently poor progress in restructuring the irrigation sector, irrigation, as well as due to limited funds, only in 1999 the reform could materialize in technical assistance to Irrigation Water Users Associations (IWUA), through a 1-year project that resulted in a preliminary legislative framework (Governmental Emergency Order (GEO) No. 147/1999), which allowed the establishment and registration of IWUAs.

In 2002, there were 4.3 million farms (small and large), divided into 14.5 million plots. The average size of a farm was 3.1 ha, and the average size of a plot was 0.95 ha. Nearly 50% of the total were farms with an area of less than one hectare. In particular, within irrigated systems, the return of properties and their fragmentation inherently led to the quasi-impossibility of organizing and managing water distribution. While large areas of land could be irrigated in the past by relatively simple watering schemes, after the property restitution process, a lot of small landowners had to be supplied with irrigation water.

Statistical data from the National Statistics Institute (2002) indicate that the irrigation infrastructure was not located proportionally to the different types of agricultural producers. Thus, the irrigable area reported was 1.5 million ha (10.8% of Romania's agricultural area), of which 532.000 ha (35%) were managed by individual agricultural producers and 979.000 ha (65%) were managed within the entities with legal status. Forty-two percent of the total irrigable area was concentrated with 1,295 commercial companies with an average size of 488 ha. Agricultural companies managed only 16% of the total irrigable area.

In 2004, a new law for land improvement systems was given. This new law on land improvements (Law No. 138/2004) replaced GEO No. 147/1999 and detailed the new tasks of National Society of Land Improvement (NSLI) and IWUAs. NSLI was transformed into two separate entities. First, the National Land Improvement Administration (NLIA) was responsible for basic land improvement activities, including system management. The new NSLI had tasks related to maintenance and repairs of land improvement systems. Shortly after, the IWUA's attributions have been reviewed and adapted to land improvement organizations that are responsible for irrigation, surface drainage, land drainage, flood defenses and soil erosion, or a combination of these activities. With this change, IWUA's had to re-register as IWUOs (Irrigation Water Users Organizations). The same law has provisions on the establishment of IWOU's federations and attributions.

The application of this law also created some administrative problems. In accordance with the provisions of this law, IWUAs were now responsible for water distribution and maintenance of the irrigation systems on their territory, although in many cases the infrastructure itself was still owned by the state. This concession arrangement was set up to protect the existing infrastructure until it was clear which systems would be rehabilitated and until the IWUAs could become strong enough to finance their maintenance and repair work at their own systems.

The mechanism for the allocation of subsidies for irrigation and its essence, defined within Law No. 138/2004, stipulate that this subsidy is determined each year as "a maximum ceiling per hectare" corresponding to the contracted area, which is unique at the national level and applicable across the country, regardless of the location of each irrigation water supply point within a particular irrigation system. According to this law, the allocation and distribution of the state subsidy is based on the following principles:

- The beneficiary of the subsidy is the water user who must be a member of an IWUO;
- The subsidies are granted only if the request for water supply involves the irrigation of at least 20% of a surface served together;
- The subsidy for irrigation is actually paid only if the IWUO has paid its own contribution to the annual tariff;
- The subsidy will be used only for the payment of electricity and for the exploitation, maintenance and repair works for the irrigation systems.

The grant ceiling covered the following categories of expenditure:

- 80% of the annual tariff (expressing maintenance costs for the water supply and distribution network of the water supplier (NLIA));
- 90% of the operating costs of the irrigation water supplier or delivery charge (NLIA);
- 100% of the energy costs (part of the IWUO operating costs) at the Pumping Station level to the electricity company.

Few observations can be made to the current subsidy mechanism:

- The existing subsidizing mechanism favors commercial IWUOs compared to farmers.
- The Irrigation Grant is defined as a sum of money for the area contracted between NLIA and IWUO each year, but there is no analysis (reporting) of the actual irrigated area and irrigated crops.

- The existing subsidizing mechanism does not stimulate any improvement in the management activity of the main irrigation water supplier (NLIA) or the achievement of higher results, including the reduction of water losses and the timely, quantitative and qualitative delivery of irrigation water required.
- The current subsidizing mechanism is restrictive for crops with higher water demand and is "friendly" to field crops such as cereals. This mechanism does not encourage the introduction of crops of high economic value to use irrigation water more efficiently.

Despite institutional reforms, the use of irrigation water has remained limited over the years. Between 2000 and 2006, a maximum utilization rate of 25% was achieved in 2002, within southern and southeastern Romania, and a similar percentage was reached in 2003. Both years were considered dry. For all other years of the range, the degree of use in the irrigation systems of different territorial branches was below that value. For 2007, almost 290,000 ha were initially contracted for irrigation, most of them being on the territory of IWUOs. In May 2007, when it was already clear that the year would be very dry, the government increased the subsidy ceiling and eventually irrigated some 350,000 ha. Data pertinent to previous years show that the actual irrigated percentage of the total contract area didn't exceed 50%. The main constraining factor in the use of irrigation water was its price despite the subsidies granted.

Another constraint was the lack of watering equipment. The Ministry of Agriculture and Rural Development (MARD) estimated that in 2004 the watering equipment was sufficient only for the actual irrigation of 350,000 ha. Farmers who used these equipment were mostly commercial, large-scale farmers associated with IWUOs with small farmers who practiced subsistence farming. These commercial farmers could afford to purchase watering equipment and apply irrigation for crops of high economic value (MARD, 2009, 2016).

Once Romania joined the EU, it had to comply with community policies and directives on agriculture and rural development and in this context to adapt its legislation. EU policies and directives in the field of agriculture and rural development are geared toward market liberalization, decentralization of decision-making and implementation of support programs, as well as a strengthening of farmers' status, aimed at transforming the agriculture sector into a more competitive one, considering that development should not be at the expense of the environment.

However, irrigation sector is known to be on a continuous decline. The total irrigated area compared to the total area arranged for irrigation, represented 11% in 2004 and 5% in 2013. The total irrigated area in NLIA administration decreased from 8% in 2004 to 0.5% in 2013 while the total irrigated areas in IWUOs ownership decreased from 26% in 2004 to 16% in 2013. As a result, for 95% of all irrigation facilities in 2013, the costs were incurred only for security, maintenance and repairs without the respective facilities being put in operation.

13.4 Brief Actual SWOT Analysis of Irrigation Systems in Romania

The Romanian MARD recently conducted a SWOT [1]analysis of the irrigation systems in Romania. In principle, the weaknesses of these irrigation systems exceed their strengths. At national and international level, potential risks regarding the irrigation systems created a state of uncertainty regarding the future of these systems.

Thus, in this report developed by MARD, we can notice the following strengths:

- The functioning of IWUA;
- Irrigation used mainly by commercial farms;
- The expertise of farmers in using irrigation;
- Water price in viable arrangements;
- The investments were made in this field starting with 2004 (approx. 80 million euro).

The weaknesses of irrigation systems in Romania were divided into several categories like: economic, technical, management, perceptive and organizational. The culture structure, irrigation systems design for large exploitations, old infrastructure, water losses from irrigation systems, energy consumption, and lack of cooperation from farmers are only a part of the weaknesses that characterize the Romanian irrigation systems (Romanian Court of Auditors, 2014; Romanian Government 2018).

In this report, presented are several opportunities (unconvincing) as well as several threats that can easily affect the current state of these systems in a negative sense.

13.5 Irrigation within Sustainable Agricultural Water Management

Romanian-specific terminology states that land improvement works include in particular irrigation and drainage arrangements but also soil erosion control. The land improvement works involve land, water and plant management, are energy users and suppliers, have a strong impact on land management and respond to climate change by mitigating their effects and by creating microclimates.

Understanding the interactions between agriculture and water and, in a broader context, the land–water–energy–climate nexus is crucial. Agriculture is inextricably linked to ecosystems, climate, energy and water in particular (Strzepek and Boehlert, 2010; Konikow and Kendy, 2005; Milly et al., 2005; Rosegrant et al., 2009). There is a particular need for the development of relevant climate–energy–economy models based on a specific approach to ecosystem services for a proper land-use analysis.

Land improvements represent an important component of water management in agriculture and have widespread influences in all components of the land–water–climate–energy nexus. These land improvement works provide important ecosystem services including aquifer recharge, flood retention, carbon fixation, soil organic matter accumulation, soil nutrient recycling and support for flora and fauna diversity through habitat creation. Integrating these benefits into agricultural water management requires breaking the barriers between engineers, ecologists, agronomists, economists, hydrologists and climate researchers, respectively, applying valid climate–energy–economy models as well as land-use models.

Improving the understanding of these ecosystem services provided by land improvement systems and the relationships developed within the land–water–climate–energy nexuses as well as the implementation of land improvement arrangements adaptable to climate change will reduce the actual existing pressure on basic resources.

Water deficit and excess water (water stagnation on land) are both causes of land degradation. There are a number of technical options that address the impact generated by these causes in agriculture and that have been considered quite often as adaptation strategies (Fleskens et al., 2005). The land improvement works represent an important category of these technical options. However, the limitations of such an approach must be well understood (Vincent et al., 2013) and interactions with ecosystem service provision become even more important (Dale and Polasky, 2007). The attention that must be paid to the ecosystem services provided by these land improvement works should be increased considering that this approach has been neglected quite a lot.

Agriculture water management has important interactions with these services, such as carbon sequestration (Follett, 2001; Lal, 2008) and greenhouse gas emissions from turbid soils (Wessolek et al., 2002; Kluge et al., 2008). These dynamic interactions between the principles of sustainable land management and the working of decision-makers must also be better understood (Fleskens and Hubacek, 2013).

At the European level, there is a large volume of data on land improvement systems and the types of degradation to which they respond but these data are scattered, fragmented and sometimes incomplete, especially when we talk about the complexity, functioning and services of land improvement works as well as interactions with the nexus of food–climate–energy–ecosystems.

Land improvement works should also be part of sustainable land management, which is defined as the use of water and land resources, including soil, water, animals and plants, for the production of goods to meet human needs, while ensuring, in the long term, the productive potential of these resources and maintaining their ecological (environmental) functions (UN Earth Summit, 1992).

The production of food requires very large quantities of water. It is estimated that only crop evaporation can reach 6,700 km^3 of which 18% is irrigation water (Siebert and Doll, 2010). Irrigation consumes about 70% of the water taken by humans from the surface and deep sources. However, this huge amount of water contributes significantly to global food production (Turral et al., 2010). The differences between the production obtained in irrigated agriculture and those in nonirrigated agriculture were emphasized by Siebert and Doll (2010) who estimate that in the case of irrigated cereal areas, the production is about 4.4 t/ha compared to 2.7 t/ha obtained in the case of nonirrigated areas. Over 40% of the cereals come from the irrigated parts of the world, with the cereal production registering a decrease of more than 20% if they would abandon these arrangements (Siebert and Doll, 2010).

Understanding the regional characteristics is a key element for the decision-makers involved in planning, designing and operating climate-adaptable land improvement systems but also in establishing/mapping new relevant policies in this sector.

Agriculture is one of the largest contributors to climate change through greenhouse gas emissions and indirectly through the conversion of natural ecosystems into agricultural land. Also, agriculture has the greatest potential for mitigating the effects of climate change, especially by sequestering carbon in soils (IPCC, 2007). The application of some measures of conservation of agriculture, the re-establishment with native vegetation of the abandoned agricultural areas represents only a few measures with high potential in the face of the challenges generated by the climatic changes.

Irrigation in arid and semi-arid areas plays an important role in achieving this goal. The addition of water to the soil contributes to the development of plants and the carbon deposition in the soil (Entry et al., 2004). The extension of irrigated areas and the conversion of water-bearing areas from rainfall to forest areas will contribute to increased productivity and at the same time to carbon fixation in soil (Entry et al., 2004), while the application of long-term irrigation will contribute to increasing the amount of carbon in the soil by adding biomass (Wu et al., 2008). However, there are some voices who argue that irrigation has not reached such a level of development that it makes a high contribution to carbon fixation (Schlesinger, 1999). The emissions generated as a result of the operation of the pumping stations may in some cases exceed the amount of carbon fixed in the soil due to irrigation application. Furthermore, groundwater in arid areas contains significant amounts of dissolved Ca and CO_2 that will be released into the atmosphere with the use of these water sources in irrigation. Rice fields, although effective in combating floods, contribute to climate change through methane emissions. Tyagi et al. (2010) proposed in this respect complex irrigation and drainage systems.

Improving water productivity and meeting environmental requirements will reduce the impact of irrigation arrangements through lower water consumption required for system exploitation and crop development and in parallel with maintaining water flows needed to support riparian ecosystems. Water productivity has sometimes been defined by the phrase "more production per drop," which means maximizing agricultural production per unit of water used. In a broader sense, considering the multiple ways of using water resources, the expression "more value per drop" or "more jobs per drop" was also used.

De Fraiture et al. (2013), however, proposed the use of the expression "more benefits per drop" or "net income per unit of used water" thus to include all ecosystem services provided by water management in agriculture. Integrating these benefits (food production, aquifer recharge, flood retention, biodiversity and carbon fixation), however, requires inter- and multi-disciplinary collaboration between researchers and policy makers in fields such as engineering, ecology, agronomy, economics, hydrology and climatology.

Identifying a balance between maximizing food production through the intensive exploitation of water and soil resources and conserving these resources to ensure sustainable development and preserving biodiversity is a future challenge. Most research on ecosystem services provided by irrigation systems is limited to the study of rice fields. It is therefore necessary to continue and encourage the study of good practices in water management in agriculture for the development of ecosystem services in parallel with the production of sufficient quantities of food.

13.6 Perspectives of Irrigation in Romania

From the data presented by Smedema (1995, 2000), it follows that, worldwide, about 1,100 million ha with agricultural holdings do not have a water management system. In these areas, the productivity reaches up to 45%. Irrigation covers an area of 270 million ha and contributes to obtaining 40% of production using about 70% of the water resources extracted from river systems worldwide. Sixty million ha with irrigation also have drainage systems.

According to Romanian National Strategy for Sustainable Development, having the year 2030 as the horizon of time, Romania is far behind other European countries in terms of irrigation systems' development. This situation should be studied from the perspective of climate changes, given that drought currently affects more than 50% of agricultural land and only 12% is covered by viable irrigation systems.

In 2016, it was developed and approved by the National Rehabilitation Program of Main Irrigation Infrastructure, valuing around 1 billion euros and which aims to adapt Romanian agriculture to climate changes and to reduce their effects on agricultural production.

The general objective of the program is to rehabilitate the main irrigation infrastructure that will lead to the increase of the functional surface from the economically viable and marginally viable surface for irrigation to 70% in 2020 and to 90% in 2030, and consists in the rehabilitation of the main irrigation infrastructure from 86 viable arrangements belonging to the public domain of the state, with an area of approximately 1.8 million ha, by the end of 2020.

The specific objective of the program is to increase the efficiency of the main pumping stations (fixed and floating) and re-pumping, to eliminate the water losses by infiltration from the irrigation channels belonging to the public domain of the state and to eliminate the degradation occurring in the hydrotechnical constructions belonging to these systems.

The restoration of the existing irrigation capacities is a basic measure for the development of the agricultural sector in Romania. The rehabilitation of the main irrigation infrastructure will lead to an increase in the operating efficiency of the irrigation facilities with direct reflection in reducing the tariff/1,000 m^3 pumped water, which will create greater possibilities for farmers to use irrigation water.

Investments in the rehabilitation of the main irrigation infrastructure will generate macroeconomic effects consisting mainly of obtaining a net increased income compared to the situation before the rehabilitation as a result of:

- improving lands' productivity, currently with a deficit of humidity, saltiness, acids, etc.;
- improving culture plan structure, by using valuable and cost-effective plants;
- increasing average production per hectare.

The farmers will feel the effects of the measures to reduce the tariffs for irrigation, in the sense of reducing the costs of production and implicitly of increasing the profitability of the agricultural activities.

The rehabilitation of the main irrigation infrastructure will lead to savings of water and energy resources, by reducing losses and applying an efficient management of resources. The social impact of the rehabilitation will be materialized by carrying out the rehabilitation action that will reduce the vulnerability of the human communities to drought by avoiding affecting a population of 1,097,433 inhabitants from 190 localities.

Ecological impact will result from the rehabilitation of some existing works whose ecological impact was evaluated at the time of execution, without generating changes in the initial ecological impact.

Thus, the MARD, under the condition that the government assumes, by decision, the National Program for the Rehabilitation of the Main Irrigation Infrastructure, will receive during 2016–2020 the amount of 1.015 billion euros necessary for the rehabilitation of the main infrastructure.

Irrigation systems, under the current climatic conditions and under the pressure of new challenges regarding the availability of water resources, have the obligation to produce increasing quantities of food on the basis of a lower consumption of water. Innovative practices in the field can bring an economic advantage in parallel with reducing the negative effects on the environment (quantity of water extracted from sources, energy consumption, polluting elements) (Faures et al., 2007). Water-efficient practices have the potential to increase economic viability and environmental sustainability in irrigated agriculture without the need to reduce water consumption.

Innovative irrigation technologies are generally seen as technologies that improve the efficiency of water use in addition to other benefits. The policies of the EU place a special emphasis on this kind of technologies and have high expectations from them. The European Commission highlights "technological innovation in the field of water, considering that water efficiency will be a factor of increased importance for competitiveness" (CEC, 2008). European Parliament believes that solutions must be found in "environmentally friendly technologies that facilitate the efficient use of water" (EP, 2008).

Higher water use efficiency depends on proper management of agricultural practices in addition to superior technology. However, major companies in the field have preferred to focus on modern technologies in the idea that they will only benefit. Perry et al. (2009) stated, however, that a modern irrigation system, based only on the implementation of superior technologies, can be as unproductive and useless as a faulty management of a classical irrigation system. In case of incorrect application, irrigation technologies can cause significant losses based on high investments and an economic indicator of low water productivity (Battilani, 2012).

The limitations encountered in the implementation of technologies needed to improve the efficient use of water are due to several essential factors:

- Perception of irrigation equipment as exclusive benefit providers;
- The desire of farmers to increase their net income at the expense of improving water productivity;
- Lack of technical recommendations in the field;
- Farmers not knowing how to anticipate specific practices in the field of irrigation to evaluate the efficiency of these systems.

Pereira et al. (2012) stressed that farmers generally do not have assistance in developing and adopting sustainable environmental practices in parallel with maintaining financial and social-economic objectives. Decision support systems aim to improve the efficiency of water use on farms, but few are widely applied as they require certain skills for use.

For a decision support system to be successful, it must include a few elements:

- Must provide farmers for daily irrigation with a simple, real-time, easy-to-use, free system;
- To be able to provide specific solutions to as many cultures as possible;
- To analyze the economic benefits, the relevance for the next irrigation and to calculate the profitability of the irrigation (Levidow et al., 2014).

In order to exploit the full potential of water efficiency measures, it is necessary to disseminate their benefits as widely as possible, the specific training of farmers and the coupling of technological solutions with operational practices to increase the economic performance of farms (Tollefson and Wahab, 1994).

13.7 Conclusions

In recent years, new modern methods in the field of irrigation have been developed. Their occurrence is related to the need to minimize the negative consequences generated by the anthropogenic impact on water and land resources. These methods are especially dedicated to water deficit management, soil pollution control and soil erosion control. Together with the traditional methods that have been used so far, the new methods form the basis of the sustainable development of a society capable of responding to the challenges of the present.

Irrigation is a vital element of modern agricultural systems as they increase soil fertility and ensure high and stable agricultural productivity over time. The sustainability of agricultural areas is ensured by a control of the processes of hydrological, bio-chemical, and hydro-geological nature, as well as of other processes that take place in these landscapes. The main purpose of these control measures is to control the flows of matter and energy within agricultural systems and to minimize losses.

A major problem to be answered by these measures is the management of an increasingly intense competition for water resources against the need to maintain a clean and sustainable environment. In the agricultural sector, we can see that an increase in productivity will be definitely based on the implementation of new technologies in the field of irrigation, on a more concrete calculation of the water requirement for plants, on significant improvements in the design, implementation and on operation of irrigation systems. This will be able to answer the current problems in the irrigation sector—problems that need to be solved as efficiently and as quickly as possible (Den Toorn, 1993; Kuroda, 1995; Shady, 1999).

The modernization of irrigation systems and the adoption of financially, socioeconomically and environmentally compatible management systems require a holistic approach in trying to increase agricultural production, to improve water resource conservation techniques, to prevent soil salinization as well as water stagnation, respectively, to protect the environment (Malek-Mohammadi, 1998; Dudley, 1999).

Looking from the perspective of sustainable water management in agriculture we can see that most irrigation systems have developed in a step-by-step conception over the decades. In many of these systems, some structures are aging and/or damaged. In addition, irrigation systems must withstand social, economic pressure and be able to respond positively to changes in the requirements and needs. As a consequence, the infrastructure in these irrigated areas must be renewed or even replaced in order to continue the sustainability of agricultural production. These processes depend on a number of factors that need to be well coordinated: new and modern technology, environmental protection, institutional strengthening, economic and financial analysis, and quality research and human resources development. However, many of these factors are associated with the risks and uncertainties arising from climate change, market and international trade.

Note

1 SWOT – Strengths, Weaknesses, Opportunities, Threats.

References

Alexandrescu A. (1970) Histrian landscape in antiquity (in Romanian). *Pontica* 3:77.

Ammianus M. (1982) *Roman history (In Romanian)*. E.S.C., Bucharest, Romania.

Aricescu A. (1977) *The army in Roman Dobrogea (In Romanian)*. Military Printing House, Bucharest, Romania.

Avram A., Bounegru O. (1986) New contributions at Histria aqueducts issue (In Romanian). *Studies and Research in Ancient History* 37(3):262–265.

Barbulescu M. (1984) *Spiritual interferences in Roman Dacia (In Romanian)*. Ed. Dacia, Cluj-Napoca.

Battilani A. (2012) Sustainable knowledge-base-d irrigation management: The IRRINET package. http://ec.europa.eu/environment/archives/greenweek2012/. (accessed on May 26 2019).

Bleahu M. (1963) Observations on Histria area evolution in the last three millenniums (In Romanian). *Probleme de Geografie* 9:45–56.

Blidaru V., Wehry A., Pricop G. (1997) *Irrigation and drainage arrangements (in Romanian)*. Interprint, Bucharest, Romania.

Bodor A. (1957) Contributions on agricultural issue in Dacia before Roman conquer (In Romanian). *Studies and Research in Ancient History* 1–4:137–148.

Botzan M. (1979) Observations from the last century on some antique constructions from Dobrogea (In Romanian). *Pontica* XII:171–179.

Botzan M. (1980) Hydrotechnical constructions in antique Dobrogea (In Romanian). *Hidrotehnica* 23(10):70–71.

Botzan M. (1984) *Waters in the life of Romanian people (In Romanian).* Ed. CERES, Bucharest, Romania.

Botzan M. (1989) *The beginning of hydrotechnics on Romanian territory (in Romanian).* Technical Printing House, Bucharest, Romania.

Botzan M., Albota M. (1980) *The age of agricultural irrigation in Romania (In Romanian).* Terra Nostra IV, Bucharest, Romania.

Branga N. (1980) *The urbanism of Roman Dacia (In Romanian).* Ed. Facla, Timisoara, Romania.

Canarache V. (1951) *About Histria supply with drinking water (In Romanian).* SCIV II (2), Bucharest, Romania.

Cantor L.M. (1967) *A world geography of irrigation.* Oliver&Boyd. Edinburgh, Scotland.

Cazacu E., Dobre V., Mihnea I., Pricop G., Rosca M., Sarbu E., Stanciu I., Wehry A. (1989) *Irrigations (in Romanian).* CERES Printing House, Bucharest, Romania.

CEC (2008) *Follow up communication on water scarcity and droughts in the European Union COM (2007) 414 final [SEC(2008) 3069].* Commission of the European Communities, Brussels, Belgium.

Conea I. (1935) *Lovistei country (In Romanian).* S.R.R.G., Bucharest, Romania.

Dale V.H., Polasky S. (2007). Measures of the effects of agricultural practices on ecosystem services. *Ecological Economics* 64:286–296.

De Fraiture C., Faryap A., Unver O. (2013). Integrated Water Management Approaches for Sustainable Food Production. Background paper for World Irrigation Forum, 28.09-5.10.2013, Mardin, Turkey. Online la www.icid.org/wif1_iwrm_bgpap.pdf (accessed at November 11, 2015).

Den Toorn W.H. (1993) Irrigation development. Project or process? Land and Water International, No. 77.

Dudley N.J. (1999). Integrating environmental and irrigation management in large-scale water resource systems. In: Mahendrarajah S., Jakeman A.J., McAleer M. (eds.): *Modelling change in integrated economic and environmental systems.* John Wiley and Sons Ltd., New York.

Dumitriu-Snagov I. (1979) *Romanian countries in the XIV century (In Romanian).* Cartea Romaneasca, Bucharest, Romania.

Entry J.A., Sojka R.E., Shewmaker G.E. (2004). Management of irrigated agriculture to increase organic carbon storage in soils. *Soil Science Society of America Journal* 66:1957–1964.

EP (2008) Addressing the challenge of water scarcity and droughts in the European Union. European Parliament Resolution of 9 October 2008, Addressing the Challenge of Water Scarcity and Droughts in the European Union (2008/2074(INI)).

Faures J., Svendsen M., Turral H. (2007) Reinventing irrigation. In: Molden D. (ed.): *Water for food, water for life: A comprehensive assessment of water management in agriculture.* Earthscan and International Water Management Institute, London; Colombo (Capitolul 9).

Fleskens L., Hubacek K. (2013). Modelling land management for ecosystem services. *Regional Environmental Change* 13:363–366.

Fleskens L., Stroosnijder L., Ouessar M., De Graaff J. (2005). Evaluation of the on-site impact of water harvesting in southern Tunisia. *Journal of Arid Environments* 62:613–630.

Follett R.F. (2001). Soil management concepts and carbon sequestration in cropland soils. *Soil and Tillage Research* 61:77–92.

Giurescu C.C. (1973) *Contributions to the history of Romanian science and technology in the 15th century – beginning of the 19th century.* Scientific Printing House, Bucharest, Romania.

Halbac-Cotoara-Zamfir, R., Eslamian, S. (2017). Functional analysis of regional drought management. In: Eslamian S. and Eslamian F. (eds.): *Handbook of drought and water scarcity, Vol. 3: Management of drought and water scarcity,* Taylor and Francis, CRC Press, Boca Raton, FL, pp. 305–328.

Hoffmann H. (1981) *A superior solution in the evolution of plants for the processing of cereal seeds: Floating mill, studies and communications on the history of popular civilization.* Terra Nostra, Sibiu, Romania, p. 125.

IPCC. (2007). Climate Change 2007: Synthesis Report. Contribution of Working Groups I, II and III to the Fourth Assessment Report of the Intergovernmental Panel on Climate Change. IPCC, Geneva, Switzerland.

IPCC. (2012). Managing the risks of extreme events and disasters to advance climate change adaptation. A Special Report of Working Groups I and II of the IPCC. Cambridge University Press, Cambridge, UK and New York, USA.

Irimie C. (1968) Traditional types of popular technical solutions in southern areas of Hunedoara, Sargetia V., Museum of Dacian and Roman Civilization Deva, Romania, p. 531.

Kluge B., Wessolek G., Facklam M., Lorenz M., Schwärzel K. (2008). Carbon loss and CO_2-C release of drained peatland soils in northeast Germany. *European Journal of Soil Science* 59:1076–1086.

Konikow L.F., Kendy E. (2005). Groundwater depletion: A global problem. *Hydrogeology Journal* 13, 317–320.

Kuroda M. (1995) The role of advanced technologies in irrigation and drainage systems in making effective use of scarce water resources. General Report. Proceedings of the ICID Special Technical Session on the Role of Advanced Technologies in Irrigation and Drainage Systems in Making Effective Use of Scarce Water Resources. September, Rome, Italy.

Lal R. (2008). Carbon sequestration. *Philosophical Transactions of the Royal Society B: Biological Sciences* 363:815–830

Levidow L., Zaccaria D., Maia R., Vivas E., Todorovic M., Scardigno A. (2014). Improving water-efficient irrigation: Prospects and difficulties of innovative practices. *Agricultural Water Management* 146:84–94

Malek-Mohammadi E. (1998). Irrigation planning. Integrated approach. *Journal of Water Resources Planning and Management* 124(5):272–279.

Mateescu C.N. (1975) Remarks on cattle breeding and agriculture in the middle and Late Neolithic on the Lower Danube Considérations sur l'élevage et l'agriculture au Néolithique moyen et final du Bas-Danube. *Dacia. Revue d'Archéologie et d'Histoire Ancienne* 19:13–18.

Metz B., Davidson O.R., Bosch P.R., Dave R., Meyer L.A. (2007). IPCC. Climate Change 2007: Mitigation, Contribution of Working Group III to the Fourth Assessment Report of the IPCC. Cambridge University Press, Cambridge, United Kingdom and New York, USA.

Milly P.C.D., Dunne K.A., Vecchia A.V. (2005). Global pattern of trends in streamflow and water availability in a changing climate. *Nature* 438: 347–350

Mindru R. (1969) *Short overview on land improvement history in Romania (in Romanian)*. Ministry of Agriculture, Terra Nostra Handbook, Bucharest, Romania.

Ministry of Agriculture and Rural Development (2009) Project of irrigation rehabilitation and reform. Economic Analysis of the Irrigation Sector. A World Bank report.

Ministry of Agriculture and Rural Development (2016) *The national program for the rehabilitation of main infrastructure from irrigation systems*, Bucharest, Romania. www.madr.ro (accessed on July 15, 2018).

National Statistics Institute (2002) Agricultural census. Preliminary results, Bucharest, Romania. www.insse.ro (accessed on November 12, 2018).

Nedelea D. (1971) Aspects from popular technique in water supply of vegetable crops. *Terra Nostra* 2:31–44.

Nicolau C., Vaisman I., Plesa I., Ceausu N., Muresan D., Popescu I. (1970) *Land improvement (in Romanian)*. E.D.P., Bucharest, Romania.

Parry M.L., Canziani O.F., Palutikof J.P., van der Linder P.J., Hanson C.E. (eds). 2007. Intergovernmental Panel on Climate Change (IPCC). Climate change 2007: Impacts, adaptation and vulnerability, Contribution of Working Group II to the Fourth Assessment Report of the IPCC. Glossary, pp. 869–883, Cambridge University Press, Cambridge, UK.

Pereira L.S., Cordery I., Iacovides I. (2012). Improved indicators of water use performance and productivity for sustainable water conservation and saving. *Agricultural Water Management* 108:39–51.

Perry C., Steduto P., Allen R.G., Burt C.M. (2009) Increasing productivity in irrigated agriculture: Agronomic constraints and hydrological realities. *Agricultural Water Management* 96:1517–1524.

Petrescu-Dambovita M. (1978) *Short history of pre-Roman Dacia*. Ed. Junimea, Iaşi, Romania.

Romanian Court of Auditors (2014) Report on the performance regarding the implementation of land improvement reform regulated by Law 138/2014, Bucharest, Romania.

Romanian Government (2018) *Romania 2030 national strategy for sustainable development*. Paideia, Bucharest, Romania.

Rosegrant M.W., Ringler C., Zhu T. 2009. Water for agriculture: Maintaining food security under growing scarcity. *Annual Review of Environmental Resources* 34:205–222.

Schlesinger W.H. (1999). Carbon sequestration in soils. *Science* 284(5423):2095.

Semenescu M. (1983) Quays and other hydro-technical constructions realized in the past in some Romanian harbors. *Hidrotehnica* 28(9):271–275.

Shady A.M. (1999) Water, food and agriculture. Challenges and issues for the 21st century. Keynote address. In: *Proceedings of the 17th ICID Congress*. September, Granada, Spain.

Siebert S., Doll P. (2010). Quantifying blue and green virtual water contents in global crop production as well as potential production losses without irrigation. *Journal of Hydrology* 384: 198–217. Doi: 10.1016/j.jhydrol.2009.07.031.

Smedema L.K. (1995). The global state of drainage development. Grid, IPTRID Network Magazine, No. 6.

Smedema L.K. (2000). Global drainage needs and challenges. In: *Proceedings of the 8th International Workshop on Drainage*, New Delhi, India.

Strzepek K., Boehlert B. (2010). Competition for water for the food system. *Philosophical Transactions of the Royal Society B: Biological Sciences* 365:2927–2940.

Telegut M. (1971) A peasant irrigation installation in Banatului plain (in Romanian). *Tibiscus* 1:152–156.

Tollefson L.C., Wahab M.N.J. (1994) *Better research-extension-farmer interaction can improve the impact of irrigation scheduling techniques*. World Bank, Washington, DC.

Turral H., Svendsen M., Faures J.M. (2010). Investing in irrigation: Reviewing the past and looking to the future. *Agricultural Water Management* 97(4):551–560.

Tyagi L., Kumari B., Singh S.N. (2010). Water management — a tool for methane mitigation from irrigated paddy fields. *Science of the Total Environment* 408(5):1085–1090.

UN Earth Summit (1992) Environment and sustainable development: Implementation of Agenda 21 and the Programme for the Further Implementation of Agenda 21. http://daccess-dds-ny.un.org/ (accessed on November 12, 2015).

Vincent K., Cull T., Chanika D., Hamazakaza P., Joubert A., Macome E., Mutonhodza-Davies C. (2013). Farmers' responses to climate variability and change in southern Africa – is it coping or adaptation? *Climate and Development* 5:194–205.

Wessolek G., Schwärzel K., Renger M., Sauerbrey R., Siewert R. (2002). Soil hydrology and CO_2 release of peat soils. *Journal of Soil Science and Plant Nutrition* 165:494–500.

Wu L., Wood Y., Jiang P., Li L., Pan G., Lu J., Chang A.C., Enloe H.A. (2008). Carbon sequestration and dynamics of two irrigated agricultural soils in California. *Soil Science Society of America Journal* 72:808–814.

V

Water Scarcity and Irrigation in India

14

Water Scarcity in India: An Evaluation Prior to Independence

Shovan K. Saha
Institution for Hygiene and Environmental Sanitation, New Delhi

Achintya Kumar Sen Gupta
Institute for Hygiene and Environmental Sanitation, New Delhi World Health Organization

Rina Surana and Prerna Jasuja
Malaviya National Institute of Technology

Saeid Eslamian
Isfahan University of Technology

14.1 Introduction

The **Indus Valley** was home to one of the world's first large civilizations. It began nearly 5,000 years ago in an area of modern-day Pakistan and Northern India.

There were more than 1,400 towns and cities in the Indus Valley. The biggest were Harappa and Mohenjo-Daro. Around 80,000 people lived in these cities.

(BBC)

DOI: 10.1201/9781003353928-19

Abundant availability of sweet water through millennia was one of the three necessary preconditions that led to incidence, growth and establishment of the Indus Valley Civilization. The other two are availability of large tracts of fertile land and stable politico-military situation. During her long history, India was dominated by Hindu or Vedic, Buddhist and Islamic cultures followed by British colonial rule from which India gained independence in 1947 (Saha, 2019). Subsequently, till date, India represents a multi-culture nation that officially practices secularism. The prosperity attained and sustained during the Hindu or Vedic era was rooted in a uniquely vast and deep pool of knowledge-base that evolved and accumulated over several millennia. The resultant culture and way of living essentially ensured well-being of both the society and the natural environment. Subsequently, variously interpreted and translated versions of some of that inherited reservoir of knowledge are found in the present era in the form of Vedas, Upanishad, Puranas (ancient scriptures) and the two epics Ramayana and Mahabharata. Thus, it is reasonable to believe that an efficient irrigation system existed throughout that period that contributed significantly to India's great prosperity and in turn, to her great fame as a country of abundant riches. Possibly so much so, that a large number of voyagers sailed across the oceans and ventured to reach India through her ports for trade and have glimpses of the famed land of abundance. Unmistakably, they associated the final part of their voyage across a smaller ocean that led them to the shores of India. Thus, the ocean around the Indian subcontinent – though smallest among all oceans of the world – was easily identified as the 'Indian Ocean'. India's landscape has been historically characterized by several major rivers and their numerous tributaries crisscrossing the vast land mass. Therefore, access to the groundwater was relatively easy, enabling the farmers and non-farmers alike to utilize groundwater in required amounts for irrigating their fields as well as meeting domestic consumption needs. With evidence found in the remains of Indus Civilization such as wells, public bath and brick-lined drains, one allows oneself to believe that the irrigation system during the long Hindu and Buddhist era, consisted of irrigation channels possibly fed directly by rivers and tanks created by holding rain water by check dams as well as wells. The wells were frequently equipped with Persian wheels.

Though the place of origin of this ingenious device is not clearly known, on the basis of description found in the Panchatantra[1] dated third century BC, Vishwanath observed (2008, 2009), "The Persian Wheel is perhaps actually the Indian Wheel".

The southern part of Indian subcontinent referred to as the Deccan plateau or simply as the Deccan, is distinguished from the northern region in terms of heavier annual precipitation, undulating topography and rainfed non-perennial rivers. Since ancient times, local communities therefore built anicuts at key locations on the rivers and devised channels to fill a series of tanks located carefully to form a cascade. Tanks were also built utilizing the natural topography which too were connected to the elaborate tank irrigation system (Figure 14.1). The communities, usually led by experienced men, managed the system well enough to ensure equitable distribution of water all the year round, as well as prevented droughts (Irshad and Eslamian, 2017). Until 1970s, the share of tank irrigation in South India was 92% which declined to 53% by 1990s (Sivasubramaniyan and Gopalakrishnan). The anicut canal across river Kaveri (Cauvery) in Tamil Nadu is believed to have been built between 100 BC and 100 CE as it is mentioned in Buddhist literature. The canal has been repaired and maintained through centuries and is still functioning. Soon after the management of the larger tank irrigation systems was taken over by the Public Works Department (PWD) during colonial rule in 1800, the efficacy of the system declined.

A firm conclusion, however, is that during the first 4,500 years of India's recorded history, the irrigation system in India evolved possibly from a somewhat simple concept to a highly sophisticated one that was efficient, sustainable and stable over the millennia.

During the Islamic era spanning over five and a half centuries,[2] though the irrigation system remained fundamentally unchanged from the preceding era, it was rejuvenated, augmented and improved. For example, the Western Yamuna Canal (WYC) – built before the Islamic era – was repaired by Firoz Shah Tughlaq (reigned 1351–1388) and utilized as the major conduit transporting water from snow-fed river Yamuna to the agricultural fields few hundred kilometres away. He is also credited for digging wells, *baolis* (usually rectangular deep stepped wells) and new canals

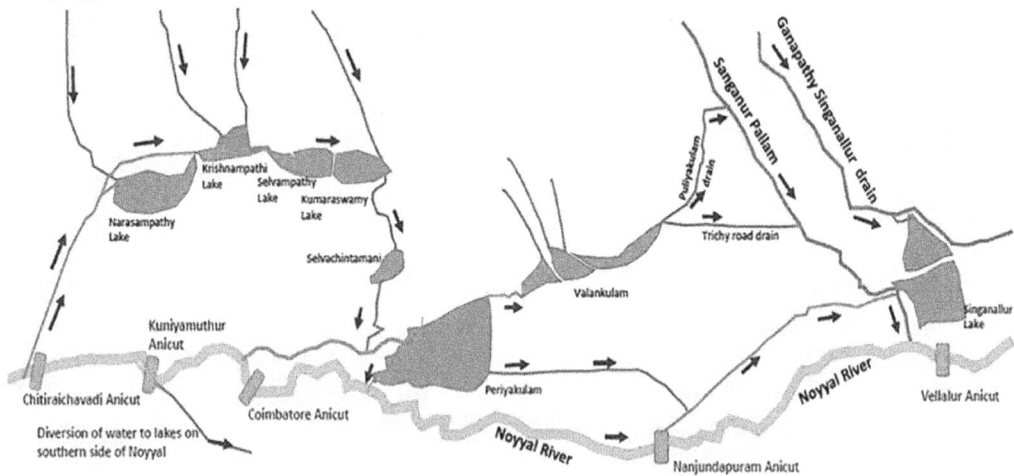

FIGURE 14.1 River Noyyal in the Coimbatore District, Tamil Nadu has got a number of anicuts to divert water to interconnected reservoirs for irrigation. The overflow water from these reservoirs gets connected to the river Noyyal by channels. This is a virtual sketch developed by field survey (map not to scale) (integrated analysis of water management and Infrastructure in Coimbatore – Report by Alyssa Neskamp, Dr. Stefan Liehr. Ing., Marius Mohr).

across the sprawling Tughlaq Empire, evidently to increase the agricultural production and collect larger amounts of revenues through taxes on land and farm produce. Powered by abundant supply of water from snow and heavy-monsoon-fed perennial rivers as well as a good irrigation system, both agricultural products and wealth accumulated to create huge reserves. In spite of repeated invasions by plunderers from far and near over centuries, early representatives of colonial rulers quickly discovered that India is a rich store house of resources.

While the area of irrigated land during the colonial period increased from 13 to 22 mha (between 1900 and 1947) and major canals were built – for example, Ganges canal (built 1854) fed by river Ganges – millions died of starvation due to famines at intervals of 2–3 years between 1860 and 1878. In the remaining period three famines occurred causing nearly 20 million deaths. Evidently much of the irrigated land was used for cultivation of poppy for producing opium exported to China. Irrigated fields particularly in Bihar and Bengal, traditionally used for paddy cultivation, were forcibly used for indigo cultivation in colonial India. Through a tyrannical cyclic loan arrangement, farmers were compelled by English and Scottish planters to cultivate indigo instead of paddy and other crops cultivated traditionally. Between 1788 and 1810 the share of indigo imported from India in the world market (predominantly European Market) rose from 30% to 95% giving huge profits to the exporters (Prasad, 2018). The direct implications of this policy consisted of degradation of fertile land, sharp decline of food crop production leading to frequent famines and millions of death (Box 14.1).

Towards the end of British period (1757–1947), several contemporary technologies and practices were introduced in selected industries and infrastructure operations and management of resources such as forest and water at the cost of traditional sustainable practices. Some of them were introduction of piped water supply in cities replacing traditional water management practices, long distance travel by railways and decline of transportation system by rivers and waterways. Some other trends during that period such as continued domestic population growth and exploitative taxation policies, regional and international geo-political disturbances such as WW I (1914–1918) and WW II (1939–1945), including siphoning of resources to England at the cost of starving millions in India to death contributed majorly to the general economic decline as reflected by massive unemployment, poverty, famines, erosion of physical and social infrastructure across the country (Mukerjee, 2010).

Box 14.1   Famines in India during British Rule

Twelve major famines took place in India from 1765 to 1943. Usually, a few million lives were lost in each, adding to an estimated figure of 77 million. The first known as Bengal Famine took place in 1769–1770 when 10 million or about one-third of then population of Bengal perished. The worst in terms of loss of life was the Indian Famine of 1896–1897 covering a major part of the subcontinent including Bengal, Madras, Bombay, United Provinces (present-day Chennai, Mumbai, Uttar Pradesh, respectively) and parts of Punjab. Ironically, the last one is also identified as Bengal Famine 1943–1944 followed by an epidemic of cholera, causing loss of life totalling to 3 million.

The following pages present an evaluation of India's irrigation status starting with pre-Harappan era of Indus Valley Civilization to Vedic times to pre-Islamic rule. Subsequently, the status of irrigation during Islamic and colonial era is considered. The entire discussion focuses on the fundamental characteristics of the irrigation systems which have been – like other ancient civilizations of the world – segregated from dimensions of demographic, technological, economic as well as socio-cultural boundaries of societies.

The status and evaluation of the irrigation system during the post-independence era that is 1947 onwards, are discussed in the subsequent chapter.

14.1.1 Study Limitations

Country-level studies on similar scale are usually undertaken over a period of 2 years or more, with the support of a sponsor and a team of researchers for a vast subject such as the present one. The present work, however, conducted without such support systems had to depend heavily on secondary sources of data.

14.2 Water Management Systems in Ancient India

The Rig Veda,[3] one of India's most ancient and revered literature, reminds the humans of the presence of water everywhere in nature, and its mystic role in making civilization happen in the following expression:

> *The water which is created in the universe, the water which flows in the springs, rivers, the water which comes from digging of wells, canals, the water which is self-created in the form of waterfalls, who enters into the ocean and who is pure and full of light, who is full of divine characteristics, help me in this world and be received by me.*

> *(Saxena, 2012)*

Water is among the precious sources of life after air and all ancient civilizations flourished around this core resource. The earliest human settlements developed on river banks, besides lakes and on the sea coast. While they settled there, they had to introduce first agricultural activities for their own survival and sustenance of community and later cottage and other industries for domestic and economic purposes, which is a basic aspect of social management.

Geographically, Ancient India consisted of not only present-day India but also entire Pakistan and parts of Afghanistan. Therefore, the discussion on development of civilization and water development of Ancient India includes the regions mentioned above representing a vast area, where one of the earliest civilizations of the world flourished.

1. Mehrgarh, which began in the Pre-Pottery Neolithic period (circa 7,000 BC or even earlier), is located in the Bolan Pass (west of Indus River Valley) on the Kachi plain of Baluchistan, Pakistan, on the principal route between what is now Afghanistan and Indus Valley (Kenoyer, 1991). The habitation got developed in seven phases. The site was abandoned between 2,000 and 2,500 BC, when the inhabitants got in contact with Indus Civilization. Mehrgarh had an access to river water for survival and development for thousands of years. Evidence shows that they had developed farming techniques and harnessed rainwater along with water from the river.

Studies show that Mehrgarh had a gradual development from a village society to a regional centre that covered an area of around 200 ha in its prime time. The inhabitants lived in houses and were involved in hunting and domestication of animals like cows, sheep and goats. Their farming products included cereals like barley and wheat. During the early period, these people used stone and bone tools including polished stone axes, flint blades and bone-pointer. By 6,000 BC, these people started working on hand-made pottery and introduced potter-wheel. Some fine terra-cotta figurine and pottery with exotic geometrical designs had been excavated from the site. Studies also show the evidence of textile and use of cotton even in that period. They also used ornaments made of beads, seashells and semi-precious stones like Lapis Lazuli (Kenoyer, 1991).

Indus Valley Civilization, which peaked around 2,500 BC, is spread out in an area of around 1,260,000 km² in Afghanistan, Pakistan and India (Kenoyer, 1991). There is an Indus Valley site on the Oxus river at Sortughai in northern Afghanistan and another at Alamgirpur around 28 km from Delhi, India. To date, over 1,052 cities and settlements have been found, mainly in the region around the Hakra-Ghaggar river and its tributaries. Out of these settlements, major urban centres like Harappa and Mohenjo-Daro, as well as Lothal, Dholavira, Ganweriwala, Kalibanga and Rakhigiri are quite famous.

Unfortunately, the Indus Civilization has still not been properly understood, more so, because no expert has been able to decipher the language the community had used to express their means of subsistence and the causes of their sudden disappearance beginning around 1900 BCE (Jansen, 1989). In fact, we are yet to understand the language they had used and how they had administered themselves both socially and economically.

In some of the excavations in the Indus Civilization area, systematically planned city with proper design of houses, road system, drainage and other facilities have been unearthed. Evidence of proper administration and management is visible. Studies show the habitants adopted rainwater harvesting (RWH) as well as river training methods as per site requirements. The Harappan irrigation system was one of the architectural feats. Instead of using canals or waterways all year round, they would instead merely use the flood season to their advantage. From Harrapan city we observe there is a huge wall surrounding the city with proper gates and watch towers at regular intervals.

Square seals or coins made of powder of white stone, ivory, clay and metals with pictures of animals, manlike figures and write-up have been unearthed from these sites. Due to lack of knowledge, it is difficult to make out what for these were used. Historians believe that seals might have been used for trade and rituals. Earthen pottery wares with designs and paintings were quite common and had been excavated from most of the sites.

The community mostly depended on farming, domesticating animals and trade including cottage industry. Crops grown those days mostly include peas, sesame seeds, dates, cotton, wheat and barley. Apart from sheep, goat and cattle they had domesticated water buffalos for intensive agricultural activities.

A brief description of the water and sanitation facilities of Mohenjo-Daro and Dholavira is given below.

1. The Mohenjo-Daro Water Supply: The people in the town were provided with fresh water drawn from wells, vertically built with specially designed wedge-shaped bricks. This well design was probably invented by the Harappan people. There must have been around 700 wells, one in every third house in Mohenjo-Daro. There was effluent drain network from every house in Mohenjo-Daro. The excavation of Great Bath of Mohenjo-Daro also shows the presence of a functional 'bathing platform' with a pool measuring around 12 m×7 m and 2.4 m deep. This was a brick-lined pool located in the central courtyard. There were group of rooms around the Great Bath for ablution. The construction of actual basin is a technical masterpiece which shows the very high standard of Harappan Engineering (Haraniya, 2017).

2. Deep into the arid stretch of Kutch, Gujarat, there is an island named Dholavira amidst the salt pans. More than 4,000 years ago, Dholavira was one of the largest cities of its time. It had a continuous occupation of more than 1,200 years. It existed between 2650 BC and 1450 BC. Dholavira seems to have evolved into a sophisticated planned city with a castle, middle town, lower town, bailey, stadium, water reservoirs and planned drainage system as per life and time of Harappan people. A regular house in Dholavira consisted of four rooms, a spacious courtyard, a bathroom and also a kitchen.

   Dholavira had a unique water management system. The main sources of water were two seasonal streams and groundwater. City administrations harnessed the fresh water and created a complex and effective water management system consisting of combination of channels and reservoirs of stones. As many as 16 reservoirs have been located and one of them was bigger than 'Great Bath' of Mohenjo-Daro (UNESCO, 2014; Singh, 2019) (Figure 14.2).

14.3 Irrigation Practices during Vedic Period

The earliest mentions of irrigation methods are found in *Rig Veda*, where a number of *slokas* are dedicated to this practice. *The Veda* mentions about irrigation from well system where *Kupa (wells)* once dug are stated to be always full of water. The *varata (rope)* and *cakra (wheel)* were used to pull *kosa (pails)* of water. The water was drained through *surmi susira* (broad channels) and from there into *khantitrima* (diverting channels) into fields (Biradar, 2015).

In third century BCE, Kautilya's (a celebrated Indian Saint-Politician) *Arthashastra* mentions about irrigation using water harvesting systems. Panini, the Indian scholar in the fourth century BCE, mentioned that rivers like Sindhu, Suvastu, Varnu, Sarayu, Vipas and Chnadhaga were trapped for irrigation purposes.

Cultivation of a wide range of cereals, vegetables and fruits was common and animal husbandry happened to be another main occupation for their livelihood. The cropping sequencing was prescribed and practiced rigorously. Preparation of manure from cow dung was also practiced during that period (Mulage, 2017).

In 1st century BC at Sringaverapura near Allahabad, U.P. had a sophisticated water harvesting system using flood water of Ganga. A 800 feet long, 60 ft. wide and 12 feet brick lined tank was constructed to collect overflow of river Ganga. The natural slop of the land was used to bring water. Two deep earthen tanks were used for settlement of silt. The settled water in the tank was used by the community during dry season.

(Source: C.P.R. Environmental Education Centre, Ministry of Environment and Forest, GOI)

14.3.1 *Araghatta* or the Persian Wheel

In Sanskrit language the word *araghatta* has been used in the ancient texts to describe the Persian Wheel. The word *ara-ghatta* comes from combination of two words, *ara* meaning spoke and *ghatta*

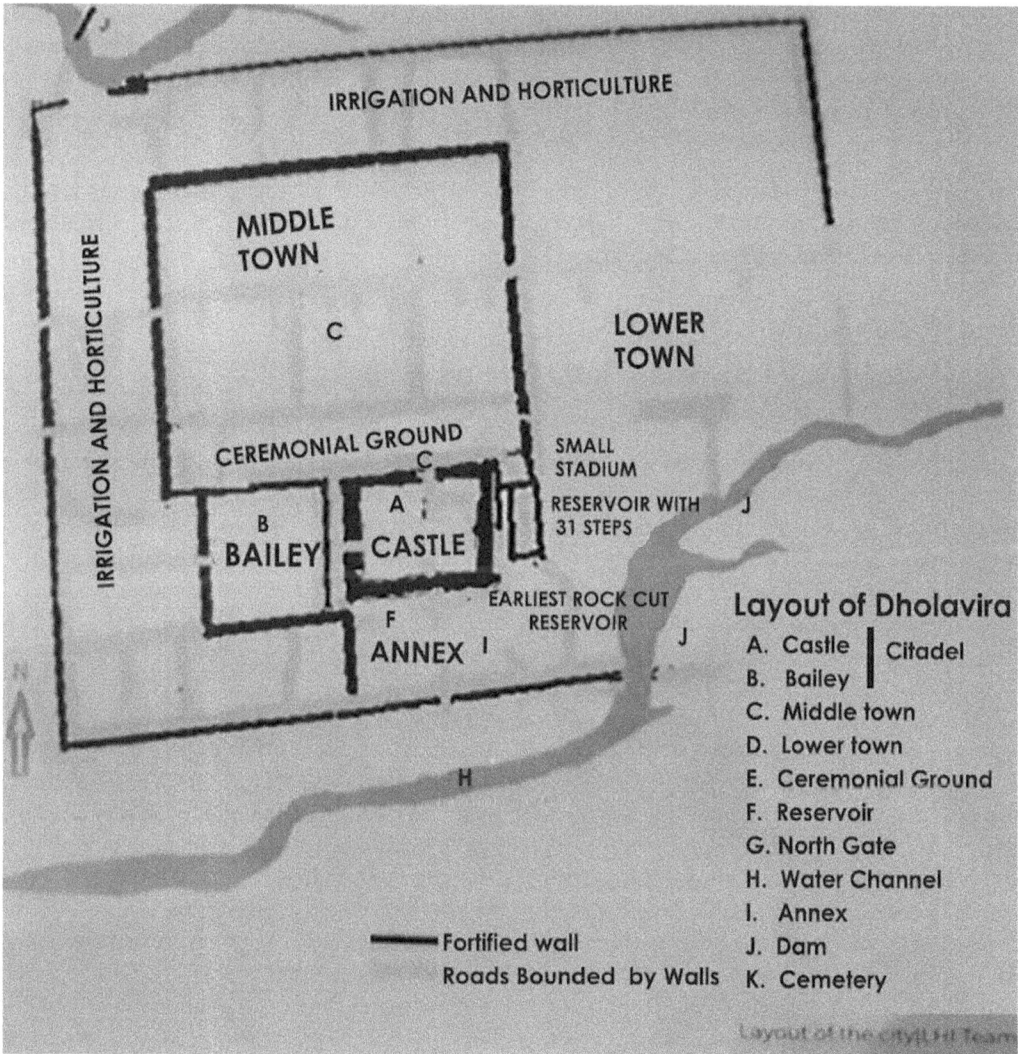

FIGURE 14.2 Layout of Dholavira, Indus Valley Civilization. Location of dams on the water channels show the river training activities and reservoirs show the water management activities carried out by the community (article by Haraniya Krutika – "Dholavira & Its Story of a Civilization", August 2017).

meaning pot. It is mentioned in the Panchatantra (third century BC) and Rajatarangini (12th century CE) as *cakka-vattakka* or the *ghati yantra*. The word *araghatta* is also called *rahat* or *reghat* in North India in later period and is being named even now. Usually men, bullocks, elephants or camels were used to convert the circular motion into vertical motion as used in the Persian Wheel. During Delhi Sultanate, this technology was re-introduced in Punjab province (Ahmed, 2018).

14.4 Irrigation Practices during Mauryan Period

1. The Mauryan Empire (322–185 BC) helped in categorizing soils and made metrological observations for the agricultural use. They tried to implement three types of irrigation systems by erecting dams across rivers/streams, by digging canals and by digging wells. Supply of water was regulated

and supervised by village headman. From *Arthasastra* we understand water tax was regularly collected from the farmers who were using irrigation water. Government agencies used to maintain various reservoirs, tanks, canals and wells. During the reign of Chandragupta Maurya a 100 ft. dam was built near Girnar to provide irrigation (*https://www.historyforexam.com/201f28/09/irrigation-system-of-ancient-india.html*).

The Greek diplomat Megasthenes (300 BC) in his book *Indika*[4] mentioned that greater part of the country were irrigated to bear two crops in a year consisting of cereals, millets, different sorts of pulses and rice apart from fruit trees all over the country (Mulage, 2017).

14.5 Irrigation Facilities and Its Development in South India during Chola Period (850–1250)

Since there are no snow-fed perennial rivers available in South Indian states of Tamil Nadu, Karnataka, Telangana and Andhra Pradesh and seasonal rains are sufficient, tank irrigation was more prevalent in these states. The people in South cultivated a wide range of crops namely rice, sugarcane, millets, black pepper, various grains, areca, cotton, sandalwood, plantation trees, etc. Systematic ploughing, manuring, weeding, irrigation and crop protection were practiced for a sustainable agriculture. During 100 BC–100 AD, a big anicut was first constructed across the river Cauvery at Kallanai, Thanjavur district, Tamil Nadu, to improve the irrigation system, which is still functional (Dharshivkar, 2007). Anicuts have also been built across the rivers Tambaraparani, Chittar and many other rivers. The anicuts in the Kanyakumari district had been built 1,000 years ago.

Due to royal patronage and with increased role of temples and village assembles, highly planned water control system in most of the villages were developed to control and manage water systems for the community benefit. Committees like *eri-variyam* (tank committee) and *tota-variam* (garden committee) were active. Apart from these, the temples with their vast resources of manpower, money and land contributed towards management of water resources and irrigation processes (Perundevi, 2019).

The system of collective holding of land got disintegrated into individual plots during the Chola period (875–1279 AD). Slowly each plot holder used to develop his own irrigation system as per the requirement. The Cholas also had bureaucrats who used to oversee the distribution of water, particularly where tank and channel system were introduced for dry lands (Mulage, 2017).

During Chola period a large number of very large water tanks were built for irrigation purposes. King Rajendra Chola built a huge tank named *Solagangam* in his capital city Gangaikonda Cholapuram and was described as 'liquid pillar of victory'. About 25.5 km long (16 miles), it was provided with sluices and canals for irrigation in the neighbouring areas. Some other very large tanks built for irrigation were Viranameri near Kattumannarkoil in South Arcot District by King Parantaka Chola. Other famous lakes are Madurantakam, Sundra-cholapereri and Kundavai-Pereri (Gopal and Shrivastava, 2008).

14.6 Two Other Irrigation Systems of Ancient India

14.6.1 'Phad'[5] System of Khandesh, North Maharashtra

'Phad' is a shallow pool of water on agricultural fields created by plugging their outlets. This system – as per folk information and a few researchers – was initiated sometime in 300 BC with boulder system and is in use even now with permanent structures in this part of the country. The irrigation system is developed by diverting water from the river by constructing overflow type of weir known as *Bandhara* (i.e. Bund). The irrigated land is generally divided into three or four parts known as *Phad*, each one given a separate name by the community for easy reference. It is an irrigation system mainly on tributaries of the river Tapi, namely, Panzara, Girna and Burai and the rivers Kan and Jhamkheli in Maharashtra. The total irrigation process functions as overflow system and excess water flows back to the river.

It is a totally people-managed water distribution and management system. These *Bandharas* are either constructed by village heads or government authority who collect revenue from the water users. There are chains of *Bandharas* in the river system and as many as 112 units were functional on the rivers and tributaries mentioned above in 2006–2007 survey (Dharshivkar, 2007).

The salient features of the *Phad* system are as follows:

1. It has a chain of *Bandharas*.
2. It has to have levelled fields with shallow river basin.
3. It has no gates or valves at main tributary but has a control outlet to the field.
4. The inflow of water is controlled only at entry point and entry point of each *Phad*.
5. The excess water runs back to main source/river.
6. Proper care and distribution of water are managed by water-using body.
7. The system can work when there is flow in the river flooding or no flooding.
8. The beneficiaries have a controlling/management body.
9. Fishing is not possible.

14.6.2 Overflow Irrigation System of Bhagalpur, Bihar

A series of parallel canals were constructed from the river Ganges to the river Damodar in Eastern part of India to carry the floodwater of river Ganges. Studies show that this system was created around 2,300–2,500 years ago. The shallow canals (depth around 1.5 m) were having a width between 50 and 13 m from head to tail end. Records show that all the six canals were not of same length and varied between 100 and 50 km. Floodwater from the main river and canals was diverted into the agricultural fields by cutting the riverbank or providing notch on the canal bank. This was sealed just by putting obstruction when water was not required. The diverted water was muddy and carried lots of silt, which enriched the soil. The agricultural fields were used mainly for mono-crop paddy cultivation. There was no valve/gate in the system. As water was standing in the field for quite some time, fish cultivation was also practiced. The system could be operated whenever there were floods in the river Ganges. The records do not show any controlling mechanism for running of the system. The system was abandoned in late 19th century (Dharshivkar, 2007).

14.7 Irrigation Systems during Medieval India

As per the studies, in the early 12th century, Prithviraj Chauhan got constructed the WYC for increasing irrigation potential in North India. These systems got a boost up during the Medieval Period. During Sultanate period (1309–1388 AD), Firoz Shah Tughlaq constructed a double system of canals from the river Yamuna to Sutlej. These are referred as *asrajwahas* in Indo-Persian historical texts. He further renovated the WYC constructed during Prithviraj Chauhan's era (1178–1192), for increasing irrigation potentials covering more areas under cultivation for growing grains and fruits (Chauhan and Ram, 2023).

Baolis or step wells were constructed in Delhi region during Sultanate era (1206–1526). During Mohammad-bin-Tughlaq period the Satpula (seven spans) were constructed to regulate water supply for irrigating the area falling outside the city. Built across the southern wall of *Jahanpanah*, the dam was several metres above ground level. The seven spans had sluice gates to control water in an artificial lake.

During Shahjahan's period (1628–1658 AD), a number of canal systems were introduced to improve crop and horticultural potential in Northern India, especially for Shahajanabad, the newly built capital in Delhi. Ali Mardan Khan, the Persian artisan introduced a canal system from Khizrabad to Safidon (from present-day Karnal to Hisar) which later got silted up. Ali Mardan also developed a system to bring Yamuna water into the Palace and connected the same with the canal from Najafgarh area (Chauhan and Ram, 2023).

A *dighi* was a square or circular reservoir about 0.38 m by 0.38 m with steps to enter. These were created in Shahjanabad area. These *dighis* were fed by canal system and the community used water for domestic purposes.

14.8 Irrigation System during Colonial Era

In 1800, some 800,000 ha of land were irrigated in India. By 1900, due to building of canal system in Uttar Pradesh, Bihar, Punjab, Assam and Orissa the irrigation potential in the country increased to 1.62 m ha. The biggest developments were the Ganges canal from Haridwar to Kanpur and new network of canals in Punjab. With considerable increase in irrigation it was expected that the food production would increase many fold, but unfortunately it did not happen so, as it was slowly realized that irrigation potential was created more to do with commercialization of crop rather than for family consumption (Mulage, 2017). Production of opium and poppy was introduced in the country at a very large scale in eastern and northern regions of India for exporting them to China. Apart from these, as Industrial Revolution had been completed in England, commercial crops like tea, cotton, jute, sugarcane, groundnuts, indigo, etc. were also cultivated in different parts of India to meet increased demands of British industries, for which Indian farmers were getting just peanuts whereas the British coffers were getting filled up (Mulage, 2017). As farmers were forced to grow commercial crops to meet the needs of their masters and to meet ever-increasing land revenue demands by *Zamindars* (landlords) and State agencies, family consumption crops production suffered drastically, resulting in famine conditions, more so, during low rainfall seasons in the country. Therefore, all these exploitation of farmers in India saw massive famines over the period between 1850 and 1905.

After witnessing millions of deaths due to famine, British India started major irrigation projects in the country and by the year 1947, the irrigated area had increased to 22 million ha from 13 million ha in 1900. Meanwhile, irrigation canal projects from groundwater sources like wells and ponds were also introduced in South Indian states, which helped in increased agricultural return. In between with two World Wars, farmers in India were busy in increasing food production to meet the demands of other countries with a result that famine conditions prevailed in the country. With partition of the country in 1947, India lost valuable agricultural land and irrigation potential. With massive influx of population, India slowly became a food-scarce country, which needed to be tackled in war footing subsequently.

For about three millennia ending with the advent of colonial rule in 1757, India was the largest economy of the world. Traditionally evolved water management system sustained a high level of prosperity in the country up to early 19th century when India's population was about 200 million. By then, Presidency Towns Kolkata (Calcutta), Mumbai (Bombay) and Chennai (Madras) founded by the colonial rulers were three most important cities of India in terms of their trade transactions as well as administrative centres for the regions.[6] The proportion of people living in towns and cities in the mid-18th century could be 5% or less. Also, by this time Industrial Revolution took deeper roots in Europe and continued to contribute to the rapidly growing prosperity of the constituent countries through mass production and manufacture of newly invented products. Maddison estimated in 2007, between 1500 and 1820, while India's GDP grew from 60,500 to 111,417 million USD, that of UK grew from 2,815 to 36,232 million USD during the same period. Other European countries were better than UK but far behind India. However, during the same period, the per capita GDP of India decreased from 600 to 550 USD while that of England grew from 685 to 3,274 USD (Lee 1984). Also, between 1700 and 1950, India's share of the world economy declined from 24.4% to 4.2% and from 1700 to 1900 India's "share of global industrial output" had shrunk from 25% to 2%.

14.9 Irrigation Systems of Rajasthan and Tamil Nadu: Representing Significantly Different Eivirons and Cultures

Varied environmental settings across India contributed to the emergence of varied cultures of India. As such, the States or Provinces represent the two types of variations, though cultural boundaries often cut across administrative boundaries. The states Rajasthan (the westernmost State sharing boundary with Gujarat and

Pakistan) and Tamil Nadu (the southernmost State of the subcontinent protruding into the Indian Ocean) represent distinctly different cultural practices determined by low and heavy annual precipitation, respectively. Scanty rainfall and semi-arid climatic conditions in the first taught the communities about different devices of RWH, suitable crops and low-rise high-density built-form in the settlements. Apart from storing rainwater for farming, every medium to large size mansions (*havelis* in local language) had an underground tank for collecting rainwater to be consumed by the households. In the southern region of India, abundant precipitation on the undulating Deccan plateau naturally created lakes and a system of perennial rivers flowing into the Bay of Bengal on the east or the Arabian Sea on the west. In a way, similar systems consisting of carefully laid network of canals connected with a series of water tanks ensured availability of sufficient water for irrigation of the paddy fields and other crops. For domestic consumption, easy access to wells and ponds provided sufficient water. Wells and ponds supplying drinking water were not used for any other purpose.

14.10 Irrigation Practices of Rajasthan

The scarcest natural resource in the arid State of Rajasthan is water – the crucial determinant for life, livelihood, and development activities. The State Water Policy (P.2, 2010) observed, "Rajasthan has just 1.16 percent of total surface water available in India, whereas it has 5.5 percent of the human population, 18.70 percent of the livestock and more than 10.4 percent of the geographical area of the country. Precipitation, which is scanty, is the only source of renewable fresh water. Out of the 14 rivers Chambal is the only perennial river in the State".

The tough living conditions due to the hot and arid climate coupled with water scarcity make living in this parched State a challenge. The people, who settled here, adapted their lifestyles to brave the harsh climate and unforgiving living conditions. Since ancient time in Rajasthan, camels served as the all-purpose workhorse engaged in drawing water from wells, ploughing the fields, carrying stones for building fort cities on hill tops, meeting the demand for domestic consumption of milk and more importantly, long distance travel across the desert.

The community at large owed its sustainability to a culture that evolved over centuries and nearly perfectly harmonized with the water-scarce ecosystem. The royal nobles, financially well-off families and community as joint venture mostly built community water supply systems. In majority of cases these were operated and maintained by the community. Numerous legends and stories based on battles and sacrifices of the independence loving kings, queens and the commoners represent the other perhaps more important dimension of Rajasthan's culture. Unlike many other ruling communities, the rulers of princely states of Rajasthan could never be subjugated either by the Islamic dynasties including the Mughals nor by the British. Legends of valour and heroism arising out of Rajasthan were captured by literary icons of Bengal in eastern region of India such as Abanindranath Tagore of the celebrated Tagore family of Bengal.

For irrigation of agricultural fields as well as meeting needs of domestic consumption, available rainwater was carefully harvested and stored by building appropriate devices according to local conditions such as topography and soil quality. Generally, crops typically suitable for the region were cultivated. Check dams, wells, stepped wells, sluices, streams and canals, streams and canals, overflow reservoirs and more represent the wide range of water harvesting devices used. Percolation zones and percolations sinks aimed at recharging aquifers were also parts of the system. They were owned and managed by the community and unwritten rules and codes aimed at creating the then 'best practices' were followed for collective well-being of the community. The codes were for both use and maintenance of the water supply system.

14.11 Traditional Rainwater Harvesting (RWH) Structures in Rural Rajasthan

Though the average annual rainfall in Rajasthan is around 600 mm, the annual rainfall in Thar Desert area is reported to be less than 200 mm. Groundwater availability in the Thar Desert area is low, as it is highly saline as well as contains high concentration of chemicals like fluoride, chloride, nitrate, etc. making it unfit for agriculture, animal husbandry or human consumption. Moreover, increased impact of both the human and livestock population is a constant threat to land, surface and groundwater resources.

There were quite or few community-based RWH structures used mainly for drinking water purposes. Some common RWH structures found in Western Rajasthan and Thar Desert such as *taanka*, *kund*, and *baori* are listed in Table 14.1 (Refer to Annexure 1, Table 14.1: Types of Rain Water Harvesting structures traditionally built in Rajasthan).

There were a few common ground rules for all these community-based RWH structures constructed and maintained in the villages. These are as follows:

- Strict community control ensured that catchments were left clean and undisturbed, water bodies are not sullied, if they are meant for household use.
- The villagers have to contribute towards construction and maintenance of the system. Family contribution generally decided by the village committee. In some cases, richer individuals and families contribute the capital cost for the benefit of the community on philanthropic basis.
- Distribution of water is controlled by the village committee.

This was a system as well worked out as any integrated water resource management scheme of today involving stakeholder participation in decision making, creation and maintenance of the resource and its catchment, its careful consumption and conservation and strict bans on its pollution. In some areas of Rajasthan, the above-described system is still in use.

14.12 Water Management in Jodhpur and Jaipur, Two Historic Cities of Rajasthan

As examples of indigenously developed good water management in two cities located in arid and semi-arid environment, the cases of the Jodhpur the celebrated fort city built in 15th century and Jaipur built in early 18th century and present capital of Rajasthan are briefly presented here.

14.12.1 Jodhpur

Founded in 1495 AD, the city of Jodhpur has a huge network of surface and groundwater bodies. Most of these water bodies in the Jodhpur area were mostly natural but were improved by the community members and royal families for more than 500 years. While selecting the site for locating the city, the planners had given a serious thought about its local water potential and important physical features like *nadis*, *talabs* (ponds), canals and lakes to introduce important groundwater features like wells *baories* and *jhalaras* including step wells. The *Chonka-Daijar* Plateau, 30 km long, 5 km wide (at the centre) with heights ranging between 120 and 150 m above ground level served as the catchment for 50 functional surface water bodies like *nadis*, *talabs*, tanks and lakes and also indirectly to about 154 groundwater bodies – wells, *baolis* and *jhalaras* (Centre for Science and Environment, 1997; Surana, 2019; Figures 14.3 and 14.4).

As many as 46 *talabs*, water reservoirs situated in a valley or a natural depression, were constructed between 500 and 300 years ago in Jodhpur. These were meant for community water storage and were preserved from any sort of pollution. Five large lakes were built around Jodhpur, in the outskirts of the city. The oldest one named Balsamand was constructed in 1126 AD. The five lakes together have the capacity to hold 700 million cubic feet of water at a given time. These lakes, still in existence, are fed by canal system, many of which are still functional. Some of these lakes are still being used as storage reservoirs for drinking water system in the city.

Unfortunately, the construction and maintenance of *talabs*, located within the city, started deteriorating around 1897–1898, when the centralized public water supply systems were introduced and by

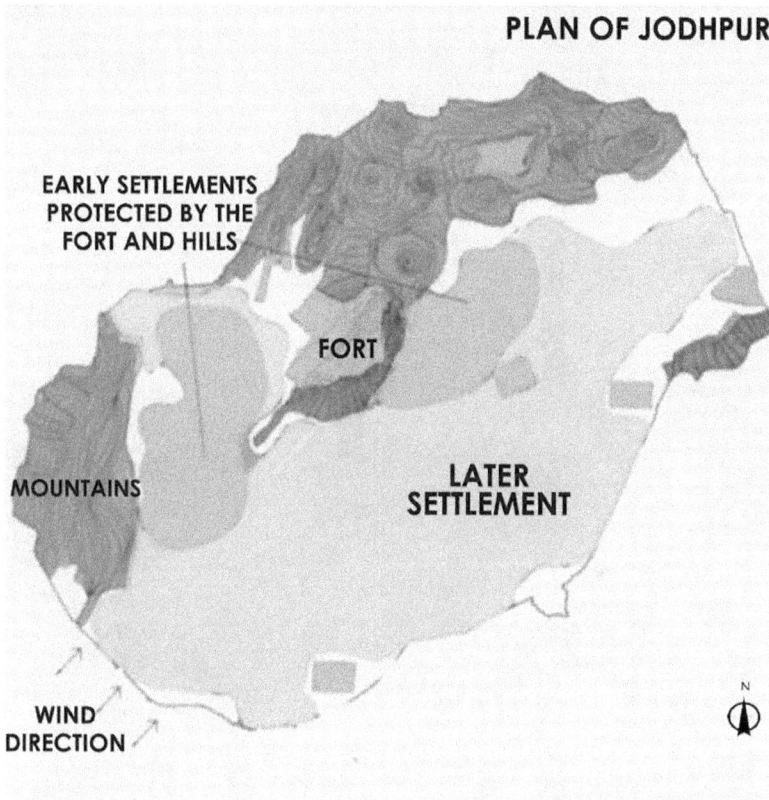

PLAN OF JODHPUR

EARLY SETTLEMENTS
PROTECTED BY THE
FORT AND HILLS

FORT

MOUNTAINS

LATER
SETTLEMENT

WIND
DIRECTION

N

FIGURE 14.3 Plan of Jodhpur ("Water Resources and Development of Human Settlements: The Case of Udaipur, Rajasthan", unpublished PhD thesis, School of Planning and Architecture, New Delhi).

1960 almost whole of the old system had collapsed. Most of these water bodies are highly in dilapidated condition and getting polluted by rapid urbanization.

14.12.2 Jaipur

Popularly referred to as the 'Pink City', all the buildings were painted pink, symbolizing a colour of good hospitality, to welcome Queen Victoria and the Prince of Wales in 1876. The city of Jaipur, the capital of Rajasthan since 1949, had a population of 3 million in 2011 and is estimated to have a population of over 4 million in 2021. Carefully located, planned and built, Jaipur started functioning in 1727 for a population of less than 1,00,000 persons occupying an area of about 600 ha (authors' estimate). Jaipur was located at the northern end of a gently contoured parcel of land, close to the Nahargarh Hills on the north and rocky outcrops of the Aravalli Ranges on the east. The rainwater from the Nahargarh Hills was channelized to form a water body Man Sagar to serve as the reservoir of potable water for Jaipur. By mid-19th century, as the city population grew, supply of water from Man Sagar was supplemented by groundwater drawn by wells dug in Amanishah river bed. By 1900 Jaipur had a population of over 1,60,000 persons that nearly doubled by 1951 and doubled again by 1971 (Figures 14.5 and 14.6).

With the advent of colonial dominance in Rajasthan in 1817–1818 scarcity of water was addressed by replacing the traditional community-managed system with introduction of a centrally controlled system administered by the PWD (Centre for Science and Environment, 1997). In response to increasing water scarcity through 19th century, the Ramgarh reservoir was constructed at a distance of about 30 km on river Banganga which started water supply in 1931. The dependence on groundwater continued to increase, as

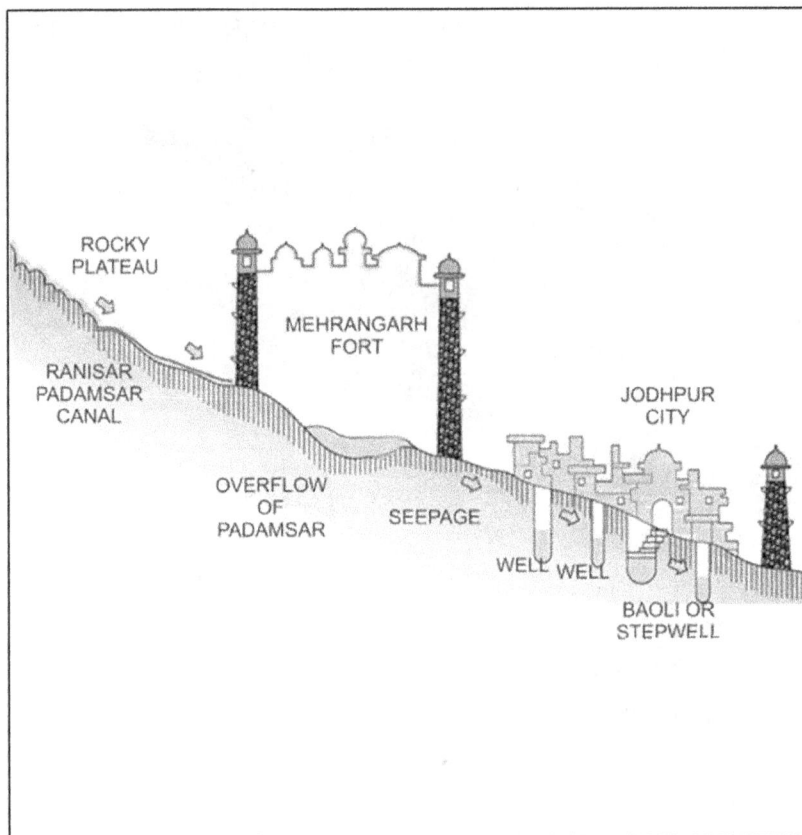

FIGURE 14.4 Schematic section of the water system of Jodhpur. The natural slope illustrates the catchment has been used for harvesting of water in reservoirs which also percolates to step wells and tanks (Surana, R, 2019, "Water Resources and Development of Human Settlements: The Case of Udaipur, Rajasthan", unpublished PhD thesis, School of Planning and Architecture, New Delhi. Centre for Science and Environment (CSE), "Dying Wisdom", New Delhi, 1997).

the erstwhile reservoirs and channels got heavily silted due to centuries of neglect and non-maintenance making them unusable; unabated spread of built-up area often occurred on natural drainage channels within city limits resulting in increasing runoff and declining recharge of groundwater and city floods.

Thus, the fragile but sustainable environment–human habitat relationship that emerged in Rajasthan on the basis of centuries-old accumulated community experience at the ground-zero was disrupted and replaced by an alien system, operated by the aliens. Evidently, though the new system appeared to resolve the water scarcity, it was hardly sustainable.

14.13 Rainfed-River-Rich Tamil Nadu

India is dotted with tens of thousands of tanks, mostly man made to conserve water to meet community water needs as well as for irrigation purposes. Tank irrigation in India is concentrated in

- Coastal Tamil Nadu and Andhra Pradesh;
- South Central Karnataka, Telangana and Eastern Vidarbha;
- North East Uttar Pradesh in the farmer kingdom of Awadh; and
- Rajasthan, east of the Aravalli mountains.

FIGURE 14.5 A view of terraced gardens (top) and Chand baori step well in Abhaneri near Jaipur (bottom) (Surana, 2019).

Tanks or reservoirs were the most important source of irrigation in South India. As observed by an 1856 study of irrigation in South India, the presence of more than 53,000 tanks in 14 districts of the Madras Presidency were recorded. Records also show that in 1883, around 50% of the cropped area was irrigated by tanks, built by the Chola and Pandya Kings. Most of the tanks were built by the kings to meet the irrigation and drinking water needs of the people. The construction and maintenance was, however, done entirely by the community (Centre for Science and Environment).

Water bodies in Tamil Nadu fall under three categories: lakes or eris, tanks or *kulams* and ponds or *kuttais*. A eri was a large earthenware tank dug out of the ground with the dug-out mud making the side walls or bunds. A *kulam* was built with bricks (and occasionally granite) and was attached to a temple, giving it the name *kovil kulam* or temple tank. A *kuttai* was a small pond. In Tamil Nadu there are a few tanks that irrigate more than 1,000 ha and one of which serves 6,000 ha. Rectangular tanks are constructed above Uranis, natural springs, to collect the water coming out from the ground. A winding staircase is provided for people to go down and collect the water for their domestic use.

Irrigation by tanks, known by *eris*, has been practiced in approximately one-third areas of Tamil Nadu. *Eris* have played an important role in maintaining ecological harmony as flood control measure, soil erosion prevention and wastage of runoff during heavy monsoon period. It also helps in recharging the groundwater. Paddy cultivation in Tamil Nadu solely depended on *eris*. Before the British rule, the *eris* were maintained by the community members with special budget to run the *eris* and maintain the irrigation channels sluices, inlets, etc. With land tenure system introduced by British, the maintenance funding of *eris* suffered completely and eventually collapsed (Centre for Science and Environment).

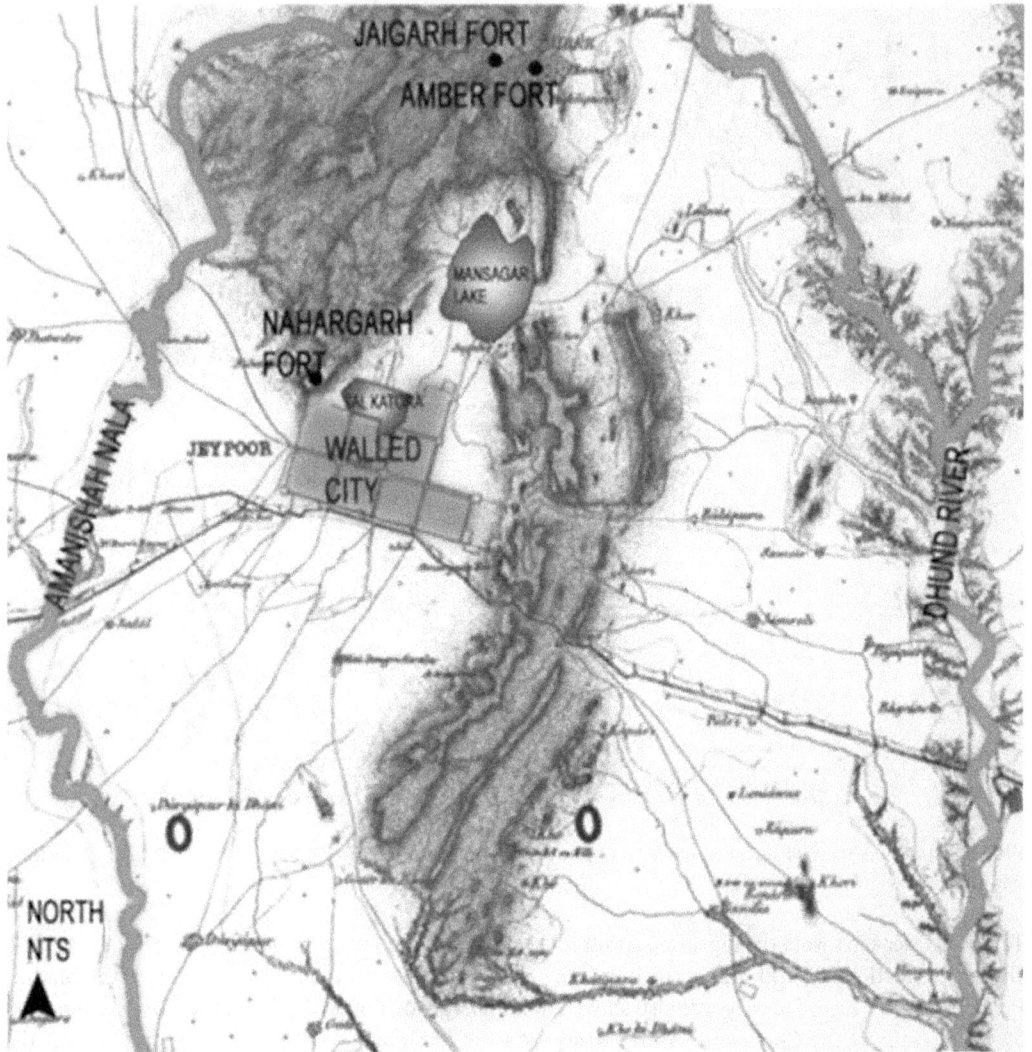

FIGURE 14.6 Runoff from the Nahargarh Hills north of Jaipur city was channelized to create the reservoir Man Sagar to supply water to the city. Note: Each sector of Jaipur (Jeypoor in the map) is about 1 km×1 km (Surana, 2019).

Traditionally, the village community decided what crops to grow and where to grow them, how long the sluice gates of the *eris* should be kept open, how much water should flow to each field and so on.

Tanks can be classified into system and non-system tanks. System tanks receive plenty of water as they catch the overflow from a reservoir, nearby stream and the runoff from around their catchment. They help farmers to raise more than one crop. Non-system tanks depend entirely on rainfall and can support only one crop. The tanks used a system of canals and sluice gates to control and transport the water. Several tanks linked by canals were also built in the watershed areas (Common Pool Resource, 2018).

Anicuts were small or medium dams built across rivers to divert water into irrigation channels. The Grand Anicut or Kallanai was built in the second century AD by Karikala Chola. It was made of stone and situated on the river Cauvery where the River Kollidam branches off. Anicuts have also been built across the Tambaraparani, Chittar and many other rivers. The anicuts in the Kanyakumari district are said to have been built 1,000 years ago.

14.14 Temple Tanks of South India

In South India the ancient tradition of establishing a tank alongside a temple prevails. Since every village has a temple, it also has a temple tank. These tanks were constructed to harvest water. They captured rainwater and runoff. Sometimes, channels were constructed to bring water from a nearby steam or river.

The temple tanks are known as *kovil kulam* in Tamil Nadu, *kulam* in Kerala, *kalyani* in Karnataka and *cheruvu* or *pushkarini* in Andhra Pradesh. The water from the temple tank was mainly meant for the ritual bath of the deity and to provide water for the flowering plants. Devotees also washed their hands and feet or even bathed in a separate tank maintained for the purpose before entering the temple. The temple tank was the focal point of several religious activities like the *theppam* or float festival, for the offering of prayers to one's ancestors and mediation.

The temple tanks vary in size and shape and are a masterpiece of engineering. Corridors and long flights of steps surround them. They have intricate inlet channels for bringing water from a stream or river and outlets that carry away the excess water. Some have natural springs in their bed and others have wells that can be accessed when the tank is dry. In ancient times the temple tanks always had water, even when all other sources had dried up.

During the ancient period – possibly up to the medieval time – construction industry was a major employer of a wide range of workers and professionals. They represented unskilled labourers to extraordinarily skilled designers, sculptors and craftsmen. Construction of a temple and its complex, complete with *Gopurams* (towering gates to temple complexes), water tanks, accommodation for the priests and workers, not to mention the iconic images of Gods and Goddesses could take several generations of dedicated work. In fact, additions and alterations of the temple complexes and palaces continued through the dynasties ruling for decades and even centuries. The King of Chola dynasty is believed to have constructed the Meenakshi Temple in the third to fourth century BCE. However, the dynasty reigned for about 1,500 years up to 1279 CE.

14.15 Conclusions

Archaeological remains of the Indus Valley Civilization as well as recommendations found in India's earliest known scriptures such as the Rig Veda clearly establish that the Indian society was well aware since millennia, about the critical role of water adequacy in sustaining prosperity for individuals as well as for the communities in the long term. Possibly on the basis of experiencing natural climatic fluctuations leading to highly variable water availability, communities across the subcontinent evolved remarkably efficient and elaborate irrigation systems that ensured reasonably equitable availability of water throughout the year, in all regions and sub-regions and kept water scarcity at bay. Essentially, those systems consisted of small, medium and large natural or manmade water bodies, strategically located to create a cascaded system connecting the water tanks by channels with control devices that monitored flow and distribution of water in required amounts and returning the excess water if any, to the main stream or river. Flood control was the other important function of these water management systems. Evidently, appropriate management of the irrigation system could be possible as the entire community was not only involved in establishing the system but also participated in its operations and management to ensure that the system was operated in a highly coordinated fashion, with sufficient efficacy. Needless to say, the subsystems required to ensure a stable and resilient manmade water management system such as quick communication system were established and maintained through centuries. Further, there must have been dependable devices in place to ensure good management of the interface of the natural hydrological system and the manmade ones in case of perturbations such as droughts and floods.

In the 18th century, India's culture and financial resources, though based mainly on agriculture, were highly developed and India had a flourishing economy, which proved to be a 'Golden Goose' to the British administration. During the colonial rule spanning nearly two centuries, though considerable expansion of the irrigation network in the subcontinent was built, adding significantly to the area under irrigated farmland, it heavily benefitted the economy of the UK at the cost of decline of the well-being of the native community of colonial India. As noted in the beginning of the chapter, a dozen famines occurred in India between 1765 and 1943. The far-reaching adverse impacts of replacing the community-managed irrigation system by a centrally controlled one, combined with population growth, multiple cropping and urbanization, have been serious leading to difficult-to-predict floods and droughts. Granted that the traditional practice of community-managed irrigation systems had been very successful at the regional level, they could be well utilized even within a larger system that would link them for inter-regional optimization of the resource. However, this possibility was not considered by the colonial rulers due to a variety of reasons, beyond the scope of the present discussion.

Annexure 1

TABLE 14.1 Types of Rain Water Harvesting (RWH) Structures Traditionally Built in Rajasthan

Name	Location	Description
Kuis/beri	Western Rajasthan. Location based on local knowledge. The *beries* are found infrequently and are scattered.	*Kuis/beri* is a large naturally occurring underground *taanka*-shaped reservoir. A large community-owned *beri* (underground reservoir) can hold more than 400 cum of water and a smaller one around 100 cum in capacity. Mostly found in the impermeable rock areas and are lined with naturally occurring clay. *Kuis/beris* are 10–12 m deep pits dug near tanks to collect seepage. Most *beries* have not been maintained for many years and have silted up. Desilting involves the community working together to remove all the silt. Presently, a concrete cap in the access area and hatch is provided for the safety of water stored.
Nadi	Ponds storing water from the adjoining natural catchment during rains in the Jodhpur district. Site selection as per local knowledge.	In dune areas, the *nadis* are of 1.5–4.0 m and those in sandy areas between 3 and 12 m. The storage capacity is for a few months. Almost all the villages do have one or more such ponds. These are mainly being used to provide drinking water for the human beings as well as animals. Community members take care to keep the area clean especially before the onset of rainy season. Since government agencies are providing alternate safe sources of drinking water, these ponds are neglected. These *nadis* do require regular desilting and upkeep of the embankments. Upkeep of *nadis* does help in reduction of soil erosion and groundwater recharge.
Taanka	Jodhpur and Bikaner districts. A *taanka* (small tanks) traditional RWH structure at household level. In case, rain failure, *taankas* are also filled by water from other sources.	This is basically a household storage tank to meet the drinking water needs during dry season. The *taanka* is a cylindrical underground water tank made primarily from local materials and cement mortar. In general, a *taanka* has a storage capacity of around 20,000 L. which is good enough for a family to meet their drinking water needs during dry period. In arid region most houses have got a small taanka to meet their emergency water needs. The *taanka* system is also found in Dwarka in Gujarat.
Kunds/Kundis	Western Rajasthan and some areas of Gujarat.	*Kunds/Kundis* is a circular underground well with a saucer-shaped catchment area that gently slopes towards the centre potion which is a well. A wire mesh across water inlets prevents debris to enter. Water is disinfected by lime and ash. These RWH structures are mostly owned by rich people.
Baoris or *Bers*	Rajasthan.	*Baoris or Bers* are quite deep community wells built for storage of drinking water. Majority of them were built by banjaras, the mobile trading community.
Jhalaras	Rajasthan and Gujarat. Oldest *Jhalara* in Jodhpur city, which had eight of them, was built around 1660 AD.	*Jhalaras* were human-made community storage tanks, mostly rectangular in design, having steps in three or all the four sides. These *jhalaras* collect subterranean seepage of a *talab* or a lake located upstream. This water from *jhalaras* was used for community bathing and religious rites.

(Continued)

TABLE 14.1 *Continued* Types of Rain Water Harvesting (RWH) Structures Traditionally Built in Rajasthan

Name	Location	Description
Khadin	Constructed in Jaisalmer district of Rajasthan in 15th century.	*Khadin*, also called a *dhora*, is a rainwater harvesting structure used for agricultural purposes. It has a long (100–300 m) earthen embankment with spillways or sluice, built across lower hill slopes lying below gravelly uplands. It has got great similarity with irrigation methods of the Ur (present Iraq) around 4500 BC and the Nabateans of the Middle East. Similar systems were also adopted in Nager desert 4,000 years ago and in Southwestern Colorado 500 years ago.
Vav/Vavdi/Baoli/Bavadis	Gujarat and Rajasthan. The step wells were constructed mainly in four periods: (i) Pre-Solanki Period (8th–11th centuries); Solanki period (11th–12th century CE); Veghela period (mid-13th to end 14th century CE); and the Sultanate period (mid-13th to late 15th centuries CE).	The traditional step wells, known as *Vav* or *Vavdi* in Gujarat and *Baoli/Bavadis* in Rajasthan, were highly decorative strong structures built by nobles usually for strategic and/or philanthropic reasons. These were constructed for community use, and located in strategic places like trade routes, for social gatherings and on military routes as well. Sculpture and inscriptions in the step wells demonstrate their importance to the traditional, social and cultural lives of people.

Source: 1. CSE Website, Rainwaterharvesting.org.
2. GoI, 2018, C.P.R. Environmental Education Centre, Ministry of Environment and Forest, Government of India. 2018.

Notes

1 Panchatantra is a collection of ancient Indian fables designed to teach children about character of individuals in the real world and how to deal with them while achieving one's goal.
2 Defeat of Prithviraj Chauhan, the Hindu ruler of Delhi in 1192 by Muhammad Ghori in the Second battle of Tarain is usually taken the beginning of Islamic era in India. In the Battle of Plassey 1757 (150 km north of present-day Kolkata) defeat of Nawab Siraj-ud-Daula by Robert Clive is marked as the beginning of colonial rule in India. Thus, Islamic rule was for 675 years (Majumdar, RC et al., 1956, *An Advanced History of India*, MacMillan & Co., London).
3 The Rig Veda is an ancient Indian Collection of Vedic Sanskrit hymns. It is one of the four sacred canonical texts of Hinduism, known as the Vedas, Period 1500–1200 BC.
4 *Indika* is an account of Mauryan India by the Greek writer Megasthenes. The original book is now lost, but its fragments have survived in Latin Later Greek and Latin works. The earliest of these works are those by Diodorous Siculus, Strabo (Geographica), Pliny and Arrian (Indica)
5 Phads are low-cost small-diversion irrigation system developed by the community on the small rivers, prevailing in the north-western part of Maharashtra.
6 Chennai (Madras) in 1639 and Kolkata (Calcutta) in 1690 were founded by the British. Mumbai (Bombay) was already inhabited since many centuries first as a fishing village subsequently growing as a trading settlement and ruled by different dynasties. It came under British control since 1661. These three cities were patronized by the British rulers in terms of establishing industries, administrative institutions such as the high courts, universities and more. New Delhi was built between 1913 and 1937 as the new capital city of India. Till 1960s four key Indian metropolises included Kolkata, Mumbai, Delhi and Chennai.

References

Ahmed R., 2018, "A Study of Agricultural Production for Medieval India", *Research Directions*, Vol 9, No.12, pp. 116-120.

BBC, n.d., https:// HYPERLINK "http://www.bbc.co.uk/bitesize/topics/zxn3r82/articles/z9mpsbk"www.bbc.co.uk/bitesize/topics/zxn3r82/articles/z9mpsbk Biradar, V. B., 2015, "Irrigation system in ancient Karnataka", *Indian streams Research Journal*, Vol 5 , No. 7, pp. 1-5.

Centre for Science and Environment, 1997, *Dying Wisdom*, Centre for Science and Environment, New Delhi, India.

Chauhan, M. K. and Ram, S. 2023, "Rehabilitation of canal irrigation schemes in India: a qualitative analysis", *Water Policy*, Vol. 25, No. 1, pp. 59-68.

Common Pool Resource, 2018, *Environmental Education Centre*, Ministry of Environment and Forest, Traditional Water Harvesting Systems in India.

Centre for Science and Environment Website, Rainwaterharvesting.org. https:// HYPERLINK "http://www.rainwaterharvesting.org/rural/Traditional3.htm"www.rainwaterharvesting.org/rural/Traditional3.htm

CSE Website, Rainwaterharvesting.org. https://www.rainwaterharvesting.org/rural/Traditional3.htm

Dharshivkar, M., 2007, "The Two Ancient Water Systems in India", Presented at National Seminar on Water and Culture, Hampi, Bellary, Karnataka, 25–27 June, 2007, India.

Gopal L., Shrivastava V. C., 2008, *History of agriculture in India, up to c. 1200 A.D.*, Ed. by Chattopadhyaya, Concept publishing company and Centre for studies in Civilizations, New Delhi, pp. 501

Haraniya K., 2017, "Dholavira and the Story of a Civilization", Shrijan: Reviving the Embroideries of Kutch: file:///c:users/Sony/Desktop/Dholavira and its story of Civilization.html.

History for Exam, n.d., https://www.historyforexam.com/201f28/09/irrigation-system-of-ancient-india.html

Irshad, S. M., Eslamian, S., 2017, Politics of Drought Management and Water Control in India, in *Handbook of Drought and Water Scarcity, Vol. 3: Management of Drought and Water Scarcity*, Ed. by Eslamian, S. and Eslamian, F., Taylor and Francis, CRC Press, Boca Raton pp. 447–460.

Jansen, M., 1989, "Water Supply and Sewage Disposal at Mohenjo-Daro" *World Archeology, the Archeology of Public Health*, Vol. 21, No. 2.

Kenoyer J. M. 1991, "The Indus Valley Tradition of Pakistan and Western India" Journal of World Prehistory Vol. 5, No. 4, pp. 331–385.

Lee, S. J., 1984, European Population Growth 1500-1800 in *Aspects of European History 1494-1789*, Routledge, Newyork, pp. 12.

Mukerjee, M. 2010, *Churchill's Secret War: The British Empire and the Ravaging of India during World War II*, Penguin, Tranquebar Press, Chennai.

Mulage, B. S. 2017, "History of Agriculture System in India: A Legal Perspective" *International Journal of Humanities Social Sciences and Education*, Vol. 4, No. 7, pp. 25–30.

Perundevi, B., 2019, "Water Governance through Lakes in medieval Tamilnadu- A Study" *Think India Journal* Vol. 22, No. 14, pp. 9893-9896.

Prasad, R. 2018, "Indigo—The Crop that Created History and then Itself became History" *Indian Journal of History of Science*, Vol. 53, No. 3, pp. 296–311.

Saha , S, 2019, Lecture on "Role of City Planners in City Building in India: Past, Present and Future", presented in Lovely Professional University Jalandhar, India.

Saxena, C. 2012, "The Concept of Water in Rigveda" *International Journal of Social Science& Interdisciplinary Research*, Vol. 1, No. 8.

Singh, R. 2019,"Water Management and Conservation Practices in Indus Valley Civilization" , *International Journal of Applied Social Science*, Vol.6, No. 9, pp 2195-2199.

Sivasubramaniyan, K., Gopalakrishnan, S., *Ancient Engineering Marvels of Tamil Nadu*, Madras Institute of Development Studies, India Water Portal.

Surana, R., 2019, "Water Resources and Development of Human Settlements: The Case of Udaipur, Rajasthan", Unpublished PhD thesis, School of Planning and Architecture, New Delhi, India.

UNESCO, 2014, "Dholavira: A Harappan City" United Nations.

Vishwanath, S. 2009, "Persian Wheel in India End of an Era". http://base.d-p-h.info/en/fiches/dph/fiche-dph-7866.html

Vishwanath, S., Acharya, A., 2008, "The Persian wheel: The water lifting device in Kolar, Karnataka" in India Water Portal. https://www.indiawaterportal.org/articles/persian-wheel-water-lifting-device-kolar-karnataka

15

Water Resources and Irrigation Practices of Contemporary India

Shovan K. Saha
Institution for Hygiene and Environmental Sanitation, New Delhi

Achintya Kumar Sen Gupta
Institute for Hygiene and Environmental Sanitation, New Delhi
World Health Organization

Rina Surana and Prerna Jasuja
Malaviya National Institute of Technology

Saeid Eslamian
Isfahan University of Technology

15.1 Introduction

Immediately after attaining independence from colonial rule in August 1947, India faced severe shortage of food and shelter, massive unemployment, a shattered economy reflected by low Gross Domestic Product and more importantly, abysmally inadequate infrastructure that could contribute to improve the situation. The worst affected was the food supply sector with low stock of staple food grains such as wheat and paddy. The Indian University Commission in its report submitted in 1949 observed "the country's position with regard to food production was pathetic. India with 70 percent of its population engaged in agriculture, imported food grains at a high cost of foreign exchange" (Indian University Commission, 1949).

The multipurpose river valley projects operationalized within about a decade after 1947 hardly made a difference. Nearly two decades into the post-independence era, driven by persisting food shortage, in 1965, the Green Revolution (GR) modelled after Norman Ernest Borlaug's successful experiment in Mexico, was launched by the Government of India in order to stem the situation. It essentially consisted of using bioengineered seeds, heavier irrigation, use of chemical pesticides, multiple cropping and development of infrastructure to ensure access to regional markets (Box 15.1).

Massive improvement of irrigation system was identified as the second most important input of the GR, while the first was to introduce hybrid crop varieties for higher yield. The goal was to significantly raise the irrigated farmland and reduce dependence on difficult-to-forecast monsoon rains. This included not only rejuvenating the existing irrigation system but also the construction of new multipurpose hydroelectric projects aimed at reaching hitherto un-irrigated or poorly irrigated land particularly in semi-arid and arid regions of Rajasthan, Gujarat, Andhra Pradesh, Orissa and Madhya Pradesh as well as others. Due to lack of surface water availability and inadequate rainfall in and around semi-arid regions, over-exploitation of groundwater has been recorded, resulting in the creation of dark zones.

Out of the four types of irrigation methods used, viz. Well irrigation, Reservoir or Tank irrigation, Canal irrigation and multipurpose river valley projects, the third accounts for 42% of the total irrigated land (Varshney, 2015). Evidently, on the whole, the irrigation system thus created were far beyond the direct control or monitoring by small farmers – who were usually either landless or with small holdings.

Box 15.1   Green Revolution in India

In 1943, as Bengal and Bihar, two major eastern states of India, were experiencing the infamous famine, Rockefeller Foundation initiated the Mexico Agricultural Program essentially consisting of introducing a modern scientific approach to agriculture. Led by Norman Ernest Borlaug, by the mid-1950s the dramatic improvement of agricultural productivity was achieved in Mexico and the pioneering approach was being applied in less developed countries around the world, namely in Latin America, Africa and Asia. Borlaug's prescription included using bioengineered high yielding variety strains of grains, synthetic fertilizers, strong insecticides and pesticides and appropriate irrigation among other inputs. This radically innovative approach earned great fame as the Green Revolution (GR) and earned Borlaug the Nobel Prize in 1970. In India, MS Swaminathan led the application of the GR concept in 1965 that enabled the government to avert a serious food crisis. But by the 1970s, criticism against the same grew louder and louder. The major criticisms were:

- Rapid decline of shifting cultivation, conversion of forest land to agricultural land, mono cropping and overgrazing leading to serious imbalance of ecosystem;
- Use of pesticides and insecticides in heavy doses for high yielding crops causing long-term loss of natural soil fertility;
- The toxic elements entering the food chain causing health hazard; and
- Increased water logging due to increased irrigation particularly for rice-based cropping system.

All these need a further review of GR in favour of introduction of the Second Green Revolution of Green Agriculture (GA) on a large scale. Currently, GA is gaining popularity across India on a local level.

Source: Based on a number of reports including GoI publications.

Thus, the age-old traditional practices and methods of irrigation known to farmers were by and large replaced by large-scale canal-fed distribution systems that had to be monitored by trained technical personnel.

Further, as the population of India grew naturally and due to unexpected influx of people from neighbouring regions who were driven by politico-economic compulsions, as observed by the World Bank (2019), water availability in India has reduced from 3,000 to 4,000 m³ in 1950 to 1,000 m³/person in 2019 (World Bank, 2019). The groundwater extraction in India is highest in the world, accounts for about 25% of total water extracted worldwide. The inequity of water availability and the skewed pattern of urbanization in favour of largest cities are known and discussed in formal, semi-formal and informal meetings since the 1970s by lay citizens, government representatives and technical experts resulting in recommending varied strategies. Considering the potential reserves of water, with appropriate management and application of technological prowess, India should possibly be a water surplus nation. The possibility of redistribution of water across the subcontinent by means of a network of canals interlinking the overflowing rivers causing devastating floods in surplus regions to nearly dried up streams in the drought regions, has been considered during the colonial rule towards the end of that regime. Subsequently too, such proposals have been under discussion from time to time. Issues such as environmental impact, displacement of people, challenges of coordinated operation, inter-state conflicts regarding sharing water and more prevented consistent and full implementation of projects under the scheme. Usually such projects need a minimum gestation period of a decade. For completion of all key link canals that would reduce the imbalance of river water distribution across the country, it could take at least two decades or more. Nevertheless, the promising aspects of the megaproject seem to have compelled the current government to initiate follow-up action within weeks of assuming power in May 2014. While this mega-initiative may be underway to its realization on ground, other localized efforts towards depressing and stabilizing the demand curve and enhancing the rate of supply of water must be attempted.

Apart from sufficiency, since the 1970s, the deteriorating quality of water has been a source of risk to the health of the citizens in several regions of India both in rural and urban areas. In the rural areas, consequent to introduction of the GR, poorly managed application of synthetic fertilizers and excessive withdrawal of groundwater required for hybrid varieties of crops not only led to severe adverse impact on soil but also contributed to pollution of groundwater reserves. In the urban areas, disposal of untreated domestic and industrial waste in rivers and lakes led to serious pollution of surface water sources. In addition, rapid increase of the impervious surface as well as over-withdrawal of groundwater caused steady drop of water table and also to its pollution. Groundwater quality in many a pocket, is showing presence of ingredients not desirable for domestic use. This is the result of over-pumping and uncontrolled exploration of groundwater resources. Ambitious schemes such as the Ganga Action Plan initiated in 1985 with proposed allotment of Rs. 70,000/ million or 1 billion USD (@Rs70/ per 1 USD) have not yet resulted in any significant improvement in the water quality of river Ganga. Though guidelines to prevent pollution of surface as well as groundwater are in place, adequately strong control measures are missing.

In a new-normal post-Covid-19 era, radical interventions would be required to sustain the population of India within an acceptable comfort zone in terms of access to safe and sufficient water.

This chapter first evaluates the irrigation scenario during the post-independence era, considers the options for the future and suggests an approach towards a stable irrigation system that would sustain one of the basic needs of all Indian citizens.

15.2 Surface Water

Though due to partition of India it lost geographical area and water potential substantially, India, with 2% of the land mass, is the seventh largest country in the world. It supports about one-sixth of world population and possesses around 4% of world's water resources (Jain, 2019). With this,

India is rather a water-rich nation. The vast rivers of India helped in physical, financial and cultural growth of the country. Amongst rivers, 12 of them may be classified as major rivers covering an area of 253 mha of catchment and 46 as medium rivers with combined catchment capacity of 24.6 mha (Centre for Science and Environment, n.d.). There are four groups of rivers in the country. These are (i) Himalayan rivers, (ii) Deccan rivers, (iii) Coastal rivers and (iv) rivers of the inland drainage basin.

The three major rivers, namely the Indus, the Ganga and the Brahmaputra along with their tributaries form the Himalayan rivers. These river systems which are fed by rain, snow and glaciers melt. They carry more than two-thirds of total water available in India. Majority of the rivers and their tributaries originate from neighbouring countries and are trans-boundary rivers. India is located either upstream or downstream of the river systems and has to face occasional flooding during torrential rains in the catchment, which is beyond their approach.

Major rivers in the Deccan Group the Mahanadi, the Godavari, the Krishna, the Narmada, the Tapi and the Cauvery are rain-fed and carry much less silt than the Himalayan rivers. Coastal rivers of India are, in general, smaller in lengths and catchment areas. Apart from this, the river Subernarekha (with 1.9 mha catchment) is the largest of several other medium rivers in the country (Figure 15.1).

Other than rivers and canal system spread out in the country, there are other inland water resources like reservoirs, lakes, tanks, bheels, oxbow lakes, brackish water, etc., which cover almost 7 mha of the country. More than 50% of these water resources are spread out in the states of Orissa, Andhra Pradesh, Gujarat, Karnataka and West Bengal (Table 15.1).

The three major issues concerning variability in water resources in India are as follows:

- India has large temporal variability in water availability.
- There is a regional mismatch between water availability and much higher demands, which is ever increasing.
- Withdrawal of water from surface and subsurface water bodies is increasing and slowly becoming unsustainable (Jain, 2019).

During monsoon in India, more than 70% precipitation takes place within 4 months of the year and the rivers carry between 70% and 75% of their annual flow during that period, many a times, inundating the flat plains and flooding. During rest of the 8 months, majority of rivers in India account for balance 25%–30% of flow causing negligible or dry runs during some summer months. Groundwater also shows some such rise and fall, with some delay during dry and post-monsoon periods. Subsequently, this large variability in water availability gives rise to host of problems such as floods and droughts.

Apart from this, water availability in India varies from place to place and region to region simultaneously due to diversity of geo-morphological characteristics. Thus, one often finds draught conditions in some river basins or regions of India, whereas some others are suffering from flood conditions during the same period. There is a need to address both types of variability at the same time as most of the control measures are similar in nature.

Along with growth in population, changes in lifestyles and economic activities are occurring very rapidly in the country. With these rapid increases, the demand for water is also increasing proportionately. Apart from domestic water demand, the agricultural water demands, which accounts for around 85% of annual water demand, and industrial water needs are recording substantial increase. As there is no major trend in annual rainfall, the gap between demand and supply is becoming wider and many regions/basins are facing scarcity.

To meet increasing water demands, progressively larger quantities of water are being withdrawn from both surface and subsurface water resources, thereby causing adverse impact downstream of many rivers, which are showing no flow during summer days. Even rapid fall of groundwater table shows a delicate situation in many regions in the country. This is highly detrimental to the environment.

FIGURE 15.1 River Basins of India (article by Dr. Sharad K Jain, "Water Resources Management in India", National Institute of Hydrology, Roorkee 247667, India).

15.3 Groundwater

Groundwater, the most preferred source of water, is mostly available in the country. To meet the needs of vast rural population spread out in over 700 districts of the country, groundwater happens to be most dependable source of water. Groundwater has contributed significantly towards India's economy and is the most precious natural resource. It meets around 85% of India's rural water needs, 50% of urban water needs and around 50% irrigational water needs of the country. With increased over-dependence on groundwater, un-planned and non-scientific development of groundwater resources, mostly led by individual initiatives, adverse impacts are observed in terms of long-term decline of groundwater level,

TABLE 15.1   Surface Water Resources Potential of the River Basins of India

S. No.	River Basin	Catchment Area (km²)	Avg. Water Resources Potential (bcm)	Utilizable Surface Water Resources (bcm)
1	Indus (up to border)	321,289	73.31	46
2	Ganga	861,452	525.02	250
	Brahmaputra	194,413	537.24	24
	Barak and Others	41,723	48.36	
3	Godavari	312,812	110.54	76.3
4	Krishna	268,948	78.12	58
5	Cauvery	81,155	21.36	19
6	Subarnarekha	29,196	12.37	6.8
7	Brahmani and Baitarani	51,822	28.48	18.3
8	Mahanadi	141,589	66.88	50
9	Pennar	55,213	6.32	6.9
10	Mahi	34,842	11.02	3.1
11	Sabarmati	21,674	3.81	1.9
12	Narmada	98,796	45.64	34.5
13	Tapi	65,145	14.88	14.5
14	West-flowing rivers from Tapi to Tadri	55,940	87.41	11.9
15	West-flowing rivers from Tardi to Kanyakumari	56,177	113.53	24.3
16	East-flowing rivers between Mahanadi and Pennar	86,643	22.52	13.1
17	East-flowing rivers between Pennar and Kanyakumari	100,139	16.46	16.5
18	West-flowing rivers of Kutch and Saurashtra including Luni	321,851	15.1	15
19	Area of inland drainage in Rajasthan	36,202	NA	NA
20	Minor river basins draining to Myanmar and Bangladesh		31	NA
	Total		1,869.35	690.1

Source: Centre for Science and Environment, n.d., India Water Facts_Water Resources of India.

quality deterioration and increased energy consumption for lifting water from deeper levels. On the other hand, there are some regions in the country where groundwater development is in low key, in spite of the availability of groundwater resources, and on the other hand, in the canal command areas water logging and soil salinity along with rise in groundwater table are prevalent (Jha and Sinha, 2009). Due to the occurrence of diversified geological formation with considerable lithological and chronological variations, complex tectonic framework, climatologically dissimilarities and various hydrochemical conditions, India has a highly diversified hydrogeological set-up. The major geological formations are:

1. Consolidated formations represented by Igneous and Metamorphic rocks with major rock types consisting of granites, charnockites, quartzites and associated phyllite, slate, etc.; basalt and associated igneous rocks;
2. The semi-consolidated rock formations are represented by Mesozoic and Tertiary period with major rock types represented by limestone, sandstone, pebbles and boulder conglomerates;
3. The unconsolidated formations belong to Pleistocene to recent period and are represented by major rocks such as boulders, pebbles, different grade of sands and silt-clay. These rocks form the major potential aquifer zone (Jha and Sinha, 2009) (Figure 15.2, Table 15.2).

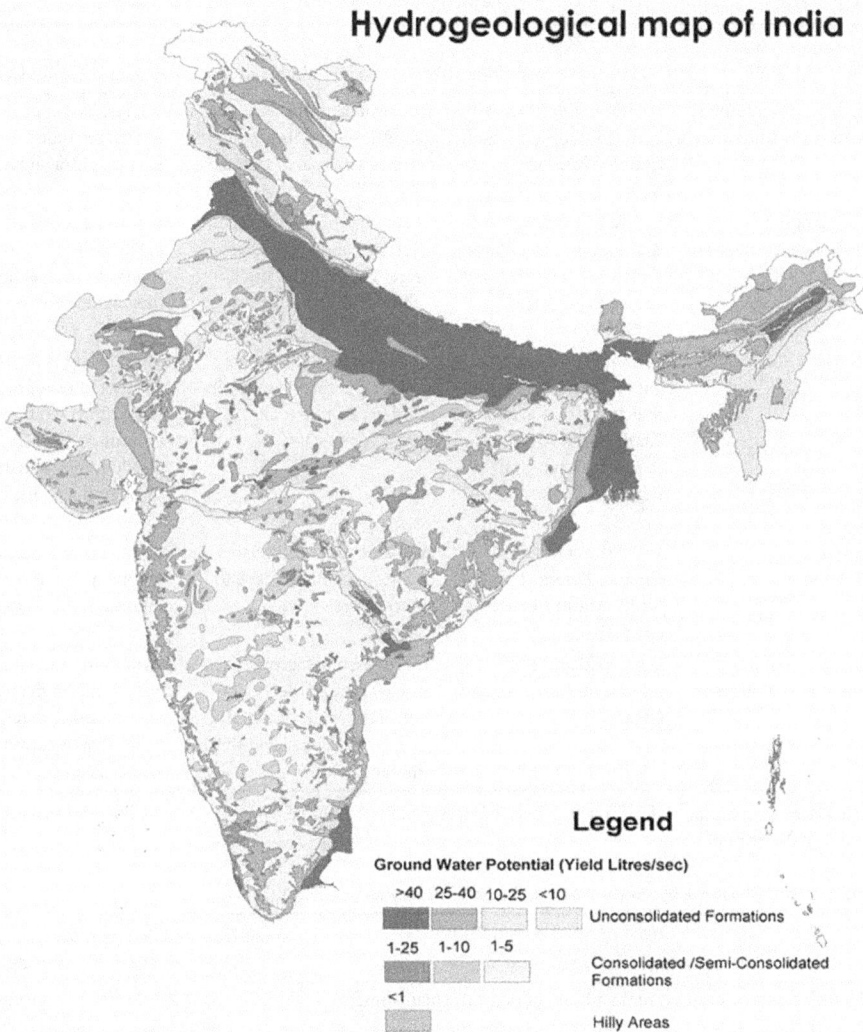

FIGURE 15.2 Hydrologeological map of India (article written by B. M. Jha and S. K. Sinha "Towards Better Management of Ground Water Resources in India", Central Ground Water Board, Government of India).

As per the assessment done by Central Ground Water Board (CGWB), Ministry of Water Resources, Government of India, the annual replenishable groundwater resource was estimated to be 433 billion cubic metre (bcm) out of which 399 bcm was available for development (2006 report). During the same period groundwater draft for the country as a whole was estimated to be 231 bcm of which 92% was utilized for irrigation and balance 8% for domestic and industrial use. CGWB had further carried out groundwater resources availability and status of its utilization in 5,723 units, spread out in all the regions of the country. The findings of the study are enclosed in Table 15.3.

The Central Ground Water Authority has notified 20 severely critical or over-exploited areas in the country for regulation of groundwater development and management (Jha and Sinha, 2009). Although groundwater use has provided the much-needed food security and drinking water needs of the people, over-withdrawal has caused many a problem such as increase in cost of pumping, investment going waste due to decline of water table and even dry run of different stretches of rivers. Over-pumping of groundwater has also resulted in contamination due to chemicals, namely, arsenic, fluoride and iron,

TABLE 15.2   Region-Wise Physiographic and Geomorphologic Settings

S. No.	Region	States/Area Description	Formation Details	Expected Groundwater Yield
1	Northern mountainous terrain and hilly areas	The mountainous terrain in the Himalayan region from Kashmir to Arunachal Pradesh	Rocks such as granites, slates, sandstone and lime stone from Paleozoic to Cenozoic Age	1–40 lps. No scope of storage. Recharges Indo-Gangetic-Brahmaputra alluvial plains
2	Indo-Gangetic-Brahmaputra alluvial plains	Vast plains of Ganga and Brahmaputra river basins more than one/fourth of country's land area (850,000 km²)	Thick piles of sediments of Tertiary and Quaternary Age. The vast and thick alluvial fill (<1,000 m at places)	Extensive and highly productive groundwater reservoir. Deeper aquifers in this region offer good scope of further exploitation
3	Peninsular shield area	South of Indo-Gangetic plains covering states of Maharashtra, Karnataka, Tamil Nadu, Andhra Pradesh, Orissa and Kerala	Consolidated sedimentary rocks, Deccan Trap basalts and crystalline rocks	Groundwater is restricted to weathered residuum and interconnected fractures. Ground Water (GW) development is mainly through dugwells. Yield varies from 2 to 10 lps
4	Coastal Area	Covers the states of Gujarat, Kerala, Tamil Nadu, Andhra Pradesh, and Orissa	A thick cover of alluvial deposits of Pleistocene to Recent age.	Potential multi-aquifer system but has a risk of seawater ingress. Yield range between 5 and 25 lps
5	Cenozoic fault basin and low-rainfall areas	Covers arid and semi-arid regions of Rajasthan and Gujarat	Three discrete faulty basins, the Narmada, the Purna and Tapti Valleys contain valley-fill deposits 50–150 m in depth	Scanty rains and GW occurrence in deep aquifer tapping fossil water. Purna valley GW is extremely saline. Yield 1–10 lps

Source: Jha, B. M. and Sinha, S. K., "Towards Better Management of Ground Water Resources in India", Central Ground Water Board, Government of India.

TABLE 15.3   Groundwater Exploitation Status in the Country as per CGWB

S. No.	Status of GW Development	No. of Units Studied with the Percentage	Categorization
1	GW development exceed more than 100% of natural replenishment	839 (14.7%)	Over-exploited
2	GW development between 90% and 100% of utilizable resources	226 (3.9%)	Critical
3	GW development between 70% and 100% of utilizable resources and long-term decline of water level	550 (8%)	Semi-critical
4	GW development below 70% of utilizable resources	4,078 (72.8%)	Safe
5	Units having salinity in GW	30	

Source: Jha, B. M. and Sinha, S. K., 2009, "Towards Better Management of Ground Water Resources in India", Central Ground Water Board, Government of India.

salinity, etc. Eventually, India is going to face a situation when in a normal monsoon year, India will just be able to meet its water demand but in a below-normal monsoon period, the situation will be precarious in some of its regions.

15.4 Water Availability and Future Water Demand

The per capita availability of water at national level is reducing at a very fast pace. This is mainly due to increase in population, which is quite high. In 1951, the annual per capita availability of water was 5,177 m³, which has come down to as low as 1,508 m³ in 2014. Table 15.4 gives further projection of

TABLE 15.4 Estimated Per Capita Availability of Water

S. No.	Year	Estimated Annual Per Capita Availability of Water in m³
1	1951	5,177
2	2001	1,820
3	2014	1,508
4	2025	1,465
5	2050	1,235

Source: Indian Council of Agricultural Research (ICAR), 2019, India's per capita water availability to decline further, New Delhi.

TABLE 15.5 Water Demand in Billion Cubic Metres for Various Sectors from 2010 to 2050

Sector	Standing Sub-Committee of Ministry of Water Resources			National Commission on Integrated Water Resource Development		
Year	2010	2025	2050	2010	2025	2050
Irrigation	688	910	1072	557	611	807
Drinking water	56	73	102	43	62	111
Industry	12	23	63	37	67	81
Energy	5	15	130	19	33	70
Others	52	72	80	54	70	111
Total	813	1,093	1,447	710	843	1,180

Source: India Environmental Portal, n.d., Water Resources Development of India: Critical Issues and Strategic Options. https://www.adrindia.org/adri/india-water-facts#.

expected per capita water availability in the country in near future (Indian Council of Agricultural Research (ICAR), 2019) ().

Given the potential increase in water demand due to rise in population and lack of conservation measures, very soon India will face stressed conditions if the per capita water availability reduces further and comes down between 1,000 and 1,100 cum.

Total utilizable water resources in India are assessed as 1,123 bcm. Total annual net groundwater availability has been estimated as 399 bcm. A total storage capacity of 212.78 bcm has been created in the country through major and medium projects with additional projects under implementation in 2006 contributed 76.26 bcm, while projects under consideration during that period was expected to contribute 109.77 bcm. The irrigational potential of the country has been estimated at around 139.9 million hectares (mha). With inter-basin sharing, this may be increased by 35 mha. The CGWB estimates that by introducing groundwater recharging measures like artificial recharging and rainwater harvesting (RWH) the availability may be enhanced by 45 mha (Jha and Sinha, 2009). In 2006, water demand was estimated by (i) Standing Sub-Committee of the Ministry of Water Resources, Government of India, and (ii) the National Commission for Integrated Water Resources Development (NCIWRD). Details are given in Table 15.5.

The estimates by Ministry of Water Resources (MoWR) indicate that the industrial water demand will rise by five times in 2050 and that for energy by 16 times, while the drinking water need will double and irrigation needs will increase by 50%. Therefore, every effort needs to be made to fill this resource gap.

India invested a major chunk of its investment for the development and expansion of its irrigational facilities, along with consolidation of its existing systems. The main part of the strategy was to increase the food grains production. With sustained and systematic development of irrigation since 1951, the combined potential of all the three irrigation methods, namely major, medium and minor, have increased from 22.6 to 98.84 mha.

15.5 Suggested Remedial Measures

To take up corrective measures it is essential to look into water crisis in totality, i.e. surface and subsurface water resources together and not in isolation. Three distinct actions are required, namely reduce demands, conserve water and move water across geographies (Jain, 2019). Since agriculture is the largest consumer of water resources hovering around 80%, all efforts need to be made to increase the efficiency of its use. Present efficiency of water utilization in agriculture sector is merely 40%–50%. With mere 20% increase in efficiency of water utilization in irrigation sector, demands of water for industrial and urban use could easily be met. A higher water efficiency may be achieved by adopting measures like drip irrigation, sprinklers and other water-saving measures that must be integrated with cropping pattern and demand for crops in the market. Needless to say, for doing so, farmers, especially the young and prospective farmers and the community at large, need to be involved in the whole process including achieving progressively better awareness about key aspects of contemporary agricultural science. These measures need to be adopted first in the regions where water availability is low and eventually across the subcontinent.

15.5.1 Utilization of Floodwater

As discussed earlier, 70% of water flow in the rivers in India takes place within a period of 4 months creating flood like situation almost every year. Hence, it is necessary to conserve this excess flow of water, which is estimated to be 500 billion cubic metres (bcm), and use it during dry season. This storage of water may be done on the ground or below ground.

Ground-level storage will certainly cover lots of precious land and create numerous problems like displacement of people, forest submergence, threat to biodiversity, environmental issues, etc. To conserve water below ground, suitable hydrogeological features are necessary and suitable large-scale managed aquifer recharge system needs to be created. It is equally important to control groundwater withdrawal measures especially in the regions where groundwater withdrawal is more than its recharge capacity and water table is declining.

Groundwater recharge of the aquifer needs to be planned at both macro-level and micro-level. Currently, macro-scale information on recharge potential is available, whereas recharge activities could only be planned at local/micro-level. Hence, for recharge planning, it will be necessary to identify and obtain geological data at local level and delineate recharge sites. It is also equally important to identify the de-saturated aquifers and determine their water-holding capacity to estimate recharge planning at local level.

Villagers and farmers may do their bit of water conservation activities by construction of check dams and farm ponds to store rainwater (Raghavan and Eslamian, 2021). This will help in meeting local water demands as well as recharging the groundwater aquifers. However, if a large number of such structures are created at the upstream side, there may be some water shortage in the downstream areas more so in the low-rainfall regions. Therefore, it will be advisable to go for such large-scale interventions at the river-basin scale so that impacts for such interventions at upstream are known and help in development of proper river basin plans.

In spite of its overriding importance, urbanization and systematic and coordinated urban development strategy at the national or sub-national levels have received little attention in India. Except for the policies spelt out in the Report of the National Commission on Urbanization in 1988, no serious official output is available on the subject till date. The Jawaharlal Nehru National Urban Renewal Mission (JNNURM) launched in 2005 was aimed at strengthening the infrastructure and images of 65 cities selected based on their typical identity. The Smart City Mission launched in 2015 aimed at transforming 100 cities into "smart cities" by digitizing every action related to city development and management. Both missions have their own pitfalls that they virtually failed to recognize the human component of Indian towns and cities including the most populous cities that act as powerful magnets drawing millions from all over the country for employment and better quality of life.

In most metropolises and megacities of India, water supply is a critical issue. And further, with increasing paved area in them and little attention paid to the location of natural drainage channels running through them, the run-off generated by monsoon contributes to flash floods in them as well as to post-monsoon floods.

15.5.2 Loss of Reservoir Storage Due to Sedimentation

Annually India is losing around 0.8%–1% of its created storage of more than 300 bcm due to sedimentation (Jain, 2019). As it is impossible to create new storage every year, it becomes necessary to control the sedimentation flow or de-sludge the accumulated siltation by developing technologies as per site requirements. The technologies adopted in India so far include (i) flushing out stored sediments using low-level outlets by having low-level spillway crests; (ii) spillway gates may be raised in the early stages of wet season to allow sediments to exit the reservoir without settling; (iii) physically removing sediments deposited in the existing reservoirs.

15.5.3 Flood Management

Out of country's total geographical area of 329 mha, about 40 mha are flood-prone out of which around 32 mha can be provided with reasonable degree of flood protection. Till March 2004, 16.46 mha of flood-prone area has been provided with a reasonable level of flood protection by providing embankments, drainage channels, town protection work and raising platforms etc. Central Water Commission has developed a nationwide Flood Forecasting and Warning System on inter-state water basins. Flood forecasts are being issued through 173 stations of which 145 are river-level forecasting stations and 28 are inflow-forecasting stations on major dams/reservoirs throughout the country (India Environmental Portal). These reliable advance information systems do help in loss of life, property and structural damages appreciably.

15.6 Water Quality Issues and Environmental Factors

Water pollution, both surface and groundwater, is slowly acquiring serious dimensions in the country. Situation is so precarious in some stretches of the river is that though river flows by the urban centre, one has either to bring water from a distance of hundreds of kilometres or go for subsurface water flow or both to meet domestic water needs of the inhabitants. Biological, toxic organic and inorganic pollutants already contaminate more than 70% of surface water sources and large part of groundwater reserves. There are some areas in the country where water available is not suitable for any use, namely domestic, agriculture or industry. There are many sources for water pollution. These include:

- Dumping of untreated and partially treated municipal waste into water bodies;
- Partially treated and untreated industrial waste discharge into surface water bodies and injecting into groundwater;
- Drainage from residues of agricultural fertilizers, insecticides and pesticides;
- Degradation of catchment areas;
- The hidden chemicals stored in soils and sediments with the potential to be remobilized by land-use change, dredging or climate change;
- Intensive livestock production helping increase in solid and liquid waste production;

- Destruction of wetlands, the natural filter of water;
- Mining and other extraction activities including mine water pumping, ore washing and effects of mine trailing;
- Over-exploitation of groundwater in coastal and semi-arid regions causing an increase in chemical contents and salinity; and
- Oil spills from oil transport systems in the major rivers.

Water quality criteria are available for various categories of beneficiaries ranging from drinking water, outdoor bathing, propagation of wildlife and fisheries, irrigation, industrial cooling to controlled waste disposal. The water quality standards developed by Central Pollution Control Board (CPCB) are listed in Table 15.6.

Based on the long-term water quality data generated over the years, it has been estimated that around 6,750 km of river length is severely polluted (Biological Oxygen Demand [BOD] more than 6 mg/L), a length of 8,550 km is moderately polluted (BOD between 3 and 6 mg/L) and 29,700 km length is relatively clean (BOD less than 3 mg/L) (World Bank, 2019). In most of the river stretches, the three identifiable pollutant indicators, namely BOD, dissolved oxygen and total coliforms are not adequately able to indicate overall health of the water body.

Agriculture: Agriculture happens to be one of the biggest contributors of waste in the water bodies. Indiscriminate use of pesticides and insecticides as well as fertilizer beyond the soil and crop needs is contributing to this problem. High level of fertilizer use has been associated with increased eutrophication in many inland water bodies. Switching to organic farming will certainly reduce this problem, but proper education to the farmers and price control measures and appropriate legislation will certainly help to improve the situation and stop indiscriminate use.

Industry: Water pollution by industries is quite significant in the country. Of the total pollution load, 40%–45% is contributed by the processing of industrial chemicals, while 40% of the total organic pollution (expressed in BOD) arises from the food industries followed by industrial chemicals and the pulp and paper industry (Jain, n.d.). As per rules, all the industries are expected to treat their waste before discharge into water bodies or drains but rarely do so as per guidelines.

Private landowners and many industrial units in India have absolute ownership of groundwater beneath their land and they can extract any amount of groundwater without regard to the impact on

TABLE 15.6 Designated Best Use Water Quality Criteria

Designated Best Use	Class of Water	Criteria
Drinking water source without conventional treatment but after disinfection	A	Total Coliforms Organism MPN/100 mL shall be 50 or less pH between 6.5 and 8.5 Dissolved oxygen, 6 mg/L or more Biochemical oxygen demand, 5 days, 20°C, 2 mg/L or less
Outdoor bathing (Organized)	B	Total Coliforms Organism MPN/100 mL shall be 500 or less pH between 6.5 and 8.5 Dissolved oxygen, 5 mg/L or more Biochemical oxygen demand, 5 days, 20°C, 3 mg/L or less
Drinking water source after conventional treatment and disinfection	C	Total Coliforms Organism MPN/100 mL shall be 5,000 or less pH between 6 and 9 Dissolved oxygen, 4 mg/L or more Biochemical oxygen demand, 5 days, 20°C, 3 mg/L or less
Propagation of wild life and fisheries	D	pH between 6.5 and 8.5 Dissolved oxygen, 4 mg/L or more Free ammonia (as N), 1.2 mg/L or less
Irrigation, industrial cooling, controlled waste disposal	E	pH between 6.0 and 8.5 Electrical conductivity at 25°C micro mhos/cm, Max. 2,250 Sodium absorption ratio, Max. 26 Boron, Max. 2 mg/L

Source: CPCB, Government of India. www.india.gov.in/official-website-central.

aquifer and environment. This create over-pumping in majority of cases disturbing the groundwater quality and lowering the groundwater table. Industrial symbiosis, in which unusable waste from one industry becomes input for the other industry, encouraging reductions and replacement of toxic chemicals and "polluter to pay", will certainly help in reducing the pollution load.

Domestic: It is estimated that about 38,254 mld (2009) of wastewater gets generated in Class I and Class II (up to 50,000 population) urban centres in the country. This accounts for almost 70% of total urban population of the country. Less than 35% of waste generated in the urban centres gets treated for safe disposal. CPCB studies show that many of the existing sewage treatment plants are only partially functional due to many reasons. The main cause pollution of rivers and other water bodies is the discharge of untreated and partially wastewater from the urban centres. Under the National River Action Plans, initiatives have been taken by the central and state agencies to control or reduce river water pollution (Kamyotra and Bhardwaj, 2011). Since long-distance transportation of wastewater for centralized treatment costs enormously, efforts are now being propagated to go for decentralized treatment based mostly on natural processes. After proper treatment, it is possible to use this wastewater for pisciculture, irrigation, forestry and horticulture. Municipal wastewater can be recycled for irrigation or for industries or thermal power as utilities in cooling towers and boilers.

We have a success story of using municipal wastewater for irrigation as well as pisciculture for many decades in the country without going for any conventional treatment. The biggest wastewater treatment process involving pisciculture and agriculture activities along natural treatment process has been developed over the years at East Kolkata Wetlands (EKW) to treat around 1,000 mld of wastewater from the Kolkata Metropolitan area. These 12,500 ha of area has got 254 sewage-fed fisheries, agricultural land, garbage disposal sites and some built-up area. EKW have been designated as a "Ramsar Site" in November 2002.

15.7 Selected Irrigation Practices and Their Applications in India

Various irrigation technologies have been adopted in this country for improving the yield and quality of crops. Irrigation techniques have differed from crop to crop as well. It also depended on the type of source of water.

15.7.1 Surface Irrigation

It had been adapted in majority of areas in which water is distributed over the soil surface by gravity flow. The irrigation water, in general, passes through level or graded furrows or basins using siphons, gated pipe, or turnout structures and is allowed to advance across the field. Surface irrigation is most suited for flat land slopes and medium to fine texture soil type, which allows lateral spread of water. This technique is simple and does not require much of technical skill, excepting a basic knowledge of water management depending on type of crop. Surface irrigation allows full utilization of rainwater. There are mainly four types of surface irrigation, namely basin irrigation, border irrigation, furrow irrigation and wild irrigation (Jondhale et al., 2017).

15.7.2 Drip Irrigation

This method is also called micro- or trickle irrigation. In this method water at a slow rate spreads over the soil, drop by drop, to irrigate limited area around the plant. Drip irrigation saves water by reaching directly to the root zone with the help of some irrigation components namely pump unit, control head, main and sub-main lines, laterals and dippers. The rate of flow in general is kept between 2 and 20 L/hour. The soil moisture is kept at an optimum level with frequent irrigation. This method can reach an efficiency of 90%–95% (Indian Agricultural Research Institute, n.d.).

Drip irrigation has been extensively used in the country. As on March 2012, more than 2.5 million ha of agricultural area is under drip irrigation in the country. This type of irrigation is mostly suitable for all orchard and vegetable crops. Plans for further extension of this irrigation are in hand.

15.7.3 Sprinkler Irrigation

In this method of irrigating water is sprayed or sprinkled through the air like rain drops. The spray or sprinkling device may be permanently set in a place or moved on a trolley. Or, this may be temporarily set and moved to a new area as per needs. Water under pressure is carried and sprayed over the crop. Due to spraying method refreshing effect occurs on plant. Sprinklers mostly operate through electrical or hydraulic technology. The main components are pump unit, mainline, lateral line, sprinklers and nozzles that may be rotating or fixed type.

Sprinkler irrigation saves around 30% of water. Around 3.6 million ha of irrigated land is under sprinkler irrigation in the country as on March 2012. Sprinkler irrigation is mostly applied for oil seeds, cereals and vegetable crops (Irrigation Engineering, n.d.).

15.7.4 Smart Irrigation

It is the latest method of irrigation. It is reported to be completely automated and cost-effective but needs huge initial investment. For Smart Irrigation a proper study of available water, soil, topography, climate, crop type, soil infiltration rate and capital cost need to be made. Smart Irrigation system uses sensors, tunnels, GPRS system and wireless network. It irrigates automatically as per site requirement, without waiting for the presence of an operator. Though it consumes less time and is cost-effective, the initial investment is quite high and needs good technical support for its initial running. To begin with high-skilled human resource development will be a need. It has been adopted in a very limited scale in India (Jondhale et al., 2017).

15.8 Climate Change and Its Impact on Water Resources

Climate change means human-induced change to the global climate system. In its 2007 report, the Intergovernmental Panel on Climate Change (IPCC) had stated, "Global GHG (Green House Gas) emissions due to human activities have grown since pre-historic times, with an increase of 70% between 1970 and 2004". It further explained, "Global atmospheric concentration of CO_2, methane (CH_4) and nitrous oxide (N_2O) have increased markedly as a result of human activities since 1750 and now far exceed pre-industrial values determined from ice cores spanning many thousand of years" (Intergovernmental Panel on Climate Change (IPCC), 2008).

This increase has resulted in increased average temperature of the global ocean, sea level rise and decline in glaciers and snow cover. There is increased frequency of droughts, cold days and cold nights. At the same time, hot days, hot nights and heat waves have become more frequent. Analyses done by the India Meteorological Department and the Indian Institute of Tropical Meteorology, Pune show the same trends for temperature, heat waves, glaciers, droughts and floods, and sea level rise as shown by IPCC, although the magnitude of changes do vary. Monsoon rainfall is increasing along the west coast, north Andhra Pradesh and north-west India. It is reported to be decreasing over east Madhya Pradesh and adjoining areas, north-east India and parts of Gujarat and Kerala (Sengupta, 2012).

> Glacial Water: Changes in climate is already affecting many mountain glaciers around the world. The run-off of Himalayan rivers is expected to be highly vulnerable to climate change because warmer climate increases the melting of snow and ice earlier and faster shifting the timing and distribution of run-off. These changes will affect the availability of fresh water.

Indian Himalayas has nearly 9,575 glaciers and it is estimated that these cover an area of about 38,000 km². Almost 67% of the Himalayan glaciers have retreated in past few decades due to warming effects. A study by the Geological Survey of India has revealed that the glaciers are receding at an alarming rate (Tangri, 2000). Apart from receding glaciers, another alarming problem is formation of lakes due to melting of the glacial rapidly. A sudden discharge of large volumes of water with debris from these lakes caused glacial-lake outburst floods (GLOFs) in valley downstream causing destruction of valuable natural resources and flash floods.

Fresh Water (Surface and Groundwater): Water resources will be coming under increasing pressure in the Indian subcontinent due to changing climate, which affects water demand, lack of water during high demand and quality. This will certainly affect the life of the community and may create crisis and disputes over water resources. Ganga–Brahmaputra basin is certain to get affected due to receding glaciers and GLOFs formation causing damage to human life and property including environmental degradation.

Most of the rivers in southern peninsular are fed through groundwater discharge and supplemented by monsoon rains. Therefore, these rivers have very limited flow during non-monsoon seasons. With fluctuation of monsoon period and intensity, the flow in these rivers is bound to get affected.

The demand for groundwater is expected to increase in the future due to declining surface water availability, increase variability in precipitation and reduced summer flows in snow dominated basins. Climate change will certainly affect the groundwater recharge rates and GW level. In many aquifers, the spring recharge shifts to winter and summer recharge will decline. In high altitudes, thawing of permafrost will cause changes in groundwater level and quality. Climate change will also cause the vegetation changes, which will have an impact on GW recharge (Sengupta, 2012; Table 15.7).

Water Quality and Health: Studies have shown climate change induces regional weather changes causing heat waves, extreme weather conditions, rise in temperature and increase/decrease in precipitation. All these extreme weather conditions are directly or indirectly involved in change in microbial contamination pathways in certain vulnerable sectors, transmission dynamics, agro-ecosystems, hydrology and socio-economic factors affecting health of the community members.

TABLE 15.7 Select Reports on the Impact of Climate Change on Water Resources in India during the Next Century

Region/Location	Impact	Reference
Indian subcontinent	Increase in monsoon and annual run-off in the central plains	Lal and Chandra (1993)
Odisha and West Bengal	A rise in 1 m in sea level inundates 1,700 km² of agricultural land	IPCC (1992)
Indian coastline	A rise in 1m in sea level is likely to affect a total area of 5,763 km² and put 7.3 million people at risk.	JNU (1993)
All India	Increase in potential evaporation across India	Chattopadhyay and Hulme (1997)
Central India	Basin located in the comparatively dry region is more sensitive to climate change	Mehrotra (1999)
Kosi basin	Decrease in the run-off by 2%–8%	Sharma et al. (2000)
Southern and Central India	Soil moisture increases marginally by 15%–20% in monsoon months	Lal and Singh (2001)
River basins of India	A general reduction in the quantity of available run-off, increase in Mahanadi and Brahmini basins	Gosain et al. (2006)
Damodar basin	Decreased river flow	Roy et al. (2003)
Rajasthan	Increase in evapotranspiration	Goyal (2004)

Source: Sharma, A., and Sarkar, A., (eds.), 2007, Impact of Climate Change on Water Resources. Jalvigyan Sameeksha (Hydrology Review) 22 (Indian National Committee on Hydrology and Institute of Hydrology, Roorkee).

TABLE 15.8 Health Concerns and Vulnerability Due to Climate Change

Weather Events	Impact on Human Health
Warm spells, heat waves and stagnant air masses	• Heat strokes, affecting mainly children and elderly • Increase in respiratory diseases • Cardiovascular illnesses
Warmer temperatures and disturbed rainfall patterns	• More exposure to diseases and an increase in vector-borne diseases
Heavy precipitation events	• Increased risk of diseases related to contaminated water and to unsafe food. Poor sanitation aggravates the situation
Draughts	• Malnutrition and starvation affect mainly children and the elderly • Reduced crop yields affect farmers' income including psychological stress of non-payment of debts
Intense weather events (cyclones, storms)	• Loss of life, injuries, life-long handicaps • Damaged public health infrastructure • Loss of property, land, displacement and forced migration
Sea level rise and coastal storm	• Loss of livelihood, the disappearance of land and massive migration causing social conflicts affecting mental health

Source: WHO-South East Asia Region – How Is Climate Change Affecting Our Health – A Manual for Teachers. World Health Day, 7 April 2008.

Warmer temperature will raise the risk of flooding, increasing diarrheal diseases. In addition, algal blooms could occur more frequently as warm temperature helps this cause, particularly in areas with polluted water.

Climate change may induce the risk of some infectious diseases, particularly those diseases, which strive during warm areas spread by vectors such as mosquitoes and rodents as climate is more favourable for their growth.

Food production, especially cereal crops, is expected to be severely affected by climate change. A change in climate will affect temperature, rainfall patterns, soil moisture and soil fertility. Crop pests may thrive in this climatic condition creating widespread malnutrition conditions within the community.

Climate change may also contribute towards social disruption, psychological stress, economical loss, decline and displacement of populations due to effects on agriculture, water resources and extreme weather conditions. All these issues will affect developing countries more due to continued uncontrolled population growth and already fragile environment (WHO Publication, 2008) (Table 15.8).

15.9 Irrigation Systems of Rajasthan and Tamil Nadu

Being a vast land inhabited continuously through millennia, India developed into a country with over 300 languages and over 1,600 dialects, multiple faiths and cultures. Rajasthan and Tamil Nadu are two of the 15 major states (provinces) of India located in western and southern parts of the country, respectively. They vary considerably from each other not only in terms of geoclimatic, physiographic and landscape characteristics but also traditional cultural practices that evolved. Rajasthan is set in a hot semi-arid to arid region and Tamil Nadu in undulating terrain, dotted with water bodies and rain-fed rivers crisscrossing a landscape which is far greener than Rajasthan (Table 15.9).

The State of Rajasthan, founded in 1949, occupies a significantly large part of the Thar Desert in north-west India (Earth Eclipse, n.d.). The climate here is commonly identified as "hot desertic" or arid to semi-arid having highly variable rainfall, consisting of a landscape made of vast and shifting sand dunes, seasonal rivers and thorny bushes. In spite of such harsh conditions, deep water table interrupted by rocky outcrops of Aravalli Ranges, the medieval cities of Rajasthan flourished rather well. While

TABLE 15.9   Comparison of Selected Features of Rajasthan and Tamil Nadu

Aspect	Region			Remarks
	Rajasthan	Tamil Nadu	India	
Area in '000 sq.km	342.2	130.0	3,280	Rajasthan is the largest state in terms of area
Population in 2011, in million	68.5	72.1	1,210.5	
Percent population living in urban areas	24.87	48.4	31.1	Tamil Nadu is one of the most urbanized states of India
Environmental setting	Hot semi-arid to arid	Coastal humid	Diverse	
Annual average precipitation in mm	600	930	Diverse	
Crops grown	Barley, mustard oilseeds, pulses, cotton, tobacco, groundnut	Paddy, maize, pulses, palm oil, sugarcane, coconut, banana, mango	Diverse	
Sources of irrigation	24% canal, 73% tube wells and open wells	26% canal, 17% tanks, 55% wells	24% canal, 65% groundwater	
Irrigated land	25% or 6.66 mha under irrigation		47% or 65 mha of farmland, rest dependent on monsoon	

Source: Various reports and Internet sites in public domain.

grains and lentil varieties requiring limited water were cultivated, red chillies were generously used in cooking to counter the adverse health effects of poor quality of drinking water.

15.9.1 Groundwater Scenario of Rajasthan

Since major part of Rajasthan is not served by any perennial river, people had to solely depend on meagre rainfall that was there in the western part of Rajasthan and adapted various water harvesting structures, which had been discussed in detail in the previous chapter. Even in western Rajasthan open wells had been constructed to a depth of 200–250 ft to harness groundwater and to meet the daily needs of the people. Manmade reservoirs, big and small, were constructed by studying the topography and local knowledge to accumulate the water that drained out during occasional rains to meet the agricultural and animal demand. These were mostly created with a community effort with or without supports from Royal families or as a philanthropic work by some noble person (Box 15.2).

After independence, Government of Rajasthan created Ground Water Department in the 1950s, to take up scientific investigation and exploration of groundwater sources to meet minor irrigation and drinking water needs. The department started geo-hydrological investigation in various districts of Rajasthan and engaged various types of drilling equipment to meet the water needs.

The geo-hydrological map of Rajasthan shows that the state is characterized by heterogeneity in groundwater conditions. The state can be divided into three geo-hydrological units, namely unconsolidated sediments, semi-consolidated sediments and consolidated rocks. The unconsolidated sediments are of two types – alluvial sediments and aeolian deposits. The alluvial deposits are confined to Barmer, Jalore and Jodhpur districts, consisting of sand, clay, gravel and cobbles. Valley fills have been reported from Jhunjhunu, Ajmer, Bhilwara and Udaipur districts. The aeolian sediments constitute one of the major aquifers east of major fault, east of Bikaner. It occupies an area of 1,400 sq.km. The aquifer thickness is 40–80 m. The yield of wells ranges from 100 to 150 m³/hour.

Box 15.2   Rediscovering the Johads and Reviving a River

In 1986, the villagers of Bhaonta-Koyala of Alwar district, Rajasthan, took the help of Tarum Bharat Sangh, an NGO to protect forests and repairing the Johads. They studied and mapped the natural drainage system and chose the best sites for building the Johads. Today, the village has a total of 15 water harvesting structures including one 244 m long 7 m tall concrete dam in the upper catchment of Aravalli to stop water from draining out. After construction of the dam, water levels of wells even up to 20 km downstream have started rising appreciably and retained throughout the year. This has helped increase in cultivation and mill products. The Arvari river, which had dried up completely, started reviving after the dam got constructed. By 1995, the Arvari had become a perennial river. Hence, 70 odd villages in the Arvari basin formed a protection committee of the villagers to help in afforestation in the forest area, restriction in tree felling, maintenance of the Johads and allowing grazing in the restricted areas (Centre for Science and Environment, n.d.).

Box 15.3   Indira Gandhi Canal Project

The Thar Desert, which is covered in rolling dunes for almost its whole expense covers most of the western areas of Rajasthan. The annual precipitation is on an average in between 200 and 300 mm. The Indira Gandhi Canal Project (IGCP) has been constructed in the north-western part of the state of Rajasthan covering a part of Thar Desert districts, i.e. Ganganagar, Churu, Hanaumangarh, Bikaner, Jodhpur, Jaisalmer and Barmer. IGCP is aimed to irrigate the thirsty desert land of western Rajasthan with Himalaya's water and provide drinking water to millions of inhabitants and provide irrigation potential to thousands of hectares of Thar Desert. It is a multidisciplinary irrigation cum area development project aiming to transform desert waste land into agriculturally production area.

Semi-consolidated formations include sandstones, limestones and Aur beds, covering Jaisalmer and Barmer districts. The dugwells in Jaisalmer limestones yield 13–68 m³/day. The yield of wells in Lathi sandstone varies from 50 to 150 m³/hour.

The consolidated rocks includes gneiss, granites, schist, phyllites, marble and Vindhyan sandstones, limestone, quartzite and basaltic flows, mostly restricted to eastern part of the state. The yield prospect is limited unless the well is located near major lineaments or any other weak planes. The groundwater quality is in general poor (brackish to saline) at deeper levels.

Around 140 blocks have been designated as over-exploited, 50 blocks are in critical conditions and 14 blocks have been identified as at semi-critical condition in Rajasthan. More than 39,000 km² area has been identified for artificial recharge. Groundwater in almost all the districts of Rajasthan is having more or less problems of salinity (EC > 3,000 μs/cm at 25°C), fluoride (>1.5 mg/L), iron (>1.0 mg/L) and nitrate (>45 mag/L) along with chloride (>1,000 mg/L) in some selected districts. Even with these water quality standards, as water is scare in Rajasthan, people have to depend on groundwater source for their daily use. Hence, there is a big emphasis on RWH structures in Rajasthan. Roof-Top RWH has been made mandatory in state-owned buildings of plot size more than 300 m² with effect from 2006. For violation of building bye-laws, punitive measures, viz. disconnection of water supply, have also been made. The government has made the provision of compulsory installation of RWH system in all newly constructed and existing buildings and government offices (Central Ground Water Board, 2006).

Origin of this canal is from Harike barrage situated in Punjab. The Feeder off-take from Harika is 204 km long with 170 km length in Punjab and Haryana and balance 34 km in Rajasthan. This canal enters in Rajasthan at Hanumangarh district. From the tail of Indira Gandhi Feeder, 445 km long Indira Gandhi Main Canal starts, which passes through Sri Ganganagar, Bikaner districts and ends at Mohangarh in Jaisalmer district. The project has envisaged utilization of 7.59 MAF surplus water of Ravi-Beas rivers, which forms part of Rajasthan's share (Figure 15.3).

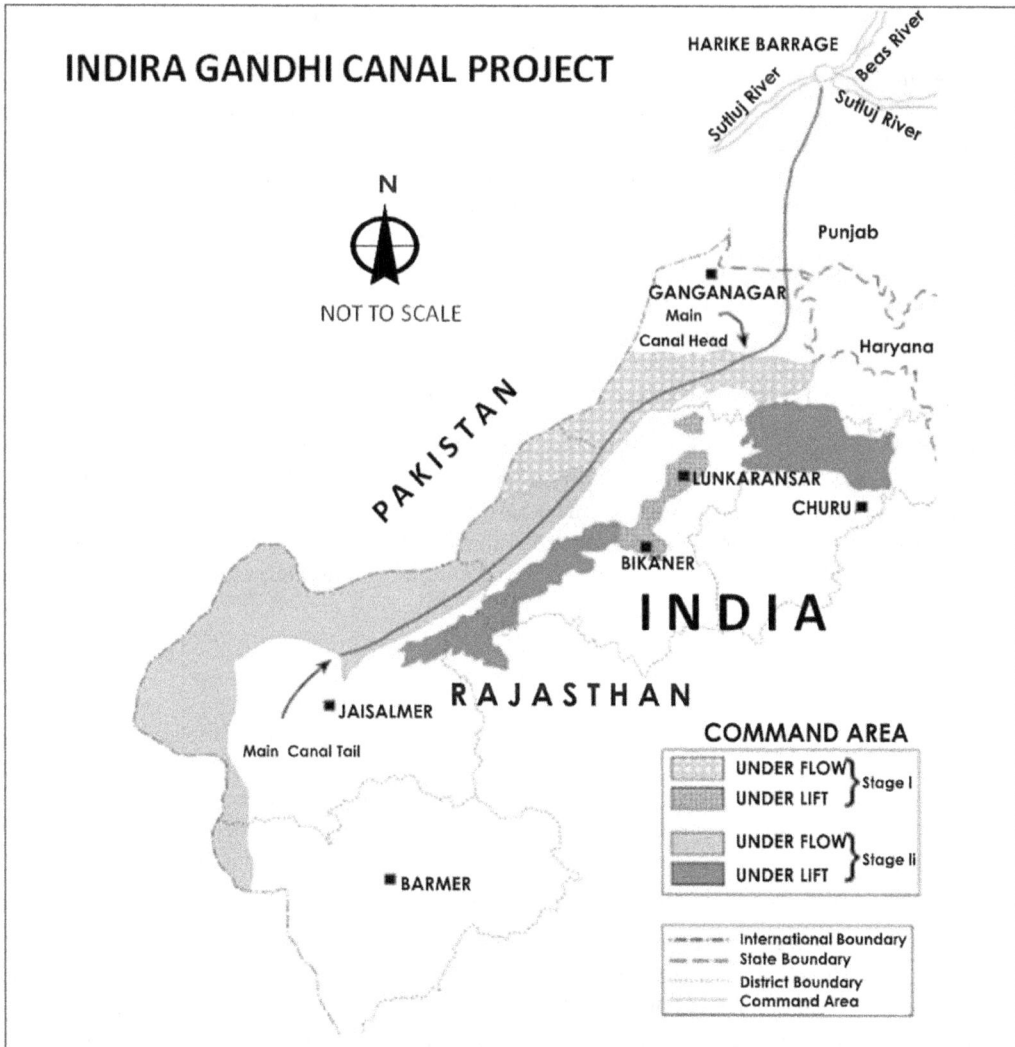

FIGURE 15.3 Location map of Indira Gandhi Canal Project (authors, based on course material for training of graduate students).

The main objectives of this project envisaged in the 1960s were as follows:

- To provide irrigation facilities in desert area to meet the increasing demand of agricultural products.
- To provide water for drinking and industrial uses.
- Drought proofing of the area and improving living conditions.
- To meet the needs of drinking water, fodder, etc. for the animal wealth in the region.
- To provide opportunities for employment and overall development of the area.

The whole project, being huge in nature, had been divided into two stages with many sub-projects as per site requirements. The 393 km long canal portion (Feeder 204, Main Canal 189) from Harike barrage to Pugal in Bikaner distt., i.e. up to 620 RD of Main Canal, with its distribution system (excluding Sahwa lift) is called stage I. Original work of this stage has been completed and major portion of this stage (except Kanwar Sain lift) has been transferred to Water Resources Department for operation and maintenance.

Work of stage II of Indira Gandhi Nahar Project (IGNP) is under progress. The area downstream to RD 620 of Main Canal with its distribution system (including Sahwa lift) is in stage II. Length of the Main Canal in this stage is 256 km, which extends from Pugal to Mohangarh. Work of Main Canal was completed in 1986.

As per project estimate of year 1993 and subsequent decisions of State Agencies, total Cultivable Command Area (CCA) of the project was 19.63 lac hectare (5.53 lac hectare in stage I + 14.10 lac hectare in stage II). In view of reduced availability of water for irrigation in project, state government in year 2005 took decision to complete canal construction works in 16.17 lac hectare of CCA. This area has been opened for irrigation after completion of canal construction works. However, Command Area Development and Water Management Programme works in lift schemes of stage II is remaining, where works for development of pressure irrigation system have been initiated.

15.9.2 Pressure Irrigation System in Lift Schemes of IGNP Stage II

In lift schemes of stage II, sprinkler irrigation system has been adopted for efficient and optimal use of precious water. A pilot project of sprinkler irrigation was taken up in 27,449 ha in the year 2007–2008, in which all works to be executed by department have been completed. All the constituted Water User Associations have started irrigation with sprinklers in most of the area. Ministry of Water Resources, Government of India, has approved and provided partial funding to establish sprinkler irrigation system in remaining area of stage II lift schemes.

15.9.3 SCADA System

SCADA system has been installed in project for effective control on water regulation and distribution in canals through latest techniques. The quantity of water flowing in Main Canal and other important canals off-taking from Main Canal is available at main controlling places of project. This is also available on Internet. Extension of this system on some new locations is under process (Indira Gandhi Nahar Department, n.d.).

With completion of most of the construction work as above, IGNP has benefited the state as below.

Total irrigation area in Rajasthan has increased with assured water supply from the canal system bringing in benefit to the state and community. Crops of wheat, mustard, paddy, groundnuts, sugarcane and cotton flourish with available canal irrigation.

The irrigation in hot arid land of western districts of Rajasthan has helped in improving the micro-climatic conditions. It has resulted in minimizing the dedicating effects of temperature and strong winds on biomass production and settlements.

Drinking water from this canal system is being supplied to various villages, towns and cities of Bikaner, Jodhpur, Sri Ganganagar, Hanumangarh, Jaisalmer, Jodhpur, Churu, Nagaur and Barmer.

Water for power generation is being supplied to various power projects of Suratgarh, Barsingsar, Guda, Ramgarh, Giral, Rajwest, etc. and various industries.

Elimination of drought conditions in Thar Desert area, with greenery.

Remarkable improvement in socio-economic conditions of the people and increase in all economic activities.

Rise in groundwater table in the command area at around 0.8 m/year.

Massive forestation along the canal, roads and newly settled areas has helped in reducing the intensity and impact of blown sand. Pastureland development and sand dunes stabilization works have been carried out so that the supply of fodder can be made available to the livestock (Indra Gandhi Nahar Department, n.d.).

With all these very high achievements of irrigation system in the desert area, there will be obviously some hindsight of the story and that includes:

Water logging in some areas of the stage I of command area due to over-irrigation.

Increase in salinity of soil in some areas.

Increase in vector-borne diseases in some areas due to water logging.

The main factors responsible for rapid rise of water level are liberal use of canal waters for irrigation and groundwater recharge due to Ghaggar flood inflow. The other factors responsible may be as below.

1. Seepage of canal water
2. Over-irrigation by cultivators
3. Absence of natural drainage and out-falls in the area
4. Continuous pounding of water in Ghaggar diversion depressions.

In most of the areas, the problem is likely to be eased with the reduction of availability of water due to opening of more and more areas in stage II, since the share of water of stage II is also hitherto used in this area (Hussain and Mohammad, 2018).

15.9.4 Overall Water Management Initiatives in Rajasthan

As stated above the massive irrigation project aimed at rapid enhancement of irrigated farmland in Rajasthan, 650 km long Indira Gandhi Canal system along with extensive network of canals started functioning in 1961 (Box 15.3). As of 2010–2011, nearly 74% of irrigated land is served by groundwater and 24% by canals. The new centrally managed system not only contributed significantly to increase agricultural output of Rajasthan but also to the success of the GR initiative of India and that of the region in particular. However, in the process, the monitoring of irrigation according to ground conditions at the micro-level was no longer under the purview of the stakeholders and the community of farmers (Figure 15.4).

On gaining independence in 1947, the state took possession of all water resources. The ideal image of urban life now included piped water supply, wet flushing latrines and a sewer connection. This along with the unprecedented urban population growth increased the water consumption as well as the wastewater and sewage generated which no Urban Local Bodies (ULBs) were capable of treating and safely disposing them. Groundwater was mined, facilitated by the power subsidy policy of the government. Society and communities became indifferent to the fate of urban water bodies, which the cash-strapped ULBs started using as convenient dumping sites for untreated sewage and industrial effluent (Figure 15.5).

Many changes in the citizens' attitude towards water took place over time in Rajasthan (Figure 15.5). Initiated during the British rule and further strengthened especially after the Independence of India from the British in 1947, have taken place in Rajasthan. With independence from British rule in 1947 and formation of the Republic of India and the State of Rajasthan in 1949, water bodies were acquired by

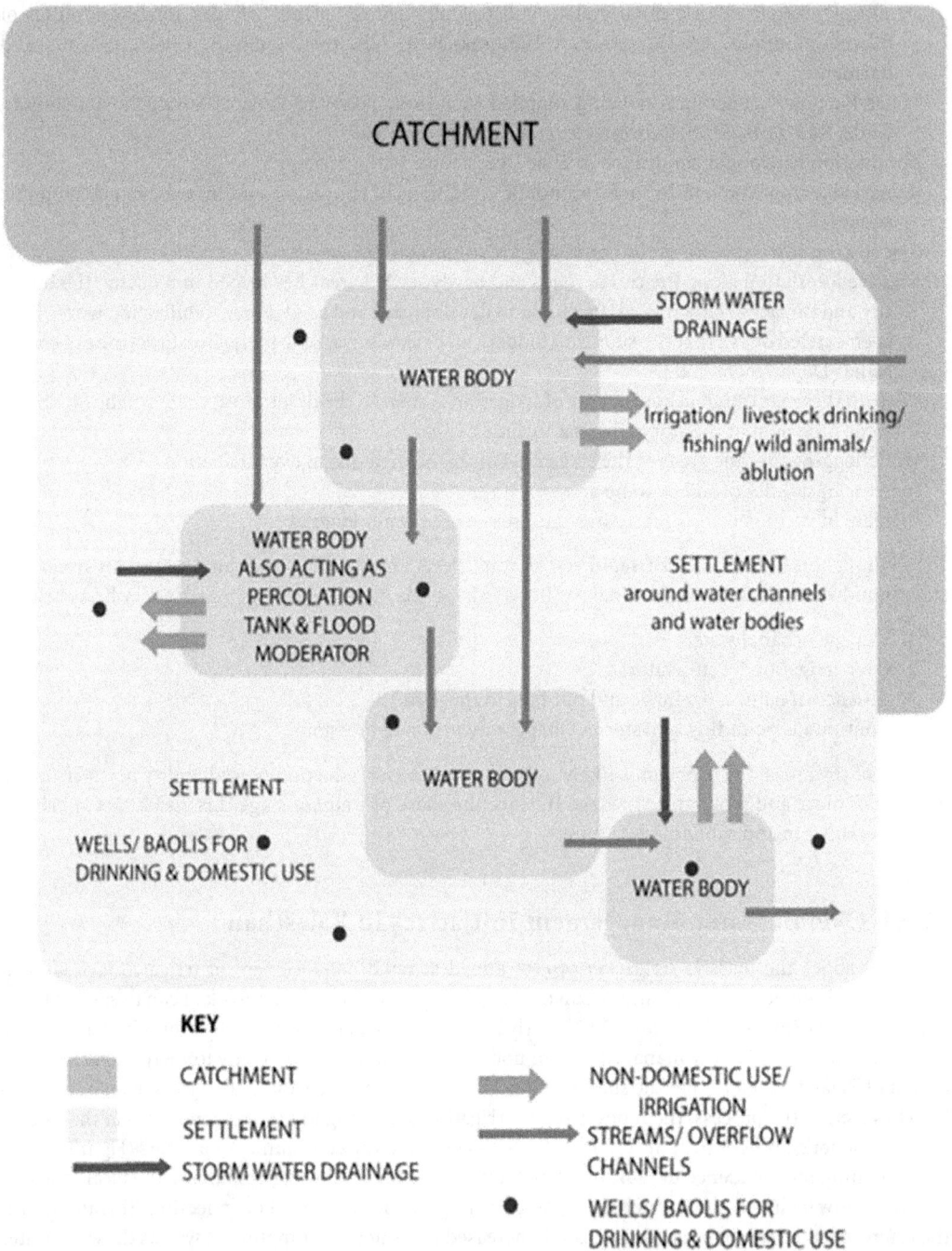

FIGURE 15.4 Schematic and linear layout of relationship between catchment, settlement and water body (pre-independence, pre-industrial and prior to piped water supply) (Surana, 2019).

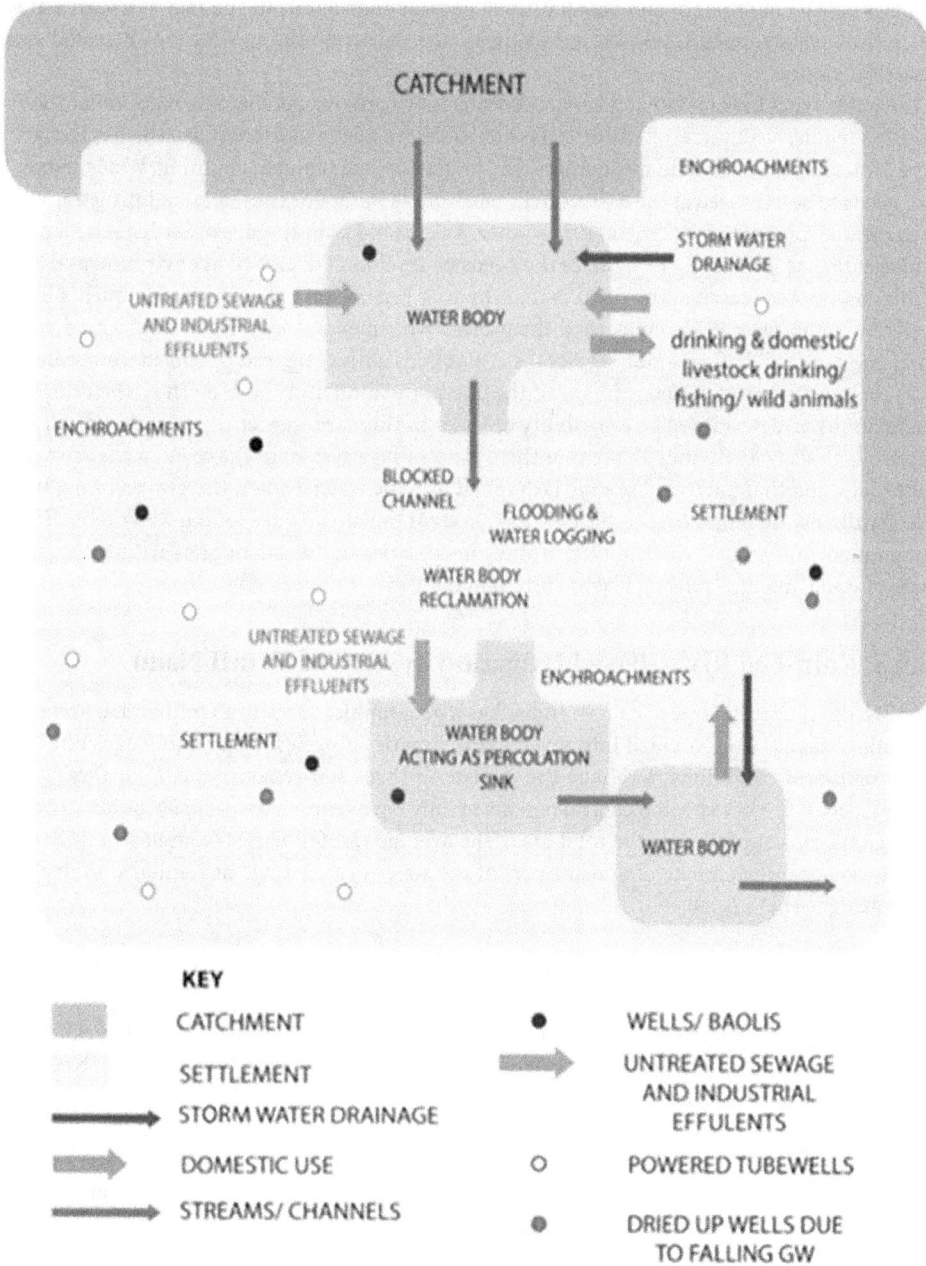

FIGURE 15.5 Schematic and linear layout of relationship between catchment, water body and urban settlement after 1947, continuing till date, having industries and partial piped water supply (Surana, 2019).

the government and also the onus to develop and maintain them. Water supply in India and Rajasthan became solely a state obligation and not a collective responsibility of state and society.

The ideal image of urban life now included piped water supply, wet flushing latrines and a sewer connection. This along with the unprecedented urban population growth increased the water consumption as well as the wastewater and sewage generated which no ULB was capable of treating and safely disposing them. Groundwater was mined rapidly, facilitated by the power subsidy policy of the government. Society and communities became indifferent to the fate of urban water bodies, which the cash-strapped ULBs started using as convenient dumping sites for untreated sewage and industrial effluent.

The water crisis in this state is a very real deterrent to progress as the state reels under the setback of droughts regularly. Physical planning so far has had three major shortcomings which is likely to become more critical in this state than most others. The first is a lack of regional planning. Water resources' planning needs to be managed at the basin/catchment level as water does not follow administrative boundaries so regional planning is of critical importance. The second is an absence of Settlement Planning policy for the state that sufficiently recognizes the needs of development as well as environmental boundaries, in this case, water scarcity. The third is that the first-generation master plan documents in Rajasthan prepared in the early 1970s concerned themselves with means and ways of augmented water supply under the head of "Utilities and Services" or "Public Utilities" instead of considering water availability as the most important determinant of the city population and the rest. Thus, the critical aspect of sustainability as determined by availability of water in this case was hardly considered. All subsequent master plans also clearly state that either there is no sewage system in the town or the sewage system is inadequate, and that untreated sewage is dumped into the water bodies, thereby polluting them as well as jeopardizing the remaining recharge possibilities of the already deepening water table (Hussain and Mohammad, 2018). Since the late 19th century or so, cities in Rajasthan grew without considering the limits of water, along a trajectory of unsustainability.

15.9.5 Rain-Fed River-Based Irrigation System of Tamil Nadu

Tamil Nadu, the tenth largest state in India, has a geographical area of 13 million ha. The coastal state in South India has been divided into seven agro-climatic subzones for planning agricultural development. Semi-arid conditions dominate the climate in three sub-regions: north, northeast coastal and southeast coastal. The delta and central regions mainly have semi-arid to dry sub-humid climates. These five regions consist of 97% of the total area. The average rainfall varies from 865 to 3,127 mm among sub-regions, and the climate of a major part of the state is categorized as semi-arid to dry sub-humid (Amarasinghe et al., 2009).

15.9.6 Irrigation Potential in Tamil Nadu

Studies show total Net Sown Area of the state has declined from 6.3 mha in 1971 to 4.8 mha in 2005 significantly due to various reasons. Source-wise contribution to Net Irrigated Area (NIA) shows that though groundwater irrigation expanded rapidly between 1971 and 2005, canals and tanks, which were the main sources of irrigation in the 1970s and 1980s, contributing to about two-thirds of the total NIA, declined rapidly.

Canal irrigation commands, which have lost more than 140,000 ha between 1971 and 2005, account for only 29% of the total canal NIA. Tank irrigation, which contributed to one-third of the total NIA in 1971, has lost more than half of its NIA by 2005. Although low rainfall in three consecutive years explains the short-term variation, there seems to be a consistent declining trend in the recorded NIA under tanks during the last few decades. A 1996–1997 estimate shows there were 39,202 tanks in Tamil Nadu (Table 15.10).

Though the net tank-irrigated area has decreased, much of that has been replaced by groundwater irrigation. This pattern is prominent in the northeast coastal and central regions, where the NIA under tank commands has decreased by 243,000 and 103,000 ha, respectively, and NIA under canal-irrigated areas has decreased by 38,000 and 10,000 ha, respectively. Over this period, the NIAs under groundwater in the two regions have increased by 303,000 and 195,000 ha, respectively, and have offset the loss of surface-irrigated area. Secondly, groundwater irrigation has also spread well outside surface command areas, which has affected overall agricultural production (Amarasinghe et al., 2009).

TABLE 15.10   District-Wise Number of Tanks in Tamil Nadu, 1996–1997

District	No. of Panchayat Union Tanks	No. of Public Works Department Tanks	No. of Ex-zamin Tanks	Total
Chengalpattu	1,733	1,207	756	3,746
North Arcot	2,084	1,169	482	3,735
South Arcot	1,766	757	79	2,602
Salem	549	188	-	737
Dharmapuri	1,579	101	154	1,834
Coimbatore	64	59	-	123
Thanjavur	491	685	-	1,176
Pudukkottai and Trichy	5,334	268	214	6,394
Madurai	3,391	771	331	4,493
Ramanathapuram	1,333	1,508	7,367	10,208
Tirunelveli	965	686	445	2,096
Kanyakumari	1,074	984	-	2,058
Nilgiris	-	-	-	-
Total	20,413	8,903	9,886	39,202

Source: Palanisamy K, Easter K W, 2000, *Tank Irrigation in the 21st Century – What Next?* Discovery Publishing House, New Delhi.

Although groundwater development has been able to attain NIA at the present level, it has affected 85% of the net groundwater resource in the state. CGWB studies have shown that out of the 385 blocks in the state, 142 are over-exploited and 33 blocks are categorized as critical, where the stage of development is between 90% and 100% in both pre- and post-monsoons, and 57 are semi-critical, where the stage of development is between 70% and 100% in either pre- or post-monsoons (Central Ground Water Board, 2006).

15.9.7 Water Crisis of Chennai

According to 2011 census records, Tamil Nadu was the most urbanized state of India with nearly half (48.4%) of the 72.14 million population living in urban settlements (Census, 2011). In the next census scheduled to be held in 2021, Tamil Nadu may have a population of 84 million with more than half (42 million) living in urban areas. Chennai (founded as Madras in 1639 by the East India Company) had a population of 4.6 million in 2011 and estimated to be inhabited by nearly 6 million in 2021.

The coastal city of Chennai, earlier known as Madras, was a fishing village with some trading initiated by Portuguese and Dutch merchants in the 16th century. By 1640, the British East India Company obtained permission from Vijayanagara Rulers to establish a Fort and factory and some residential colonies in the village, Madraspatnam, which was renamed as Madras. Slowly, this colony became "Madras Presidency" and military garrison by the British Rulers to control southern part of India. Today with over 8 million people, Chennai is the fourth largest city of the country.

Although there are three rivers that flow through the Chennai region and discharge into Bay of Bengal, the township, till the mid-19th century, was receiving water from various shallow wells and tanks located within and outside the city boundary. By 1870, the first piped water supply scheme was installed for Chennai region by constructing a masonry weir across Kortalayar river at Tamaraipakkam, 160 km north-east of the city. The water was first diverted into Cholavaram lake with a capacity of 881 mcft and later to the Red Hills lake having a capacity of 3,300 mcft, by channel. Subsequently, water is channelled to Kilpauk from where it was distributed within city boundary. The system is still functional, though storage capacities have reduced drastically.

The annual average rainfall in the Chennai region is around 1,200 mm. If water is harvested and stored in various reservoirs, there is enough water to meet needs of the Metropolitan City. There are four major reservoirs with combined capacity of 11,057 mcft in and around Chennai. Apart from the two mentioned earlier, the other two are Poondi (capacity 3,231 mcft) and Chembarambakkam (capacity 3,645 mcft). These four reservoirs together lose around 5 mcft daily due to evaporation.

The city supplies around 1,710 mld of water per day. Two desalination plants of 100 mld capacity each, using sea water, have been commissioned. Water supply schemes from Krishna river (500 mld) and Cauvery water from Veeranam Tank (140 mld) have augmented Chennai water supply in the last few years. Even then a large population in Chennai suburbs and slums are still depended on private water tankers every day to meet their daily water needs. Packaged drinking water is in great demand in City even now.

As per the records of Water Resources Department, only 19 of the 29 major water bodies in the city periphery can be rejuvenated as the others have been lost due to encroachments. Once rejuvenated, additional storage capacity of 1,000 mcft can be added. If the four primary reservoirs are de-silted by 1 m another 500 mcft could be added as well. Studies further show that there are around 3,600 tanks in and around Chennai metropolitan area. If these tanks are properly preserved and networked, around 80,000 mcft water could be harnessed from these sources.

Three years of deficient monsoon rainfall between 2016 and 2018 led to a huge water crisis in Chennai. In fact on 19 June 2019, Chennai City Officials declared "Day Zero", on the day when almost no water was left in all the four major reservoirs of the city. Though there was deficient rainfall for a few consecutive years, encroachments of the natural catchments, over-exploitation of groundwater resources are equally responsible for creating this crisis. Tanker water supply to the city population was the only alternative and people had to pay a big cost to that resulting in conflicts. While preparing the Second Master Plan of Chennai (2005–2026) in 2005, the First Master Plan of Chennai (1976–1996) was reviewed. It was observed that the water "supply of 2,271 pcd still remains a distant dream" and the anticipated capacity of Sewage Treatment Plants was 1,177.5 mld by 2001 against the available capacity of 481 mld (Chennai Metropolitan Development Authority, n.d.).

The rampant and unchecked urbanization has led to shrinking lakes and wetlands over time. The authorities had given permissions for construction without adhering to land-use planning regulations. This has resulted in the area under water bodies shrinking by 2,389 acres between 1979 and 2016. Illegal construction was allowed to encroach upon much-needed water sources and close to water bodies, having a devastating impact on urban ecology (Natarajan, 2019). To improve the water resources for Chennai, the state agencies are engaged in de-siltation and rejuvenation of various existing water reservoirs as well as adding new water reservoirs in a large scale. The augmentation of water supply system by adding more desalination units is also under progress. Constructions of RWH structures in individual households are also being pursued religiously. All these measures are expected to meet future water demand of Chennai.

15.10 Conclusions

Water scarcity in post-independence India has been essentially a manmade phenomenon resulting from policies with limited imagination of the foreseeable future, promoted through decades that eventually heightened water scarcity which had already set in since early 19th century. Much of the traditional wisdom on this subject has been lost or forced out by contemporary methods that promised tangible benefits to the agricultural sector at the cost of destruction of natural ecosystems including displacement of population due to submergence of settlements. Constitutionally, water has been under the purview of the state or the provincial governance. The ideology of state governments changed dramatically as the members occupying ruling party and opposition benches in the state legislature exchanged places as desired by the electorate and the crafty king makers. Therefore, the top-down decision-making mechanism practiced generally throughout the country promoted sets of policies that rarely matched with those of the neighbouring States and Districts in terms of priorities.

As the administrative boundaries of States and Districts often cut across the natural boundaries of river drainage basins and watersheds, the issue of sharing and management of water assumed great complexity in post-independence India. Thus, evolving a sensible, sustainable national water management system requiring inter-district and inter-state coordination for location and operation of irrigation infrastructure including multipurpose hydroelectric projects hardly ever happened. Effective intervention by the National government has been difficult in this context as in cases of several other issues.

The first GR of India was a timely escape route adopted by the Government of India that prevented a possible national famine much worse than that of 1943. Subsequently, the number of Agricultural Universities grew from 17 in 1949 to 75 in 2019, while the population grew from 0.36 billion in 1951 to 1.25 billion in 2011. Their commendable research activities covered not only application of bioengineering and biogenetics but also smart cropping patterns required in the present era. However, the need of evolving a vision of Indian development based on coordinated strategies of development of water resources, urbanization pattern and trends of population growth, industries and related infrastructure, as well as the diverse and fragile environmental systems must be addressed urgently. Uncontrolled withdrawal from groundwater reserves to meet the requirements of advanced agricultural practices ushered in by the first GR especially during low-rainfall years combined with continued urbanization led to rapid depletion of the groundwater reserves. Qualitatively, disposal of untreated excess water from agricultural fields and effluents from industries laden with chemicals into abandoned or dried up wells led to serious pollution of groundwater in several regions. The laws and regulatory control measures designed to monitor and stop such activities evidently, did not serve the purpose. Experience shows India consistently failed to implement well-intended and well-expressed policies and programmes introduced from time to time or enforce provisions of the existing laws and regulations aimed at protecting non-renewable resources including rivers, lakes and aquifers. The enabling tools led by sound governance principles were either circumvented or were not sensitive enough as they took too long to respond with effective measures within a real-world timeframe to resolve or address prevailing issues.

A positive view of melting of the Himalayan ice caps, increasing flooding of rivers and rise of sea level is that they are good trends! It is as if Mother Nature or Planet Earth is responding to the progressively greater demand for water generated by the growing population of India. She seems to be saying "Store the water from rain and flood creatively to meet your demands, and move away from the coast and you will see yourselves in a comfortable and sustainable state for decades even centuries to come!" Fortunately, like the tip of and ice berg, the accumulated wisdom of Indian civilization is hardly visible to the "naked eye" as exemplified by the contemporary water managers of Rajasthan and Tamil Nadu and their strategies. Expert Committees, bureaucrats as well as elected representatives of the people and others seem to be entangled in a mess of fuzzy ideas and procedures inherited from the recent past that allowed them to lead the development of water resources of India only into blind lanes or cul-de-sacs. The fact that water must be first managed regionally, river-basin-wise, rather than according to administrative boundaries of governance needs to be sufficiently recognized and understood.

As the stranglehold of Coronavirus (Covid-19) seems to tighten its grip slowly at the time of writing these lines during the middle of October 2020, it is increasingly clear that even with eventual loosening of its grip, civilization will coexist with Covid-19 like with other scary health hazards in the past. Culture will subtly mask itself once again in an innovative way to move forward with its evolution. Perhaps the most important lesson learnt from the Covid-19 experience is the psychological readiness of the society to accept similar (or not so similar!) challenges from unexpected quarters, adapt to their implications at a minimum social cost.

In the present context, India may be required to attain water self-sufficiency at the earliest for a safe journey into the future.

References

Amarasinghe U A, Singh OP, Sakthivadivel R, Palanisami K, 2009, *State of Irrigation in Tamil Nadu: Trends and Turning Points*, International Water Management Institute, Sri Lanka.

Census, 2011. Government of India, India. https://www.census2011.co.in/census/state/tamil+nadu.html

Central Water Commission, 2014, "Basinwise assessment of Water resources", https://cwc.gov.in/sites/default/files/nwauser/basinwise-asssmnt-wr9.pdf

Central Water Commission, 2017, "Annual Report 2016-2017", https://cwc.gov.in/sites/default/files/CWC_AY_2016-17.pdf

Central Water Commission, 2021, "Water related statistics Report 2021", https://cwc.gov.in/sites/default/files/water-and-related-statistics-2021compressed-2.pdf

Central Ground Water Board, 2006. Dynamic Groundwater Resources of India (Accessed in March 2004). cgwb.gov.in/documents/DGWR2004.pdf

Chattopadhyay, N. and Hulme M., 1997, Evaporation and potential evapotranspiration in India under conditions of recent and future climate change, *Agriculture and Forest Meteorology*, Vol 87, pp. 55-73.

Chennai Metropolitan Development Authority, n.d., "Review of First Master Plan of Chennai", HYPERLINK "http://www/"http://www. cmdachennai.gov.in/SMP_main.html

Gosain, A.K., Rao, S. and Basuray, D., 2006, Climate change impact assessment on hydrology of Indian river basins. *Current Science*, Vol.90, No.3, pp. 346-353.

Goyal, R.K., 2004, "Sensitivity of evapotranspiration to global warning: a case study of arid zone of Rajasthan (India)", *Agricultural Water Management*, 69, pp. 1-11.

Gupta R. K., Bakshi S. R., 2008, "Studies in Indian History: Rajasthan Through The Ages The Heritage Of Rajputs" , Sarup & Sons, pp. 143

Gupta, P. P., "Economic Water Scarcity: The Beginning of the End of Growth Story" Retrieved from https://adriindia.org/images/newsletter/15427130456-Economic-WaterScarcity-The-Beginning-of-the-End-of-Growth-Story.pdf23-29, Accessed 1 Apri 2021

Hussain A., Mohammad T A, 2018, "Indira Gandhi Canal Project Environment and Changing Scenario of Western Rajasthan: A Case Study", *International Journal of Academic Research and Development*, Vol. 3, No. 4, pp. 15–19.

IPCC, 1992, "Global Climate Change and the Rising Challenge of the Sea. Supporting document for the Intergovernmental Panel on Climate Change", World Meteorological Organization and United Nations Environment Programme, Geneva.

Imam S., 2020, "Indira Gandhi Canal: Afforestation and its Impact on Forest Cover in Rajasthan (India)", *International Journal of Science and Research*, Vol 9, No. 10, pp. 434-446.

India Environmental Portal, 2009, "Water Resources Development of India: Critical Issues and Strategic Options", "http://www.indiaenvironmentportal.org.in/water-Assessment.pdf"www.indiaenvironmentportal.org.in/water-Assessment.pdf

Indian Council of Agricultural Research (ICAR), 2019, India's Per Capita Water Availability to Decline Further, New Delhi, India.

Indian University Commission, 1949, Report of the Indian University Commission, India, Intergovernmental Panel on Climate Change (IPCC), 2008, "Climate Change 2007, Synthesis Report". p. 5.

Jain S K, 2019, "Water Resources Management in India- Challenges and the Way Forward", *Current Science*, Vol. 117, No. 4, pp. 569–576.

Jain S K, 2017, *Water Resources Management in India*, National Institute of Hydrology, Roorkee, India. www.nihroorkee.gov.in.

JNU, 1993, "Impacts of greenhouse inducted sea-level rise on the islands and coasts of India", School of Environmental Sciences, Jawaharlal Nehru University (JNU), New Delhi, India.

Jha B M, Sinha S K, 2009, *Towards Better Management of Ground Water Resources in India*. Central Ground Water Board, Government of India, Haryana, India.

Jondhale A S, Bhosale V P, Takate V S, 2017 "Irrigation Systems and Its Methods", *International Journal for Research and Applied Science and Engineering Technology*, Vol. 5, No. V, pp. 1550–1552

Kamyotra J S, Bhardwaj RN, 2011, "Municipal Wastewater Management in India." India Infrastructure Report, pp. 299–311.

Lal, M. and Chander, S., 1993, "Potential impacts of greenhouse warming on the water resources of the Indian subcontinent", JEH, 1(3), pp. 3-13.

Lal, M. and Singh , S.K., 2001, "Global warming and monsoon climate", *Mausam*, Vol. 52, pp. 245-262.

Mall R. K., Bhatla R., and Pandey S. N., 2007,"Water resources of India and Impact of Climate Change", *Jalvigyan Sameeksha*, Vol. 22, pp. 157-176.

Mehrotra, R., 1999, "Sensitivity of runoff, soil moisture and reservoir design to climate change in central Indian river basins, Climatic Change", Vol. 42, No.4, pp. 725-757.

Ministry of Agriculture & Farmers Welfare, 2014, "National Mission on Micro Irrigation", https://pmksy.gov.in/microirrigation/Archive/IES-June2014.pdf

Natarajan A, 2019, https://chennai.citizenmatters.in/chennai-water-crisis-causes-responsibility-solution-10629

Palanisamy K, Easter K. W., 2000, *Tank Irrigation in the 21st Century – What Next?*, Discovery Publishing House, New Delhi.

Raghavan S, Eslamian S, 2021, "Rainwater Harvesting: The Indian Experience – Traditional and Contempora", in *Handbook of Water Harvesting and Conservation, Vol. 2: Case Studies and Application Examples*, Ed. by Eslamian, S. & Eslamian, F., John Wiley & Sons, Inc., New Jersey, pp. 359–372.

Roy, P.K., Roy, D., Mazumdar, A. and Bose, B., 2003, "Vulnerability assessment of the lower Ganga-Brahmaputra-Meghna basins" In: Proceeding of the NATCOM-V&A Workshop on Water Resources, Coastal Zones and Human Health, IIT Delhi, New Delhi, June 27- 28, 2003.

Sengupta A K, 2012, *Resources and Infrastructure: Climate Change and Disease Dynamics in India*. TERI, New Delhi, pp. 313–343.

Sharma, K.P., Moore, B. and Vorosmarty, C.J., 2000, Anthropogenic, climatic, and hydrologic trends in the Kosi Basin, Himalaya, *Climatic Change*, Vol. 47, pp. 141-165.

Singh, R., 2006 "Water conservation", http://117.252.14.250:8080/jspui/bitstream/123456789/4605/1/46-Water%20Conservation.pdf , pp. 477-487

Surana , R, 2019, "Water Resources and Development of Human Settlements: The Case of Udaipur, Rajasthan", unpublished Ph.D. thesis, School of Planning and Architecture, New Delhi, India.

Tangri A K, 2000, *Integration of Remote Sensing Data with Conventional Methodologies in Snowmelt Run-Off in Bhagirathi River Basin, UP Himalaya*. Remote Sensing Application Centre, Lucknow, India.

WHO Publication, 2008, *How is Climate Change Affecting Our Health*. WHO, Switzerland.

World Bank, 2019, https://www.worldbank.org/en/news/feature/2019/03/22/helping-india-manage-its-complex-water-resources

16

Water Scarcity in India: Irrigation Practices

Bhasker Vijaykumar
Bhatt
Facile Maven Pvt Ltd, Surat

Neerajkumar
D. Sharma
*En-vision Enviro
Technologies Pvt Ltd*

Symbols

AIS and LUS	All India Soil and Land Use Survey
BCE	Before Common Era
bcm	Billion Cubic Metre
BPL	Below Poverty Line
CCA	Culturable Command Area (in a canal network)
CGWB	Central Ground Water Board, Government of India
CPHEEO	Central Public Health and Environmental Engineering Organization
CWC	Central Water Commission
DAP	Drought-prone Areas Programme
DDP	Dessert Development Programme
FRL	Full Reservoir Level
FWSI	Fulkenmark Water Stress Indicator
GCA	Gross Command Area (of a canal network)
GW	Giga Watt (of electricity)
IMD	Indian Meteorological Department, Government of India
INR	Indian National Rupee
IWMP	Integrated Watershed Management Programme

DOI: 10.1201/9781003353928-21

IWDP	Integrated Wasteland development Programme
KLBMC	Kakrapar Left Bank Main Canal
km	Kilometre
km²	Square kilometre (of area)
KRBMC	Kakrapar Right Bank Main Canal
LPH	Litre Per Hour
m	Metre (measurement of length)
m³	Cubic metre
MCM	Million cubic metre (of storage)
mcm	Million Cubic Metre
MI	Micro-Irrigation
MIC	Minor Irrigation Census
mm	Millimetre
MoA	Ministry of Agriculture
MoUD	Ministry of Urban Development, Government of India
MoWR	Ministry of Water Resources, Government of India
MT	Metric Tonne
MW	Mega Watt (of electricity)
NCERT	National Council for Educational Research & Training
NCIWRD	National Commission for Integrated Water Resources Development, India
NIR	Near Infra-Red (electromagnetic waves)
NMMI	National Mission on Micro-Irrigation
NWDA	National Water Development Agency
RS	Remote Sensing
UN	United Nations
UNEP	United National Environmental Programme
UNICEF	United Nations Children's Fund
URBMC	Ukai Right Bank Main Canal
VIS	Visible (spectrum of electromagnetic waves)

16.1 Introduction

Water is a proven essential for all forms of life. Water as a resource has a high degree of unevenness regarding its distribution on the earth. About ten countries on the globe possess approximately 60% of the available freshwater. These countries are Brazil, China, India, Russia, Canada, Indonesia, United States of America, the Democratic Republic of Congo and Columbia. These countries also have enormous local variations and, with growing urbanization globally, the ecosystems are facing changes (UNICEF, 2016). It has raised sincere environmental concerns towards issues observed. Human activities have affected the ecosystems' hydrological, biological and chemical function and seeking a preservation approach towards ensuring an adequate supply of water for the entire population on the earth. Major concerns related to water are its quality, quantity and availability. Considering the facts as produced by several studies at the global level, the United Nations Environmental Programme (UNEP) has included water crisis under Agenda 21 as a challenge to the global environment communities (United Nations, 1992). Deliberations regarding rational use and management of the resources related to ocean water as well as freshwater on the globe are included in detail under item 17 and 18 of Section II (United Nations, 1992). The global balance of water reveals that about 2.5% freshwater lies in the deep and frozen Antarctica and Greenland. Only about 0.26% flows in the form of rivers-streams on the surface, lies in lakes and in the soils and shallow aquifers that may provide access. Again, the uneven distribution acts as additives in the complexity and many resources are located far from the human settlements (Singh et al., 2016).

The geography of India is such that comparatively larger amounts of freshwater are available. The water stills are in the form of natural dip-forming ponds, lakes, reservoirs formed due to embankments of several dams and so on. It has also enriched the groundwater resources. Surrounded with the sea on three sides of the nation, ample quantity of precipitation is received during the monsoon season that lasts about 4 months in a year in most parts of the country. The country is located in the northern hemisphere near the equator, and it helps in considerable evaporation from the surface sources. The cycle of water seems well balanced until a hike in the population in a recent century and rapid urbanization, creating pressure on the natural resources in existence.

As per the decadal recording agency in India (known as Census of India), the country had a population of 446.2 million in the year 1961, which was surpassed by a population count of 1.2 billion persons by the latest census year 2011 (Chandramauli, 2013). India has a geographic area that comprises about 2.45% of the land area of the globe and has about 4% of available water resources. It provides shelter to about 16% of population of the world (NCERT, 2015). It is perceived that the inhabitant tally must be totalling more than 1.4 billion persons by the year 2026 (Gupta, 2019). The growth rate of population in the country is declining slowly since the 1980s. However, the net population is increasing. The rising population poses a high demand for domestic, agricultural and industrial water supply. Especially, the increasing urbanization asks for ample and regular domestic water supply with essentially, water-centric lifestyles. Rising demand for water at several ends is resulting in making available water to have a notion as a scarce resource.

Further in the chapter, discussions are made to introduce the diversified geography of India, the demographical shifts and trends along with water resources in the country, condition of the groundwater and depletion. The scarcity of water in general is addressed with climatological variety in the nation. The irrigation practices with historical-traditional approach in some parts of India are highlighted along with modern programmes of interlinking of rivers. The man-made irrigation by means of unique approach of group wells is elaborated along with recently adopted approach of micro-irrigation towards consumptive use of irrigable water. Prior to conclude the overall study, a national programme of integrated watershed management is also discussed.

16.2 Geographical Diversities in India

India has a high diversity in terms of its landscape, culture, languages, religion, economic activities, and so on. The landscape includes ranges of snow-capped mountains to desserts, flat planes, hilly areas and plateaus. The country comprises most of the Indian subcontinent situated on the Indian plate. The country has a long coastline to the extent of about 7,000 km in length and covers on its three sides. The coastlines have the Arabian Sea on its West end and the Bay of Bengal on the East end.

The northern plains, eastern regions and central parts of the country are occupied by the fertile Indo-Gangetic plains where the Deccan plateau has its spread in the southern part. Thar desert having rocks and sand dunes are occupying lands in western India. The peak Himalayan range has its spread in the northeast and eastern parts of the country. The climate in the country ranges from the equatorial in the southern parts to tundra in the Himalayan peaks. The highest point of the nation is situated at the peak of Kanchenjunga (8.598 m), and the lowest point is at Kuttanad (−2.2 m) in Kerala. Rivers of the Ganga and Brahmaputra are the longest in the country. Largest surface water in a lake is in the Chika lake. Geographically, India is the seventh largest country on the globe. The country is bordered with nations of People's Republic of China (3,380 km), Bangladesh (4,053 km), Myanmar (1,463 km), Nepal (1,690 km), Bhutan (605 km), Afghanistan, Sri Lanka, Maldives and Pakistan (2,912 km). The borders of the nation have an extension in the Arabian Sea till the Lakshadweep islands on the southwest. And, the borders extend to the islands of Andaman and Nicobar in the southeast (Ministry of Home Affairs, 2020)

Political diversity of the country is denoted by 28 administrative states and eight union territories. These states established under the 'States Reorganization Act, 1956' have their independent government that operates under guidance and assistance of the central government (Ministry of Law and Justice,

1956). The union territories are independent of the state government and administered by the central government irrespective of its geographic locational belongingness.

Geographically, the nation is divided into seven regions. These regions are the northern mountains (Himalaya range), Indo-Gangetic plains, Thar dessert, Central highlands and Deccan plateau, the East Coast, the West Coast, the bordering seas and islands. Major mountain ranges include for the Aravalli, Eastern Ghats, Himalayas, Pataki, Vidhya, Sahyadri (Western Ghats), Satpura and Karakoram. The Himalayas form the country's northeastern border and separates the country from the rest parts of Asia.

The Indo-Gangetic plains ('Basin 2a' as in Figure 16.1) are the floodplains of the Ganga-Brahmaputra and the Indus rivers. The plains run along with the alignment of the Himalaya mountain range. The plains have

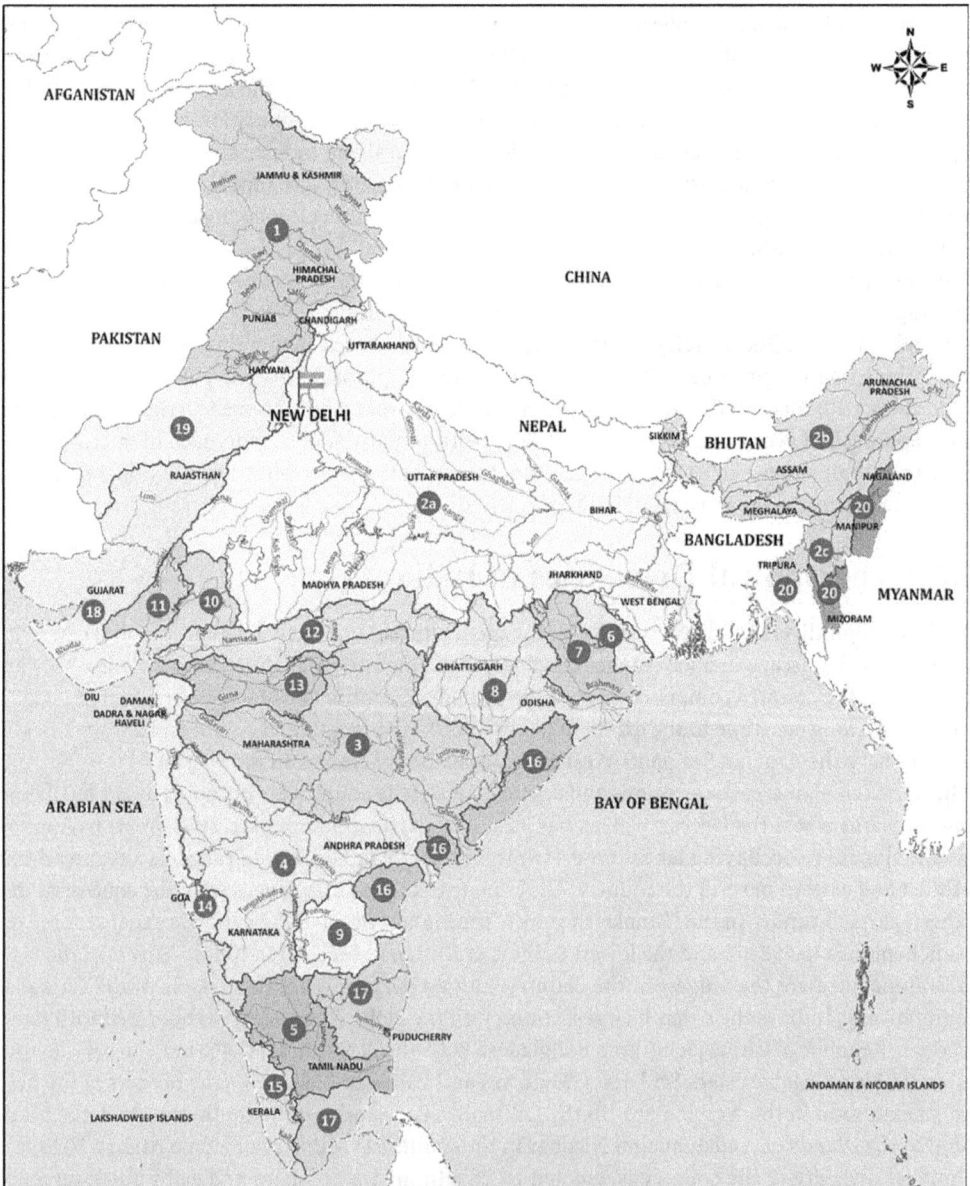

FIGURE 16.1 Drainage basins of rivers in India (PMF IAS, 2016).

their spread from the Jammu-Kashmir in the west and extend till Assam in the East having an area of about 700,000 km². They have variation in their length through breadth by several hundred kilometres. They have the world's most extensive spread of alluvium belt formed by the deposition of silts of three major rivers having several tributaries. The plains are mostly flat in terrain and treeless, making it ideal for irrigable land by a system of canals. The plain is also rich by means of the groundwater resource. It has intense farming and major crops include rice and wheat grown in rotation. Other crops such as maize, sugarcane and cotton also are cultivated. Also, the plains are identified amongst the densely populated places across the globe. The cities of Uttar Pradesh, Delhi and Punjab are highly populated since the past several decades.

An area of about 600,000 km² forms a rectangular tract that extends from Ahmedabad to Kanpur (forming southeast to the eastern boundary), from Kanpur to Jallandhar (forming north to the northeast border) and from Jallandhar to Kutch along the western border of the country is identified as dessert and semi-dessert region. The region has reported low rainfall ranging from 350 to 700 mm per annum. Extreme deserted areas receive rainfall even lower than 350 mm annually in the region. There is almost no area covered with the canal network.

Another chronic drought-prone zone forms a rectangular tract spreading over rain shadow parts of the Western Ghats. In fact, the region is situated on the east of the Western Ghats and has a spread of about 300 km width. It includes parts of land from Andhra Pradesh and eastern Karnataka. The area under the region that is about 370,000 km² has a character of high erratic mean annual rainfalls lesser than 750 mm.

In addition to these two regions, there are several scattered pockets that become prone to droughts with several reasons. About 100,000 km² of area lies in these scattered pockets. Henceforth, of the total land area in the country, about 1 million km² of lands have a tendency to receive the inadequate and highly variable rainfall.

16.3 Demography and Water Resources in India

Today, India has the second largest population on the globe. It is one of the oldest civilizations dating back to third and second millennia BC, which, over a period of time, expanded to northern and eastern India (Central Intelligence Agency, n.d.). Geographically, the country is the seventh largest nation with an administrative area of 3.28 million km² (nationsonline.org, 1998).

The population in the urban pockets in India is increasing. An examination is illustrated graphically (in Figure 16.2 below) showing the shift of citizens from rural hinterlands to the urban pockets. The trend shows that the urban population was reported to be the highest, only in the state of Maharashtra and Mizoram (higher than 35% of the total population) back in the year 1991. Subsequently, during 2001 and 2011, census enumerations reported for the addition of Tamil Nadu, Kerala, Karnataka, Goa, Gujarat, Delhi and Punjab to cross the mark of 35% urban population in the state (Chandramauli, 2013) (Figure 16.2).

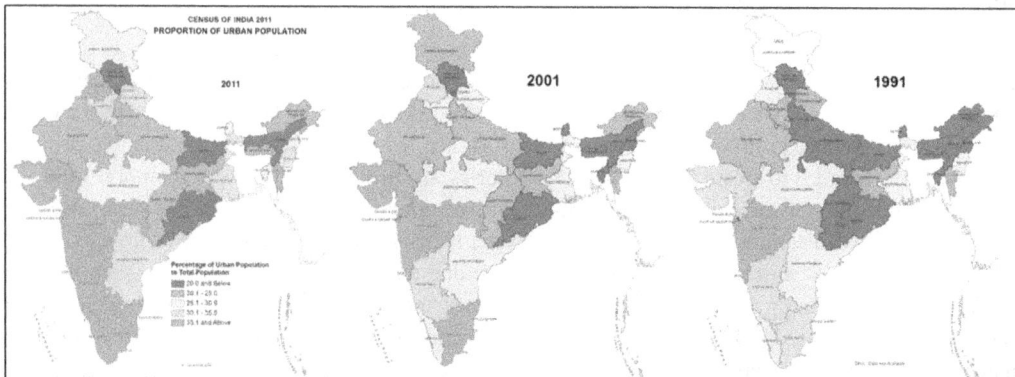

FIGURE 16.2 Urban population trends in India (Chandramauli, 2013).

As per the prevailing guidelines of Central Public Health and Environmental Engineering Organization (CPHEEO) in the country, all the urban local bodies are to ensure a minimum supply of potable water to individuals at a rate of 135 L/day. In addition to it, the Ministry of Urban Development is targeting to enhance the systems to ensure continuous water supply round the day and year (Ministry of Urban Development, 2009). Earlier, about 40–45 years ago, the urban centres were able to cater to the water supply needs on the continuous supply basis; however, with increasing population and deteriorating condition of infrastructure, the intermittent system of supply came in to practice.

With enriched water resources in the nation, India has developed a system of canal-based irrigation. The canals take water from dams and weirs constructed on rivers. Water made available by nature every year, by means of precipitation reaches an extent of about 4,000 km³. It enriches a network of about 10,360 rivers and its tributaries having a length longer than 1.6 km each. The river catchments receive a mean annual rainfall to the extent of 1,869 km³ out of which only about 32% is put to use. In addition, there is high spatial variation in the precipitation distribution with its duration limited to monsoon months only. Ganga, Brahmaputra and the Indus rivers have huge catchment areas. These catchments, however, account for about 60% of total surface water available in the country and cover only about one-third area of the country in the northern region. In the central part of the country, Narmada and Tapi rivers have larger basins. And, in the southern region of the nation, Godavari, Krishna and Kaveri are the major rivers. Optimum capacity is achieved as anticipated for these rivers towards harnessing of available water (NCERT, 2015). Figure 16.3 shows the major rivers in India.

Water resources in India are monitored by the Ministry of Water Resources (MoWR), Government of India. State governments and departments thereof work under the guidelines and policies of the ministry. Also, as per the constitution of the nation, the water resources are identified as a public good within an administrative state and managed by the government of a state. The ministry takes care of the overall planning, coordination, availing guidance as well as the formulation of policies in the water resources sector. Following the policies and funding assistance guidelines, the ministry also attempts to provide technical guidance for planning, scrutiny of proposal reports, clearance for funding and monitoring of irrigation, flood-control and the major-minor multipurpose projects and financial assistance thereof.

The ministry recently had to formulate the national water development perspective along with carrying out a study on water balance towards the identification of potential inter-basin transfers. At times, disputes arise among state governments in relation to the consumption of water from an inter-state river. The ministry intervenes and mediates for resolution and assists for the accomplishment of ambitious, inter-state projects. For basins of the Ganga and the Brahmaputra, the MoWR has prepared the master plans for flood control. The ministry also governs the operations related to groundwater (Department of Water Resources, 2017).

16.3.1 Groundwater and Depletion

The groundwater resource is considered to be a preferred one for application in many sectors in the country. The MoWR, Government of India, through its division called Central Ground Water Board (CGWB) manages the resource of water (Department of Water Resources, 2011). The board operates with a vision to develop and manage groundwater resources in a sustainable manner across the nation. The primary function of the board is to create and maintain inventories, study the utilization on a temporal scale and evaluate for possible consumptive use through the preparation of policies and availing guidelines. The board also is active in actions pertaining to groundwater recharge tasks by the state governments.

The groundwater resources used in the country are due to their almost universal availability, low capital cost for harnessing and dependability. Such a character of the resource has led to exploitation of the resource and resulted in depletion in terms of quality and quantity to a large extent. The limitation of aquifers in terms of recharge capacities and other environmental factors is testimony for the careful use of the resources. The aspect is seldom addressed by individual users. A study of the CGWB mentions that the annual replenishable groundwater in the country is estimated to be about 433 billion cubic

FIGURE 16.3 Major rivers in India. Courtesy: Google Earth Pro, 2018.

metres (bcm) out of which about 399 bcm of water is identified as available for development for different utilization (Jha & Sinha, 2008). Groundwater sources are utilized to the maximum by the irrigation sector. Being a major consumer, the irrigation sector reports the use of groundwater to the extent of 92% for an annual withdrawal. The development of the groundwater has a large variation in terms of recharging in different parts of the country. Some state governments have imposed regulations making groundwater recharge mandatory for building approvals for a variety of projects. In a way, the development of groundwater as a resource is observed to be an extremely complex proposition. It is established in the country, due to highly uneven distribution and utilization, a single management strategy for groundwater resource cannot be formed covering all the parts and sectors. However, initiatives are taken up in various places in different activities. Best practices are illustrated, encompassing several cases by the ministry (MoWR, 2019).

An assessment was carried out in the year 2017 by the Central Ground Water Board (2017). As of the March 2017, the study identified that the extraction of the groundwater was 297 bcm (about 63% of the total availability in the country) where largest extraction was carried out in the irrigation sector. The extraction pattern was found to be non-uniform across the nation. The assessment also noted a comparative analysis for the recent (the year 2017) and previous (the year 2013) exercise results. The assessment was focusing on two aspects: dynamic resources where replenishment is possible and static resources where aquifers do not allow for recharging. The assessment was carried out for depths up to bedrock or 300 m from the ground level in the alluvial areas and up to 100 m depth where there were hard rocks. Scope for identifying the potential for augmentation of aquifers was also attempted. The balance was extracted using a formula that stated as:

$$\text{Inflow} - \text{Outflow} = \text{Change in storage (of an aquifer)} \tag{16.1}$$

That is,

$$\Delta S = R_{Rf} + R_{Str} + R_c + R_{Swi} + R_{Gwi} + R_{Tp} + R_{Wcs} \pm VF \pm LF - GE - T - E - B \tag{16.2}$$

where

ΔS=Change is storage; R_{Rf}=Rainfall recharge; R_{Str}=Recharge from stream channels; R_c=Recharge from canals; R_{Swi}=Recharge from surface water irrigation; R_{Gwi}=Recharge from groundwater irrigation; R_{Tp}=Recharge from Tanks & Ponds; R_{Wcs}=Recharge from water conservation structures; VF=Vertical flow across the aquifer system; LF=Lateral flow along the aquifer system (through flow); GE=Ground Water Extraction; T=Transpiration; E=Evaporation; B=Baseflow

Here, in the absence of statistical information for all the parameters at all the blocks, major components with data were considered for budgeting with reasonable assumptions. A lumped parameter estimation approach was undertaken to keep a possibility to update the assessment as the data may be made available from other sources towards a refined assessment (Central Ground Water Board, 2017). The findings were reported for identification of overexploited areas to have a concentration in (i) Haryana, Punjab, Western Uttar Pradesh and Delhi where despite having abundant replenishable resources, there is over-exploitation in withdrawals; (ii) Rajasthan and Gujarat, where self-water recharge is limited, and resource is stressed due to arid climate; and (iii) the peninsular India in the southern part of Andhra Pradesh, Telangana, Karnataka and Tamil Nadu where there are crystalline aquifers leading to less groundwater availability.

The study was carried out in 6,881 administrative blocks of India. It was a joint exercise by organizations under the directives of the central government towards an assessment of groundwater status. As a result, 17% blocks were identified as 'overexploited' towards the extraction exceeding the annual recharging. The extraction in these areas was noted to cross 100% of groundwater (Central Ground Water Board, 2017). The assessment also revealed that 5% blocks are 'critical' and 14% as 'semi-critical' in terms of groundwater extraction. About 1% area is remarked as 'saline' and unsafe for groundwater extraction except for the specific application of the water with such a quality. Still, 63% of areas were declared as 'safe' for utilization of groundwater having replenishing abilities. The report identified a scope of future assessments for carrying out water balance studies with more accuracy, characterization of aquifers with estimation of parameters, inclusion of best practice case studies with assessment and management of groundwater, a continuous monitoring of groundwater resources to make temporal data available, introduction of mechanisms to have an automatic groundwater resource assessment, research on aquifer stream interaction at several instances, groundwater modelling and predictive simulation exercise.

The central government, through the Ministry of Water Resources, has taken the initiative to prepare and implement district-wise groundwater recharge plans where the blocks were identified to follow over-exploitation in practice (Central Ground Water Board, 2015). These blocks in districts are applicable to many administrative states of the country. The funds are made available through national financial devolution plans implemented through annual outlays.

16.4 Climate and Water Scarcity

Climatic regions in India are classified by Koeppen's classification (Secure IAS, 2018). It is an empirical classification based on mean monthly and annual precipitation and temperature data. The data are recorded at several instances and maintained in a repository of the Indian Meteorological Department (IMD) of India. The climate in the country is split between temperate and tropical. A transitional zone also exists between these two. The temperate climate region includes the areas of the Himalaya, northwestern region, arid lowlands of Rajasthan (Thar), the region of moderate rainfall (Gujarat, Madhya Pradesh, part of Maharashtra, Telangana and Andhra Pradesh). Tropical India includes the regions of very heavy rainfall of the northeastern parts, region of heavy rainfall (Odisha, Bihar, West Bengal), the Konkan coast (on the west), the Malabar coast (on the southwest) and Tamil Nadu. These regions face temperatures from less than freezing temperature to peaks of 48°C. The rainfall has variations from absolutely no rain to a little higher than 2,000 mm per annum.

A research by Jain (2011) suggests for domestic water requirements in India and annually per capita availability and utilizable surface water in India for the years of 2025, 2050 and 2065 based on other studies of past trends. The table below illustrates the statistics (Table 16.1).

Jain (2011) using UN medium variant for population estimation stated that the population of India will increase to the extent of 1,459, 1,692 and 1,718 million persons with a rise of 9.21%, 28.47% and 29.56%, respectively, for the years of 2025, 2050 and 2065 on the base year of 2011. The increased population will pose a requirement for domestic water supply to the tune of 66.90, 118.58 and 122.59 bcm for the respective years of 2025, 2050 and 2065. He also stated that the rural demands would be seeking about 89.67% of domestic water supply by the year 2065. As per his projections, it is estimated that the urban population in India will keep on increasing and will reach about 65% by the year 2065. Currently, the available surface water is about 1,544.38 mcm per annum, and the population increase by the year 2065 will reduce the per capita water availability to 1137 mcm. If it happens so, then as per the Falkenmark Water Stress Indicator (FWSI), the nation is sure to enter the water-stressed status, and the stress will become severe over a period of time (Damkjaer & Taylor, 2017).

As per the FWSI, water availability higher than 1,700 m³/capita/year is defined as a threshold value and if it is higher than th, water will pose for shortage only irregularly or at times.at Lower values will generate water scarcity in various levels of severity. A value less than 1,000 m³/capita/year is assumed to impose a limitation on the economic development, human health and well-being. Yet, if the value is

TABLE 16.1 Domestic Water Demands and Availability

Years	1951	2001	2011	2025	2050	2065
Population (millions)	361	1,027	1,210	1,459	1,692	1,718
Percentage Urban	0.17	0.27	0.31	0.45	0.60	0.65
Percentage Rural	0.83	0.73	0.69	0.55	0.40	0.35
Per capita surface water availability (MCM/year)	5,410	1,902	1,614	1,339	1,154	1,137
Per capita utilizable water (MCM/year)	1,911	672	570	473	408	402
Norm—Urban area (lpcd)	-	135	135	220	220	220
Norm—Rural area (lpcd)	-	70	70	70	150	150
Demand—Urban (BCM)	-	-	-	48.17	81.52	89.67
Demand—Rural (BCM)	-	-	-	18.73	37.05	32.92
Total (BCM)	-	-	-	66.90	118.58	122.59

Source: Jain (2011).

lower than 500 m³/capita/year, there will be severe constraints concerning water application for sustenance (Jain, 2011). The study reveals the limitation of natural resource and warns India to be on the verge of entering water scarcity state in the near future. In addition to water scarcity, as discussed earlier, parts of the country also face droughts and are prone to the hazard.

Droughts are a result of deficit rainfall, and it is a complex phenomenon (Rath et al., 2021). As a hazard occurring naturally, unlike other events, it begins slowly and has a long duration covering a vast area of the earth's surface. In India, its occurrences upset the food security by devastatingly affecting the agricultural economy that is majorly dependent on the monsoon. Every year, the central government has to declare a few or many administrative districts as 'drought-affected' due to depleted groundwater resources and sparse rainfall. In India, annual rainfall of an extent of 200 mm in the dry regions does not create a concern for agriculturists as the agricultural activities are adopted with such a meagre amount of precipitation. However, if such amounts, if received over a duration of few years, in a region where the annual average of rainfall is observed to be about 500–800 mm, it can cause failure of crops and subsequently may result in a drought condition (Figure 16.4).

The IMD defines drought as a situation when an area receives less than 75% of its annual mean rainfall in a year (Shewale & Kumar, 2005). The IMD classifies the droughts in the country as (i) severe drought, i.e., the annual rainfall deficiency exceeds 50% of the normal rainfall and (ii) moderate drought— the deficiency is between 25% and 50%. Rao empirically explored several aspects and impacts of droughts in the Indian context and suggested space-based technology for monitoring and remedial actions (Rao, 1996). A diagrammatic representation showing causes of droughts and impacts identified with indicators resulting in a shortage of food and water due to the type of drought occurrence that is reasonably established based on rational studies.

To study drought, different parameters can be used by employing space technology that is developed so far by several researchers. As the drought occurrence has a classification, it needs a different methodology to be employed for understanding a particular type and observe it happening (Figure 16.5).

A course of remedial action may be worked out based on it. The remote sensing applications, with their technological establishment and constant upgradation, are allowing to study natural events by availing temporal data. It has certain limitations for a micro-level assessment; however, for a larger region, the technology provides opportunities to understand the indicative nature. Vulnerability information in relation to drought parameters can be studied by different methodology using space technology (Jayaraman et al., 1997). These are narrated in Table 16.2.

The Brahmaputra and Ganga-Barak river basins together account for about 60% of the average water resources potential per annum; however, due to regional terrestrial constraints as well as and place–time variations in the availability of water, about only 4% of the total potential water in the basin of Brahmaputra and about 48% in the Ganga basin is put to use. The National Commission for Integrated Water Resources Development (NCIWRD) had projected that 9 out of 20 river catchments do not have adequate water for industrial and other uses (Central Water Commission, 2019). River basins like that of Sabarmati and Indus do not make available any residual water left for industrial and other such uses even in the year of 2010. The circumstances are expected to worsen by the year 2050 by the time the basin of the Ganga river being fed by the east-flowing rivers such as Mahanadi and Pennar river that are facing problems for water availability (Singh et al., 2016).

Water is essential, not only for palatable use and household purposes but also for agricultural-crop production and industrial applications. The demand for safe water increases with the expansion of population to sustain life, sanitation, energy production and to run industries. Based on the discussion here, the water utilization for food is a major form of water use by the society, and so far, by and large, on a global scale, the agricultural sector uses the maximum of available water. Amplification in urbanization and industrialization along with changing zonal climatic characteristics may be considered among reasons for the propagation for water scarcity. The next section highlights the irrigation practices in India along with its impacts.

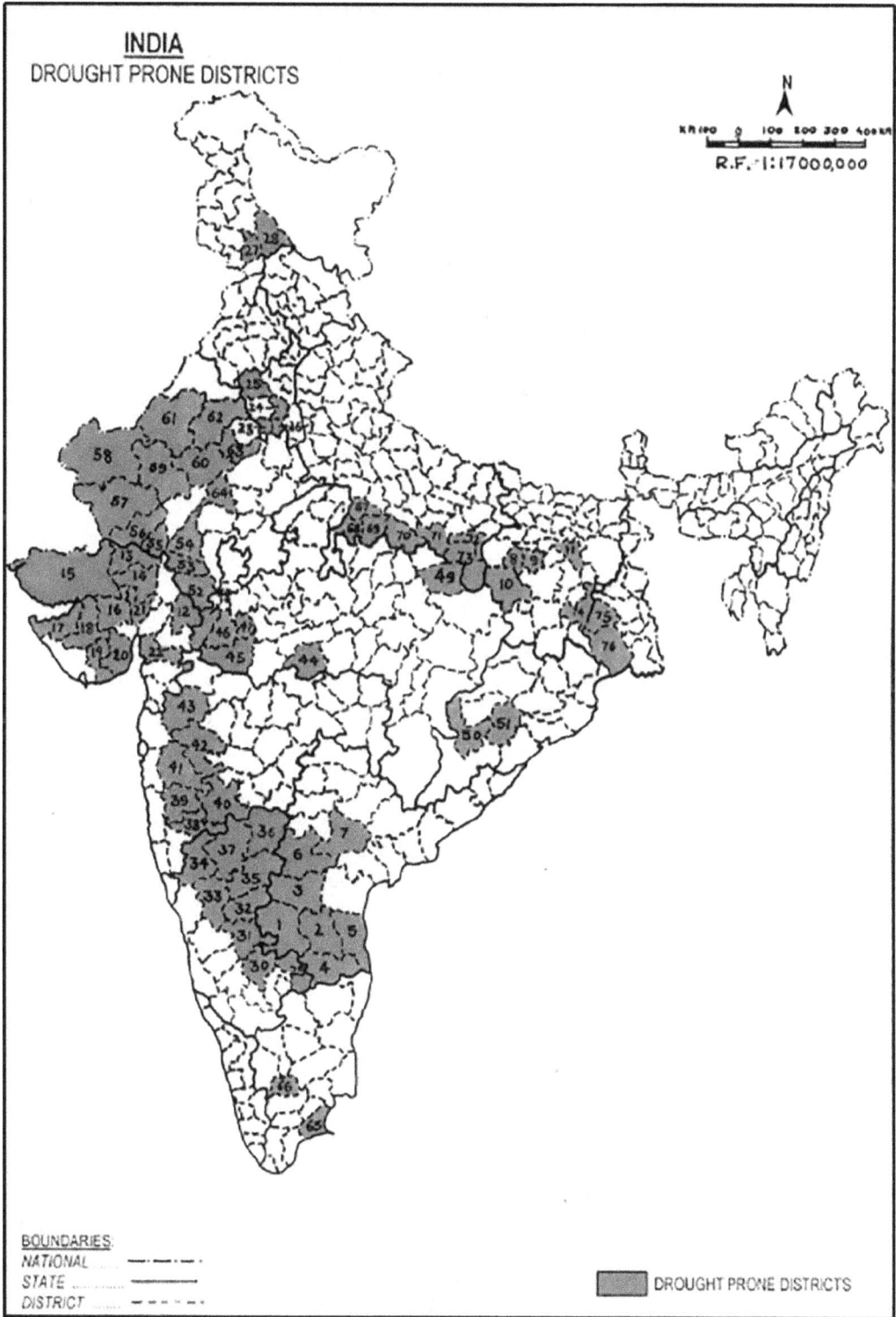

FIGURE 16.4 Drought\-prone areas in India (Directorate of Economics and Statistics, Government of Gujarat, 1995).

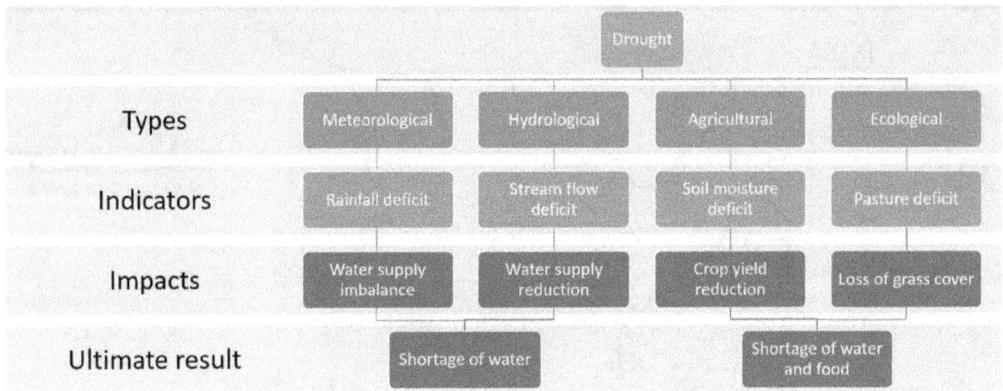

FIGURE 16.5 Types of droughts and their impact (Rao, 1996).

TABLE 16.2 Remote Sensing Applications in Drought Assessment

Drought Parameter	Methodology	Vulnerability Information
Vegetation index	VIS and NIR Absorption	Crop water stress
Aridity index	Optical and Thermal data	Evaporative demand of atoms
Precipitation	Microwave, Radar	Stress severity
Geomorphology	Optical and Microwave data	Groundwater targeting
Soil moisture	Optical and Microwave data, 3D Modelling	Recharge stress detection

Source: Jayaraman et al. (1997).

16.5 Irrigation Practices

Since the independence of India in the year 1947, most of the irrigated land of greater India was allocated to Pakistan. In the year 1951, by the time when the first 5-year plan (sectoral financial assistance) was begun, the country had an approximate population of 300 million and the cultivable area under irrigation was about 23 million ha with a production of 90 MT food grains. Since then, a large number of irrigation schemes have been completed ensuring a firm supply of water for the community involved in agricultural activities. As of today, about 113 million ha land is being irrigated in India out of which about 58 million ha is covered under major and medium surface irrigation scheme. About 15 million ha area is irrigated through the minor surface scheme, and about 40 million ha have the minor groundwater schemes. Today, the nation is producing about 250 MT of food grain to suffice the requirement of citizens across the country (Mazumdar, 2016).

It is estimated that the overall performance efficiency of irrigation projects in the country is too low, having an average of about 35% in the case of major and medium irrigation projects. Mazumdar (1984) found that the average of project efficiency for three major river-valley irrigation projects have a variation from 18.6% to 38.8% only. Much of the water was lost in transit and application on the fields along with extremely poor water management of water at the farm level (Mazumdar, 1984). In the year 1992, the Planning Commission of India recognized three major deficiencies accountable for poor functioning of irrigation projects, namely: (i) unlined channels; (ii) lack of land consolidation and improper levelling, sizing of the irrigated land; and (iii) improper water management at farm level beyond canal outlets. The commission also noted the prevalent policy of charging farmers for drawing water from the canal. The farmers at the head end had a tendency of drawing water as much possible without bothering about farmers at the tail end of a canal (Planning Commission of India, 1992). There is, till date, a complete absence of volumetric measurement-based water supply at the farm level and pricing in accordance for the supplied water for irrigation. In addition, over the years, the focus of the central government towards investments in the irrigation sector is shifted due to the demand for the development of other infrastructure (Mazumdar, 2016).

16.5.1 Historical Perspectives on Irrigation Systems in India

The irrigation practice history in India has its traces to prehistoric times. Ancient Indian writers and scripters have mentioned wells, canals, dams and tanks. Those mentions depicted the then systems to be of small in size having minor works and allowing for operations by individuals for irrigating little sized land parcels. In the southern part of India, peninsular region, during the reign of the Chola Kingdom, the construction of Grand Anicut allowed for perennial irrigation in the period of the second century. The system was developed for the water of the Cauvery river. By the fifth-millennium BCE–fourth-millennium BCE, cotton was cultivated. The Indus cotton industry was well developed and used to practice for spinning of cotton (Tripathi, 2016).

India has a long history of civilization with the development of irrigation channels in an early stage. The Harappa and Mohenjo-Daro civilization remains have revealed shreds of evidence of canal under practice for multiple purposes. However, the section here discusses traditional irrigation practiced and developed in villages and tribal areas of various parts of the country with a timeline prior to modern technological age (Debnath, 2016).

16.5.1.1 Kuhl System of Irrigation

It was developed as a traditional irrigation system in the village called Mumta in the state of Gujarat. In the system, a group of users forms an association. The leader of the association is known as 'Kohli'. His primary role is to build and rebuild dams/barriers across springs and monitoring the water distribution among the members of the association. In the system, a temporary head well was constructed near a gorge to divert the water in the canals to the fields. An average distance of 3 km from canals to the first field was observed. With the slope of the ground, the water from one stream to the other flows and ultimately reaches the gorge downstream. The width of the outlet was controlled to ensure control of the water flow between two successive fields. All the users of the association contributed to the construction of the system, and there was a provision for penalty for the offenders. Upon observing the absence of a user during construction activity, the water supply was denied. During a period of water scarcity, the Kohli was empowered to decide on the quantity of water allocation for each of the fields under command.

16.5.1.2 Zabo System of Irrigation

The system was very much famous in the region of Nagaland, a northeastern state of India. This region, as discussed earlier, lies in the eastern Himalayas. Terrace farming was an important practice adopted in the region. The hilltops have remained under protected forests in the form of a sacred grove. The tribal citizens practiced maintaining the watershed by maintaining of these forests. After precipitation, the water was received in the terraces successively, and the land was irrigated. Water was stored in a pond-like structure located approximately in the centre of the terrace. Over a period of time, water flowed downward to the paddy fields in the foothills. As a combination, bamboos were used to distribute water in the terraces to increase the coverage of the irrigated land of the sloping terrain.

16.5.1.3 Bamboo Drip Irrigation in the Himalayas

The tribes of 'Khasi' and 'Jayantia' used the system to irrigate the plantation of black pepper. Bamboo was used to tap water from the springs and streams flowing on the hills. The bamboo allowed for an outflow of 20–80 drops per minute at the point of application. The inlet for water in bamboo was set up at the hilltops allowing for a flow of water at farms in the foothills. The bamboo intake sizes were manipulated at different stages to control the flow of water. The internodes in the bamboo were removed to form a pipe-like continuity to be employed for the fabrication of the system. A unique-smooth axe, called 'Dao', was used to smoothen the bamboo channel.

16.5.1.4 Ahar-Pyne System

This system was followed in the state of Bihar. Water collection canals were known as 'Pyne' that allowed for the flow of water to the tanks called 'Ahars'. These Ahars were used to recharge the groundwater resources as, in general, it could absorb a good amount of water. The practice was lost in the transition of time; however; recently, attempts were made to revive it in some parts such as Jehanabad of the state.

16.5.1.5 Inundation System of West Bengal

The region of West Bengal has been prone to floods since ages. The riverbanks were allowed to get inundated during the floods. Advantageously, the floodwaters carry a huge load of silt. Once the flood was stopped, the riverbanks were employed for cultivation. The system was abolished during the British Rule due to the construction of embankments in the flood plains.

16.5.1.6 Traditional Irrigation System of Kerala

The traditional irrigation system in the state of Kerala in India was developed with a high response to the ecology and cultural practices. The system in practice was eco-friendly, socially accepted, flexible, less costly and constructed using local knowledge, resources and engaging citizens. Agricultural activities in Kerala are rainfall dependent. 'Keni' named after a very shallow well in the Wayanad district was dug in the valley. It had small diameter walls strengthened by driving of hard, hollow, seasoned tree trunk. Tree trunks used for the usage were from 'Choondapana' (Caryota urens), 'Aanjili' (Artocarpus hirsutus), 'Nelli' (Emblica officinalis) and similar trees. Even today, where some of the wells are of age more than 500 years, exists and being used by the citizens by lifting water (Joseph, 2016).

In Kerala, there are a large number of ponds and tanks employed for domestic and irrigation purposes. These ponds, small in size and shallow in depth, constructed in the farms facilitate the growth of a large variety of plants, herbs and hefty trees with several species on the surrounding moist banks. It also allows for the growth of algal and other aquatic organisms in the water that helps in purification of the water and assists in percolation of water in recharging of aquifers. In early days, several ponds were interconnected forming a network (known locally as 'eri') cascading system. It helped in draining away of floodwaters.

Tunnels were dug in the hills of laterite strata, known as 'Surangam' or 'thurangam' majority of which existed in the Kasargod district of the state. Trickling water from the wall of the tunnels formed a flow with virtue of gravitational forces. At the end of the tunnel, the open mouth (known as 'Mathakkam') allows for the collection of water for diverting it to open pit/tank/ponds. Apart from the ponds and tunnels, people of Kerala practiced construction of pits to store water from springs, check dams ('Katta'), sub-surface check dams, water harvesting structures at the foothills ('Pallam'), barrages and diversion weirs and so on.

16.5.2 Irrigation through Canals in India

In India, irrigation through open channels in the form of scientifically designed canal is prevailing since the early British Rule period. Some of the earmarking systems are discussed here. However, almost all the river basins, as discussed earlier, have been used for surface water extraction for irrigation purposes.

16.5.2.1 Great Ganges Canal of Northern India

As discussed earlier, the northern plains of the Ganga are among the largest fertile agricultural land across the globe. The plains have deposition of alluvial soils that are suitable for irrigation and responds well for artificial watering of the crops. Most of the plains are under the administrative boundary of the Uttar Pradesh state. With a large number of perennial streams flowing on a mildly sloping terrain, the lands here generally are favourably providing water bearing stratum. Groundwater resources in the plains are available within 5–15 m depths from the ground surface.

Back in the year 1880, an Indian famine commission was formed due to consecutive droughts occurring at an interval of every 5 years with one severe event at an interval of 12 years. Famine event in the year 1838 affected almost all the parts of the country. Sir Proby Thomas Coutley, English Engineer and palaeontologist working for the then East India Company, initiated and designed for the construction of a canal system that was the largest in the world. During the period of January 1848–April 1854, canals having a length of 1,052 km, consisting of main canal and branches were constructed. The system designed many cross-drainage structures including complex designs for aqueducts, viaducts and so on. During that period, there was no precedence of any work of this nature and magnitude ever attempted anywhere in the world. With a very limited technical manpower and indigenous techniques, the task was accomplished. Post construction, it was identified that the bed slope of the canal was a little higher and was causing erosion in the canal bed. A remedial action was performed in the year 1885. The system is still protecting the citizens from famine and droughts as well as yielding returns on the capital spent for the project. Principles used for designing the system were adopted as: (i) to keep the direction of the canal along the highest lands or watershed to be carried with; (ii) to avoid all interference with the natural escapes for the flood waters by turning the heads of all drainage lines; (iii) not admitting drainage water into the canal channel except in the immediate vicinity of the canal works; and (iv) whenever a natural drainage interference is observed along the canal alignment, an efficient escape portion was to be provided for the drainage water to get directed away from the course of canal. Application of these principles for establishing a system of artificial irrigation by means of canal network has improved the natural drainage of the country (Goyal et al., 2016).

16.5.2.2 Canal Network in the Tapi Basin

The development of irrigated agriculture in the Tapi basin, Ukai-Kakrapar Scheme, the biggest irrigation scheme in Gujarat, was taken up by the Government of Gujarat and completed in two stages, the first stage in 1957–1958 and the second in 1971–1972 (partially). The total service area (culturable command area-CCA) under the scheme in its final stage of development as planned (1966) was to be 3,79,940 ha (Revised area 3,56,480 ha in 1978) with a total irrigation intensity of 109% under Kakrapar and 103.5% under Ukai. Irrigation network in 3,59,280 ha (revised area 3,37,200 ha in 1978) has been completed.

Agriculture development in the command area has been fairly rapid since 1958 after the completion of the Kakrapar System and has gathered greater momentum after the completion of Ukai dam in 1971. Fertile land, good rainfall, large storage at Ukai, good marketing facilities, rapid development of agroindustries, and progressive nature of farming community in the command area have provided the necessary climate for accelerating the pace of agriculture development still further.

Organized irrigated agriculture in the command area started in the year 1957–1958 when Kakrapar weir at FRL 155.00 (subsequently raised to 160.00 in 1971) was constructed across Tapi near Kakrapar village 80 km from Surat in the erstwhile Bombay State along with two canals, Right Bank Main Canal 64 km long (K.R.B.M.C) and Left Bank Main Canal also 64 km long (K.L.B.M.C) to provide irrigation in 2,27,540 ha—about 1,66,190 ha under K.L.B.M.C and 61,350 ha under K.R.B.M.C. As there was no storage reservoir upstream, the scheme functioned purely as a diversion scheme mainly for growing seasonal crops. In order to improve the irrigation intensities in the above area and also to bring an additional area of 152,400 ha under irrigation, the Government of Gujarat constructed 69 m high storage dam at Ukai 23 km upstream of the pick-up weir with a storage capacity of 0.71 million ha metre live storage 0.142 million hectare metre dead storage (Tapi Embankment Division, 2010). The benefit from this scheme started flowing progressively from the year 1972. Under the scheme, the capacity of Kakrapar Right Bank Canal at head was increased from 32.5 to 70.18 m³/second to carry additional water as surplus to its own discharge. The discharge of Ukai Right Bank Canal (U.R.B.M.C) taking off from K.R.B.M.C at 46.6 km, from its head to irrigate 67,400 ha in the area lying between Narmada and Kim. The entire system was designed for a cropping pattern different from the one for which the Kakrapar

System was originally designed. According to the revised cropping pattern (1966), the combined capacity required was worked out as 70.18 m³/second (2,480 Cusecs) and for U.R.B.M.C 45 m³/second. The bed gradient for the K.R.B.M.C was kept as 1 in 10,000 (Bhatt, 2015).

16.5.3 River Interlinking

The concept of interlinking of rivers is not novel. About 130 years ago, it was conceived by Sir Arthur Cotton; however, then it was to facilitate trade. Back in the year 1970, K. L. Rao, an engineer and a former minister in the irrigation ministry of India, attempted to revive the plans for interlinking of rivers in India. The purpose of the plan was to link the basins of the Ganga and the Brahmaputra (water surplus area) to the central and southern regions of India (the water deficit regions) by forming a national grid of surface water resources. By the year 1980, the national government formulated 'National perspectives for water development' for the country. From the year 1982 onward, the National Water Development Agency (NWDA) initiated to study the water balances in rivers and feasibility to check for storing, linking and transferring of water by dividing a proposed project of interlinking into Himalayan and the Peninsular regions (Bansal, 2014). It was abandoned thereafter. Later, again in the year 1999, the idea of interlinking of rivers was revived by the commission of Integrated Water Resources Development.

The national river interlinking is said to have proposed development of 30 links. It will be connecting 37 rivers of national importance. Also, it is envisaged to connect 3,000 storage dam reservoirs. Below figures show the two proposed components of river interlinking regions.

With rising water woes, in the year 2002, the then President of India, Dr A. P. J. Abdul Kalam mentioned about interlinking of rivers to overcome the situation of a water crisis (Bansal, 2014). It was a result of a legal application to the Supreme Court and subsequent ordering to the central government to initiate working on the interlinking of major rivers in the country. The then government through appointing a task force with a deadline till the year 2016, where no intermediate outcomes were produced, and as a result of it, the central government appointed a 'Special Committee for interlinking of river' under the MoWR in the year 2012 by the virtue of another order from the Supreme Court. The ruling government earlier than the year 2014 was not in favour of the project, however, later the newly formed government earmarked INR 100 crores in its budget for the year 2014–2015 to expedite the project and prepare a detailed project report (Figure 16.6).

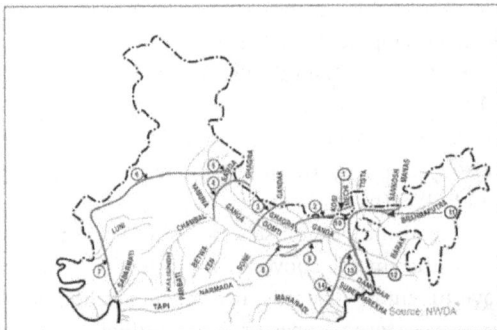

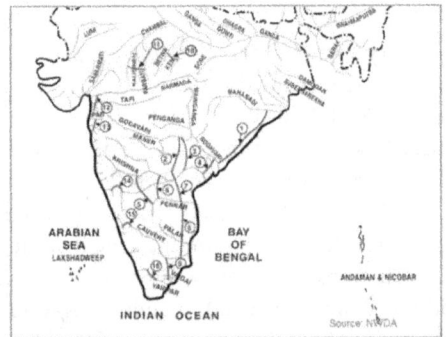

FIGURE 16.6   Proposed interlinking of rivers in India (Bansal, 2014).

The project of interlinking of rivers envisages conserving monsoon precipitation resulting in inflows for irrigation, hydropower generation and flood control. The linkages of Himalayan rivers are proposed from the Ganga and the Brahmaputra to transfer water for the western, central as well as water-deficient eastern part of the country through 14 links. The peninsular component is being explored for proposing 16 links to connect the rivers in Southern India. Here, the construction of many large dams and major canals is envisaged. The Mahanadi and the Godavari are expected to play a key role for supplying its surplus water to the deficient regions by the construction of large-sized canals connecting Parbati, Betwa, Kalisindh and Chambal rivers.

The project anticipates providing an additional 35 million ha available for irrigation in the water-scarce peninsular and western region of India. About 25 million ha of land will be made available through surface irrigation and rest 10 million ha through groundwater. It shall create additional employment, boosting of crop outputs and income from farming. It also will enhance potential navigation and fisheries activities. Also, the benefits of new capacities of hydropower generation are derived to the extent of an additional 34 GW. Such capacities will be curbing the demands of availability for potable water supply (Figure 16.7).

The total cost of the interlinking project is expected to be INR 560,000 crores at the 2002 price levels having INR 16,000 Crores allocation for over 35 years down the line. Some critic notes about the river interlinking project highlight a thought towards an absence of a complete scientific analysis to arrive at a surplus and deficit quantum of water in basins and sub-basins of rivers traversing in the said regions. Also, some rivers in the Himalayan region are flowing across international boundaries and may bring in a different paradigm of challenges in the project implementation. The components of social, environmental and biodiversity costs are yet to be explored at a detailed level as some forest land parcels are anticipated to be at a higher risk. River dynamics, individual quality of a river, will add for another challenge for the quality of water when mixed with other rivers that shall be inclusive of important levels of silt and pollutant loads.

16.5.4 Irrigation by Group Wells

A prevailing and widely followed practice by small and marginal farmers (having landholding of about 2 ha area or around) in India is based on sharing water for the purpose of irrigation by means of wells constructed and owned jointly. The wells are classified into three categories as (i) dug wells; (ii) shallow tube wells (with depths lesser than 70 m); and (iii) deep tube wells (depths higher than 70 m). The fifth Minor Irrigation Census (MIC) in the contour was conducted in the year 2013–2014. The report claims that the groundwater accounts for 94.5% of minor irrigation schemes with an increase in use at the national level against a decrease of surface water schemes. It reported 8.78 million dug wells irrigating 16.8 million ha, 5.9 million shallow wells irrigating 22.2 million ha and 5.77 million tube wells irrigating 24.2 million ha of farmlands. The report identified the existence of 28,10,307 group wells and 1,74,42,592 individual farmer-owned wells along with publicly owned wells at a tally of 2,68,985 making the total reach a huge number of 2,05,21,884 wells employed for irrigation across the country (Ministry of Water Resources-River Development & Ganga Rejuvenation, 2017).

The practice of group wells is not only promoting water-sharing efficiently but also involves equity among users. The normal practice adopted is of forming a group in size of two to ten farmers to share water through informally formed norms among them. Some groups also practice the sharing of water on a rotational basis. Some groups share water through channels having an equal share of water extracted from wells. The administrative states of Tamil Nadu, Rajasthan, Andhra Pradesh, Haryana, Madhya Pradesh and Maharashtra were identified as the top-listed states having the highest number of group wells in the MIC (2013–2014).

The fifth census of minor irrigation schemes has reported 98.7% of the groundwater minor irrigation schemes to be under the possession of private ownership. With such huge ownership, the structures for minor irrigation are also under private ownership and maintained by people, resulting in a maximum

PROPOSED INTER BASIN WATER TRANSFER LINKS

HIMALAYAN COMPONENT

1. Manas-Sankosh-Tista-Ganga
2. Kosi - Ghagra
3. Gandak - Ganga
4. Ghagra - Yamuna **
5. Sarda - Yamuna **
6. Yamuna - Rajasthan
7. Rajasthan - Sabarmati
8. Chunar - Sone Barrage
9. Sone Dam-Southern Tributaries of Ganga
10. Ganga - Damodar - Subernarekha
11. Subernarekha - Mahanadi
12. Kosi - Mechi
13. Farakka - Sunderbans
14. Jogighopa-Tista-Farakka (Alternative to 1)

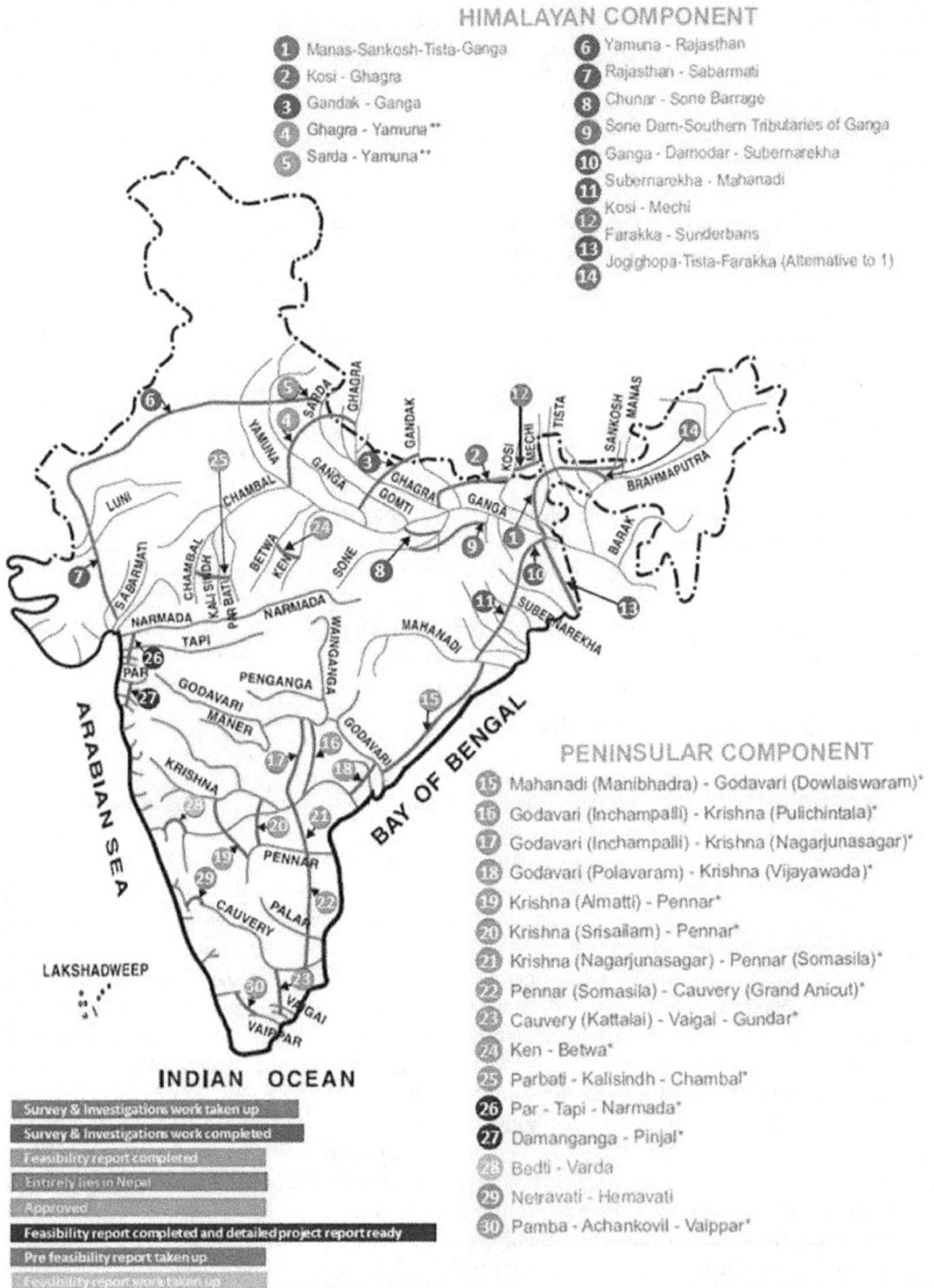

Survey & Investigations work taken up
Survey & Investigations work completed
Feasibility report completed
Entirely lies in Nepal
Approved
Feasibility report completed and detailed project report ready
Pre feasibility report taken up
Feasibility report work taken up

PENINSULAR COMPONENT

15. Mahanadi (Manibhadra) - Godavari (Dowlaiswaram)*
16. Godavari (Inchampalli) - Krishna (Pulichintala)*
17. Godavari (Inchampalli) - Krishna (Nagarjunasagar)*
18. Godavari (Polavaram) - Krishna (Vijayawada)*
19. Krishna (Almatti) - Pennar*
20. Krishna (Srisailam) - Pennar*
21. Krishna (Nagarjunasagar) - Pennar (Somasila)*
22. Pennar (Somasila) - Cauvery (Grand Anicut)*
23. Cauvery (Kattalai) - Vaigai - Gundar*
24. Ken - Betwa*
25. Parbati - Kalisindh - Chambal*
26. Par - Tapi - Narmada*
27. Damanganga - Pinjal*
28. Bedti - Varda
29. Netravati - Hemavati
30. Pamba - Achankovil - Vaippar*

FIGURE 16.7 Proposed inter-basin water transfer links in India (India water portal, National Institute of Hydrology, 2014).

outreach for irrigation purposes. The census identified that the development of the minor irrigation schemes is based on the capacity of individuals and seek financial support from the financial institutions to fully utilize the available potential of groundwater. Reduction in open channel flow of water for micro-irrigation was reported; however, it remains the major mode of conveyance of water from a source to the field of application. A major share of energy utilized across the nation is the electricity and diesel for irrigation purposes, and there is a scope for shifting the sources of energy to solar and windmill which have begun to employed at a lower scale (Ministry of Water Resources-River Development & Ganga Rejuvenation, 2017).

16.5.5 Micro-Irrigation

The micro-irrigation (MI) scheme, for the first time in India, was launched in March 2005 and was restructured in the shape of a national programme named as the 'National Mission on Micro-Irrigation' (NMMI) in June 2010. It consisted of sprinkler and drip systems focussing on precise irrigation to the crop plants' root zones targeting for minimization of irrigation water demands. The MI has increased the efficient use of water as well as the distribution of fertilizers and minerals in the form of fertigation. The Central Government of India is well promoting these systems by providing cost subsidies directly to the applicant-user at the farmer level.

The irrigation by dripping refers to a technique having an application of water in a relatively small quantity having the rate of discharge at about or a little less than 12 L/hour (LPH). In the system of drip irrigation, a mechanism is employed using a network consisting of plastic pipes fitted with emitters to supply drops of water to the root zone of the plants in a farm-field. It had emerged as one of the most suitable irrigation systems for most of the soil compositions in India. It leverages specific merits as (i) increasing the water efficiency (leading to about 85%); (ii) improved yield of the crop; (iii) uniformity and improved crop produce quality; and (iv) minimization in the soil structure damage.

Furthermore, the sprinkler method of irrigation includes a mechanism to spray water in the air to make it fall on the ground surface of a farm-field as in rainfall. The spraying is created by making the flow of pressurized water being released through small orifices/nozzle openings. It is identified as the most suitable mechanism for cultivation practices on sloping grounds having uniform/non-uniform undulations. The system can provide a gross saving in the water supplied for agricultural application to the extent of about 40% in comparison to other surface irrigation methods. Compared to the traditional irrigation practices, the trials of sprinkler system in different parts of the country, with regard to the water-saving aspect, has revealed it to be about 70% along with increase in the crop to an extent of 57% in different agro-climatic conditions (Singh et al., 2016).

16.5.6 Integrated Watershed Management

Watershed is a geo-hydrological division of land, flowing to a common merging point by a system of streams. On the surface of the earth, all land parcels belong to one watershed or the other. A watershed could be grouped into a number of classes depending upon the way of classification. Generally, common modes of making watershed categories are based on the size, drainage, shape and land-use pattern. The means of classification could also consider the base of drain or river size, a point of interception and the drainage density with its distribution. The All India Soil and Land Use Surveys (AIS and LUS) of the Ministry of Agriculture, Government of India, have developed a system for watershed dilation like water resource region, basin, catchments, sub-catchments and watershed (IDFC, 2012). The usually identified five stages of watershed delineation are mostly based on a geological area of the watershed, and are as listed herewith:

a. Macro Watershed (>50,000 ha)
b. Sub-watershed (10,000–50,000 ha)

c. Mili-watershed (1000–10,000 ha)

d. Micro watershed (100–1,000 ha)

e. Mini watershed (1–100 ha)

In India, since 2009–2010, the Ministry of Rural Development has initiated a nation-wide scheme of Integrated Watershed Management Programme (IWMP). It is a flagship programme of the ministry and under implementation by the Department of Land Resources. The programme is a result of a merger among three programmes, namely, Dessert Development Programme (DDP), Integrated Wasteland Development Programme (IWSP) and Drought-prone Areas Programme (DAP) for the development of rain-fed or degraded areas in the nation. The IWMP is targeting the soil, vegetative covers and water through conserving, harnessing and redeveloping the depleted state of these natural resources. The tasks performed in the programme focused on soil runoff prevention, harvesting mechanisms for rainwater and groundwater recharging means by the construction of entities. To an extent, the project has resulted in an increase in crop production, multi-cropping and crop-rotations based on the availability of water, introduction to diversified agro-based activities, creation of sustainable livelihoods and increase in family incomes.

The project components under the IWMP have addressed ridge-area treatment, soil-moisture conservation, local drainage line treatments, raising of nurseries, afforestation, horticulture activities, development of pastures, rainwater harvesting wells and so on. The project implementation duration had a variation from 4 to 7 years, depending on the scale of addressing a component. The ministry sanctioned 6,622 projects across the nation for micro-watersheds covering an area of 31.29 million ha. It sought central assistance to the extent of INR 8,240.61 Crores up to the year 2013–2014 (Press Information Bureau, 2014).

An exemplary case is discussed here that happened in the Ahmednagar district of Maharashtra State in Central India (Lobo & Samual, 2005). A place located at the foothills of the Sahyadri hills, Hiware bazaar is a village having a spread across 977 ha. The average precipitation here ranges from 300 to 400 mm/annum. The area often was declared as drought-affected area. The lower availability of water for irrigation and cultivation in fields had the worst impact on the citizens of Hiware bazaar. It resulted in actions such as migration to larger cities, and preparation and sale of illicit alcohol became a significant source of income. There was a severe rise in alcoholism and crime among the citizens. Over 90% of families lived Below the Poverty Line (BPL).

By the year 1990, Popatrao Pawar, an elected leader in the constituency, initiated the drawing up of a watershed management plan for rural development based on primacies as set by the citizens themselves. It identified the needs in order of having safe drinking water, water for irrigation purpose, subsequent employment, youth education and health of all. The project was funded by the central government at INR 960,000 with a contribution from the participant citizens of INR 350,000. The water-led transformation then took place, and as a result, the constituency is now among the top agro-based income-generating sites, has safe drinking water and facilities for sanitation, literacy to an extent of 95%, abrupt reduction in the BPL and landless families, high levels of groundwater, increased areas under irrigation and increased cropping intensity, rise in fodder availability, increase in primary workers and so on (IDFC, 2012).

The timeline of events over the period with very affirmative results may be considered as one of the best illustrations for a watershed management programme identification, planning and implementation. It suggests that the key to a successful watershed management programme approach involves participation by the government, representatives of the people as well as all involved citizens. There are several such cases of successful implementation where citizens have been benefitted from the results of the watershed management programme and curbed the water crises to a larger extent and adding to the prosperity amongst them.

16.6 Conclusions

Deliberations through the chapter suggest that there is a wide variety of aspects concerning irrigation and water resources in India. There exist shifts of geological, geographical and climatic zones in different parts of the country. Traditional practices of irrigation were developed with an eco-friendly approach in a way to conserve natural water resources; however, with increasing urbanization and water demanding activities, the existing resources are not sufficient. The government has taken initiatives to promote micro-irrigation practices to foster consumptive use of water for balancing the agriculture and urban water demands. The conservation and management of water is emerging as a pressing need of the time prompting action towards sustainable development in the era of increasing population, climate change, industrialization and intensifying agricultural practices to maintain a balance. If the existing trend is to continue, the predicated population increase and demands by the year 2065 will earmark the country as a water-scarce nation and it will be difficult to sustain economic activities in India.

The Central Government of India has created a national water portal and also applied online database management system for creating records through e-governance. However, at regional and local levels, there is a need for a mechanism of advanced water information and management. In addition to it, looking at the diversified use of water in several sectors, inter-sector water transactions shall be promoted through evolving a mechanism for the purpose. It will seek for an integrated approach through the national and state water, agricultural, industrial and urban development organizations. Also, efforts are needed towards reduction in the water supply–demand imbalances that prevail in the urban and rural areas and activities thereof. An improvement in the existing water infrastructure for the promotion and utilization of water consumptive methods and rational systems may assist in improving the scenario. Research activity needs to have widespread promotion across the nation by providing financial assistance and opportunities. Such studies shall be performed using modern equipment and sensible-sophisticated instruments with enabled interaction with servers and software for simulations. Fundamental education about water to change the behavioural aspects concerning awareness and responsible use of water will definitely play a key role in retaining the available water resources in a sustainable way.

Acknowledgements

The authors express their gratitude to the organizations of the Government of India, Government of Gujarat, IMD-GoI, water resources institutions and the Institution of Engineers (India) for providing information and records. Authors are also thankful to the members of the managements of APIED, Vallabh Vidyanagar, Gujarat-India and GIDC Degree Engineering College, Abrama-Navsari, Gujarat-India for sparing necessary resources and motivation for the current work.

References

Bansal, S. (2014). National River Linking Project: Dream or Disaster? Retrieved February 22, 2020, from https://www.indiawaterportal.org/articles/national-river-linking-project-dream-or-disaster

Bhatt, B. V. (2015). Appraisal of kelia irrigation systems : A boon for tribal part in South Gujarat Region. In *Annual Technical Volume of Civil Engineering Division Board 2015* (pp. 162–169). Surat, India: Institution of Engineers.

Central Ground Water Board (2017). *Dynamic Ground Water Resources of India*. (K. C. Naik & G. C. Pati, Eds.) (First). Faridabad, India: Ministry of Jal Shakti, Department of Water Resources, RD & GR Central Ground Water Board. Retrieved from http://cgwb.gov.in/GW-Assessment/GWRA-2017-National-Compilation.pdf

Central Ground Water Board. (2015). Artificial Recharge Plans. Retrieved February 23, 2020, from http://cgwb.gov.in/Ar-Plans.html

Central Intelligence Agency. (n.d.). South Asia : India — The World Factbook - Central Intelligence Agency. Retrieved June 12, 2019, from https://www.cia.gov/library/publications/resources/the-world-factbook/geos/in.html

Central Water Commission. (2019). *Reassessment of Water Availability in India Using Space Inputs*. New Delhi, India. Retrieved from http://cwc.gov.in/sites/default/files/main-report.pdf

Chandramauli, C. (2013). PCA Release 2011. New Delhi: Census of India. Retrieved from http://censusindia. gov.in/DigitalLibrary/data/Census_2011/Presentation/India/PCA Release 2011.pptx

Damkjaer, S., & Taylor, R. (2017). The measurement of water scarcity: Defining a meaningful indicator. *Ambio, 46*(5), 513–531. https://doi.org/10.1007/s13280-017-0912-z

Debnath, S. (2016). *Traditional Irrigation Systems in India*. Kolkata, India: Institution of Engineers.

Department of Water Resources. (2011). Comprehensive Mission Document for National Water Mission - Volume I. New Delhi, India. Retrieved from http://jalshakti-dowr.gov.in/sites/default/files/Document_of_NWM_Vol_I_April 20117821020996_0.pdf

Department of Water Resources. (2017). Mission/Vision/Functions | Department of Water Resources, RD & GR | Government of India. Retrieved February 22, 2020, from http://mowr.gov. in/about-us/functions

Directorate of Economics and Statistics. (1995). *"Previous At a Glance"*. Retrieved from Directorate of Economics and Statistics. Retrieved from https://eands.dacnet.nic.in/Previous_AT_Glance.htm

Google Earth Pro. (2018). Google Earth Pro. California.

Goyal, S. C., Singh, N., & Yadav, N. K. (2016). *The Great Ganges Canal of Northern India*. Kolkata, India: Institution of Engineers.

Gupta, N. (2019). Projected population characteristics as on 1st March 2013: 2001–2026- India | data.gov.in. Retrieved February 21, 2020, from https://data.gov.in/resources/projected-population-characteristics-1st-march-2013-2001-2026-india

IDFC. (2012). Hiware Bazar: A water-led transformation of a village | Hindi Water Portal. Retrieved February 22, 2020, from https://hindi.indiawaterportal.org/content/hiware-bazar-water-led-transformation-village/content-type-page/53529

India Water Portal. (2014). *National Insitiue of Hydrology*. Retrieved from India Water Portal: https://www.indiawaterportal.org/articles/research-reports-national-institute-hydrology-1996-2001-highlights

Jain, S. K. (2011). Population rise and growing water scarcity in India – revised estimates and required initiatives. *Current Science, 101*(3), 271–276.

Jayaraman, V., Chandrashekhar, M. G., & Rao, U. R. (1997). Managing the natural disasters from Space Technology inputs. In *47th IAF Congress, Acra Astronautica* (pp. 291–325). Great Britain: International Astronautical Federation. Published by Elsevier Science Ltd. Retrieved from http://abelo.zlibcdn.com/dtoken/e3cd1e2f087e3fde663a78270a74fe35/s0094-5765(97)00101-x.pdf

Jha, B. M., & Sinha, S. K. (2008). *Towards Better Management of Ground Water Resources in India*. Faridabad, India. Retrieved from http://cgwb.gov.in/documents/papers/incidpapers/Paper 1-B.M.Jha.pdf

Joseph, E. J. (2016). *Traditional Irrigation Systems in Kerala*. Kolkata, India: Institution of Engineers.

Lobo, C., & Samual, A. (2005). *Participatory Monitoring and Evaluation Systems in Watershed Development: Case Studies of Applied Tools. Encyclopedia of Evaluation Encyclopedia of evaluation*. Ahmednagar, Maharashtra: Watershed Organisation Trust (WOTR). Retrieved from https://wotr.org/publication/participatory-monitoring-and-evaluation-systems-in-watershed-development-case-studies-of-applied-tools/

Mazumdar, S. K. (1984). Efficiency of irrigation in the command areas of DVC, Mayurakshi and Kangsabati projects in West Bengal. In *National Seminar on Water Resources Management for Rural Development*. Sultanpur, India: KNIT.

Mazumdar, S. K. (2016). *Post Independence Scenario in Irrigation Sector in India – Need for Private Participation*. Kolkata, India: Institution of Engineers.

Ministry of Home Affairs. Border Area Development Program. 2020, Pub. L. No. No. 12/63/2014-BADP(Pt.-I), 13 (2020). India: Ministry of Home Affairs.

Ministry of Law and Justice. States Reorganization Act. 1956, Pub. L. No. 37 of 1956, 51 (1956). India: India Code Website. Retrieved from https://legislative.gov.in/sites/default/files/A1956-37.pdf

Ministry of Urban Development. (2009). A guide to project preparation, implementation and appraisal. New Delhi. Retrieved from http://cpheeo.gov.in/upload/uploadfiles/files/Guidance Note for Continuous Water Supply.pdf

Ministry of Water Resources-River Development & Ganga Rejuvenation. (2017). *5th Census of Minor Irrigation Schemes*. New Delhi, India. Retrieved from http://164.100.229.38/sites/default/files/5th-MICensusReport.pdf

MoWR. (2019). Government Initiatives | Department of Water Resources, RD & GR | Government of India. Retrieved February 23, 2020, from http://mowr.gov.in/government-initiatives

nationsonline.org. (1998). Countries of the World by Area. Retrieved March 30, 2020, from https://www.nationsonline.org/oneworld/countries_by_area.htm

NCERT. (2015). *Water Resources in India* (pp. 60–71). New Delhi: NCERT. Retrieved from http://ncert.nic.in/ncerts/l/legy206.pdf

Planning Commission of India. (1992). *Report of the Committee on 'Pricing Irrigation Water.'* New Delhi, India.

PMF IAS. (2016). Classification of Drainage Systems of India | PMF IAS. Retrieved February 22, 2020, from https://www.pmfias.com/classification-of-drainage-systems-of-india/

Press Information Bureau. (2014). IWMP - Harnessing, Conserving and Developing Degraded Natural Resources. Retrieved February 22, 2020, from https://pib.gov.in/newsite/mbErel.aspx?relid=102727

Rao, U. R. (1996). *Space Technology for Sustainable Development*. New Delhi, India: Tata McGraw Hill.

Rath, A., Samantaray, S., Raj, R. D., Swain, P. C. and Eslamian, S. (2021). Impact assignment of integrated watershed management in the micro watersheds of Sambalpur District, Odisha, India. In *Handbook of Water Harvesting and Conservation, Vol. 2: Case Studies and Application Examples* (pp. 341–358), ed. by Eslamian, S. & Eslamian, F., Hoboken, NJ: John Wiley & Sons, Inc.

Secure IAS. (2018). Classification of the Climatic Regions of India. Retrieved from https://secureias.com/climatic-regions-of-india-with-maps/

Shewale, M. P., & Kumar, S. (2005). *Climatological Features of Drought Incidences in India*. Pune. Retrieved from https://imdpune.gov.in/hydrology/Drought/drought.pdf

Singh, P. K., Nema, M. K., Jain, S. K., & Mishra, S. K. (2016). *Water Conservation and Management: Towards a Societal and Technological Initiative for Sustainable Development* (p. 204). Kolkata: ATV CVDB - The Institution of Engineers.

Tapi Embankment Division. (2010). *Action Taken Report For The Inquiry Commission Surat Flood Disaster*. Surat, Gujarat, India: Government of Gujarat.

Tripathi, K. P. (2016). *Traditional Irrigation Systems in India*. Kolkata, India: Institution of Engineers.

UNICEF. (2016). Drought in Gujarat : Need for Adaptation Measures. Towards Drought Free India, (147). Retrieved from southasiadisasters.net

United Nations. (1992). *United Nations Conference on Environment & Development*, Rio de Janeiro, Brazil. Retrieved from http://www.un.org/esa/sustdev/agenda21.htm

VI

Middle East Irrigation and Deficit Irrigation

17

Middle East and Origin of Irrigation

Ahmed Hayaty
Elshaikh
University of Khartoum

Saeid Eslamian
*Isfahan University
of Technology*

17.1 Introduction

Irrigation can be defined as the process of applying water to soil to assist and enhance plant growth and yield, and, in some cases, the quality of harvested parts (Sojka et al., 2002). Irrigation engineering is a practice as old as human communities, due to its association with agriculture and food supplies. Therefore, knowing the history of such science is vital in order to be conversant with the basics and developments of irrigation engineering through the various ages and civilizations. Human history has witnessed the rise and fall of many civilizations, in different parts of the world. Many nations contributed in varying degrees to the development and growth of human civilization (Cotterell, 1980). Societies evolved in various stages, starting with the development of paintings and writing, then the emergence of agricultural societies, followed by the rise of large cities, and ending with kingdoms and empires. Overall, all of this historical movement has contributed to a better life for man.

This chapter provides a historical overview of the Middle East civilizations and how they contributed to irrigation science and technologies. It could be considered a scientific historical review; additionally, it sheds some light on the major trends of irrigation science development and how the Middle East

DOI: 10.1201/9781003353928-23

civilizations contributed to irrigation development as one of the oldest known civilizations throughout history.

17.2 History of Agriculture

In ancient times, man depended on hunting and farming plants to feed himself. However, by the last Ice Age, around 11,000 years ago, climatic conditions had changed radically in many parts of the world (Armelagos et al., 1991; Mazoyer and Roudart, 2007). An increase in temperature and torrential rainfall caused a substantial change in plant cover. In tropical and semi-tropical areas, forests and jungles covered the plains that were once covered in ice during the Ice Age. Furthermore, such recent major changes in climate toward the end of the Ice Age resulted in the extinction of the major species of large animals. Various theories sought to give an answer to the question related to how the transition from hunting and collection to planting and agriculture had taken place. As a conclusion to such theories, we may mention here that there were two categories of human beings, based on the food supply system. The first category considered agriculture as a lifestyle since some of the human groups discovered the ability to grow plants in fields. This behavior had contributed to the provision of a permanent food supply for such groups, resulting in the settlement of such human groups in fields. The second category, however, depended on hunting and the collection of fruits and plants as their food supply. Therefore, these groups move from one region to another based on the availability of food supplies (Armelagos et al., 1991).

It is essential at this stage to know how human communities have developed. There are two types of ancient societies: pre-blogging and writing societies and post-blogging and writing societies. Ages before the discovery of writing are entitled "pre-history ages" i.e., 3,200 years BC. This classification had been approved by the archaeologists, through studying and noting discovered antiquities for very ancient times in early human history. It is to be noted that such ages were nominated based on the material employed in making the tools used by man in his life. This time period was divided into three major eras by experts. The first is the Stone Age, wherein man used stones for making tools and learned how to light fire. This era is subdivided into the ancient Stone Age and the Bronze-Stone Age, wherein metals were discovered for the first time in human history. This era witnessed the discovery of copper and the formation of bronze. The Iron Age was marked by the extensive use of iron in making different weapons and tools. This era lasted until the emergence of writing and the start of history (Zvelebil and Pluciennik, 2011).

About 9000 BC, during the so-called modern Stone Age, agricultural villages began to develop. Agricultural tools then became more diversified, and it was more likely that irrigation began to develop in limited scales in Mesopotamia due to drought and the desert climate in the south and the Anatolia Heights to the north (Martin and Sauerborn, 2013). The emergence of agriculture was not a sudden event or a brilliant invention by individuals. It was rather a gradual process. There is evidence from different regions of the world to the effect that the increase in agricultural production is related, in general, to a reduction in wildlife communities. Therefore, the shortage of food supply resources is the prerequisite condition, in the first place, for the major boom in agriculture. It was noted that considerable groups have transformed from hunting to planting following climatic changes, scarcity of available resources, and a reduction in the species of big animals due to excessive hunting.

The aforementioned factors increased population density in agricultural communities, resulting in the promotion of cultural, social, and higher living standards. As a result, various civilizations had appeared and developed. The increase in population density generated a demand for increased agricultural production. However, the most influential and vital element in the process was always water. In this respect, we consider irrigation as the most important of all significant strategic techniques learned by man. Although the influence of irrigation was not always of the utmost importance for the agricultural style, which may sometimes depend on rain as a source of water, however, it had always been of considerable significance in the emergence of civilizations, particularly in areas that depend on controllable water resources such as rivers and lakes. Therefore, it is not surprising that most of the ancient civilizations had grown on riverbanks and closer to water resources (Zvelebil and Pluciennik, 2011).

17.3 The Middle East Civilizations

The Middle East region of origin covers an area ranging from Jordan and Syria, through the eastern part of Turkey to the basins of the Tigris and Euphrates. Mesopotamia is located on the eastern side of the region known as the "Fertile Crescent". This arch represents the fertile lands of Mesopotamia and Al-Sham countries (Figure 17.1). Agricultural economies reliant on a mixture of domesticated crops and livestock became established in this region approximately 9,500–9,000 years ago (Zeder, 2008).

Subsequently, man started the transitional period from collecting food to producing it, besides the constant attempts to introduce cultivation into the region, which aimed at securing a settled life instead of the

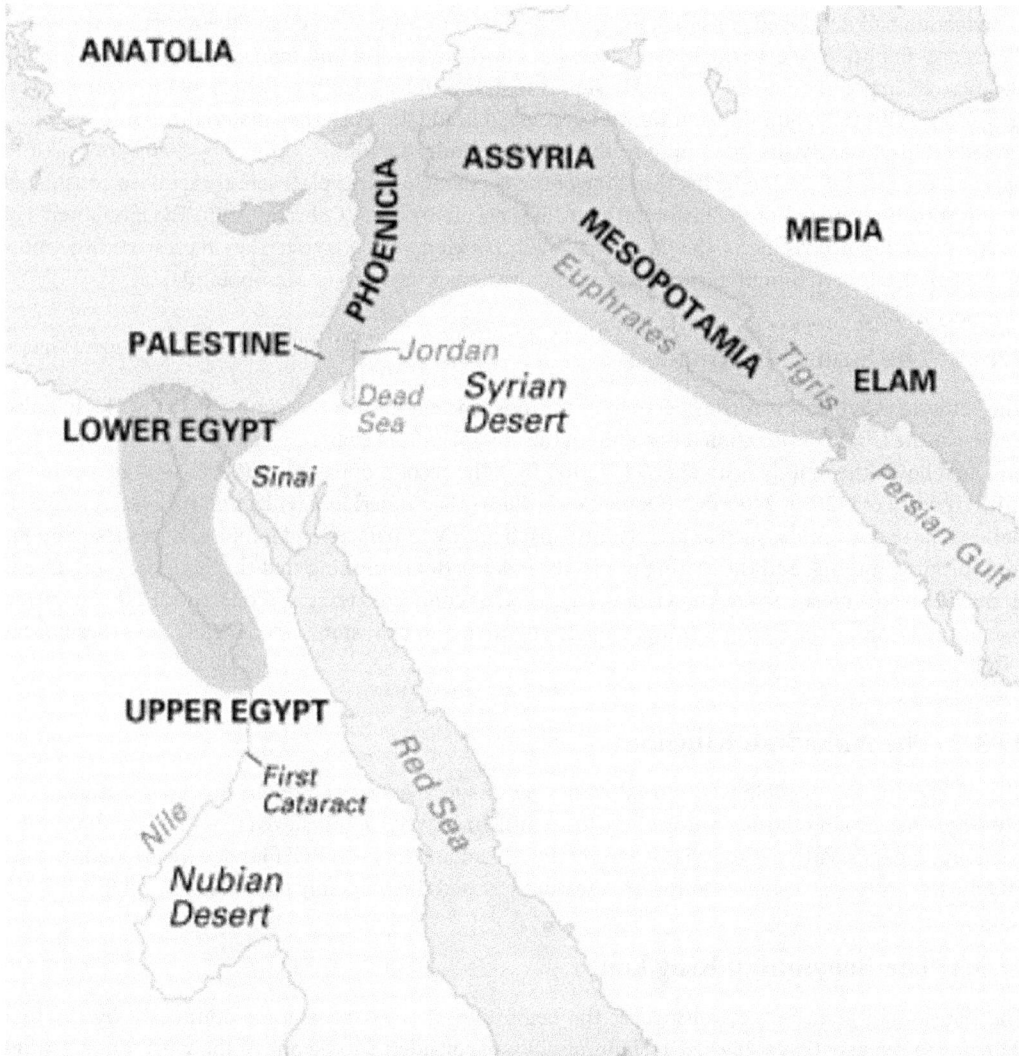

FIGURE 17.1 The Fertile Crescent (Encyclopedia Britannica, 2019).

risky nomadic life (Mays, 2010). The initials of improved agriculture started around the year 11,000 BC, at the center and origin of Middle East civilizations in the Indus Valley, Mesopotamia, and the Nile Valley.

17.3.1 Mesopotamian Civilization

The Tigris and Euphrates Valleys are described as the cradle of eastern civilization and the world. The man had inhabited the Mesopotamia area since early ages due to the availability of natural resources. There was always sufficient rainwater to support grass and tree growth, in addition to caves and forests that formed natural refuges against wildlife. Hunting was also a major source of food supply. All these resources secured human life despite the difficulties imposed by the environment. Because of the constant flow of sediments, issuing from the Tigris and Euphrates, a fertile highland in lower Mesopotamia was formed, before the rivers were deposited into the Persian Gulf. This long and relatively narrow plain has a very low slope, which makes the river courses meandering and unstable. The south area roughly corresponds to an annual rainfall of 200 mm/year; therefore, agriculture is not possible without irrigation (Cabrera, 2010).

Irrigation had developed considerably when the pioneer farmers settled in the low plains, where the Tigris and Euphrates meet. The initial successful efforts to control and manage water were driven by the increase in population and the need to supply food through agriculture, which mainly depended on irrigation. With time, human communities developed gradually until they materialized into a sophisticated civilization. By the beginning of the Ubaid period, in 6400 BC, human settlements began to appear close to what used to be the shoreline of the Persian Gulf; this place has always been considered by the Sumerians as one of the historical roots of their civilization (Cabrera, 2010). The monuments of such civilization are visible in the cities of Akkad, Babylon, and Assyria. They represented the most ancient civilization in human history, which ruled various kingdoms in Mesopotamia.

17.3.2 The Sumerian Kingdom

The Sumerian Civilization is considered to be the most ancient known civilization in the world. It is also known as the Ubaid Civilization (after a hill in the area). The Sumerian rule had been divided into three eras as follows: the first dynasty era (2800–2700 BC), the second dynasty era (2700–2600 BC), and the third dynasty era (2600–2400 BC) (Oppenheim, 2013). The Sumerian Civilization had various contributions, in various fields, such as architecture, particularly in palaces and temples. It was also famous for sculpture, mining, and the casting of metals. It is worth mentioning that the Sumerian language is deemed to be the most ancient known language as far as connected texts are concerned. It is considered one of the forms of cuneiform writing, namely engraving on clay, stone, wax, metal, and other boards (Viollet, 2010).

17.3.3 The Akkadian Kingdom

The Akkadians were people who moved from their respective homelands (in the Arab Peninsula) to Mesopotamia, around the beginning of the third millennium BC. It is believed that the Akkadians were contemporaneous with the Sumerians. Sargon the Akkadian (2371–231 BC) founded the Akkadian state, and during his reign, various arts and architecture prospered remarkably (Oppenheim, 2013).

17.3.4 The Babylonian Kingdom

The Babylonian State was established at the beginning of the second millennium BC, by the first Babylonian dynasty (2894–1595 BC). Hammurabi was considered to be one of the most famous kings of that era (1728–1686 BC). Thanks to his efforts, the country had been unified. One of his important achievements was the enactment of standard laws that prevailed in all of Mesopotamia (known as the

Hammurabi laws). Such laws were considered to be the first integrated legal system in history. The ancient Babylonian era was marked by the prosperity of human sciences and knowledge, and the extension and increase in the number of cities. The reign of King Nebuchadnezzar II was deemed to be among the most prosperous eras in history, in general. It was deemed a reign of prosperity and power for the Babylonian Civilization in particular. His era was famous for architecture: Ishtar Gate was an instance, in addition to many temples in the city of Babylon (Oppenheim, 2013).

17.3.5 The Assyrian Kingdom

Their name was derived from the name of their god, Assyria. The Assyrians were emigrants from the Arabian Peninsula and settled in Mesopotamia. Subsequently, Assyria expanded to include Syria, Palestine, and the north of Egypt, and it came to an end in 809 BC. The Assyrian Kingdom generated great kings such as Shalmaneser I, Sargon II, and Assurbanipal (Oppenheim, 2013) (Figure 17.2).

During the Bronze Age (4000–1100 BC), the urban centers were excluded from the two cities of Sumeria and Akkad in Mesopotamia. The population of cities ranged between 50,000 and 80,000 people, 90% of whom were inhabitants of cities. At the time of the initial development of such cities, almost half of the population worked on major irrigated farms around each city. Following the promotion of modern techniques and tools during the Sumerian era, Mesopotamia had witnessed a considerable development in irrigated agriculture, in such a way that only 1% of the human population was able to feed the remaining population. Thus, farmers were tempted to emphasize the efficient management of water, as noted on one of the boards, namely that it was possible to increase the return by 10% through the proper management of irrigation (Oppenheim, 2013; Viollet, 2010).

Ancient farmers also used various hydraulic structures to irrigate their lands, for example, Shadouf, waterwheels (Noria), irrigation canals, and dams. Furthermore, the ancient Sumerians had come to

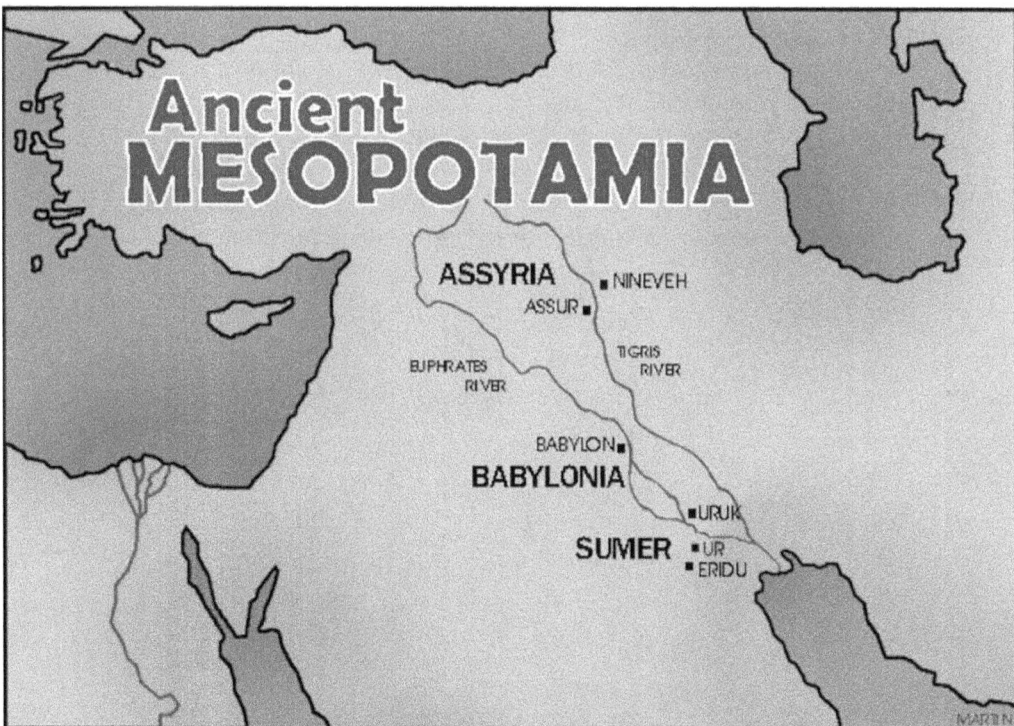

FIGURE 17.2 Locations of the ancient kingdoms in Mesopotamia (Donn, 2019).

know good operation practice for canal seepage, comprising a multi-layer, huge roll filled with thorns, papyrus, cane, and the remains of trees, and then rolled in many rows of ropes. The same method was used later by the Babylonians (Mays, 2010).

During the Ubaid Civilization in Mesopotamia, concentrated irrigated agriculture appeared at the same time in both civilizations about 8,000 years ago. The prevailing climate in the central and southern regions of Mesopotamia played a significant role in the development of the irrigation system. The flood season in Tigris and Euphrates occurs during April and May, which is a very late time for winter crops, and, at the same time, too early for summer crops. Because irrigation water security is a constant concern for a community that is entirely dependent on agriculture, a sophisticated irrigation system of dams, canals, and water reservoirs has been developed to ensure water supply for agricultural lands. Water supply for urban centers used to be secured from various sources, including canals that connected to rivers, wells, and underground water tanks. Other hydraulic techniques include water tunnels and pottery pipes, in addition to animal-powered water lifting (Waller and Yitayew, 2015).

Moreover, many dams were used for diverting irrigation water. The dam of the city of Nimrud was one of the oldest dams on Tigris, located about 15 km to the south of the existing city of Samarra. The river water is diverted through the Nehruan canal in order to irrigate an area extending to 100 km to the existing city of Jacoba. The old dam and the diversion canal not only irrigate the desert land but are also deemed as a canal connecting to the Tigris River and forming a major stream for navigation, in addition to supplying water for cities to meet daily use and domestic needs. To run this system, there was a pottery pipe water network within the city. At that time, the Nimrud Palace water tank used to be manually filled out by servants, to store freshwater. Subsequently, however, other tanks had been built and connected to a rainwater collection system (Mays, 2010).

17.3.6 Hanging Gardens of Babylon

The hanging gardens of Babylon were considered one of the seven wonders of the ancient world (Figure 17.3). There was a misconception to the effect that such hanging gardens were built during the era of the new Babylonian dynasty, allegedly built in the old city of Babylon, attributed to King Nebuchadnezzar II who ruled between 605 and 562 BC. The reason for building the gardens was said to be to please his wife, who missed living in the fertile hills of Persia and hated living in the barren

FIGURE 17.3 A conceptual drawing for the hanging gardens of Babylon (Sydalg, 2015).

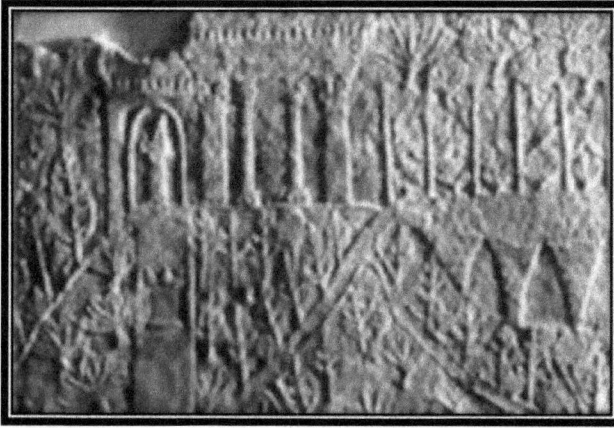

FIGURE 17.4 A wall drawing of gardens on the walls of the palace of the Assyrian king Sennacherib near Nineveh (Wikiwand, 2018).

plains of Babylon. However, there was no evidence for the existence of such gardens in the ancient city of Babylon (Mays, 2010).

More recently, in 2013, in a study conducted by Dr. Stephanie Daley from the University of Oxford, Daly decided to focus her research in the city of Nineveh, about 450 km north of Babylon, which is now represented by the northern city of Mosul, and she revealed a theory based on research spanning 20 years (Dalley, 2013).

Dr. Stephanie has succeeded in deciphering the talismans of one of the tablets displayed in the British Museum, a clay board written in the ancient cuneiform, about 2,500 years old. It was concluded that the historical description of the gardens is evident in the city of Nineveh, namely in the palace of the Assyrian king Sennacherib. This king lived 100 years before King Nebuchadnezzar II. The boards described a "palace not matched by any other palace". Moreover, the related writings also described the style of planting trees on the covered porch and the balconies of the palace, as reported in the description of the hanging gardens of Babylon (Dalley, 2013).

The archaeological inscriptions (Figure 17.4) stated that the Assyrian king Sennacherib (705–681 BC), the son of the Assyrian king Sennacherib II, had constructed a major network of canals for irrigating the gardens and parks of the city. He also transferred the capital to Nineveh. Accordingly, he started developing a modern system for water supply. He constructed a big network of canals, not only for agriculture but also for the preservation of life through supplying water to the new urban zone, which was part of the Assyrian urbanization (Viollet, 2010).

The technique used for lifting water to these gardens was described as an amazing system. Rainwater is scarce due to the desert nature of the surroundings. Therefore, it was essential to secure a water supply as well as an advanced irrigation system to water the plants on the high stands for each floor of the building. However, the system used in lifting irrigation water for the hanging gardens of Babylon is still a mystery. One assumption is to the effect that serial water-lifting tools had been used to lift water to the upper stores, comprising two gigantic wheels, one over the other (waterwheels). However, another explanation suggested the use of screws for lifting water.

17.4 Water-Lifting Techniques

The Mesopotamian Civilization had a major contribution to water-pumping engineering techniques. They had developed the Shadouf and waterwheel water-lifting systems. Additionally, they invented the screw pump, which was improved later by Archimedes. The Shadouf water-lifting technique was known in Mesopotamia since 3,000 years BC (Mays, 2010).

FIGURE 17.5 Wall paintings showing the shape of the Shadouf in the Mesopotamian Civilization (Delougaz et al., 1967; Mays, 2010). (a) Drawing showing the Sumerian Shadouf. (b) Drawing showing the Assyrian Shadouf.

FIGURE 17.6 Images showing the types of waterwheel (Mays, 2010).

Shadouf is an axial machine used for lifting water from lowlands to higher ground. It consists of a long wooden rod and an axial pivotal point. At the terminal, closer to the water source, there is a container or a skin case tied with a rope and used for collecting water. On the opposite terminal, the weight of the stone is used to support the water-lifting process (Figure 17.5).

The waterwheel was deemed to be one of the most prominent machines used for water lifting (also known as the Noria). It consisted of a big-geared wooden wheel operating in a circular movement and holding several pots, which were used to plunge into the water while the wheel circulated. Accordingly, such pots were filled in such a way as to pour water into a water basin allocated for irrigating adjusted lands. The waterwheel was operated, either by water flow power or by the movement of an animal (an ox, horse, or camel). In such a case, the animal would revolve around the axle of the last wheel, while the wheel would revolve accordingly (see Figure 17.6).

The other type of water-lifting machine is the screw pump, which consists of a hollow cylinder containing an axial screw within. The upper part of the axial screw is connected to a circular turning arm. Water is being lifted for irrigating fields through the turning of the aforesaid arm. The arm is titled on a horizontal plane at about 30°. The bottom terminal of the same is plunged into the stream from which the water is required to be lifted (Figure 17.7).

17.5 Field Irrigation Techniques

With respect to the irrigation process, the management system noted on archaeological boards for the Mesopotamian Civilization referred to the human factor and the time required for executing the irrigation works. There was also reference to the calculations related to excavation and brick construction

FIGURE 17.7 Drawing and image showing the screw pump (Mays, 2010).

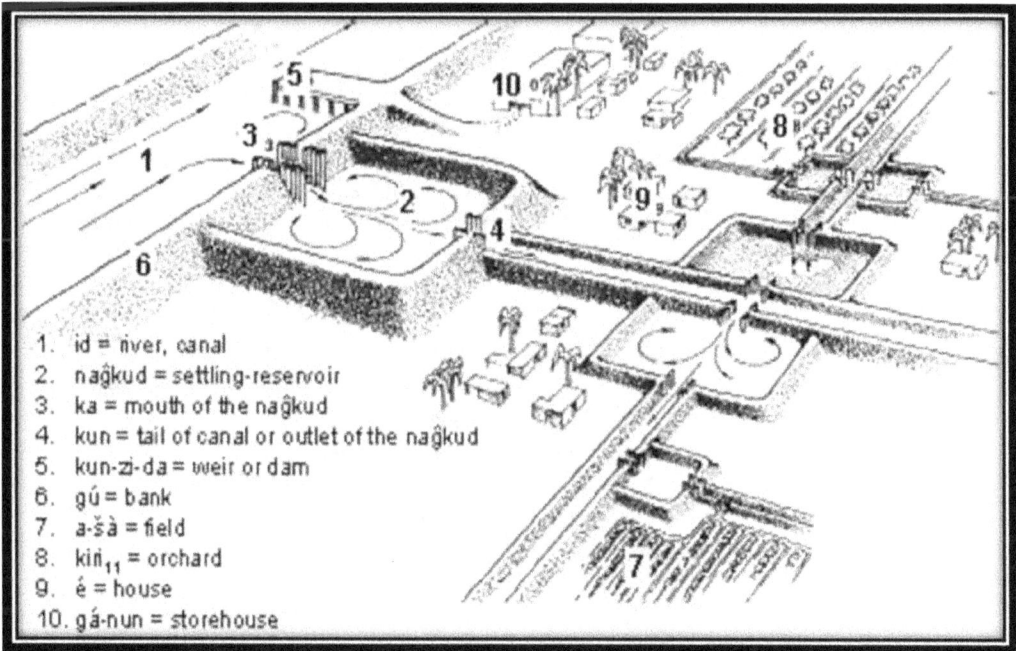

1. id = river, canal
2. naĝkud = settling-reservoir
3. ka = mouth of the naĝkud
4. kun = tail of canal or outlet of the naĝkud
5. kun-zi-da = weir or dam
6. gú = bank
7. a-šà = field
8. kiri₁₁ = orchard
9. é = house
10. ĝá-nun = storehouse

FIGURE 17.8 The different parts of the irrigation system of the Mesopotamian Civilization, as stated in Sumerian language (Kang, 1972).

works. The human cadre had two types of assignments: engineering and calculating positions, which were filled by people who could read and write. They used cane rods for measurements and lines. They also had various writings of calculations related to the irrigated land leases, as well as calculations of irrigation canals and dams. However, fieldwork was undertaken by the other category of human cadre, namely, workers, under the control and supervision of the first category. Records showed that engineers provided accurate details on the capacities and depths of canals. They further calculated the excavation quantities and the removal by workers of excavation soil (Lambert, 2007).

The Sumerian language comprised a lot of terminology for water and irrigation. Figure 17.8 showed the components of the water and irrigation network, along with the terminology of the Sumerian language for each portion of such a system. The most prominent feature of this type of basin was that its width was far less than its length. It was noted in one of the records that the width was 1 m, whereas the length was 36 m. Such dimensions were utterly tight, in such a way that they were not as efficient as storage facilities. Moreover, evaporation reduced the storage capacities of such tanks. Therefore, it was more likely that the main function of such tanks was to regulate and distribute water (Kang, 1972).

In order to lift water to the level of the main canal, barriers and dams were constructed using burnt clay bricks. Some inscriptions denoted that bricks and bitumen were used in building dams. Other references stipulated that carefully cut stones were used in building dams instead of bricks. In mathematical texts and engineering calculations, dams were exhibited as rectangular models. Other texts provided fixed ratios for the dimensions of the dam. Dams were typically built prior to the canal system, regulating water supply in both the main and small canals. For the main canal, there were established design dimensions that provided for a width of 120 m and more. This relatively vast width allowed navigation on the river into the city (Kang, 1972; Mays, 2010).

Canals were excavated and the resulting soil was used as a side landfill. A cross-section of a canal reflected that the basic shape was rectangular; however, several writings denoted that the trapezoid shape was used in the design and construction of canals. The concept of the lateral slope was introduced in the canal. Such inclination could be determined based on the horizontal distance for each length unit in a vertical direction.

The maintenance and operation of canals were considered a continuous process. Top officials to be appointed by the king would supervise the operation and maintenance processes of major canals. Moreover, it was essential to provide large groups of workers to cleanse silt from canals; usually, huge quantities of mud were removed from canals. For instance, one letter addressed to Hammurabi indicated that 1,800 m^3 of soil had to be cleansed, which meant 600 workdays (Mays, 2010). It was noted that workers mostly came from regions located on the banks of canals. Sometimes even fighters contributed to such assignments. Internal canals, however, were under the supervision of local authorities. The abovementioned practices and techniques showed the importance attached by the Sumerians to irrigation networking, including specifications and restrictions related to the construction of dams, barriers, canals, and basins. Add to this the control of products and the scheduling of irrigation. Records further emphasized that irrigation engineers and directors needed to design and maintain their irrigation systems (Kang, 1972).

One of the main causes of the deterioration of Sumerian Civilization is the decline of agricultural production. Some pieces of evidence were found in cuneiform language records referring to the prevalence of agricultural land salinity. Such records indicated that wine production from grains in 2400 BC was 2,347 L/ha. However, such production dropped to 897 L/ha in 1700 BC. Another indicator for the salinity of the soil was that the percentage of barley rose to 98% of the total planted grains in 2050 BC, since barley was known to tolerate salinity. In most cases, barley is the first product to be used for the reclamation of desert lands (Jacobsen and Adams, 1958). An additional factor was the huge amount of deposited silt in main and minor canals, in such a way that it was more feasible to abandon such blocked canals and construct new ones.

Water control was a decisive tool to secure the economic prosperity of Mesopotamian Civilization. However, the excessive exploitation of natural resources, such as land and water, had an adverse impact on the environment of such regions. Subsequently, the salinity of the soil prevailed, and accordingly, production dropped drastically. The foregoing might be among the factors that led to the decline of such a civilization, while taking into account other factors such as the political and military conditions prevailing at the time. However, the former factors constituted a source of conflict among states.

17.6 Indus Valley Civilization (Sindhu-Indus)

The Indus Valley Civilization was one of the first civilizations to emerge during the Bronze Age, which began 4,500 years ago in what is now Pakistan. It was considered one of the huge urban agglomerations, as the population was around five million (Jansen, 1989). Following the example of the ancient civilizations in gathering around the expansive valleys of rivers, this civilization was established around the Sind River and related tributaries. This civilization was also known as the Harappa Civilization, after the Pakistani city where archaeologists discovered the first signs of its existence (Waller and Yitayew, 2015). The city of Mohenjo-Daro was also one of the most prominent urban centers of such a civilization. The latter city is located on the right bank of the Sind River. The planning of this city dated back to 2450 years BC. It was located in a semi-arid environment, relying mainly on surface well water for drinking (Jansen, 1989).

Technologies that were unused in prehistoric times might have been part of the Indus agricultural regime. Developments such as irrigation and the use of bunds for impounding rainwater and soil emerge as important new technological features in this civilization. Such agricultural technology has allowed renewed settlement in different, independent, and unrelated cities (Possehl, 1997).

In general, the Indus Valley Civilization depended on irrigated agriculture, using the flooding irrigation system. The average annual flow of the Sind River was threefold greater than the Nile. Floods used to submerge expansive areas of land that produced barley, wheat, and other crops. They had also built wide irrigation canals, as well as other hydraulic systems (Waller and Yitayew, 2015). That civilization collapsed around 1800 BC, and no specific reason was known for such a fall; however, some theories attributed it to a drought caused by decreased activity of the monsoon within the region, a wave of destructive over-flooding, or both causes.

There was a written language of the Indus Civilization, but unfortunately (to date) it has not been understood or deciphered in any way. Accordingly, a considerable portion of the sciences related to such civilizations remained unknown, particularly the technical and detailed issues related to agriculture and irrigation (Samuel, 2000). It is worth mentioning that the present irrigation system in Pakistan covers an area of 16.2 million hectares, thus constituting the largest irrigation system in the world.

17.7 The Nile Valley Civilization

The birth and initial origin of this great civilization dated to 10,000 years BC. Following the example of the ancient civilizations growing around rivers, the ancient Egyptian civilization was established on the banks of the River Nile. The latter crossed the Great Desert, among the sands as a lifeline for such civilization. It extended along the Nile strip from Nubia in the south to the ends of the Nile at the Mediterranean Sea, in the north.

The Nile River is regarded as one of the most predictable rivers in the world because flood times can be predicted. Floods in Egypt are not surprising; they occur at regular intervals due to a gradual increase in water volume, unlike floods on the Tigris and Euphrates. The River Nile floods the flood plains at the start of the flood season in the summer. Water levels continue to rise until the total immersion of the southern riverbanks in mid-August. The northern areas reach such a peak several weeks later. This season is considered appropriate for planting several crops. In contrast, floods of the Tigris and Euphrates occur in April and May, which is too early since summer is dry and hot, and therefore not appropriate for planting crops (Mays, 2010). Consequently, the civilization of Mesopotamia had to develop irrigated agriculture through dams and canals to divert water from rivers to fields, whereas the Egyptian civilization relied on floodwater for irrigation purposes.

The Egyptian civilization depended totally on the River Nile in all respects of life, starting with religious rituals, continuing through daily uses, and ending with floodwater for irrigation purposes. However, the River Nile had not always been so ideal. Sometimes high floodwaters would destroy irrigation canals. In contrast, the failure of floods at certain times would impose even more serious problems than over-flooding, since the population might be exposed to famine and drought.

17.8 The Development of Field Irrigation Systems

The specific date on which the Pharaohs started the application of improved irrigation systems was still unknown, particularly canal irrigation systems, in addition to other systems that used water-lifting techniques as an improvement to the natural irrigation system, through flooding. Egyptians used water from artesian wells alongside the Nile at the time. This was done either through excavating different canals or through water-lifting techniques such as waterwheel and Shadouf (Strouhal and Forman, 1992).

According to historical studies based on photo analysis of ancient Egyptian temples, the history of the Egyptian civilization can be divided into two eras: the first era before the emergence of ruling dynasties, and the second era after the emergence of dynasties. The second era, however, marked the start of the unified Egyptian state, around the year 3100 BC (Mays, 2010). It was believed that the transitional period before dynasties witnessed major developments in field irrigation systems.

17.9 Irrigation Engineering Before the Dynasties

Egyptians had learned expansive agriculture systems from the Mesopotamian Civilization during the pre-dynastic era. During that time, their irrigation depended mainly on the annual flood of the River Nile. Floodwater was used to immerse the banks of the Nile to a height of 1.5 m during September. This peak would last about 30 days. Basins for collecting water were built along the river to irrigate the land until the fields were fully satisfied. Subsequently, such water would be drained following the completion of the irrigation process (Waller and Yitayew, 2015). Floodwater-bearing sediment served to re-fertilize lands on an annual basis, in addition to reducing or removing salinity from the soil. Following the immersion of lands by floodwater and the completion of the irrigation process at the beginning of November, farmers would start planting the winter wheat crop to be harvested during the time between April and May.

The ancient Egyptians also had a penchant for gardens, which necessitated watering all year, including during Nile depletion periods. The home gardens used to be divided into small basins by soil barriers that intersected at right angles. Water was fetched for such gardens by workers from the Nile or surface wells. This watering arrangement continued until the emergence of water-lifting techniques such as the Shadouf or the waterwheel (Strouhal and Forman, 1992).

The first recorded evidence for water management in the Egyptian civilization was traced on paintings on the royal scepter (Figure 17.9) that belonged to the king known as the Scorpion, who ruled Upper Egypt as the last of the kings before the unification of Egypt in around 3200 BC (Butzer, 1976). The lower part of the scepter head exhibited a water stream that branched out in two canals (symbolizing irrigation canals). The king held a big shovel, while his attendants stood by holding a spike of fibers and brooms. Two other workers were operating on the lower canal with hoes. This head denotes that the ancient Egyptians started the practice of some forms of water management as early as 5000 BC (Mays, 2010; Strouhal and Forman, 1992).

In Figure 17.9, this photo was interpreted to mark the start of a ceremony for the breaking of a water dam to allow water to flood over fields; otherwise, a ceremony for the inauguration of a new canal. Others believed that the photo presented the irrigation works under the supervision of the king. This document, however, showed that the transition from natural flood irrigation to developed irrigation through canals had been completed before the era of dynasties (Strouhal and Forman, 1992).

17.10 Irrigation Engineering during the Era of Dynasties

As mentioned before, the transition from natural irrigation to industrial irrigation was achieved before the era of dynasties in 3100 BC. Such developments included the management and drainage of flood-water, using water control gates and longitudinal and cross soil barriers. Such canals allowed the flow of floodwater to locations otherwise not accessible during natural flooding. Moreover, this system was used when flood levels were low. The networks of water canals had contributed even more to the transmission of water and excess water (Butzer, 1976).

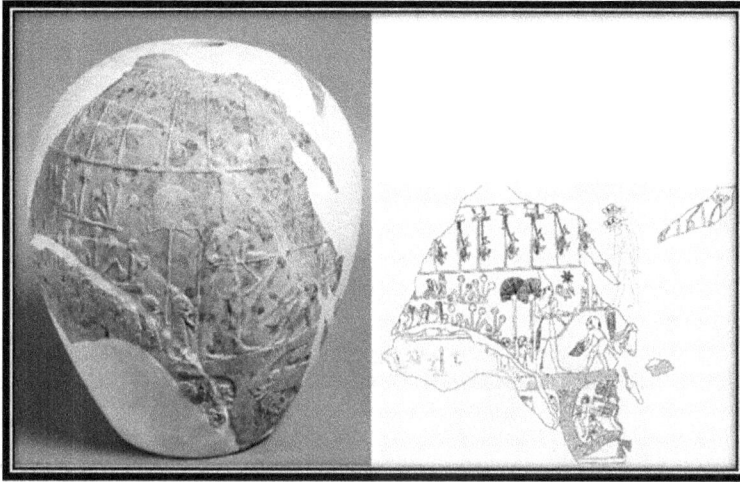

FIGURE 17.9 The head of King Scorpion's scepter, the oldest recorded evidence for irrigation management in the Egyptian civilization (Mays, 2010).

FIGURE 17.10 Basins irrigation in the Egyptian civilization (Takla, 2018).

The old irrigation system that had been developed was known as the basin system. At that time, the irrigation of basins was done and controlled on a large scale by the local authorities for each region, subject to central monitoring by the state. This type of water management addressed the control on irrigation water through basins formed by a network running parallel or perpendicular to the river. The end output of the process would be basins of different sizes (Mays, 2010).

During the flood season, beginning in August, the Nile's water would be diverted to basins in such a way that the water submerged the basins until the soil was saturated. Subsequently, water would be drained to a lower basin, toward the stream of the Nile, or else to one of the water canals (Figure 17.10). Following the completion of water drainage from the basin, the crop planting process would start. To organize and distribute fields, agricultural lands were divided into large areas ranging from 2000 Feddans in the southern Nile Valley to 20,000 Feddans at the Nile Delta, in the north (1 Feddan = 4,200 m² = 0.42 ha)

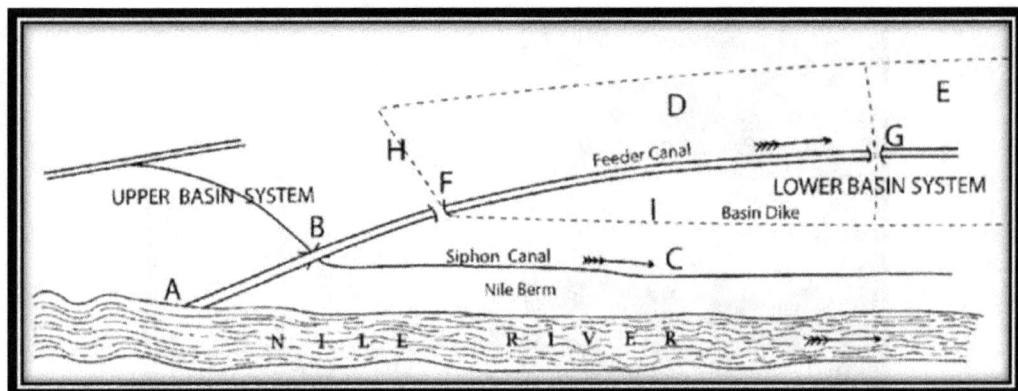

FIGURE 17.11 Components of the irrigation system in the Egyptian civilization. (Mays, 2010): (A) the beginning of the main canal, (B) the siphon of an irrigation canal from the upper region, (C) the highlands under sorghum basins, (D and E) basins, (F and G) water regulators, (H) transverse dike, and (I) longitudinal earthen barrier.

(Mays, 2010). Figure 17.11 indicates an irrigation system with basins, as basins were supplied with water through irrigation canals.

The bed level for the main canal taking water from the Nile should be in the center between the natural Nile level and the ground level on the bank, allowing direct access to such canals during the flood season. It was believed that the entry of water into such canals had been controlled to a certain extent. Such canals were not perpendicular to the river, but rather formed a curve at the beginning (as indicated in Figure 17.11), then went parallel to the Nile. The slope of the canal was in the direction of the Nile stream. However, such a slope was less than the slope of the Nile (Said, 2013).

Each canal provided water to eight basins (on average), in a straight line. The irrigation process was monitored by regulators built on the earth embankments in order to control the flow of water to the basins. There was no fixed depth range in basins; it was rather variable based on the volume of flooding from one season to another. The water remained in basins for a period of 40–60 days, following which water would be drained to low areas through a siphon to water basins in the lower regions (referred to by the letter C in Figure 17.11), instead of returning to the river. The basins were fertile and flat due to annual sediments from irrigation water that submerged such basins (Said, 2013).

During the flood season, the improved irrigation system significantly increased the annual agricultural lands. Such irrigation allowed for the full benefit of water in low floods. The same volume of water could be distributed to irrigate more basins, in addition to controlling the same during the submersion period. Thus, some of these basins might be planted for the second or third time (see Figure 17.12).

The enhancement and development of irrigation technology continued during the era of major dynasties, particularly the Roman era. After the year 1500 BC, the ancient Egyptians started lifting irrigation water by the Shadouf, which had been formerly used by Mesopotamia. Shadouf was used to irrigate small areas, particularly gardens and palaces (see Figure 17.13). Furthermore, this device provided for irrigating crops on areas adjacent to the riverbanks, as well as regions not within the reach of water canals in summer (Butzer, 1976). The waterwheel also appeared during the same era, as a tool for lifting water to irrigate more areas than was feasible with the Shadouf.

17.11 Agriculture in Fayum Area

The ancient Egyptians began agriculture in the Fayum region 5,000 years ago, around 5200–4000 BC. It is considered to be the oldest region in the world, where irrigation was used for agriculture. The Fayum region contained a large lake about 27 km north of Fayum city, which is now known

FIGURE 17.12 Wall paintings for the planting and harvesting works in ancient Egypt (Marefa, 2018).

FIGURE 17.13 Wall paintings for the use of Shadouf in lifting irrigation water (Marefa, 2018).

as Lake Qarun. It is considered one of the deepest lakes since, at certain points, it was 14 m deep. Historically, this lake was known as Moeris. During the time of the Pharaohs, it was supplied by water during the flood season, as it was lower than the sea level (−45 m). During the era of the twelfth Egyptian Dynasty, a canal was built from the Nile, and two dams were also constructed for storing water during the flood season. As indicated in the Herodotus Scrolls on the ancient history of Egypt, the lake was used to store water and supply the same to land for 6 months during the dry season (Mays, 2010).

During the early Ptolemaic era, Ptolemy II (285–246 BC) undertook the construction of a dam to control water flow from the River Nile to a canal that transmitted water to the lake. The dam was 5000 m long and 4 m high. The different openings had been closed in order to prevent water from returning to the Nile, except for one. The dam was used to maintain the level of the lake at 2 m above sea level. The water diversion system from the Nile to the Fayum region consisted of a network of relatively graded canals. It was a unique water transmission system when compared to the canal system used along the

Nile. The dam and related techniques had enabled the Ptolemaic engineers to supply water to about 325 Feddans of fertile and cultivable land. This project served to increase the fortunes of Egypt, which resulted in an increase in the population of 4.9 million, the greatest during the long history of Egypt before the 19th century (Mays, 2010).

To conclude the ancient Egyptian experience in irrigation engineering, we will find that the success of this civilization relied on the Nile's water in developing the agricultural systems, as they developed irrigation by basins. Moreover, all agricultural resources were available: fertile soil, irrigation water, a moderate climate, the developed water management prevailing at the time, the ability to forecast floods and control their consequent damage, and the irrigation system in the Fayum region. All the foregoing factors contributed to an increase in agricultural products, which, in turn, contributed to the social and cultural development of the Egyptian civilization.

17.12 Persian Civilization

Persian Civilization had emerged in the Middle East region, between the east and north of the Arabian Peninsula, and was mainly centralized in the present-day region of Iran and its surrounding areas. It was founded in 559 BC by Commander Cyrus and his son Cambyses II, when the Persian Empire gained control of vast areas. It had occupied several territories extending to ancient Egyptian Egypt (Burgan and Urban, 2009).

Through the enactment of laws and restrictions to regulate the ruling affairs within the Persian Empire, the Persian Civilization had witnessed excellence in political organization and management systems. This was undertaken by the pioneer Sassanid rulers through the chief minister known as (Ardashir). Among the most significant factors that supported the development of this civilization was the importance attached by rules to the architecture of cities, human rights values, and the various trades and sciences, to the extent that orders were issued for the translation of philosophy and medicine books, as well as references related to astronomy, from the Greek civilization (Pourshariati, 2017).

Due to the scarcity of rain in the plains and valleys, some of the smaller Aryan tribes started a gradual emigration to the Iranian Plateau, where seasonal water was available due to the melting of ice. Their former agricultural techniques required big volumes of water throughout the year. This requirement did not match the nature of the abundant water on the Iranian Plateau.

They had no choice but to use the rivers and springs that flowed from the mountains. The need to develop a new water system rose when miners faced the problem of controlling the underground water closer to the surface; and attempted to drain it through grooves and barriers. The idea developed when the ancient Iranians used the water to be disposed of by miners to develop a new basic system of irrigation, known as the "Qantas system". They operated that new system to supply the water necessary for farms during the dry season (Angelakis et al., 2016).

17.13 The Qantas System

The initial start of the system came into being closer to the Urartu Mountains, located to the north of east Anatolia, 16 km from the Iranian boundaries, and 32 km from the Armenian boundaries. From there, the system spread all around the Iranian Plateau. The most ancient Qantas system was discovered in the northern part of Iran, in the old city of Zarsh. The age of this Qantas dates to 800 BC, when the Aryans settled in the present region of Iran (Mays, 2010).

The Qantas system was a system for water collection and transmission through underground canals. It consisted of canals to be excavated under the ground, adjacent to the foot of a mountain or a plateau. Such canals would collect water from natural springs, surface water, or otherwise, obstruct and collect torrential rainwater. Subsequently, such collected water would be transmitted

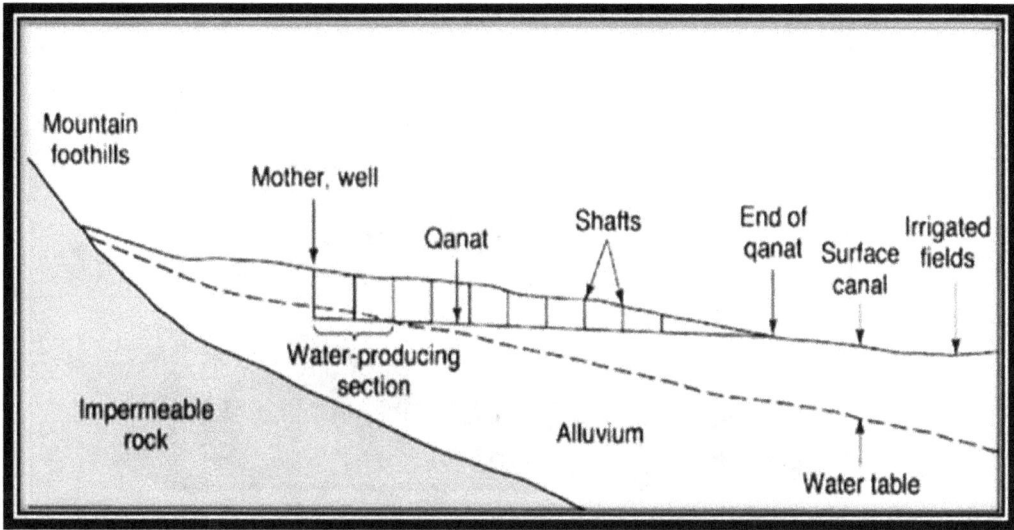

FIGURE 17.14 Components of the Qanat system to transport water from high to low lands (Mays, 2010).

naturally through the Qantas, or by gravity alone, without using any water-lifting tools. By the end of the Qantas, there would be a sedimentation lake. From there, water would be transmitted through a surface set of canals to agricultural fields and residential communities. Along the duct of the Qantas, which extended for tens of kilometers, there was a series of vertical wells (as shown in Figure 17.14), formerly used during the excavation of tunnels, which would then be used to provide lighting and ventilation.

The Qantas system operated to deliver water from different sources. In general, the process began with rain, when thunderstorms caused surface water runoff, which was often concentrated on impermeable mountains. The level of water in this type of Qantas usually increases directly upon rainfall. It might dry out when rain fails for long periods. Moreover, there were many mountains that provided water streams throughout the year, in addition to the availability of deep lakes where water remained for a long time even during the dry season. These water resources were used to irrigate fertile soils at the foot of mountains for several centuries. Water was transmitted to remote locations across mountains, plains, and valleys, as well as to farms with cultivable fertile soil and off the rock and stony soil, thanks to the Qantas system and the irrigation system, via a 10-km network (Taghavi-Jeloudar et al., 2013).

When the water flow reaches the ground level, a large basin will be excavated to collect the water. It would also serve as a sediment basin for silt and suspended materials. Usually, the canal would be lined with a flat stone to be cut from rock (Figure 17.15). At a certain distance from the sediment basin, a network of canals transmitting water to fields would branch off. With time, the underground level would drop at the source, and therefore, the water trenches had to be deepened for the water to continue flowing. After some years, the system had to be extended a little further to increase the slope and canal depth (Angelakis et al., 2016).

During this era, the tunnel excavation technique developed, despite the fact that the rock wall would sometimes be too steep and too hard. Usually, tunnels are marked by hard work and major difficulties, particularly when closer to irregular rock walls. Nevertheless, the need for tunnels was most probably associated with the development of geology. This technique started with surface excavation and continued until the required depth for the tunnel flooring, with a supporting system to strengthen the sides of the tunnel. In exposed excavations, a safe lean had to be established. Subsequently, the ceilings and walls would be strengthened.

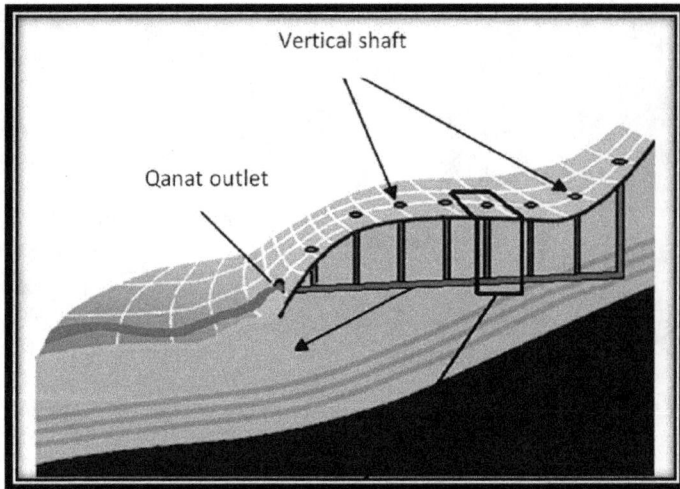

FIGURE 17.15　A diagram showing how the Qantas system works (Taghavi-Jeloudar et al., 2013).

17.14　The Dispersal of the Qantas System

Between 550 and 331 BC, the Persian Empire extended from the Sind Nile in the east to the Nile in the west, including major parts of Europe until ancient Greece. This expansion contributed to spreading the culture and technology of the Qantas system at the time. The Qantas system spread to the west from Persia to Mesopotamia, and from there to the Mediterranean and south to parts of Egypt. Furthermore, Qantas were constructed in Afghanistan, to the east of Persia, and in the settlements of the oasis on the Silk Road, in central Asia, until Chinese Turkestan. During the transfer of this technology to other civilizations, it acquired different names: "Aflaj" within the Gulf States, "Qantas" in Iran, "Karis" in Afghanistan and Pakistan, "Caner Jing" in China, and "Fujarah" in North Africa (Angelakis et al., 2016; Gorokhovich et al., 2011).

Although this system is very old, dating to 3,000 years ago, it is still working today in countries such as Iran, Oman, the UAE, and many other countries.

17.15　Conclusions

Irrigation systems have evolved significantly over time and across civilizations. As the population grows, the importance of irrigation projects grows in order to boost productivity and ensure food security. A timeline of the historical development of irrigation through the last 6,000 years of human history is considered. Many civilizations were founded in the Middle East, such as the Indus Valley, Mesopotamia, the ancient Egyptians, and the Persians. Through the history of irrigation development, it is noted that the civilizations of the Middle East have significantly contributed to the development of different technologies of irrigation engineering.

References

Angelakis, A.N., Chiotis, E., Eslamian, S., Weingartner, H., 2016. *Underground Aqueducts Handbook*. CRC Press, Boca Raton, FL.

Armelagos, G.J., Goodman, A.H., Jacobs, K.H., 1991. The origins of agriculture: Population growth during a period of declining health. *Popul. Environ.* 13, 9–22.

Burgan, M., Urban, T.G., 2009. *Empires of Ancient Persia*. Infobase Publishing, New York.

Butzer, K.W., 1976. Early hydraulic civilization in Egypt: A study in cultural ecology. *Prehist. Archeol. Ecol.* 2(3), 285–287.

Cabrera, E., 2010. *Water Engineering and Management through Time: Learning from History.* CRC Press, Boca Raton, FL.

Cotterell, A., 1980. *The Encyclopedia of Ancient Civilizations.* Mayflower Books, New York.

Dalley, S., 2013. *The Mystery of the Hanging Garden of Babylon: An Elusive World Wonder Traced.* OUP, Oxford, UK.

Delougaz, P., Hill, H.D., Lloyd, S., 1967. *Private Houses and Graves in the Diyala Region.* University of Chicago Press, Chicago, IL.

Donn, 2019. Ancient Mesopotamia [WWW Document]. URL https://mesopotamia.mrdonn.org/geography.html (accessed 11.3.19).

Encyclopædia Britannica, 2019. Fertile Crescent [WWW Document]. URL https://www.britannica.com/place/Fertile-Crescent (accessed 11.3.19).

Gorokhovich, Y., Mays, L., Ullmann, L., 2011. A survey of ancient Minoan water technologies. *Water Sci. Technol. Water Supply* 11, 388–399.

Jacobsen, T., Adams, R.M., 1958. Salt and silt in ancient Mesopotamian agriculture. *Science* 128, 1251–1258.

Jansen, M., 1989. Water supply and sewage disposal at Mohenjo-Daro. *World Archaeol.* 21, 177–192.

Kang, S.T., 1972. *Sumerian and Akkadian Cuneiform Texts in the Collection of the World Heritage Museum of the University of Illinois: Sumerian Economic Texts from the Drehem Archive.* University of Illinois Press, Champaign, IL.

Lambert, W.G., 2007. *Babylonian Oracle Questions.* Eisenbrauns, Winona Lake, IN.

Marefa, 2018. Agriculture in ancient Egypt [WWW Document]. URL https://www.marefa.org/الزراعة_في_مصر_القديمة#/media/File:Ipuy_shaduf.jpg (accessed 11.27.18).

Martin, K., Sauerborn, J., 2013. Origin and development of agriculture, in: *Agroecology.* Springer, pp. 9–48. https://link.springer.com/chapter/10.1007/978-94-007-5917-6_2

Mays, L., 2010. *Ancient Water Technologies.* Springer Science & Business Media. https://link.springer.com/book/10.1007/978-90-481-8632-7

Mazoyer, M., Roudart, L., 2007. *A History of World Agriculture: From the Neolithic Age to the Current Crisis.* Routledge, Abingdon, UK.

Oppenheim, A.L., 2013. *Ancient Mesopotamia: Portrait of a Dead Civilization.* University of Chicago Press, Chicago, IL.

Possehl, G.L., 1997. The transformation of the Indus civilization. *J. World Prehistory* 11, 425–472.

Pourshariati, P., 2017. *Decline and Fall of the Sasanian Empire: The Sasanian-Parthian Confederacy and the Arab Conquest of Iran.* IB Tauris, London.

Said, R., 2013. *The River Nile: Geology, Hydrology and Utilization.* Elsevier. https://www.elsevier.com/books/the-river-nile/said/978-0-08-041886-5

Samuel, G., 2000. The Indus Valley civilization and early Tibet. *Senri Ethnol. Rep.* 15, 651–670.

Sojka, R.E., Bjorneberg, D.L., Entry, J.A., 2002. Irrigation: An historical perspective. *Encycl. Soil Sci.* 745–749. https://eprints.nwisrl.ars.usda.gov/id/eprint/815/1/1070.pdf

Strouhal, E., Forman, W., 1992. *Life in Ancient Egypt.* Cambridge University Press, Cambridge, UK.

Sydalg, 2015. The Hanging Gardens of Babylon [WWW Document]. Thinglink. URL https://www.thinglink.com/scene/657275601245175810 (accessed 2.9.20).

Taghavi-Jeloudar, M., Han, M., Davoudi, M., Kim, M., 2013. Review of ancient wisdom of Qanat, and suggestions for future water management. *Environ. Eng. Res.* 18, 57–63.

Takla, A., 2018. Basin irrigation system [WWW Document]. URL https://st-takla.org/pub_Bible-Interpretations/Holy-Bible-Tafsir-01-Old-Testament/Father-Tadros-Yacoub-Malaty/01-Sefr-El-Takween/Tafseer-Sefr-El-Takwin__01-Chapter-39.html (accessed 11.4.18).

Viollet, P.L., 2010. Water Management in the Early Bronze Age Civilization. in: *Proceedings of the La Ingenieria Y La Gestion Del Agua a Traves de Los Tiempos*, Alicante, Spain, 30.

Waller, P., Yitayew, M., 2015. *Irrigation and Drainage Engineering.* Springer. https://link.springer.com/book/
    10.1007/978-3-319-05699-9

Wikiwand, 2018. Hanging Gardens of Babylon [WWW Document]. URL http://www.wikiwand.com/ar/
    حدائق_بابل_المعلقة (accessed 11.9.18).

Zeder, M.A., 2008. Domestication and early agriculture in the Mediterranean Basin: Origins, diffusion,
    and impact. *Proc. Natl. Acad. Sci.* 105, 11597–11604.

Zvelebil, M., Pluciennik, M., 2011. Historical origins of agriculture. *Role Food, Agric. For. Fish. Hum. Nutr.*
    41–78. http://eolss.net/Sample-Chapters/C10/E5-01A-01-01.pdf

18

Deficit Irrigation and Partial Root-Zone Drying Irrigation System in Arid Area

Abdulrasoul M.
Al-Omran, Ibrahim
Louki, and Arafat
Alkhasha
King Saud University

18.1 Introduction

The ecosystem of arid and semi-arid regions is impoverished by scarcity of water resources and soils, in particular sandy soil. Sandy soils are particularly critical for water management due to their low water holding capacity, high infiltration rate, high evaporation, low fertility levels and deep percolation, all factors that lead to lower water use efficiency (WUE), thereby decreasing crop productivity (Al-Omran et al., 2005). The water shortage and increasing demand for water in agriculture and other sectors compel the need to adoption of irrigation strategies in these regions. This may allow saving irrigation water for agricultural sector (Al-Harbi et al., 2008; Al-Omran et al., 2010). An approach to attain the objective of saving water and increasing WUE is through using DI program and partial root-zone drying system (PRD) in

DOI: 10.1201/9781003353928-24

which crops are deliberately allowed some degree of DI through the whole growth stage or at certain stages of the growth (Topcu et al., 2007; Patanè and Cosentino, 2009; Kirda et al., 2004). DI has been extensively studied on several crops (Sepaskhah and Akbari, 2005; Kirda et al, 2004; Pereira et al., 2002) and was recommended for arid and semi-arid regions (Kirda et al., 2004). Zegbe-Domìnguez et al. (2003) studied the impact of DI on tomato and found that the dry mass yield did not decrease under DI compared with full irrigation (FI). Moreover, DI can save up to 50% of irrigation water and increased WUE by 200%, with satisfactory yield. The adoption of DI requires the knowledge of crop evapotranspiration (ETc), crop response to water deficit, critical stages of growth under water deficit and economic impacts of yield reduction (Pereira et al., 2002). Agele et al. (2011) concluded that seasonal ET_c values were greater during reproduction growth stage of the crop. Amer et al. (2009) concluded that cucumber yield significantly decreased in a linear relationship with increasing water deficit. However, no significant change was observed when water was applied above 100% ETc. Mao et al. (2003) studied the effect of DI on yield and water use of grown cucumber in China and reported that WUE decreased when increasing the irrigation water applied from stem fruiting to the end of the growth stages. However, WUE increased with the increase of irrigation water from cucumber fruit setting to first fruit repining. There is increasing pressure on the agricultural sector to create ways to improve WUE by taking full advantage of available water. Drip irrigation using a deficit and/or PRD irrigation strategy and the use of plant residues or natural and industrial amendments are ways to improve the chemical and physical properties of soils to increase WUE.

DI is a strategy that aimed to decrease water applied during insensitivity stage without a significant decrease in the production of crops (English and Raja, 1996; Pereira et al., 2002). The expectation is that any yield reduction will be insignificant compared to the benefits gained from the saving of water (Eck et al., 1987). The goal of DI is to increase crop WUE by reducing the amount of water applied during the act of watering or by reducing the number of irrigation events (Kirda, 2002). DI involves the use of appropriate irrigation schedules, which were mostly derived from field trials (Oweis and Hachum, 2004), which are necessary because some crops are sensitive to water deficit during growing season changes (Istanbulluoglu, 2009). El-Mageed and Semida (2015) concluded in their study of DI on squash that no significant differences in yield between 100% ETc and 85% ETc treatments were observed.

The works on yield response factor (K_y) to water for many crops have been documented in the literature (Kirda, 2000; Moutonnet, 2000) where crops having a value of K_y lower than 1 can tolerate the water deficit. On the contrary, crops showing a K_y greater than 1 show a yield decrease more than proportional to the applied ET decrease, which means that the crop might not tolerate any irrigation deficit. Ayas and Domirtas (2009) reported that K_y value for cucumber grown in Turkey ranged between 0.196 and 1.31 depending on the water stress and growth stage, while Amer et al. (2009) concluded that these values ranged between 0.71 and 0.85 in field experiment in Egypt. The value of K_y for green bean was 1.23, while the values for safflower and eggplant were 0.97 and 1.37, respectively (Lovelli et al., 2007). In agricultural policy, the most important goal for improving WUE is to produce more food and economical yield using less water (Boutraa, 2010). Geerts and Raes (2009) concluded that the main reasons to enhance WUE are: (i) to meet poverty and human food demands, (ii) to evaluate water resources for agriculture and environment and (iii) to reduce drought stress and water losses because of evaporation rates. Sepaskhah and Ahmadi (2010) proved that partial root-zone drying is good technique for DI in agronomic and horticultural farms. In the technique of partial root-zone drying, the irrigation water can be conserved and saved up to 50% without significant reductions in yield. Qin et al. (2018), Lepaja et al. (2018), Chandra et al. (2018) and Giuliani et al. (2017) summarized the advantages of PRD like: (i) WUE and nutrient use efficiency improved, (ii) irrigation water reduced and (iii) fruit quality improved. Thereby, the partial root-zone drying irrigation is favorable and recommendable in arid and semi-arid regions.

Dorji et al. (2005) reported on irrigated pepper that irrigation use efficiency was 12%, 20.1% and 17.1% for conventional irrigation, regulated DI and PRD, respectively. They reported that PRD saved irrigation water by 50%. Al-Omran et al. (2013) performed a DI study on cucumber crop at 40%, 60%, 80% and 100% crop evapotranspiration (ETc). They reported the highest water productivity (WP) of 61.9 kg/m3 at 40% ETc, and that DI strategy increased WP and soil salinity of the field. Al-Harbi et al. (2015) conducted an experimental study for evaluating the impact of DI at 50%, 75% and 100% *ETc* on tomato using saline and non-saline water (EC 3.6 and 0.9 dS/m) under greenhouse; they reported that using saline water resulted in about 22% and 24% reduction in yield for the first and the second seasons, respectively. They recommended to use 75% *ETc* for irrigating tomato crop as a good policy for (DI) under greenhouse. In a field experimental work using surface and subsurface drip irrigation conducted by Al-Omran et al. (2005), the results showed that increasing the amendment rate of clay deposits from 1% to 2% increased the yield from 24.9 to 26.9-ton/ha and increased WUE from 2.26 to 2.76 kg/m3. The subsurface drip irrigation improved WUE to 2.89 kg/m3 compared to 2.43 kg/m3 of surface drip irrigation. In another study, under greenhouse conditions using the FI, DI and partial root-zone drying (PRD) irrigation for tomato, Akhtar et al. (2014) reported that biochar applied to soil at rate 0 and 5% by weight resulted in an increasing of soil moisture content and yield production. The results showed a significant increase in WUE by 35% and 15% for PRD and DI, respectively, compared to FI, and the fruit yield of tomato increased by 20% and 13% for FI and PRD, respectively, with biochar addition to soil compared with untreated soil.

Aladenola and Madramootoo (2013) conducted a study on the effects of varied levels of water application on WUE on bell pepper crop; they reported that WUE was 39.2, 30.7, 6.31 and 1.45 kg/m^3 for 120%, 100%, 80% and 40% *ETc*, respectively. Total yield was 26,953, 26,880, 13,605 and 5,107 kg/ha for 120%, 100%, 80% and 40% *ETc*, respectively. The results showed that the volume of irrigation water and yield relationship was linear. In addition, there were no significant differences between 120% *ETc* and 100% *ETc*. Sezen et al. (2006) investigated the effects of three irrigation intervals based on cumulative pan evaporation (18–22, 38–42 and 58–62 mm) on WUE and yield of bell pepper. They reported that when plant-pan coefficient=0.5, WUE was 7.6, 6.1 and 5.7 kg/m3 for 18–22, 38–42 and 58–62 mm, respectively. The yield values were 28.1, 22.3 and 21.6 kg/ha for 18–22, 38–42 and 58–62 mm, respectively.

The DI strategy has received very little attention in agricultural sector in arid region; therefore, the objectives of this chapter are: (i) introducing deficit (DI) and partial root-zone drying irrigation (PRD) system to arid regions, and (ii) studying the effect of DI or/and PRD at different growth stages of cucumber on yield and WUE.

18.2 Crop Water Requirements

In order to apply DI, one should determine actual crop water requirements for each crop. This can be achieved by different methods such as empirical equations, lysimeters and field methods. One of most used equation is Penman–Monteith (PM; FAO 56). Using the PM equation (FAO 56), based on climate data on the farm in both greenhouse and open field (Figure 18.1), to estimate the water needs and then calculate the total irrigation water requirements based on the quality of irrigation water and soil salinity, taking into account the values of crop coefficient K_c for each month, irrigation efficiency. The combined FAO PM method was used to calculate ET_o or using pan evaporation methods through the following equation:

$$ET_o = \frac{0.408\,\Delta(R_n - G) + \gamma\left(\dfrac{900}{T+273}\right)U_2\left(e_s - e_a\right)}{\Delta + \gamma\left(1 + 0.34U_2\right)}$$

(18.1)

where:

ET_o=Reference evapotranspiration (mm/day)
R_n=Net radiation at the crop surface (MJ/m^2/day)
G=Soil heat flux density (MJ/m^2/day)
T=Mean daily air temperature at 2 m height (°C)
U_2=Wind speed at 2 m height (m/sec)
e_s=Saturation vapor pressure (kPa)
e_a=Actual vapor pressure (kPa)

$e_s - e_a$ = Saturation vapor pressure deficit (kPa)

Δ = Slope of saturation vapor pressure curve at temperature T (kPa/°C)

γ = Psychrometric constant (kPa/°C).

- Leaching requirements:

$$LR = \left(EC_{iw}\right)/\left(2\max EC_e\right)\times\left(1/LE\right), \quad (18.2)$$

where *LR* is the leaching requirements, *ECiw* is the salinity of irrigation water (dS/m), max *ECe* is the maximum tolerable salinity of soil for pepper crop (dS/m) (max *ECe*=8.6) and LE is the leaching efficiency (*LE*=90%) (Maas and Hoffman, 1977).

- Uniformity distribution:

$$UD = \left(Q\tfrac{1}{4}\right)/\left(Q_{\text{mean}}\right)\times100, \quad (18.3)$$

where *UD* is the uniformity distribution, $Q_{1\!/\!4}$ is the mean of the lowest quarter of the observed discharge values of emitter and Q_{mean} is the average discharge of all the emitters (Karmeli and Keller, 1975).

- Storage Efficiency (K_s)

    Storage efficiency was estimated as K_s=0.91 according to (Karmeli and Keller, 1975). Then, irrigation efficiency was calculated by the following equation:

$$Ef_{firri} = EU \times K_s. \quad (18.4)$$

18.3 Crop Evapotranspiration (ET_c)

18.3.1 Pan Evaporation

The calculation of evapotranspiration was based on pan evaporation (class A pan) method according to Allen et al. (1998) as follows:

$$ET_c = E_o \times K_p \times Kc , \quad (18.5)$$

where *ETc* is the maximum daily ET in (mm), *Eo* is the evaporation from class A pan in (mm), *Kp* is the pan coefficient and *Kc* is the crop coefficient of pepper. *Kc* of bell pepper crop was recorded as *Kc-ini*=0.6, *Kc-mid*=1.15 and *Kc-end*=0.9 from FAO standard tables (Allen et al., 1998).

18.3.2 Lysimeter Method

Twelve non-weighing reinforced concrete lysimeters were used to grow the crops at the farm in open field and greenhouse (Figure 18.2). A graded gravel layer of 0.10 m was spread at the bottom of each lysimeter to facilitate drainage. Experiments were conducted using alfalfa as a reference crop, and potato, tomato and cucumber as the main crops. The irrigation water was controlled and the amount of drainage water was recorded weekly. Each lysimeter was irrigated in such a way that about 10% of water contributed to drainage. Water balance equation using the difference of soil moisture content between two irrigations by measuring changes in moisture content after and before irrigation at the root zone using a device to measure moisture (Terra Sen Dacom) at depths of 10–120 cm all year, after verifying the accuracy of moisture-sensitive, calibrated sensors with direct method (gravimetric laboratory method) with data from the sensors. The total amount of irrigation during one season is calculated by the following equation:

$$ET = P + I - Dr \pm \Delta S, \quad (18.6)$$

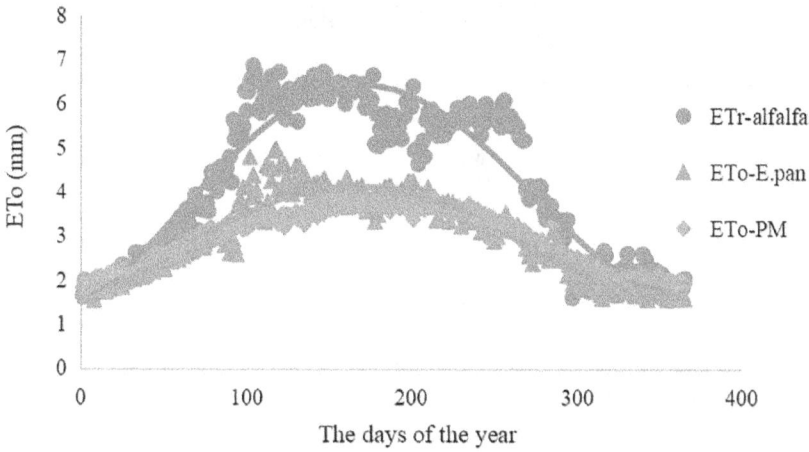

FIGURE 18.1 The daily average of ET_o according to the reference alfalfa, E-pan and FAO/Penman–Monteith in a greenhouse.

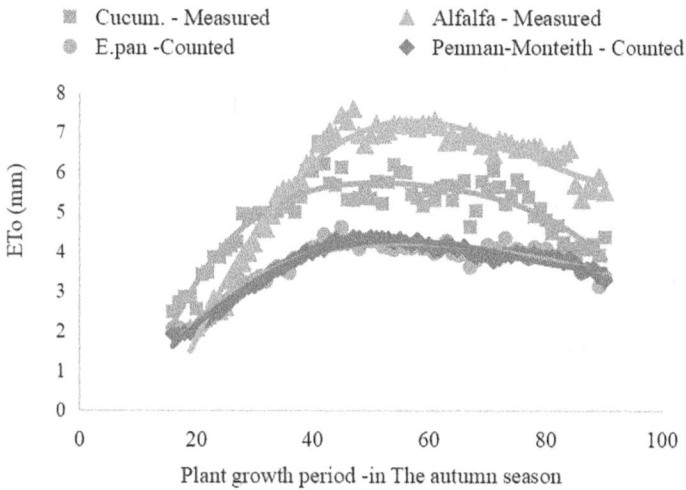

FIGURE 18.2 Comparison between ET_r for reference alfalfa measured by the lysimeter, calculated ET_o by pan evaporation method and calculated ET_o by FAO/Penman–Monteith method in a greenhouse vs. measured ETc by the lysimeter for the indoor cucumber in the autumn season.

where:

ET = Consumptive use (mm)
P = Precipitation (mm)
I = Irrigation added (mm)
Dr = Drainage (mm)
ΔS = Change in soil water content (mm).

The daily crop water requirements (mm/day) were estimated by the following equation:

$$GWR = (ET_c) / ((1 - LR) \times (Eff_{irr})).$$

(18.7)

For date palm water requirements can be determined according to Al-Omran et al. (2019) by the following equation:

Calculate the gross water requirements (*GWR*):

$$GWR = \frac{ET_c \times S_e}{(1-LR) \times E_{ffir}},$$ (18.8a)

where:

GWR = Gross water requirement (m³/ha)
ET_c = Crop evapotranspiration (m³/ha)
E_{ffir} = Efficiency (%), 90%.
LR = Leaching requirements
S_e = The percentage of evapotranspiration area.

$$S_e = \frac{\text{Shaded area per tree}}{\text{Actual area}} \times 100 = \frac{\pi R^2}{10\,\text{m} \times 10\,\text{m}} \times 100,$$ (18.8b)

where:

S_e = The percentage of evapotranspiration area
R = radius of tree (m)
Shaded area = Area of the shade of one tree measured at noon.

18.3.3 Relation between Pan Evaporation and Penman–Monteith ET_c

In this experiment, the crop evapotranspiration (ET_c) in the greenhouse was calculated by two methods: (i) pan evaporation method and (ii) estimation based on PM equation (Allen et al., 1998). The result indicated that for both ET_c there was a very good agreement between pan evaporation method and PM equation. In most cases R^2 was over 80% (Figure 18.3). Therefore, the pan evaporation method can be used instead of relying on the method of PM which needs climate data that may not be available for farmers in the field (Tiwari, 2000; Chartzoulakis and Drosos, 1997; Baille, 1994; Harmanto et al., 2004).

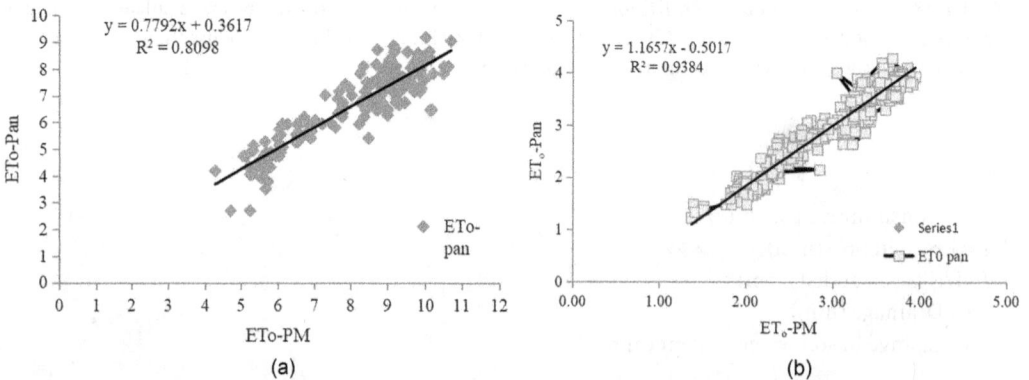

FIGURE 18.3 Correlations between *ETc* calculated using Penman–Monteith equation and pan: (a) open field and (b) greenhouse.

18.4 Date Palm Water Requirement in the Experimental Sites

Using the PM equation (56) based on climate data, the results of the study in Table 18.1 showed that the irrigation water requirements (m³/ha) after taking into account the proportion of cultivated area for each tree of the sites in Medina, Tabuk, Makkah, Al Jouf, Riyadh, Qassim, Hail and East Region, were 9,495.24, 7,340.18, 7,298.93, 8,913.59, 8,614.96, 8,568.68, 7,996.99 and 8,510.72 m³/ha, respectively, with a 100 Palm tree/ha. The total annual irrigation water requirements (m³/tree) in these sites were: 95, 73.4, 73, 89, 86, 85.7, 80 and 85 m³, respectively, as the radius of shaded area per tree is 3.5 m with effective diameter of 90%, and the rate of leaching was 12%, 8%, 13%, 12%, 14%, 11%, 13%, and 13%, respectively. Irrigation efficiency was 90%, and the average overall irrigation water requirement in all sites was 8,342.41 m³/ha/year with 100 trees/ha. These values of ET_c and crop water requirements (CWR) are attributed to the metrological conditions of each site. However, reduction in the estimated CWR to an average of 8,342 m³/ha compared to the overall average of 20,000 m³/ha as reported by many researchers (Al-Amoud et al., 2012; Ismail et al., 2014; Mihoub et al., 2015; Dewidar et al., 2015) is mainly attributed to the percentage of vegetative cover or shaded area (S_e) of the tree, as we calculated the S_e values as 0.33 of the actual area of the tree. Therefore, the practice distance of 10 m×10 m between trees in the farms of Saudi Arabia is not adequate in all sites. This area of 100 m² for each tree overestimates the crop water requirements and therefore, it must be changed to 7 m×7 m in order to have a higher vegetative cover in date palm farms.

Table 18.1 shows that the amount of irrigation water actually added by a flow meter of all study sites of the Medina, Tabuk, Makkah, Al Jouf, Riyadh, Qassim, Hail and East Region was 11,305.0, 9,463.9, 9,692.0, 11,252.75, 1,007.4, 10,035.0, 10,272.5, and 10,082.8 m³/ha/year, respectively. While these amounts added by the farmers in adjacent farms were 13,717, 12,277, 12,220, 13,340, 12,050, 12,880, 12,620, and 12,610 m³/ha/year, respectively. The increases in the amount of irrigation in adjacent farms by the farmers are mainly due to poor knowledge on irrigation requirements. Before installing the system of monitoring irrigation water in the study sites, the farmers used to add three times this amount and this might reach to 35,000 m³/ha.

18.4.1 Water Use Efficiency

WUE and irrigation water productivity (IWP) were calculated by the following equations:

$$WUE = Yield(kg)/water\ consumption, \tag{18.9}$$

$$IWP = Yield(kg)/applied\ water, \tag{18.10}$$

TABLE 18.1 Compared the Amount Water Applied in the Different Methods Sites and the Increase Water Ratio (%) Compared to Penman–Monteith Method

| Sites | Water Requirements for Different Methods (m³/ha/year) | | | | Increase in Water Ratio (%) Compared to Penman–Monteith Method | |
| | Penman–Monteith Method | Water Balance Method | Applied Irrigation Water | | | |
			Field Study	Farmer Adjacent	Field Study	Farmer Adjacent
Medina	9,495.24	-	11,305.0	13,717.00	16.0	30.8
Tabuk	7,340.18	-	9,463.9	12,277.00	22.4	40.2
Makkah	7,298.93	-	9,692.0	12,220.00	24.7	40.3
Al Jouf	8,913.59	3,515.25	11,252.8	13,340.00	20.8	33.2
Riyadh	8,614.96	-	10,007.4	12,050.00	13.9	28.5
Qassim	8,568.68	3,604.31	10,035.0	12,880.00	14.6	33.5
Hail	7,996.99	-	10,272.5	12,620.00	21.2	36.6
East Region	8,510.72	-	10,082.8	12,610.00	15.6	32.5

TABLE 18.2   Water Use Efficiency (kg/m³), Yield (kg/ha) and Water Saving (%) in the Field Study Compared to Farmer Adjacent

Sites	Field Study			Farmer Adjacent			Water Saving, %	EC_e	Yield, %
	Water Applied, m³/ha/year	Yield, kg/ha	Water Use, kg/m³	Water Applied, m³/ha/year	Yield, kg/ha	Water Use, kg/m³			
Medina	11,305.00	7,482	0.66	13,717.00	7,374	0.54	17.58	1.000	100.00
Tabuk	9,463.90	6,240	0.66	12,277.00	6,170	0.50	22.91	0.935	100.00
Makkah	9,692.00	5,406	0.56	12,220.00	5,324	0.44	20.69	4.600	97.84
Al Jouf	11,252.75	6,215	0.55	13,340.00	6,150	0.46	15.65	4.840	96.98
Riyadh	10,007.40	7,620	0.76	12,050.00	7,520	0.62	16.95	2.050	100.00
Qassim	10,035.00	6,742	0.67	12,880.00	6,531	0.51	22.09	10.950	74.98
Hail	10,272.50	6,908	0.67	12,620.00	6,708	0.53	18.60	2.600	100.00
East Region	10,082.80	8,400	0.83	12,610.00	8,520	0.68	20.04	6.030	92.69

where yield is the crop production in kg and applied water (AW) in expressed in m3 (Wang et al., 2016; Yang et al., 2017).

18.4.2   Water Use Efficiency (kg/m³), Yield (kg/ha) and Water Saving

Table 18.2 shows that the productivity per hectare ranged between 5,406 kg/ha in Makkah and 8,400 kg/ha in Al Ahasa, and WUE of palm in Medina, Tabuk, Makkah, Al Jouf, Riyadh, Qassim, Hail, and East Region in study sites was 0.66, 0.66, 0.56, 0.55, 0.76, 0.67, 0.67, and 0.83 kg/m³, respectively, while in the neighboring fields, these values were 0.54, 0.50, 0.44, 0.46, 0.62, 0.51, 0.53, and 0.68 kg/m³, respectively. The water saving values were 17.58, 22.91, 20.69, 15.65, 16.95, 22.09, 18.60, and 20.04%, respectively.

Based on the equation by Mass and Hoffman (1977), Yield% = $100 - b$ ($EC_e - a$), on the reduction of yield using saline water in all sites of the study. For the date palm tree, the threshold salinity values (a) of date palm are 4.0 dS/m and b as 3.6%. The results in Table 18.2 show that date palm production was affected by salinity in Al-Qassim site with a reduction of 25% followed by East Region farm at 7.31%. The rest areas were not affected by salinity.

18.5   Effect of Soil Amendments and DI on Tomato Yield

Table 18.3 shows the yields for different irrigation levels of fresh and saline waters during the first season. The total yield under freshwater at 100% of ET_c was significantly higher with biochar compared to the polymer treatment, i.e., biochar and polymer yields were 188.33 and 124.33 ton/ha, respectively. However, the highest yield under 80% of ET_c was recorded for the polymer and control treatments, measured at 161.67 and 157.33 ton/ha, respectively. The highest yield of 220.33 ton/ha was observed under 60% of ET_c in the mixture treatment. The yield was significantly lower by 33.15% and 14.51% for polymer and mixture treatments, respectively, while the yield was higher by 1.25% for the biochar treatment, compared to the control under 100% of ET_c. However, at the irrigation level of 80% of ET_c, the percentages of yield decrease in the mixture and biochar treatments were 5.5% and 13.76%, respectively, compared with control. In contrast, at the irrigation level of 60% of ET_c, yield increased by 28.15% and 48.87% for polymer and mixture treatments, respectively, compared with control.

Under saline water for irrigation, the yield of tomato was significantly lower by 14.44% in the polymer treatment, in contrast to the biochar and mixture treatments, whose yields were higher by 34.72% and 9.44%, respectively, compared to the control at 100% of ET_c. However, at 80% of ET_c, the yields were lower by 10.87%, 11.47%, and 12.35% in the biochar, polymer and mixture treatments, respectively, compared

TABLE 18.3 Total Fruit Yield, Water Use Efficiency (WUE) and Irrigation Water Use Efficiency (IWUE) of Tomato as Affected by the Interaction between Water Quality and DI and Soil Amendments during 2017

	Irrigation Level	Treatment	kg/m²	ton/ha	ET_c (mm)	WA (mm/season)	WA (mm/day)	WUE (kg/m³)	IWUE (kg/m³)
Fresh water	100	C	18.6	186	820.49	1,034	7.83	22.67	17.99
		B	18.83	188.33	820.49	1,034	7.83	22.95	18.22
		P	12.43	124.33	820.49	1,034	7.83	15.15	12.03
		B:P	15.90	159	820.49	1,034	7.83	19.38	15.38
	80	C	15.73	157.33	656.39	827	6.27	23.97	19.02
		B	13.57	135.67	656.39	827	6.27	20.67	16.4
		P	16.17	161.67	656.39	827	6.27	24.63	19.55
		B:P	14.87	148.67	656.39	827	6.27	22.65	17.98
	60	C	14.80	148	492.29	620	4.7	30.06	23.86
		B	12.93	129.33	492.29	620	4.7	26.27	20.85
		P	18.97	189.67	492.29	620	4.7	38.53	30.58
		B:P	22.03	220.33	492.29	620	4.7	44.76	35.52
	Least significant difference (LSD) 0.05		0.379	3.79	-	-	-	0.215	0.215
Saline water	100	C	12.00	120	820.49	1,034	7.83	14.63	11.61
		B	16.17	161.67	820.49	1,034	7.83	19.7	15.64
		P	10.27	102.67	820.49	1,034	7.83	12.51	9.93
		B:P	13.13	131.33	820.49	1,034	7.83	16.01	12.7
	80	C	11.33	113.33	656.39	827	6.27	17.27	13.7
		B	10.10	101	656.39	827	6.27	15.39	12.21
		P	10.03	100.33	656.39	827	6.27	15.29	12.13
		B:P	9.93	99.33	656.39	827	6.27	15.13	12.01
	60	C	8.63	86.33	492.29	620	4.7	17.54	13.92
		B	5.60	56.00	492.29	620	4.7	11.38	9.03
		P	10.87	108.67	492.29	620	4.7	22.07	17.52
		B:P	10.83	108.33	492.29	620	4.7	22.01	17.46
	LSD 0.05		0.234	2.34	-	-	-	0.474	0.474

to the control. At the irrigation level of 60% of ETc, the polymer and mixture treatments were higher by 25.87% and 25.48%, respectively, compared to the control. This could be due to the interaction between water and soil amendments. Moreover, irrigation using saline water led to increasing salt accumulation in the soil, which affected soil productivity adversely, as indicated by the lower yield of tomato plants with saline water, compared to freshwater irrigation. Saline water adversely influenced the yield of tomatoes compared to freshwater during the first season, under 100% of ETc, yields were increased by 55.00%, 16.49%, 21.10% and 21.07% for the control, biochar, polymer and mixture treatments, respectively. On the other hand, under 80% of ETc, the yields were increased by 38.82%, 34.33%, 61.14%, and 49.67%, respectively. Notably, at the highest level of DI of 60% of ETc, produced yields were higher by 71.44%, 103.94%, 74.54% and 103.39%, respectively, for the sequence of treatments.

The yield in the second season of 2018 is shown in Table 18.4, which lists yields for each soil amendment at each DI level of 100%, 80% and 60% of ETc using fresh and saline water. The addition of biochar to the soil tended to increase the yield. The highest yields were 339.10 and 326.69 ton/ha, obtained with biochar at 100% and 80% of ETc, respectively. The yields of tomato were higher by 23.08%, 13.25%, and 3.88%, for biochar, polymer and mixture treatments, respectively, compared to the control under FI of

TABLE 18.4   Total Fruit Yield, Water Use Efficiency (WUE) and Irrigation Water Use Efficiency (IWUE) of Tomato as Affected by the Interaction between Water Quality and DI and Soil Amendments during 2018

	Irrigation Level	Treatment	kg/m²	ton/ha	ET_c (mm)	WA (mm/ season)	WA (mm/ day)	WUE (kg/m³)	IWUE (kg/m³)
Fresh water	100	C	27.55	275.49	583	743	4.50	47.25	37.08
		B	33.91	339.10	583	743	4.50	58.16	45.64
		P	31.20	312.00	583	743	4.50	53.52	41.99
		B:P	28.62	286.19	583	743	4.50	49.09	38.52
	80	C	23.30	232.99	466	588	3.56	49.96	39.65
		B	32.67	326.69	466	588	3.56	70.06	55.60
		P	26.51	265.09	466	588	3.56	56.85	45.12
		B:P	26.51	265.10	466	588	3.56	56.85	45.12
	60	C	23.41	234.14	350	441	2.67	66.95	53.13
		B	23.73	237.31	350	441	2.67	67.85	53.85
		P	25.43	254.30	350	441	2.67	72.71	57.71
		B:P	28.51	285.10	350	441	2.67	81.52	64.70
	LSD 0.05		0.065	0.65	-	-	-	0.087	0.0588
Saline water	100	C	22.06	220.64	583	743	4.50	37.85	29.70
		B	22.50	225.00	583	743	4.50	38.59	30.28
		P	28.38	283.81	583	743	4.50	48.68	38.20
		B:P	20.67	206.74	583	743	4.50	35.46	27.83
	80	C	19.53	195.29	466	588	3.56	41.88	33.24
		B	19.70	197.00	466	588	3.56	42.24	33.53
		P	18.54	185.44	466	588	3.56	39.77	31.56
		B:P	22.23	222.30	466	588	3.56	47.67	37.83
	60	C	16.00	159.99	350	441	2.67	45.75	36.31
		B	18.73	187.32	350	441	2.67	53.56	42.51
		P	16.81	168.13	350	441	2.67	48.07	38.15
		B:P	17.42	174.20	350	441	2.67	49.81	39.53
	LSD 0.05		0.244	2.44	-	-	-	0.17	0.167

100% of ETc. On the other hand, at 80% of ETc, yield with biochar was higher by 40.21% compared to the control, while polymer and mixture treatments were higher by 13.77% and 13.78%, respectively, compared to the control. However, at the highest level of DI of 60% of ETc, the yields were higher by 1.35%, 8.61% and 21.76% for biochar, polymer and mixture treatments, respectively, compared to the control.

Therefore, in the second season, the yield data showed the same trend, i.e., yields were higher when using freshwater compared to saline. The yields using control, biochar, polymer and mixture treatments under all irrigation levels and soil amendments were higher by the following percentages: 24.86%, 50.71%, 9.93% and 38.48%, respectively, at 100% of ETc; 19.30%, 65.83%, 42.95%, and 19.25%, respectively, at 80% of ETc; and 46.35%, 26.69%, 51.25% and 63.66%, respectively, at 60% of ETc. These results are due to the high salt concentration that increased the osmotic potential in soil solution to the point that the plant has to use more energy to absorb water. Also, increasing salinity in irrigation water could lead to changes in the morphological and physiological properties of plants, such as reduction in plant leaf area, stomatal density, stomatal conductance, transpiration and net CO_2 assimilation.

Our results agree with previous findings (Romero-Aranda et al., 2001; Al-Harbi et al., 2017; Hajer et al., 2006). The saline conditions reduced the growth indices such as fresh and dry vegetative and root weights and weight at pre-harvest growth stages. Similar results were reported by Chen et al. (2009) and Wan et al. (2007), who found that the yield of oleic oil from sunflower decreases by 1.8% for every 1 dS/m increase in salinity level of the irrigation water; similarly, the yield of cucumber decreased by 5.7% per unit increase in EC_{iw}.

18.5.1 Effect of Water Quality

The interaction between water quality (fresh and saline) and DI had significant effects on total yield, WUE and irrigation water use efficiency (IWUE) during both growing seasons. Tables 18.3 and 18.4 show that the highest average yield was 171.83 ton/ha under freshwater irrigation at 60% of *ETc* during the first season, whereas the highest average yield with saline water irrigation was 128.91 ton/ha at 100% of *ETc*. The lowest yields under freshwater irrigation were 164.41 and 150.83 ton/ha, produced during the first season at 100% and 80% of *ETc*, respectively. The lowest yields under saline water irrigation were 10.35 and 8.98 ton/ha at 80% and 60% of *ETc*, respectively. Meanwhile, the values of WUE and IWUE at 60% of *ETc* tended to increase (compared to 100% *ETc*) when the amount of irrigation water decreased, with the highest values of 34.90 and 27.70 kg/m³, respectively, under freshwater irrigation, and 18.25 and 14.48 kg/m³, respectively, under saline water. The lowest values of WUE and IWUE were 20.04 and 15.90 kg/m³, respectively, under freshwater irrigation, and 15.71 and 12.47 kg/m³, respectively, under saline water irrigation. On the other hand, WUE and IWUE values at 80% of *ETc* were 22.98 and 18.23 kg/m³ under freshwater irrigation, respectively, and 15.77 and 12.51 kg/m³, respectively, under saline water irrigation.

The final average yields of tomato irrigation during the second season were affected by different levels of water irrigation, i.e., the amount of water applied was positively related to the final yield. The highest yield was 303.20 ton/ha, produced at 100% of *ETc*, while the lowest yield was 252.71 ton/ha, produced under freshwater irrigation at 60% of *ETc*. The same trend was observed with yields under saline water irrigation, which were 234.05, 200.01 and 172.41 ton/ha at 100%, 80% and 60% of *ETc*, respectively. Higher values of WUE and IWUE were obtained when the amount of irrigation water decreased. Under freshwater irrigation at 60%, 80% and 100% of *ETc*, WUE decreased in the following order: 72.26, 58.43, 52.01 kg/m³, respectively, and IWUE decreased in the following order: 57.35, 46.37, 40.81 kg/m³, respectively. Under saline water irrigation at 60%, 80% and 100% of *ETc*, WUE decreased in the following order: 49.29, 42.89 and 40.15 kg/m³, respectively, and IWUE decreased in the following order: 39.12, 34.04 and 31.50 kg/m³, respectively. These results agree with previously published reports (Al-Harbi et al., 2014; Al-Omran et al., 2013; Patanè and Cosentino 2010).

18.5.2 Water Use Efficiency

WUE can be improved by either increasing the yield or decreasing the amount of irrigation water applied, and growers usually aim to decrease the water use of crops while saving yield and quality (Kirnak et al., 2002). The tomato crop grown under DI levels of 80% and 60% of *ETc* showed higher WUE and IWUE than those grown under 100% of *ETc*. Figure 18.4 shows WUE and IWUE under DI levels of 100%, 80% and 60% of *ETc* and different soil amendments and types of irrigation water during the first season 2017. WUE values increased significantly by average values of 14.67% and 74.18% at 80% and 60% of *ETc*, respectively, as compared to 100% of *ETc*. The application of soil amendments (biochar, polymer and mixture) improved WUE. The highest WUE value of 22.95 kg/m³ was observed under biochar treatment at 100% of *ETc*, while the polymer treatment at an irrigation level of 80% of *ETc* resulted in a WUE value of 24.63 kg/m³, but the mixture treatment at 60% of *ETc* resulted in a WUE value of 44.70 kg/m³, compared with other soil amendments. This result may be related to the effect of the soil amendment to increase the specific surface area of soil and to keep water moving while increasing the water holding capacity. WUE tended to increase more under freshwater irrigation compared to saline water. The values of percentage increase in WUE values at 100% of *ETc* were 55.00%, 16.49%, 21.10% and 21.07%, respectively, while at 80% of *ETc*, the corresponding values were 38.82%, 34.32%, 61.13% and 49.66%, respectively, and at 60% of *ETc*, the corresponding values were 71.43%, 130.95%, 74.54% and 103.39%, for control, biochar, polymer and mixture treatments, respectively.

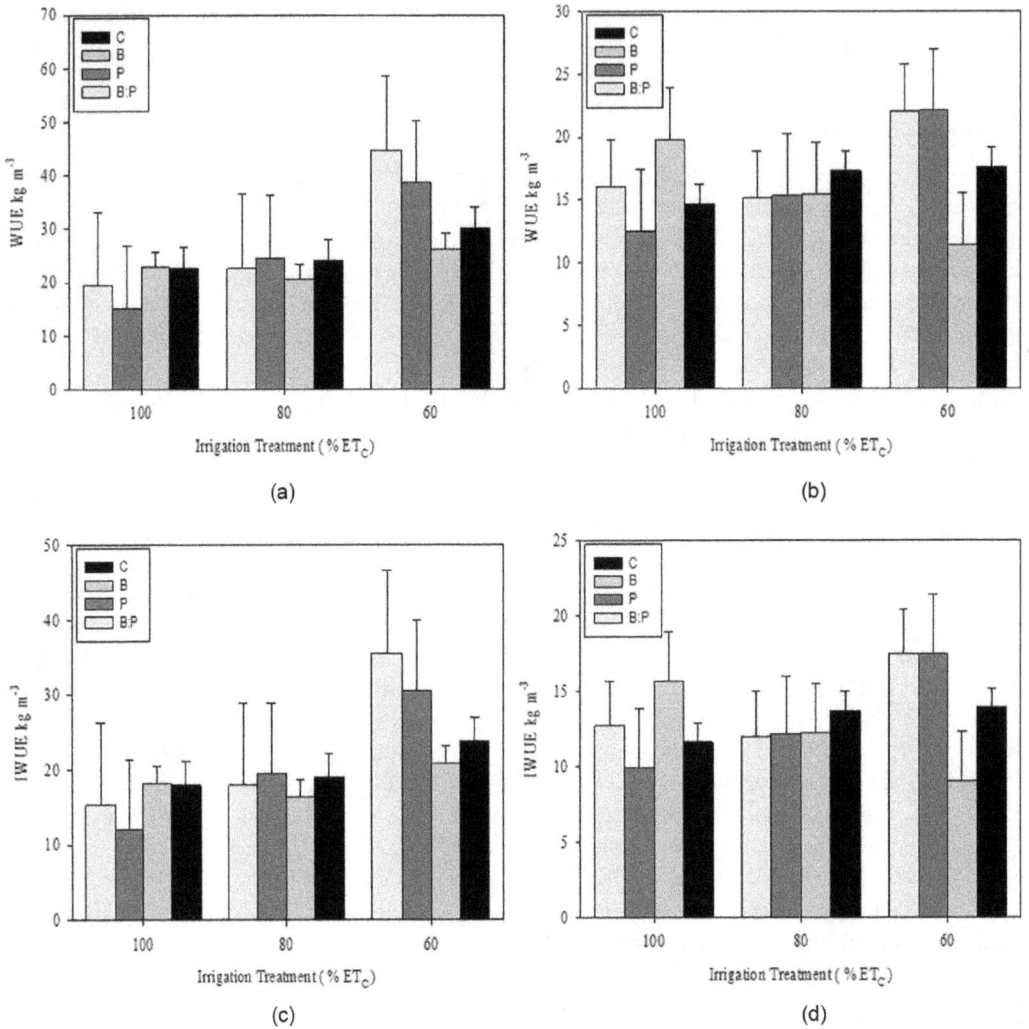

FIGURE 18.4   Effect of DI and soil amendments on water use efficiency (WUE) and irrigation water use efficiency (IWUE), (a and c, for freshwater, and b and d, for saline water).

Figure 18.5 illustrates the effect of DI and soil amendments on WUE and IWUE during the second season in 2018. Average WUE values increased by 12.99% and 39.73%, at 80% and 60% of ETc, respectively, as compared with 100% of ETc. On the other hand, WUE increased when the amount of irrigation water decreased. The highest WUE values of 72.71 and 81.52 kg/m^3 were recorded for polymer and mixture treatments, respectively, at 60% of ETc. The lowest WUE values of 47.25 and 49.09 kg/m^3 were recorded for the control and mixture treatments, respectively, at 100% of ETc. However, WUE was higher under freshwater irrigation compared with saline water, and the values of percentage increase were 24.86%, 50.71%, 9.93% and 38.43% for control, biochar, polymer and mixture treatments, respectively, at 100% of ETc; the corresponding values under 80% ETc were 19.30%, 65.83%, 42.95% and 19.25%, respectively, while under the high stress level of irrigation at 60% of ETc, the corresponding values were 46.35%, 26.69%, 51.25% and 63.66%, respectively. The results are essentially due to the accumulation of salt in the root zone, thereby increasing the osmotic pressure, which leads to reduced absorption of available water by plants. Improved WUE is attributed to the application of soil amendments that improve the soil microenvironment for crop growth, thereby significantly enhancing the crop yield and

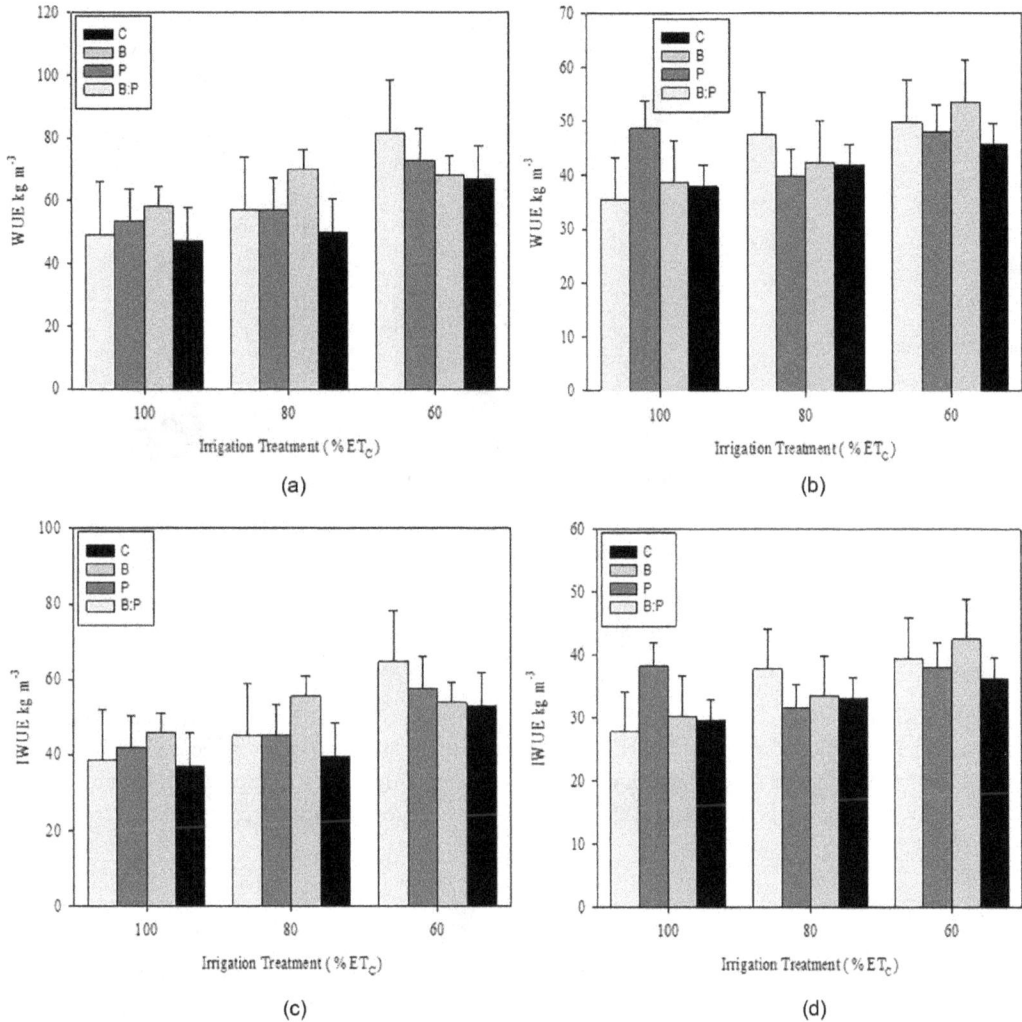

FIGURE 18.5 Effect of DI and soil amendments on water use efficiency (WUE) and irrigation water use efficiency (IWUE) (a and c, for non-saline water, and b and d, for saline water).

WUE. Generally, the application of biochar and polymer improved yield and WUE; these results are similar to previously reported findings (Uzoma et al., 2011; Usman et al., 2016; Al-Harbi et al., 2015; Qin et al., 2018; Agbna et al., 2017; Al-Omran et al., 2012; Yang et al., 2017; Islam et al., 2011; Wang et al., 2007).

18.6 Partial Root-Zone Drying System (PRD)

18.6.1 Water Requirement

Agro-meteorological conditions, air temperate, humidity and wind speed affect evaporation and AW during the four growth stages (initial, crop development, mid-season and late stage). The total evaporation amount was 87.1, 113.9, 486.2 and 176.9 mm for the first, second, third and fourth growth stage, respectively. Irrigation water requirement was calculated according to two irrigations levels 75% ETc and 50% ETc (PRD75%ET_c and PRD50%ET_c). AW for PRD75%ET_c irrigation level was 30.4, 55.9, 298.3 and 103.6 mm for four growth stages, while the AW for PRD50%ET_c level was 20.2, 37.3, 198.9 and 69.1 mm. Doorenbos and Kassam (1986) reported that total irrigation water needs are 600–900 mm per season

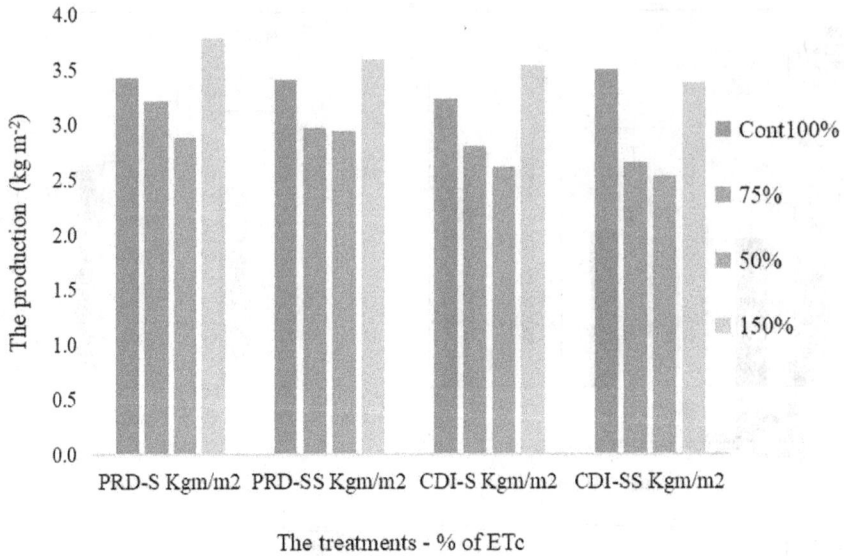

The treatments - % of ETc

FIGURE 18.6   Impact of the partial root-zone drying irrigation (PRD) system and DI using surface (S) and sub-surface (SS) irrigation method on potato productivity.

for pepper. Consequently, our findings explain that the irrigation water amounts were declined/saved by 35% and 57% for PRD75%ET_c and PRD50%ET_c, respectively. These findings agree well with the concept of PRD that irrigated the pepper plants by half of the amount of water applied (Spreer and Köller, 2005). Figure 18.6 shows the effect of PRD on crop response factor (K_y) and yield at different treatments of irrigation.

18.6.2 Yield of Sweet Pepper

Table 18.5 presents the findings of marketable yield, non-marketable yield and total yield (kg/m²) of pepper crop. Pepper yield was recorded lowest under un-amended treatments (control). The yield of control plots was 2.64 and 2.50 kg/m² for PRD75%ET_c and PRD50%ET_c, respectively. The marketable yield was 2.3, 3.2, 2.5, 3.8, 2.8, 4.3 and 3.4 kg/m² under PRD75%ET_c and 2.1, 2.7, 2.1, 2.9, 2.3, 3.4 and 2.8 kg/m2 under PRD75%ET_c for control, B4, B2, CO4, CO2, Mix4 and Mix2, respectively. The soil amended with mixture of biochar–compost produced the highest pepper yield compared with other treatments, significantly. The total pepper yield was increased by 34.9%, 9.9%, 56.8%, 17.6%, 70.3% and 37.6% under PRD75%ET_c, and by 23.3%, 3.6%, 29.4%, 80.0%, 47.7% and 28.1% for B4, B2, CO4, CO2, Mix4 and Mix2 under PRD50%ET_c compared with control, respectively. It was reported that amending sandy soils by additive materials and conditioners can result in significant changes in yield and WUE because of improving sandy soil attributes (Al-Omran et al., 2010).

18.6.3 Water Use Efficiency of Bell Pepper

Figure 18.7 illustrates the findings of WUE and IWP for all treatments under PRDET_c75% and PRDET_c50%. The addition rate of 4% of biochar, compost and their mixture improved the WUE and IWP for all treatments compared with both control and rate 2% under conditions of irrigation levels PRD75%ET_c and PRD50%ET_c. Even the lowest rate of addition increased the WUE and IWP for all treatments compared to un-amended treatment. Under PRD75%ET_c, WUE was 5.6, 7.5, 6.1,

TABLE 18.5 Combined Effects of Irrigation Level and Biochar, Compost and Biochar–Compost Mixture Amendment Materials on Pepper Crop Yield

Irrigation Level	Soil Amendment	Marketable Yield (kg/m²)	Non-Marketable Yield (kg/m²)	Total Yield (kg/m²)
PRD$_{75\%ETc}$	Control	2.335	0.304	2.638 f
	B4	3.204	0.358	3.562 c
	B2	2.54	0.361	2.901 e
	CO4	3.769	0.369	4.138 b
	CO2	2.769	0.335	3.103 d
	Mix4	4.261	0.233	4.494 a
	Mix2	3.384	0.246	3.630 c
	LSD(0.05)	0.107	0.003	0.14
PRD$_{50\%ETc}$	control	2.117	0.381	2.498 de
	B4	2.729	0.351	3.08 c
	B2	2.084	0.503	2.587 e
	CO4	2.861	0.371	3.231 b
	CO2	2.308	0.39	2.698 d
	Mix4	3.415	0.274	3.689 a
	Mix2	2.801	0.4	3.2 b
	LSD(0.05)	0.074	0.067	0.115

8.7, 6.6, 9.5 and 7.7 kg/m³ for control, B4, B2, CO4, CO2, Mix4 and Mix2, respectively. Moreover, IWP was increased by 23.3%, 3.6%, 29.4%, 8%, 47.7% and 28.1% for B4, B2, CO4, CO2, Mix4 and Mix2 compared with control (Figure 18.7). Previous studies reported that IWP was increased by 34.9%, 9.9%, 56.8%, 17.6%, 70.3% and 37.6% for B4, B2, CO4, CO2, Mix4 and Mix2 compared with control (Figure 18.7). The notable improvement of WUE was recorded under irrigation level of PRDET$_c$50% and IWP had the same trend of enhancement, significantly. IWP was 7.9, 9.8, 8.2, 10.2, pepper was increased with reducing the applied irrigation water, in particularly under PRD techniques (Aladenola and Madramootoo, 2013; Al-Harbi et al., 2014; Patil et al., 2014).

18.7 Crop Water Production Function

The relationship between crop yield and water application is called crop water production function (CWPF). The typical relationship between AW and yield for most crops is concave "curvilinear" as more of AW goes to drainage or is lost (Banihabib et al., 2016). CWPF can be divided approximately into four sections (Geerts and Raes, 2009; Foster and Brozovic, 2018). The CWPF clearly showed a linear relationship between AW and the yield, and as the level of water increases, the increase in the yield diminishes. Yield reaches a theoretical value; beyond it any further increase of AW will not increase the yield and in some cases the yield will decrease as AW increases beyond the maximum value. The shape of CWPF has very important implication for irrigation water management such as introducing the DI program (English, 1990; English et al., 2002). A useful way to express the water production function is on a relative basis, where actual yield (Y_a) is divided by maximum yield (Y_m) and actual evapotranspiration (ET_a) is divided by crop evapotranspiration (ETc). The relationship between evapotranspiration deficit ($1-(ET_a/ETc)$) and yield depression ($1-(Y_a/Y_m)$) is always linear (Doorenbos and Kassam, 1986), the slope called yield response factor of the crop (K_y). This relationship is expressed by the following equation:

$$\left(1-\left(Y_a / Y_m\right)\right) = K_y \left(1-\left(ET_a / ET_m\right)\right). \tag{18.11}$$

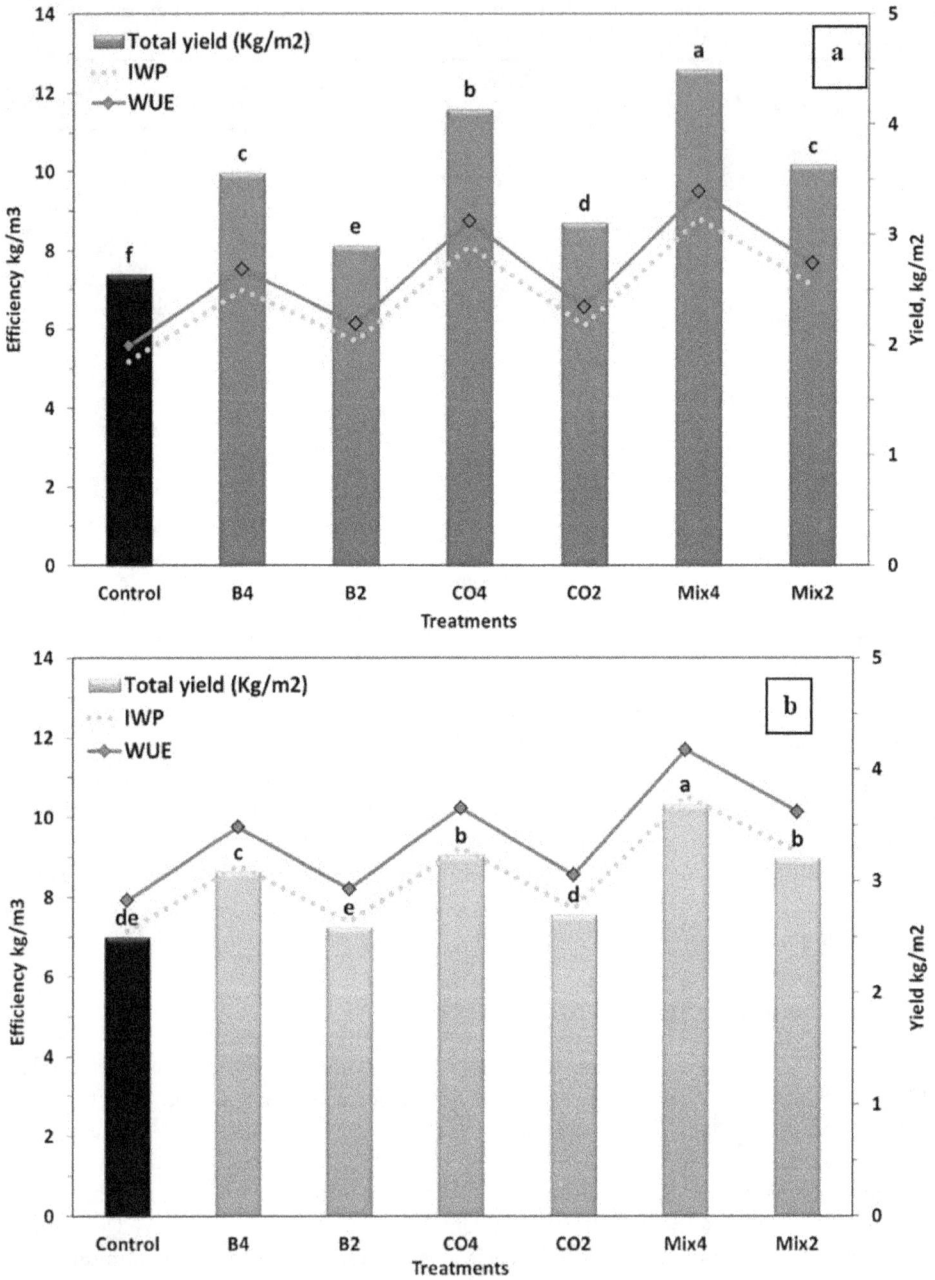

FIGURE 18.7 Impact of application rate of biochar, compost and biochar–compost mixture on IWP, WUE and pepper yield under irrigation level PRD75%ET_c (a) and PRD50%ET_c (b).

The CWPF reflects the benefit of AW in the production of dry matter or yield. It presents the relationship between the quantity of AW and the yield production. The quadratic polynomial function of Helweg (1991) was expressed as follows:

$$Y_a = b_0 + b_1 W + b_2 W^2, \tag{18.12}$$

where:

Y_a: Crop production or yield (ton/ha)
W: Applied irrigation water (m³/ha)
b_0, b_1 and b_2 are the fitting coefficients.

When yield approaches its maximum value, the slope of the WP function against water applied goes to zero; therefore, the maximum AW (W_{max}) was calculated by differentiating the WPF (Eq. 18.12) and equalized with zero, and then the maximum predicted yield (Y_{max}) can be calculated by substituting the W_{max} in Eq. (18.12).

$$\partial Y / \partial W = +b_1 + 2b_2 W = 0 \tag{18.13}$$

$$W_{max} = -b_1 / 2b_2 \tag{18.14}$$

$$Y_{max} = b_0 + b_1 W_{max} + b_2 W^2_{max} \tag{18.15}$$

18.8 Water Use Efficiency and WP

18.8.1 Open Field

The results of crop evapotranspiration (*ETc*) for each treatment and AW are presented in Table 18.6. The irrigation treatments were started by measuring evaporation from class A pan. The maximum amount of AW to the cucumber crop was 727 mm on treatment T_1_100, while the minimum AW was 267 mm for T_{12} treatment. The AW of traditional practice by the farmers in the region was 1,562 mm. The calculated *ETc* using pan evaporation method ranged between 223 and 617 mm for different treatments. WUE and crop water productivity (CWP) values increased when water amount decreased with exception of the traditional irrigation. The highest CWP value of 12.7 kg/m³ was recorded under the highest stress treatment (T_{12}). Moreover, decreasing irrigation water to level of 80% of *ETc* did not affect the growth and

TABLE 18.6 Yield, Evapotranspiration (*ET_c*), Applied Water (AW), Water Use Efficiency (WUE) and Water Productivity (WP) as Affected by DI Treatments at Different Growth Stages of Cucumber Planted in Open Field

Treatment	Average Days Per Season	Yield (kg/m²)	ET_c (mm)	AW (mm)	AW (mm/day)	WUE (kg/m³)	WP (kg/m³)
T_1–100	91	7.2	617	727	8.0	11.7	9.9
T_2–80-0	91	6.3	494	581	6.4	12.8	10.8
T_3–80-1	91	6.2	520	611	6.7	11.9	10.1
T_4–80-2	91	6.1	520	613	6.7	11.7	10.0
T_5–80-3	91	6.5	538	631	6.9	12.1	10.3
T_6–80-4	91	6.1	519	615	6.8	11.8	9.9
T_7–60-0	91	4.9	370	436	4.8	13.2	11.2
T_8–60-1	91	4.8	424	495	5.4	11.3	9.7
T_9–60-2	91	5.2	424	500	5.5	12.3	10.4
T_{10}–60-3	91	5.4	458	535	5.9	11.8	10.1
T_{11}–60-4	91	4.9	420	504	5.5	11.7	9.7
T_{12}–40-0	91	3.4	223	267	2.9	15.2	12.7
T_{13}–Traditional	91	5.8	617	1,562	17.2	9.4	3.7

yield (Table 18.6). An attempt was made to establish a relationship between water consumed and yield (Figure 18.8). According to the mathematical analysis of the CWPF, the predicted maximum yields were 7.58 and 8.96 kg/m^2 and the corresponding predicted AW quantities were 1,290 and 980 mm for summer and fall, respectively. Figure 18.9 shows comparison of the WP functions of the partial root-zone drying irrigation system (PRD) and the conventional irrigation.

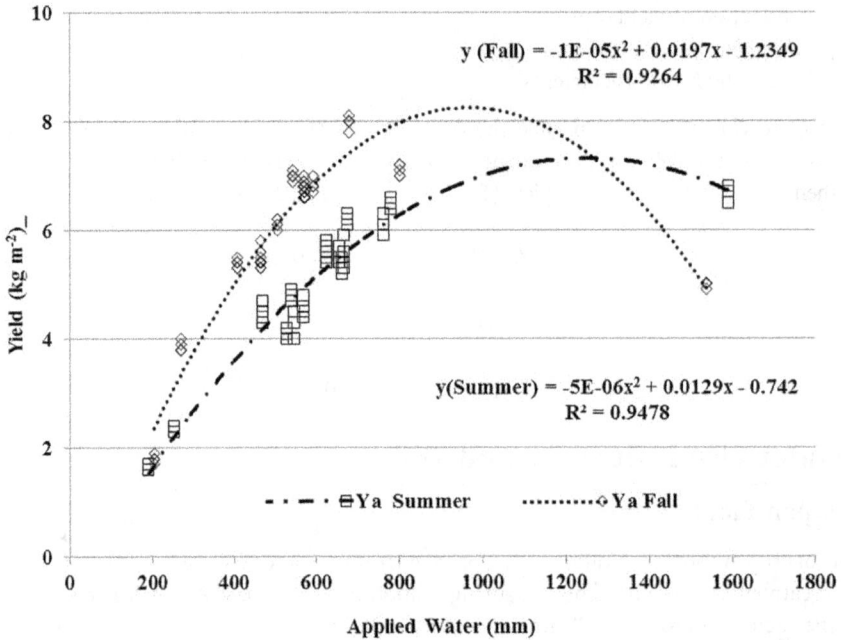

$$y \text{ (Fall)} = -1E\text{-}05x^2 + 0.0197x - 1.2349$$
$$R^2 = 0.9264$$

$$y \text{(Summer)} = -5E\text{-}06x^2 + 0.0129x - 0.742$$
$$R^2 = 0.9478$$

FIGURE 18.8　Yield as a function of AW for two seasons of cucumber (open field).

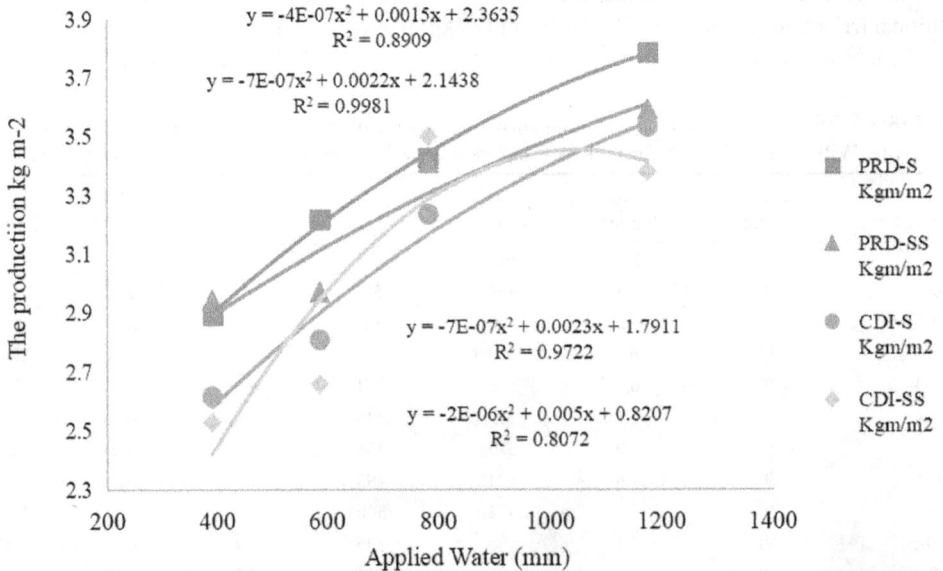

$$y = -4E\text{-}07x^2 + 0.0015x + 2.3635$$
$$R^2 = 0.8909$$

$$y = -7E\text{-}07x^2 + 0.0022x + 2.1438$$
$$R^2 = 0.9981$$

$$y = -7E\text{-}07x^2 + 0.0023x + 1.7911$$
$$R^2 = 0.9722$$

$$y = -2E\text{-}06x^2 + 0.005x + 0.8207$$
$$R^2 = 0.8072$$

FIGURE 18.9　Comparison of the WP functions of the partial root-zone drying irrigation system (PRD) and the conventional drip irrigation by surface (S) and subsurface (SS) irrigation method.

TABLE 18.7 Cucumber Water Production Function According to Applied Irrigation Water (Open Field)

Season	Crop Water Production Function	r^2	Maximum Yield (kg/m²)	Applied Water (mm)
Summer	Y (Summer) $= -5E-06x^2 + 0.0129x - 0.742$	0.9478	7.58	1,290
Fall	Y (Fall) $= -1E-05x^2 + 0.0197x - 1.2349$	0.9264	8.96	985

TABLE 18.8 Mean Yield, Evapotranspiration (ET_c), Applied Water (AW) and Water Productivity (WP) of Different Seasons as Affected by DI Treatments at Different Growth Stages of Cucumber Planted in Greenhouse

Treatments	Average Days Per Season	Yield (kg/m²)	ET_c (mm)	AW (mm)	AW (mm/day)	CWP (kg/m³)
T_1–100	108	15.0 a*	307	355	3.3	42.3
T_2–80-0	108	13.8 bc	245	283	2.6	48.8
T_3–80-1	108	13.2 d	256	295	2.7	44.7
T_4–80-2	108	14.2 b	259	299	2.8	47.5
T_5–80-3	108	14.6 ab	269	309	2.9	47.2
T_6–80-4	108	13.5 cd	260	300	2.8	45.0
T_7–60-0	108	11.4 f	184	213	2.0	53.5
T_8–60-1	108	11.7 f	204	236	2.2	49.6
T_9–60-2	108	12.4 e	210	243	2.3	51.0
T_{10}–60-3	108	12.7 e	232	267	2.5	47.6
T_{11}–60-4	108	11.5 f	213	246	2.3	46.7
T_{12}–40	108	9.1 g	123	147	1.4	61.9
T_{13}–Trad.	108	14.2 b	307	722	6.7	19.7

*Treatment means with the same letter are not significant using LSD test at 5% level.
The different letters (a–g) means significant differences.

The results in Table 18.7 were in agreement with those reported by Al-Harbi et al. (2008) and Zhang and Oweis (1999). However, Mao et al. (2003) reported a polynomial relationship between ET and yield. The study also concluded that the treatment T_1_100 had the highest yield but other treatment gave fairly good marketable yield, while economically saving water, fertilizers and pesticides. The result indicated that the WP increased with decreasing the amount of AW; the increased values were from 9.9 to 12.7 kg/m³ for T_1_100 and T_{12}_40, respectively. On the other hand, the WP of the traditional irrigation treatment recorded the lowest value (3.7 kg/m³). It was evident that over-irrigation as the traditional method used in the area lead to very high WP but yield quantity and quality decreased to be unacceptable. Similar results were reported by Ali et al. (2007), Oweis and Hachum (2004) and Zhang et al. (2004). There are many explanations reported for the reason of increasing WP with DI; some of them are that the DI can increase the ratio of yield over crop water consumption (evapotranspiration) by: (i) reducing the water loss by unproductive evaporation, (ii) increasing the proportion of marketable yield to the totally produced biomass (harvest index) and (iii) adequate fertilizer application and avoiding bad agronomic conditions during crop growth such as water logging in the root zone, pests and diseases (Geerts and Raes, 2009; Steduto and Albrizio, 2005; Pereira et al., 2002).

18.8.2 Greenhouse

The treatment T_4–80 was found to be the best treatment in terms of WP (Table 18.4); however, the traditional irrigation led to lower WP (19.7 kg/m³). Moreover, decreasing irrigation water to 40% ET caused very high WP, but decreased the final yield. Generally, the CWP values increased when water amount decreased; the maximum value of CWP was 61.9 kg/m³ for T_{12}–40 treatment, and 42.3 kg/m³ for FI treatment (T_1–100). Similar results were reported by Ali et al. (2007), Oweis and Hachum (2004) and Zhang et al. (2004). Many explanations for the reason of increasing CWP with DI are presented, some of them are that DI can increase the ratio of yield over crop water consumption (evapotranspiration) by: (i)

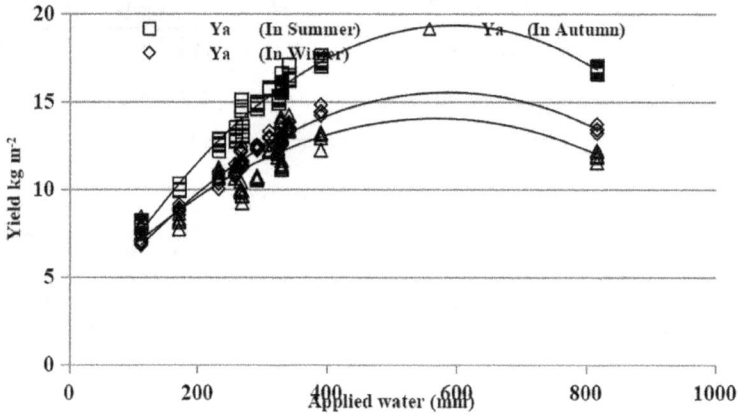

FIGURE 18.10 The relationship between marketable total cucumber yield planted on greenhouse and AW at different seasons.

TABLE 18.9 Cucumber Water Production Functions According to Applied Irrigation Water (Greenhouse)

Season	Crop Water Production Function	R^2	Maximum Yield (kg/m²)	Applied Water (mm)
Summer	$Y=-50E-06\ (AW)^2+0.0600$ $(AW)+1.491$	0.9660	19.49	600
Winter	$Y=-39E-06\ (AW)^2+0.0454$ $(AW)+2.1845$	0.9586	15.40	582
Autumn	$Y=-33E-06\ (AW)^2+0.0378$ $(AW)+3.2701$	0.7257	14.10	573

reducing the water loss by unproductive evaporation, (ii) increasing the proportion of marketable yield to the total produced biomass (harvest index) and (iii) adequate fertilizer application and avoiding bad agronomic conditions during crop growth such as water logging in the root zone, pests and diseases (Geerts and Raes, 2009; Steduto and Albrizio, 2005; Pereira et al., 2002) (Figure 18.10)

A polynomial function was fitted between yield (Y) and AW for different seasons (Figure 18.9). According to the mathematical analysis of the CWPF, the predicted maximum yields were 19.49, 15.40 and 14.10 kg/m² and the corresponding calculated AW amounts were 600, 582 and 573 mm for summer, winter and autumn, respectively (Table 18.9). These results were in agreement with those reported by Al-Harbi et al. (2008) and Zhang and Oweis (1999). However, Mao et al. (2003) reported a polynomial relationship between ET and yield. In this study, treatment T_1–100 had the highest yield, and treatments $T_{3,4,5,6}$–80 and T_{12}–40 gave fairly good marketable yield while economically saving water, fertilizers and pesticides. The result also indicated that the CWP increased with decreasing amount of AW; the CWP values were 42.3 and 61.9 kg/m³ for T_1–100 and T_{12}–40, respectively. However, the traditional irrigation treatment has the lowest value of WP (19.7 kg/m³). Although less irrigation, as in treatment T_{12}–40, led to very high WP, it also led to poor quantity and quality of yield. The results also indicated that the DI at 80% of *ETc* was more efficient in saving irrigation water with a good marketable yield compared to traditional irrigation and 100% of *ETc*. Moreover, the deficit drip irrigation helps in rationalization and preventing excessive use of pesticides and fertilizers, consequently reducing and environmental pollution.

18.9 Crop Yield Response Factor

Crop yield response factor (K_y) was determined for different treatments of DI. K_y indicates a linear relationship between the relative reduction in water consumed and relative reduction in yield (Lovelli et al.,

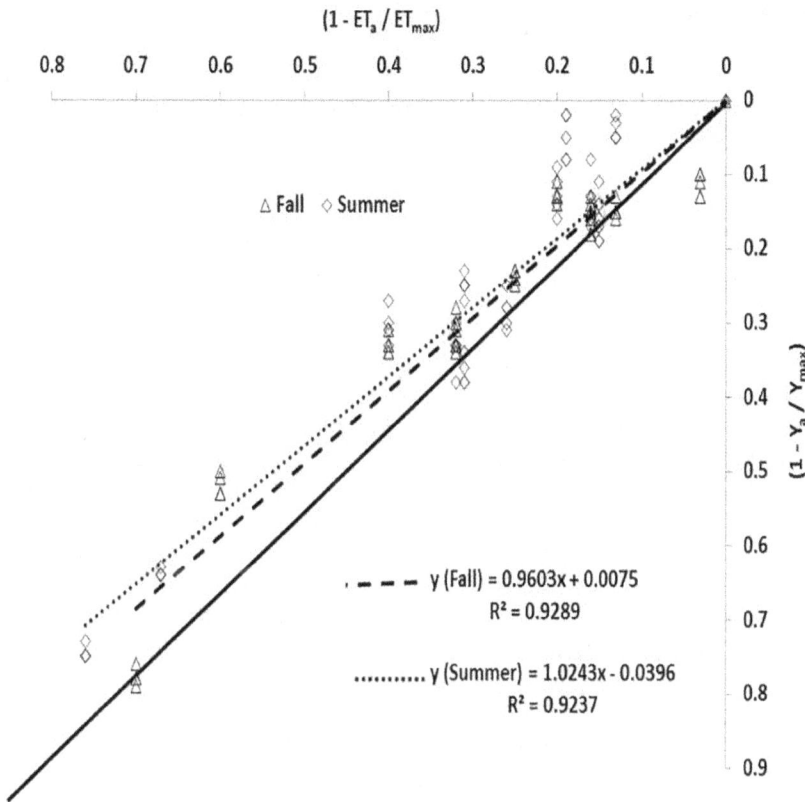

FIGURE 18.11 Relative cucumber yield decrease as a function of relative evapotranspiration decrease.

2007 and Kidra et al., 2004). Seasonal crop response factor K_y for different treatments through the open-field growth ranged between 0.96 and 1.02 for fall and summer, respectively (Figure 18.10). This means that cucumber (*Cucumis sativus* L.) grown in open field under Saudi Arabia arid conditions, cannot tolerate high severe water stress. These results were similar to those reported by Ayas and Demirta (2009) which recorded a K_y value of 1.2 for cucumber (*C. sativus* L. Maraton) grown in Turkey. On the other hand, the K_y in greenhouse ranged between 0.57 and 0.76 for fall and winter, respectively (Figure 18.10). This means that cucumber, grown in greenhouse under Saudi Arabia arid conditions, can be considered as a water-stress-tolerant crop. These results were similar to those reported by Amer et al. (2009) (Figure 18.11).

18.10 Conclusions

The management of water under water scarcity includes multiple policies. In general, policies should aim to reduce the non-beneficial water uses, particularly those related to water consumption and to the non-reusable fraction of the diverted water. However, fully exploring these concepts, mainly for farmers at field scales, requires appropriate procedures to be developed. Reduced water demand can be achieved by adopting improved farm, irrigation systems and DI. In this study DI was tested on cucumber (*C. sativus* L.) in the greenhouse and open field. It was found that FI at the early and late stage and then irrigation with 80% of *ETc* was the best treatment in terms of WP and final yield; however, decreasing irrigation water to 40% *ETc* caused very high WP but decreased the final yield. Generally, under Saudi Arabian conditions, WUE and WP values increased when the amount of AW decreased. A polynomial relationship was determined between yield (Y) and AW. However, crop yield response factor (K_y)

indicated a linear relationship between the relative reduction in water consumed and relative reduction in yield with an average of 0.65 and 0.99 for greenhouse and open field, respectively; this means that cucumber can be considered as a water-stress-tolerant crop if planted under greenhouse; on the other hand, the cucumber will not tolerate high water stress at open-field arid conditions.

References

Agbna, G.H., Dongli, S., Zhipeng, L., Elshaikh, N.A., Guangcheng, S., Timm, L.C. (2017). Effects of deficit irrigation and biochar addition on the growth, yield, and quality of tomato. *Sci. Hortic.* 222: 90–101.

Agele, S.O., Iremiren, G.O., Ojeniyi, S.O. (2011). Evapotranspiration, water use efficiency and yield of rain-fed and irrigated tomato. *Int. J. Agric. Biol.* 13: 469–476.

Akhtar, S.S., Li, G., Andersen, M.N., Liu, F. (2014). Biochar enhances yield and quality of tomato under reduced irrigation. *Agric. Water Manag.* 138: 37–44.

Aladenola, O., Madramootoo, C. (2013). Response of greenhouse-grown bell pepper (*Capsicum annuum* L.) to variable irrigation. *Can. J. Plant Sci.* 94(2): 303–310. doi: 10.4141/cjps2013-048

Al-Amoud, A.I., Mohammed, F.S., Saad, A.A., Alabdulkader, A.M. (2012). Reference evapotranspiration and date palm water use in the Kingdom of Saudi Arabia. *Int. J. Agric. Sci. Res.* 2(4):155–169.

Al-Harbi, A., Hejazi, A., Al-Omran, A. (2017). Responses of grafted tomato (*Solanum lycopersiocon* L.) to abiotic stresses in Saudi Arabia. *Saudi J Biol Sci.* 24(6):1274–1280.

Al-Harbi, A.R., Al-Omran, A.M., Alenazi, M.M., Wahb-Allah, M.A. (2015). Salinity and deficit irrigation influence tomato growth, yield and water use efficiency at different developmental stages. *Int. J Agric. Biol.* 17(2): 241–250.

Al-Harbi, A.R., Al-Omran, A.M., El-Adgham, F.I. (2008). Effect of drip irrigation levels and emitters depth on (*Abelmoschus esculentus*) growth. *J. Appl. Sci.* 8: 2764–2769.

Al-Harbi, A.R., Saleh, A.M., Al-Omran, A.M., Wahb-Allah, M.A. (2014). Response of bell-pepper (*Capsicum annuum* L.) to salt stress and deficit irrigation strategy under greenhouse conditions. In *International Symposium on Growing Media and Soilless Cultivation*, pp. 443–450.

Ali, M.A., Hoque, M.R., Hassan, A.A., Khair, A. (2007). Effects of deficit irrigation on yield, water productivity, and economic. *Agric. Water Manag.* 92: 151–161. doi:10.1016/j.agwat.2007.05.010

Allen, R.G., Pereira, L.S., Racs, D., Smith, M. (1998). Crop Evapotranspiration: Guidelines for computing crop water requirements. FAO Irrigation and Drainage paper No.56. FAO, Rome, Italy.

Al-Omran, A., Louki, I., Aly, A., Nadeem, M. (2013). Impact of deficit irrigation on soil salinity and cucumber yield under greenhouse condition in an arid environment. *J. Agric. Sci. Technol.* 15(6): 1247–1259.

Al-Omran, A.M., Al-Harbi, A.A.R., Alwabel, M.A., Nadeem, M.E.A., Al-Eter, A. (2012). Management of irrigation water salinity in greenhouse tomato production under calcareous sandy soil and drip irrigation. *J Agric. Sci. Technol.* 14(4): 939–950.

Al-Omran, A.M., Al-Harbi, A.R., Wahb-Allah, M. A., Nadeem, M., Eleter, A. (2010). Impact of irrigation water quality, irrigation systems, irrigation rates and soil amendments on tomato production in sandy calcareous soil. *Turk J. Agric. For.* 34: 59–73.

Al-Omran, A., Eid, S. and Alshemmary, F. (2019). Crop water requirements of date palm based on actual applied water and Penman–Monteith calculations in Saudi Arabia. *Appl. Wat. Sci.* 9, 69–74.

Al-Omran, A.M., Sheta, A.S., Falatah, A.M., Al-Harbi, A.R. (2005). Effect of drip irrigation on squash (*Cucurbita pepo*) yield and water use efficiency in sandy calcareous soils amended with clay deposits. *Agric. Water Manag.* 73: 43–55.

Amer, K.H., Sally, A.M., Jerry, L.H. (2009). Effect of deficit irrigation and fertilization on cucumber. *Agron. J.* 101: 1556–1564.

Ayas, S., Demirta, C. (2009). Deficit irrigation effects on cucumber (*Cucumis sativus* L. Maraton) yield in unheated greenhouse condition. *Int. J. Food Agric Enviro.* 7: 645–649.

Baille, A. (1994). Principles and methods for predicting crop water requirement in green house environments. *Cah. Options Mediterr.* 31, 177–186.

Banihabib, M.E., Zahraei, A., Eslamian, S. (2016). Dynamic programming model for the system of a non-uniform deficit irrigation and a reservoir. *Irrig. Drain. Syst.* 66(1): 71–81.

Boutraa, T. (2010). Improvement of water use efficiency in irrigated agriculture: a review. *J. Agron.* 9(1): 1–8.

Chandra, R., Jain, S.K., Kumar, M., Singh, A.K., Kumar, V. (2018). Comparative effects of deficit irrigation and partial root-zone drying (PRD) on growth, yield and water use efficiency of Rabi Maize. *Int. J. Curr. Microbiol. Appl. Sci.* 7(2): 1073–1080.

Chartzoulakis, K., Drosos, N. (1997). Water requirements of greenhouse grown pepper under drip irrigation. In II International Symposium on Irrigation of Horticultural Crops, 449: 175–180.

Chen, M, Kang, Y, Wan, S, Liu, S.P. (2009). Drip irrigation with saline water for oleic sunflower (*Helianthus annuus* L.). *Agric. Water Manage.* 96(12): 1766–1772.

Dewidar A.Z., Ben Abdallah A., Al-Fuhaid Y., Essafi B. (2015) Lysimeter based water requirements and crop coefficient of surface drip-irrigated date palm in Saudi Arabia. *Int. J. Agric. Sci. Res.* 5(7): 173–183.

Doorenbos, J., Kassam A.H., (1986). Yield response to water. FAO Irrigation and Drainage Paper No. 33, FAO, Rome, Italy.

Dorji, K., Behboudian, M.H., Zegbe-Domínguez, J.A. (2005). Water relations, growth, yield, and fruit quality of hot pepper under deficit irrigation and partial root-zone drying. *Sci. Hortic.* 104(2): 137–149. doi: 10.1016/j.scienta.2004.08.015

Eck, H.V., Mathers, A.C., Musick, J.T. (1987). Plant water stress at various growth stages and growth and yield of soybeans. *Field Crops Res.* 17(1): 1–16.

El-Mageed, T.A.A., Semida, W.M. (2015). Effect of deficit irrigation and growing seasons on plant water status, fruit yield and water use efficiency of squash under saline soil. *Sci. Hortic.* 186: 89–100.

English, M. (1990). Deficit irrigation. 1: Analytical framework. *J. Irrig. Drain. Eng.* 116, 399–412.

English, M., Raja, S.N. (1996). Perspectives on deficit irrigation. *Agric. Water Manage.* 32(1):1–14.

English, M.J., Solomon, K.H., Hoffman, G.J. (2002). A paradigm shift in irrigation management. *J. Irrig. Drain. Eng.* 128: 267–277.

Foster, T., Brozovic, N. (2018). Simulating crop water production functions using crop growth models to support water policy assessments. *Ecol. Econ.* 152:9–21.

Geerts, S., Raes, D. (2009). Deficit irrigation as an on-farm strategy to maximize crop water productivity in dry areas. *Agric. Water Manage.* 96, 1275–1284.

Giuliani, M.M., Nardella, E., Gagliardi, A., Gatta, G. (2017). Deficit irrigation and partial root-zone drying techniques in processing tomato cultivated under Mediterranean climate conditions. *Sustainablity*, 2197. doi: 10.3390/su9122197

Hajer, A.S., Malibari, A.A., Al-Zahrani, H.S., Almaghrabi, O.A. (2006). Responses of three tomato cultivars to sea water salinity 1. Effect of salinity on the seedling growth. *Afr. J. Biotechnol.* 5(10):855–861.

Harmanto, V., Babel, M.S., Tantau, H.J. (2004). Water requirement of drip irrigated tomatoes grown in greenhouse in tropical environment. *Agric. Water Manage.* 71 (3), 225–242.

Helweg, O.J. (1991). Functions of crop yield from applied water. *Agr. J.* 83, 769–773.

Islam, M.R., Hu, Y., Mao, S., Mao, J., Eneji, A.E., Xue, X. (2011). Effectiveness of a water-saving super-absorbent polymer in soil water conservation for corn (*Zea mays* L.) based on eco-physiological parameters. *J. Sci. Food Agric.* 91(11):1998–2005.

Ismail, S.M., Al-Qurashi, A.D., Awad, A.A. (2014). Optimization of irrigation water use, yield, and quality of Nabbut–Saif date palm under dry land conditions. *Irrig. Drain.* 63, 29–37.

Istanbulluoglu, A. 2009. Effects of irrigation regimes on yield and water productivity of safflower (*Carthamus tinctorius* L.) under Mediterranean climatic conditions. *Agric. Water Manage.* 96(12):1792–1798.

Karmeli, D., Keller, J. (1975). *Trickle Irrigation Design*, Rain Bird Sprinkler Manufacturing Corporation, Glendora, CA.

Kirda, C. (2000). Deficit irrigation scheduling based on plant growth stages showing water stress tolerance. In *Deficit irrigation practices*, C. Kirda, P. Moutonnet, C. Hera and D.R. Nielsen (eds). Water Report #22 FAO, Rome.

Kirda, C. (2002). Deficit irrigation scheduling based on plant growth stages showing water stress tolerance. Food and Agricultural Organization of the United Nations, Deficit Irrigation Practices, Water Reports 22, 102.

Kirda, C., Çetin, M., Dasgan, Y., Topçu, S., Kaman, H., Ekici, B., Derici, M.R., Ozguven, A.I. (2004). Yield response of greenhouse grown tomato to partial root drying and conventional deficit irrigation. *Agric. Water Manage.* 69(3), 191–201.

Kirnak, H, Tas, I, Kaya, C, Higgs, D. (2002). Effects of deficit irrigation on growth, yield and fruit quality of eggplant under semi-arid conditions. *Aust. J. Agric. Res.* 53(12):1367–1373.

Lepaja, K., Kullaj, E., Lepaja, L. (2018). Influence of irrigation management as partial root-zone drying on raspberry canes. *Bulg. J. Agric. Sci.* 24(4): 648–653.

Lovelli, S., Perniola, M, Ferrara, A., Tammaso, T.D. (2007). Yield response factor to water (Ky) and water use efficiency of *Carthamus tinctorius* L. and *Solaum melongeual* L. *Agric. Water Manage.* 92, 73–80.

Maas, E.V., Hoffman, G.J. (1977). Crop salt tolerance—current assessment. *J. Irrig. Drain. Div.* 103(2): 115–134.

Mao, X., Liu, M, Wang, X., Liu, C., Hou, Z., Shi, J. (2003). Effects of deficit irrigation on yield and water use of greenhouse grown cucumber in the North China Plain. *Agric. Water Manage.* 61(3), 219–228. doi:10.1016/S0378-3774(03)00022-2

Mihoub, A., Helimi, S., Mokhtari, S., Kharaz, E., Koull, N., Lakhdari, K., Benzaoui, T., Bougafla, A., Laouisset, M., Kherfi, Y., Halitim, A. (2015) Date palm (*Phoenix dactylifera* L.) irrigation water requirements as affected by salinity in Oued Righ Conditions, North Eastern Sahara, Algeria. *Asian J. Crop Sci.* 7(3): 174–185.

Moutonnet, P. (2000). Yield response to field crops to deficit irrigation in deficit irrigation practices. C. Kirda, P. Moutonnet, C. Hera and D.R. Nielsen (eds).Water Report #22 FAO, Rome.

Oweis, T., Hachum, A. (2004). Water harvesting and supplemental irrigation for improved water productivity of dry farming system in west Asia and North Africa. In *Proceeding of the 4th Int. Crop Sci. Congress, on the theme "Crop Science for Diversified planet"*. Brisbane, Australia.

Patanè, C., Cosentino, S.L. (2009). Effects of soil water deficit on yield and quality of processing tomato under a Mediterranean climate. *Agric. Water Manage.* 97 (1), 131–138.

Patanè, C., Cosentino, S.L. (2010). Effects of soil water deficit on yield and quality of processing tomato under a Mediterranean climate. *Agric. Water Manage.* 97(1):131–138.

Patil, V., Al-Gaadi, K., Wahb-Allah, M., Saleh, A., Marey, S., Samdani, M., Abbas, M. (2014). Use of saline water for greenhouse bell pepper (*Capsicum annuum*) production. *Am. J. Agric. Biol. Sci.* 9(2), 208.

Pereira, L.S., Oweis T., Zairi, A. (2002). Irrigation management under water scarcity. *Agric. Water Manage.* 57(3):175–206.

Qin, J., Ramirez, D.A., Xie, K., Li, W., Yactayo, W., Jim, L., Qurioz, R. (2018). Is partial rootzone drying more appropriate than drip irrigation to save water in China: A preliminary comparative analysis. *Potato Res.* 61:391–406.

Romero-Aranda, R, Soria, T, Cuartero, J. 2001. Tomato plant-water uptake and plant-water relationships under saline growth conditions. *Plant Sci.* 160(2):265–272.

Sepaskhah, A., Ahmadi, S. (2010). A review on partial root-zone drying irrigation. *Int. J. Plant Prod.* 4(4): 241–258.

Sepaskhah, A.R., Akbari, D. (2005). Deficit irrigation planning under variable seasonal rainfall. *Biosyst. Eng.* 92: 97–106.

Sezen, S.M., Yazar, A., Eker, S. (2006). Effect of drip irrigation regimes on yield and quality of field grown bell pepper. *Agric. Water Manage.* 81(1): 115–131.

Spreer, W., Köller, K. (2005). Partial rootzone drying-wer wird hier ausgetrickst? *Landtechnik* 60(1): 26–27.

Steduto, P., Albrizio, R. (2005). Resource use efficiency of field-grown sunflower, sorghum, wheat and chickpea. II. Water use efficiency and comparison with radiation use efficiency. *Agric. Forest Meteorol.* 130, 269–281.

Tiwari, K.N. (2000). *Annual Report*. Plasticulture Development Centre, Agricultural and Food Engineering Department, IIT, Kharagpur, India.

Topcu, S., Kirda, C., Dasgan, Y., Kaman, H., Cetin, M., Yazici, A., Bacon, M.A. (2007). Yield response and N-fertiliser recovery of tomato grown under deficit irrigation. *Eur. J. Agron.* 26: 64–70.

Usman, A.R.A., Al-Wabel, M.I., Al Hrbi, A., Mahmoud, W.A., El-Naggar, A.H., Ahmad, M., Abdulrasoul, A.O. (2016). Conocarpus biochar induces changes in soil nutrient availability and tomato growth under saline irrigation. *Pedosphere* 26(1): 27–38.

Uzoma, K.C., Inoue, M., Andry, H., Fujimaki, H., Zahoor, A., Nishihara, E. (2011). Effect of cow manure biochar on maize productivity under sandy soil condition. *Soil Use Manage.* 27(2): 205–212.

Wan, S., Kang, Y., Wang, D., Liu. S.P., Feng, L.P. (2007). Effect of drip irrigation with saline water on tomato (*Lycopersicon esculentum* Mill) yield and water use in semi-humid area. *Agric. Water Manage.* 90(-1–2): 63–74.

Wang, S., Kang, Y., Wang, D., Liu, S.P., Feng, L.P. (2007). Effect of drip irrigation with saline water on tomato (Lycopersicon esculentum Mill) yield and water use in semi-humid area. *Agric. Water Manage.* 90(1–2), 63–74.

Wang, D.Y., Tang, C.S., Cui, Y.J., Shi, B., Li, J. (2016). Effects of wetting–drying cycles on soil strength profile of a silty clay in micro-penetrometer tests.

Yang, H., Du, T., Qiu, R., Chen, J., Wang, F., Li, Y., Kang, S. (2017). Improved water use efficiency and fruit quality of greenhouse crops under regulated deficit irrigation in northwest China. *Agric. Water Manage.* 179: 193–204. doi: 10.1016/j.agwat.2016.05.029

Zegbe-Domìnguez, J.A., Behboudian, M.H., Lang, A., Clothier, B.E. 2003. Deficit irrigation and partial rootzone drying maintain fruit dry mass and enhance fruit quality in 'Petopride' processing tomato (*Lycopersicon sculentum* Mill.). *Sci. Hortic.* 98: 505–510.

Zhang, H., Oweis, T. (1999). Water-yield relations and optimal irrigation scheduling of wheat in the Mediterranean region. *Agric. Water Manage.* 38: 195–211.

Zhang, Y., Kendy, E., Qiang, Y., Changming, L., Yanjun, S., Hong Yong, S. (2004). Effect of soil water deficit on evapotranspiration, crop yield, and water use efficiency in the North China Plain. *Agric. Water Manage.* 64: 107–122.

19

Patricia
Amankwaa-Yeboah
*CSIR-Crops Research
Institute*

Stephen Yeboah,
*CSIR-Crops Research
Institute*

Abdul-Rauf
Malimanga
Alhassan
*University of Environment
and Sustainable
Development*

Gilbert Osei
*CSIR-Institute of
Industrial Research*

William Amponsah
*Kwame Nkrumah
University of Science
and Technology*

Bright Mayinl
Laboan
*CSIR-Soil Research
Institute*

Eric Samuel
Adu-Dankwa
*Ghana Irrigation
Development Authority*

Ian Charles Dodd
Lancaster University

Paradigms Shaping the Adoption of Irrigation Technologies in Ghana

19.1 Introduction

Agriculture is the backbone of Ghana's economy and employs a majority of the population. It continues to contribute to Ghana's economy, although the share of its contribution declined from 29.8% in 2010 to 18.3% in 2017 (MoFA, 2018), which may be attributed to the development of the oil sector and its contribution to the economy from 2011. Growth of the agriculture economy is key to overall the economic growth and development of the nation (MoFA, 2018). Yet agriculture remains largely subsistence, production has not kept pace with population growth as national and individual food self-sufficiency is challenged, and more household income is required to purchase food. The sector is mostly dominated by smallholder farmers with average farm size of about 1.2 ha and low use of the improved agricultural technologies such as irrigation.

Ghana is typical of many countries in sub-Saharan Africa where sustainable production is needed to eliminate food insufficiency and over-reliance on the food imports. This increase in production requires the sustainable and judicious use of water. On a global scale, the agricultural sector uses the vast majority of freshwater resources, and so is a logical target to manage water use and develop the techniques to produce food with less water. Irrigation has been identified as a key component to sustainably increase and stabilize the food production (Knox et al., 2010). Irrigation seeks to maximize the yield and quality while maintaining continuous year-round supply. Crop yields are consistently greater in irrigated farming than under rainfed agriculture (Kyei-Baffour & Ofori, 2007), by about 2.3 times averaged across different crops (Dowgert, 2010). However, irrigation should not be a stand-alone input, as the other agricultural inputs such as fertilizers, improved seeds and enhanced cropping systems affect its impact. In situations where food production is predominantly rainfed, adopting irrigation technologies offers the promise of greater food security and economic development by ensuring year-round agricultural production and price stability (Namara et al., 2014).

Agricultural productivity in Ghana is low and hunger and malnutrition persist, particularly in rural areas. Agriculture policy initiatives in Ghana such as the Food and Agriculture Sector Development Policy (MoFA, 2007) and the Medium-Term Agriculture Sector Investment Plan II (MoFA, 2009) overwhelmingly endorsed the need for efficient use of new production technologies such as irrigation to improve crop production. Accelerated growth and modernization of Ghana's agricultural sector require improving the performance of the irrigation sector, utilizing the country's significant irrigation potential and improving farmers' access to improved irrigation technologies. An improved irrigation sector will ensure food security and reduce the rural poverty by creating rural employment and allowing farmers to cultivate more than one crop in a year. It is thus crucial to understand current patterns in decision-making regarding irrigation technologies, the factors affecting individuals' use of these technologies and to what extent policy incentives, governance and management of irrigation infrastructure will alter behaviour to allow the production to become more efficient. This chapter explores the factors that influence the selection or adoption of irrigation technologies based on influences that are internal or external to the farm, and identifies some prospects for the promotion of irrigation technologies in Ghana.

19.2 Climatic and Agro-Ecological Settings of Ghana

Ghana is located in West Africa with Togo, Cote d'Ivoire, Burkina Faso and the Gulf of Guinea bordering to the East, West, North and South, respectively. The population is estimated to be 30,280,482 as of June 2019, and Ghana was classified as a middle-income country by the World Bank in 2010. It has 16 administrative regions and 260 districts with Accra as its national capital as well as the seat of government. The country is characterized by the different agro-ecological zones, comprising the rain forest (moist evergreen and wet evergreen), deciduous forest, transitional zone, Guinea savannah, Sudan savannah and coastal savannah (Figure 19.1a). The agro-ecological zones are defined by the rainfall, temperature, relative humidity and soil type that influence the types of crops that are cultivated in that area. Average annual temperatures in these ecological zones range between 24°C and 30°C (Asante & Amuakwa-Mensah, 2015), with temperatures as low as 18°C and as high as 40°C in the southern and northern parts of Ghana, respectively. Rainfall in Ghana generally decreases from south to north (Figure 19.1b). The southern half of Ghana (made up of the forests and transitional agro-ecological zones) is characterized by a bimodal rainfall pattern, whereas the northern half (made up of the savannah agro-ecological zones) is characterized by a unimodal rainfall pattern. This bimodal pattern comprises a major season from April to July, with a break in August and a minor season from September to November, followed by a dry season from November to February/March. The unimodal rainfall season in the north usually lasts from May to September.

19.3 Status of Irrigation Development in Ghana

Agriculture in Ghana is predominantly rainfed but is moving gradually toward an integrated irrigation system, which utilizes both the irrigated and non-irrigated production systems. The status of irrigation

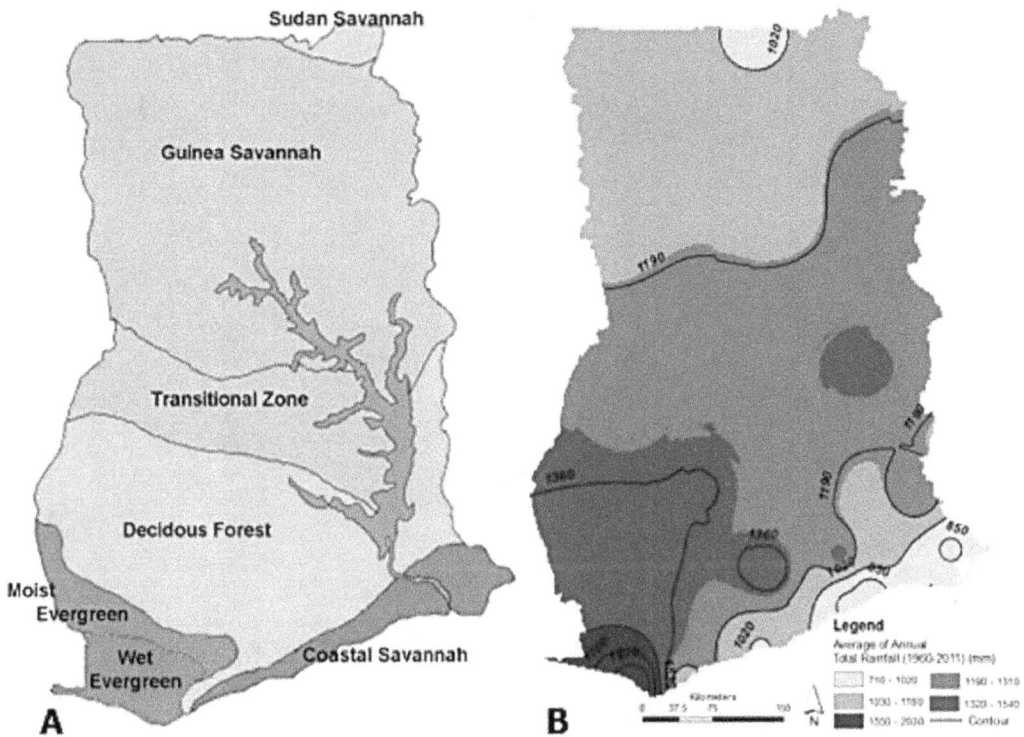

FIGURE 19.1 A map of Ghana showing (a) the agro-ecological zones and (b) corresponding spatial distribution of annual precipitation from 1960 to 2011 (Dankelman, 2008; (Owusu, 2016) (a), (Kabo-Bah et al., 2016) (b)).

in Ghana has been extensively discussed (Namara et al., 2010; 2011) with irrigated agriculture traced back to as early as 1880 in the Keta area, on land above flood level between the lagoon and the sandbar separating it from the sea (Namara et al., 2010). Agodzo and Bobobee (1994) also identified that some forms of shallow tube-well irrigation were in use in Southeastern Ghana in the 1930s. The development of formal irrigation schemes in Ghana dates back to 1960s. Construction for the Dawhenya Irrigation project was started in 1959, but available records indicate that the Asutsuare Irrigation project was completed first in 1967 (Kyei-Baffour & Ofori, 2007).

Irrigation in Ghana is often classified based on their levels of formalization (Namara et al., 2011) comprising formal, informal and large-scale commercial. Formal (public scheme) irrigation systems are mostly constructed by the government and/or in collaboration with the various donor agencies, with water typically drawn from rivers, lakes and groundwater. This type of irrigation usually comprises runoff river-diversion-based gravity-fed irrigation systems; river-pumping-based gravity-fed irrigation systems, run of the river/lake-pumping-based sprinkler irrigation systems and reservoir-based communal/sprinkler/gravity irrigation system.

As of 2019, the Government of Ghana had developed about 13,000 ha of irrigable lands from a potential of 1,900,000 ha, with 56 public irrigation schemes dotted across the country. These schemes range in size from 20 ha to over 3,000 ha (Figures 19.2 and 19.3). About 90% of the irrigation infrastructure developed under these schemes employs the surface irrigation method with some few schemes such as Weija and Akumadan irrigation schemes using the sprinkler system.

Informal (or smallholder) irrigation arises when small to medium landowners (less than 20 ha) adopt various irrigation technologies. This system mostly comprises traditional and community-initiated schemes. Although there is little data on the overall extent of informal irrigation in the country, it is estimated that this sector alone irrigates some 189,000 ha across the country, with Kumasi (the second

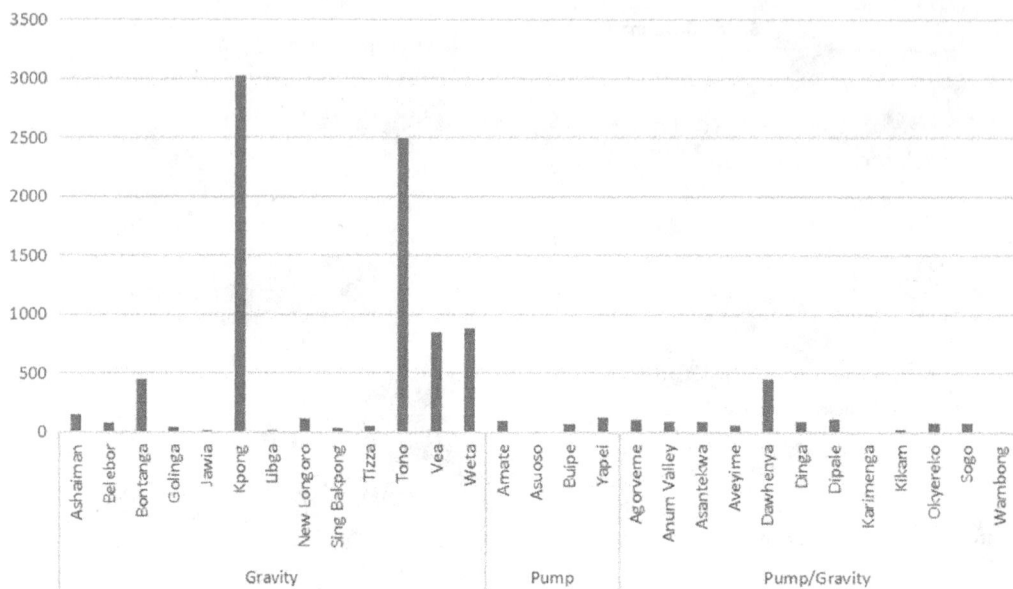

FIGURE 19.2 Operation type and irrigable capacity of some formal irrigation schemes.

largest city in Ghana) having about 12,700 smallholders irrigating more than 11,900 ha in the dry season (MoFA-GIDA, n.d.). Smallholder irrigation systems are expanding at a rapid rate, mainly fuelled by access to affordable pumping technologies and increasing the local and export markets for horticultural crops (Asante, 2013). These gravity-fed or pumped irrigation systems use water from sources such as dugouts, streams and rivers, boreholes and small reservoirs (Oteng-Darko et al., 2018).

Large-scale commercial systems (mostly applied over 20–100 ha) use high energy-consuming pumps to draw water from a river/lake, which is often stored in a reservoir for later application. Private farmers and private entrepreneurs mainly develop informal or commercial irrigation systems. These private schemes are sometimes individually developed, or access support from the Government of Ghana, International donors/partners or Non-Governmental Organizations (NGOs; Table 19.1).

Limited data on where irrigation currently exists, trends in its development, and opportunities and constraints within formal and informal schemes undermine the consensus and make it almost impossible to ascertain the extent of irrigation technology adoption in the country (Namara et al., 2011). Data to support and categorize the status of irrigation in Ghana is mostly based on government irrigation schemes and some identified commercial irrigators. Some common types of irrigation technologies found in Ghana are the sprinkler irrigation, drip irrigation, furrow irrigation, basin/flood irrigation, centre pivot (an emerging irrigation infrastructure in Ghana), pumped-hose-spray and the traditional watering can system.

There is increasing concern that less than a third of the estimated total irrigated land in Ghana lies within the well-known public schemes (Figure 19.3), and not enough is known of the location, development and management of the informal irrigation schemes that account for the remaining two-thirds of total irrigated land (Namara et al., 2011). Recently, the Government of Ghana outlined plans for constructing and rehabilitating large- and medium-scale irrigation schemes under the government's flagship programmes like the One Village One Dam (1V1D) and Planting for Food and Jobs as captured in the Medium-Term Expenditure Framework (MoFA, 2019). Since previous large-scale irrigation systems have not always produced optimal performance, utilization and the anticipated benefits, many irrigation scientists and policy actors have promoted small-scale and affordable irrigation systems to boost the food production across the country. The 1V1D aims to enable and encourage the smallholder

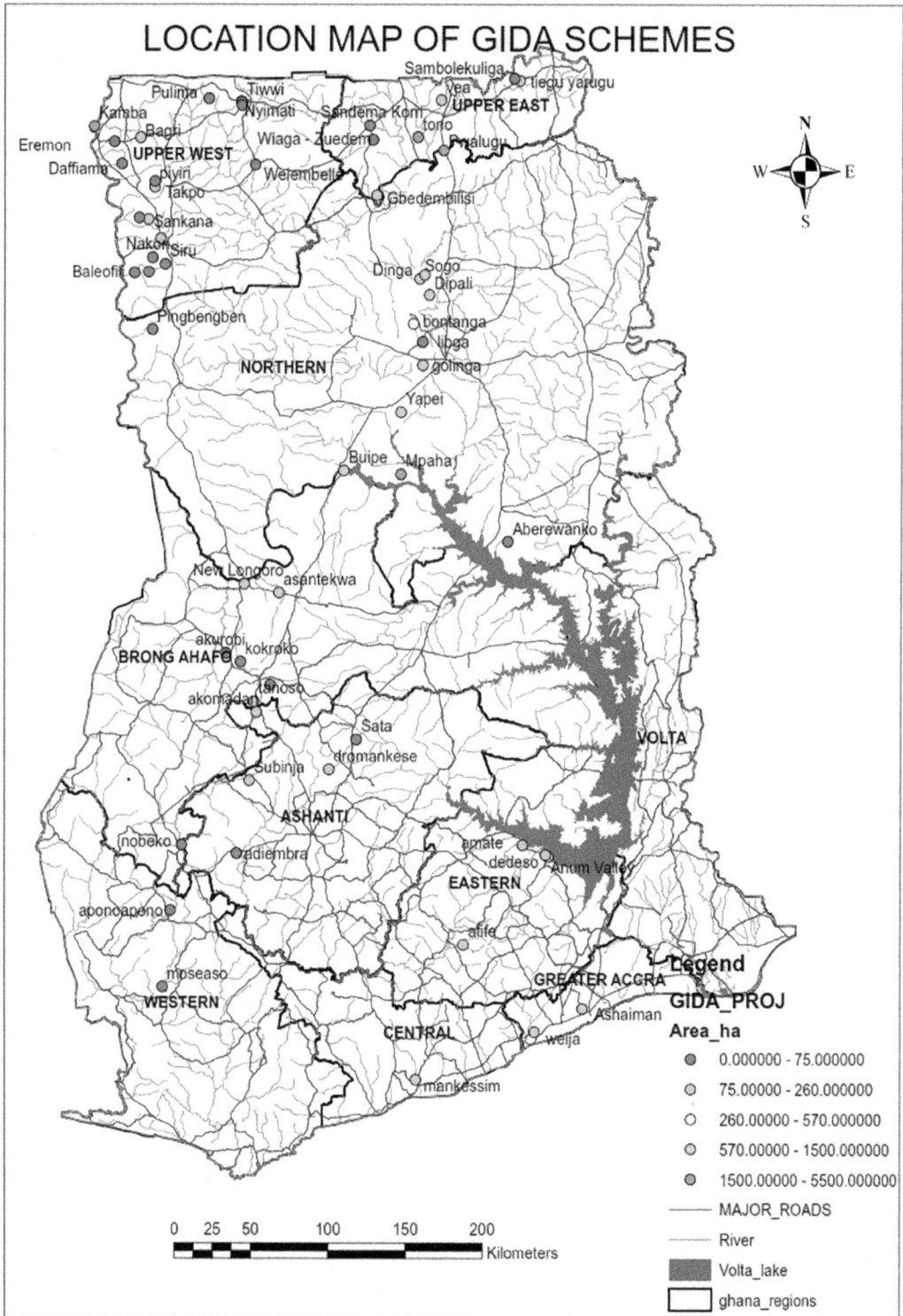

FIGURE 19.3 Some public irrigation scheme sites and their developed area (MoFA-GIDA, unpublished).

TABLE 19.1   Some Commercial Irrigation Farms in Ghana

No	Schemes	Area Developed (ha)	Area Irrigated (ha)	Irrigation System	Crops Produced
1	Golden Exotic Ltd.	1,200	1,200	Drip System	Banana
2	IWAD	250	45	Pressurized system	Assorted crops
3	VEGPRO Limited	250	205	Centre pivot	Baby corn, vegetables
4	Brazil Agro Business Group	250	250	Pump and gravity	Rice
5	Solar Harvest	750	100	Centre pivot	Maize, other crops
6	Jokopan Farms	13	13	Pump & gravity	Vegetables
7	Thai Farms	80	80	Pump & gravity	rice
8	Prairie Volta Ltd.	250	250	River pumping	Rice
9	Mawuko Farms	15	15	Pump and gravity	Vegetables
10	GADCO/WIENCO	1,200	1,200	River pumping	Rice
11	Sanford Enterprise	300	300	River pumping	Maize, other crops
12	CAISSIE Farms	300	300	River pumping	Maize, vegetables
13	AgDEVCo	5,359	225	River pumping	Maize, other crops
14	Anyako Farms	500	100	River pumping	Maize, rice

Source: Mofa-GIDA (unpublished).

irrigation in the country, with donors and policy makers also very interested in providing funds for irrigation development as a climate-smart approach for sustainable food production.

19.4 Irrigation Policy and Water Governance in Ghana

Irrigation development and governance in Ghana are outlined in the National Irrigation Policy, Strategies and Regulatory Measures (Government of Ghana Ministry of Food and Agriculture and the Food and Agriculture Organisation, 2010). The policy is aimed at sustainable growth and improved contribution of irrigation to the overall development of the Ghanaian agricultural sector as outlined in the Growth and Poverty Reduction Strategy (GPRS I and II). The policy is also designed to open up the investment space for intensified and diversified irrigated crop production in Ghana where there is clear comparative advantage (Government of Ghana Ministry of Food and Agriculture and the Food and Agriculture Organisation, 2010). The Ghanaian economy as a whole would benefit from this policy, but more specifically, it is designed to improve the livelihood of all of the existing and potential part- and full-time irrigators and the related farmer and farmer-based organizations. The policy aims to promote fairness by ensuring that private sector service providers have new opportunities to perform. The policy targets national food security, intensified and diversified production of agricultural commodities, increased livelihood options, optimum natural resource use, reduced negative environmental impacts and expanded investment space for the irrigated production (Government of Ghana Ministry of Food and Agriculture and the Food and Agriculture Organisation, 2010).

Ghana's irrigation policy has opened up the investment space for intensified and diversified irrigated cropping. The policy addresses four main issues including the low agricultural productivity and slow rates of growth, the constrained socio-economic engagements with land and water resources, environmental degradation associated with irrigation production and the lack of irrigation support. To solve these four problem areas, the policy develops "thrusts" which aim to achieve the accelerated and sustained irrigation development in Ghana (Government of Ghana Ministry of Food and Agriculture and the Food and Agriculture Organisation, 2010).

- Thrust A is about performance and growth that emphasize the productive capacity of the existing irrigation systems and respond to the new demands for irrigated production through the mix of well-coordinated public and private initiatives.
- Thrust B is about the socio-economic inclusion, which aims to remove the constraints to enhance a balanced socio-economic engagement with land and water resources.
- Thrust C covers the environmental performance of the various irrigation systems and their related practices.
- Thrust D involves extending cost-effective, demand-driven irrigation services to both the public and private irrigators.

The country is drained by three major basins and river systems, namely the Volta, South-Western and Coastal Rivers Systems covering 70%, 22% and 8%, respectively, of the total area of the country (Ghana Maritime, n.d.). The Volta River System (Figure 19.4) comprises the White, Black Volta, Red Volta and Oti Rivers (Ghana Maritime, n.d.) and contributes 64.7% out of the 38.3 billion m³ of annual runoff drained by the three major river systems in the country. The South-Western Rivers System comprises the Bia, Tano, Ankobra and Pra Rivers (Ghana Maritime, n.d.). The Coastal Rivers System comprises the Kakum/Bruku, Ochi-Nakwa, Ayensu, Densu, Odaw and Tordzie/Aka Rivers (Ghana Maritime, n.d.). According to the Ghana Maritime Agency, Ghana shares the Volta River basin with Burkina Faso, Togo, Cote d'Ivoire and Mali and also shares the Bia and Tano River basins with Cote d'Ivoire. Additional storage infrastructure development in the basin for handling ongoing challenges of climatic water stress at the entire basin scale offered great potential to increase the economic benefits to Ghana and Burkina Faso (Baah-Kumi & Ward, 2020), with no economic loss to any other riparian country. Thus, water could be released and reallocated for use in dry seasons and years.

Development of water resources in Ghana has focused on water supply for urban and rural populations, food production, environmental protection and hydropower production. The 1992 constitution paved the way to establish legislative instruments, and enact laws for water resources regulation in the country (Opoku-Agyeman, 2001), allowing secure investment in water resources. Since then, much has been done to develop a legal, political and institutional framework to manage Ghana's water resources as exhibited by the National Irrigation Policy, Strategies and Regulatory Measures (Government of Ghana Ministry of Food and Agriculture and the Food and Agriculture Organisation, 2010).

Public irrigation schemes used to be solely managed by Ghana Irrigation Development Authority (GIDA) and recently, the Irrigation Company of the Upper Regions (ICOUR). In 1987, based on a Legislative Instrument (LI 1350), Irrigation Farmers Associations (IFAs) were established on all public irrigation schemes. The Associations were registered with the Department of Cooperatives. On each irrigation scheme, a Scheme Management Unit was formed, including two farmer representatives. Some years later, farmer participation in public irrigation management was introduced. Roles and responsibility were assigned between GIDA/ICOUR and IFAs. GIDA/ICOUR were put in charge of operation and maintenance of headworks, i.e. reservoir, dam, canals and drainage, whereas IFAs were in charge of minor works, i.e. sub-drainage and canal systems, desilting of canals and maintenance of their service areas.

The regulation was amended in 2011 (LI 1995 of 2011) to make the operation and maintenance of the public schemes more public–private partnership-compliant. Under LI 2230 of 2016 all irrigation farmers were encouraged to form Water User Associations (WUAs) which were required to register with GIDA. With the WUA law in place, IFAs were dissolved. WUAs were then given the mandate for operation, maintenance and management of the public schemes in their operation zones. Thus, the WUAs were in charge of operation and maintenance of various small-scale schemes across the country. Large-scale schemes have been put under the management of private scheme management entities that oversee the operation, maintenance and management of the headworks. GIDA and ICOUR now have oversight responsibility for both public and private schemes to regulate the Irrigation subsector.

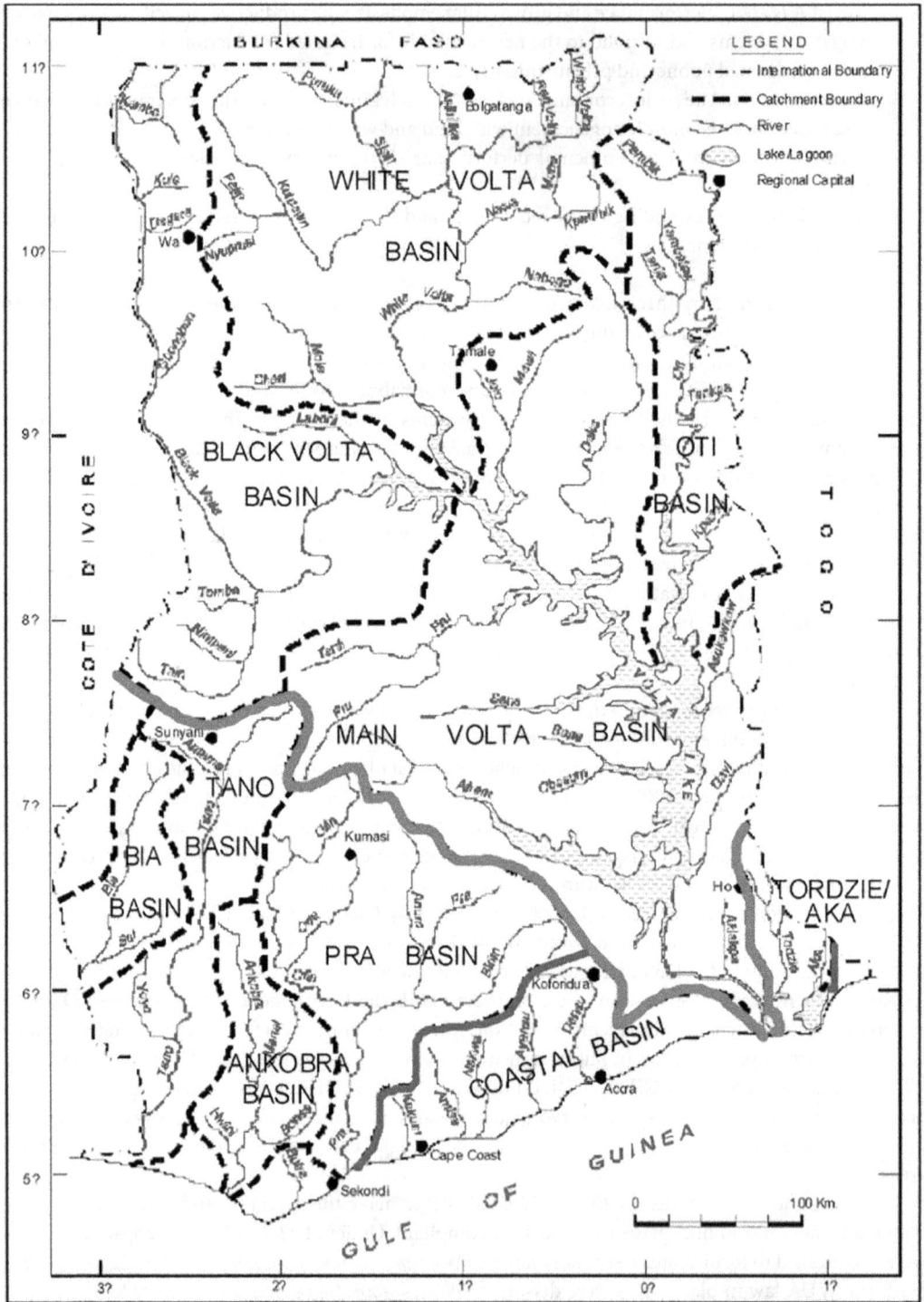

FIGURE 19.4 A map of Ghana showing the water resources and drainage basins in Ghana. The pink line indicates the delineation of the Volta Basin catchment within Ghana (Amisigo et al., 2015).

19.5 Factors that Influence the Adoption of Irrigation Technologies in Ghana

Technology plays an important role in economic development. Limited adoption of productive technologies largely contributes to the low agricultural productivity in sub-Saharan Africa (World Bank, 2007). One important means to increase agricultural productivity is by introducing improved agricultural technologies and management systems (Doss, 2006). The decision to adopt any agricultural technology depends on many factors (Mendola, 2005; Calatrava-Leyva et al., 2005), which have been categorized by Chuchird et al. (2017) as demographic, socio-economic, topographical, institutional and attitudinal. Owusu et al. (2013) also included personal, household and plot characteristics. While demographic and personal factors affect the adoption of irrigation technologies, on-farm factors such as farm size, farm hands (labour), farm location, crop type and growth duration, soil characteristics and land topography as well as some external influences determine the possibility of adopting irrigation technologies. Within this chapter, these factors are grouped into two: on-farm and external.

19.5.1 On-Farm Factors that Influence Adoption of Irrigation Technology in Ghana

19.5.1.1 Farm Size

Farm size is considered an important determinant of irrigation technology adoption in Ghana (Asante, 2013; Mugisa-Mutetikka et al., 2000). A larger farm size means that the farmer has more resources and thus a greater risk-taking ability, and is in a better position to adopt new technologies. They can also more rapidly diversify and thus operate at a profitable scale to generate returns on their investments. In Ghana small-scale farms range between 0 and 5 ha, while medium-scale farms range between 5 and 100 ha (Jayne et al., 2016). Farm size demonstrates the farmer's resources, which may determine the ability to invest in an irrigation technology, as new technologies often come with relatively high investment cost. Farm size also directly affects a farmer's access to credit which influences investments and adoption of irrigation technologies. Farm size also influences the type of irrigation that is adopted. Large-sized or commercial farms tend to adopt systems that are both technically and labour-efficient.

In Ghana, drip systems are commonly associated with small-holdings engaged in high-value crops such as pineapple, exported vegetables (Figure 19.5a) and cocoa (Figure 19.5b). Traditional rice farms readily adopt gravity-fed *sawah* systems or utilize pumping-based flood/basin irrigation systems

(a)   (b)

FIGURE 19.5 Drip irrigation adopted for (a) spring onion and (b) cocoa.

(a)   (b)

FIGURE 19.6   Surface irrigation such as (a) basin irrigation for rice production and (b) furrow irrigation in tomato production.

(Figure 19.6a). This is because rice is predominantly grown in inland valleys and lowland areas with immediate access to water. Small vegetable farms tend to adopt gravity-fed irrigation technologies such as furrow systems (Figure 19.6b).

19.5.1.2 Soil Characteristics

Soil type and quality are a major determinant of the choice of irrigation technology in Ghana, with soil fertility, water holding capacity, topography, and topsoil depth affecting crop productivity (Lichtenberg, 1989). Farmers with low soil water holding capacity were most likely to adopt sprinkler irrigation systems (Figure 19.7a and b), while those with high water holding capacity might opt for runoff systems such as the furrow or flood/basin systems (Negri & Brooks, 1990). Soil characteristics also help to assess which irrigation technology might be more profitable, hence influencing farmers' adoption of a particular technology.

Understanding soil texture aids farmers to determine the duration of irrigation and subsequently, type of irrigation system to employ. Sandy soil has poor water retention and lower water storage capacity, but high porosity and high infiltration rate. Thus, sandy soils need to be irrigated more frequently, and the regulated discharge rate of a sprinkler system is especially suitable. Farms on sandy soils were positively correlated with adopting the sprinkler irrigation, while clayey soils were negatively correlated with adoption of sprinkler irrigation (Negri and Brooks, 1990).

Drip irrigation systems operate under minimal pressure, allowing slow discharge rate and gradual release of water. Adopting an irrigation system with a high discharge rate (e.g. sprinkler irrigation) on such soils will result in water taking much longer to infiltrate, causing soil sealing, runoff and salinity issues. Loam soils with humus content have better infiltration, drainage and water retention capacity. This permits loamy soil to accommodate a wide range of irrigation systems including those with high discharge rates such as sprinklers and watering cans.

19.5.1.3 Topography

The topography of the land may also influence the choice of irrigation technology. Farmers on steep slopes are unlikely to adopt sprinkler or any surface irrigation system because of the risk of soil erosion. Fixed sprinklers on steep slopes result in non-uniform water distribution, since a section of the field is over-irrigated while the other sections receive insufficient irrigation. Even though this

FIGURE 19.7 Sprinkler irrigation in a (a) tree-crop nursery and (b) pawpaw (papaya) plantation.

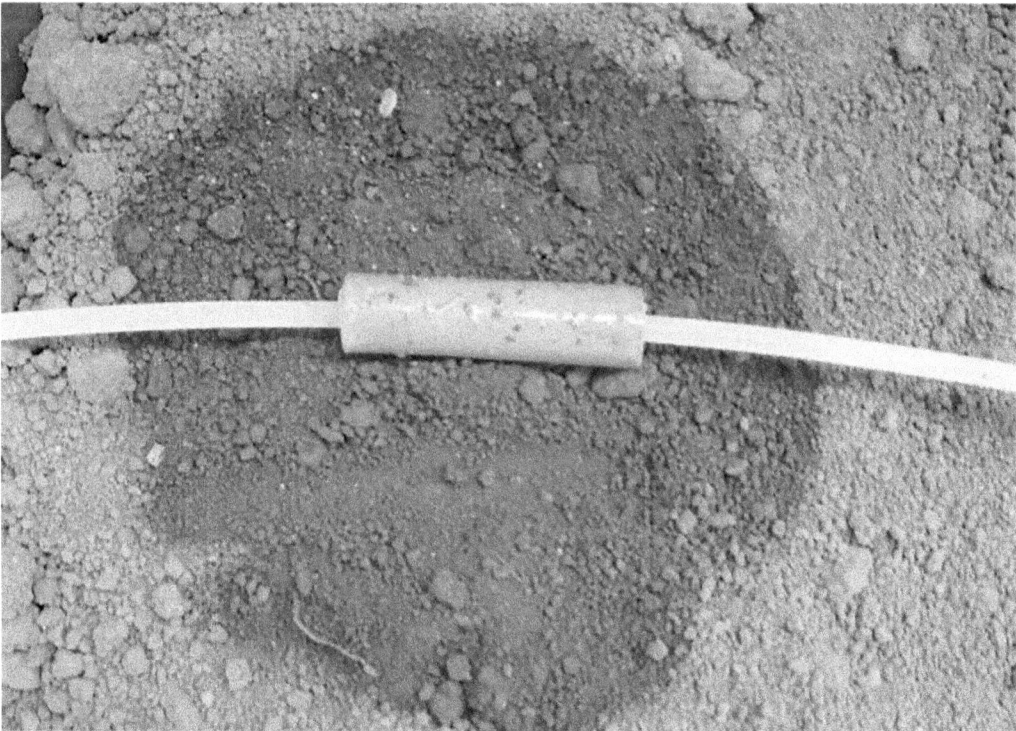

FIGURE 19.8 Water application using pitcher irrigation.

anomaly can be corrected by reducing the sprinkler riser and reducing the trajectory angle, this procedure might be cumbersome for the local farmer to undertake. Such farmers are likely to adopt drip irrigation or pitcher irrigation technology since this technology is not greatly affected by steepness of the topography. Pitcher irrigation buries a pot or a pitcher close to the plant roots and typically fills it with water manually, or by means of a flexible plastic hose using a hydraulic arrangement (Figure 19.8). Water is transported through the clay tubes into the soil according to the evaporation

FIGURE 19.9   (a) Irrigating with tail water. (b) Water application using watering can.

rates from the pitcher surface. The movement of water via evaporation depends on the soil tension (Bhatt & Kanzariya, 2017).

Sloped lands with depressions downslope are also ideal for recovering tail water (Figure 19.9a) for re-use on the crop fields. Tail water is stormwater runoff from a farm that is usually collected into a pit or other forms of reservoir for crop irrigation.

Gravity-fed surface irrigation systems are often not desirable on steeper slopes (Frisvold & Deva, 2012). Undertaking basin or furrow irrigation on steep slopes increases the velocity at which the water moves (due to gravity), detaching individual soil particles from the soil mass and transporting them down the slope. Figure 19.9a shows an example of a muddy tail water having a lot of sediments from eroded crop fields uphill. Sometimes, these tail-waters can provide nutrient dense irrigation water (from crop fields uphill) to downslope crops.

19.5.1.4 Crop Type and Length of Growing Season

Crops differ both in terms of their daily water needs and the duration of their total growing period. Thus, crops such as cassava and plantain require much more water than the crops with short growing season and low daily needs. For example, beans require 300–500 mm over an entire growing season of 75–90 days (Critchley et al., 1991). Farmers in agro-ecological zones such as the Guinea Savanna and the Sudan Savanna with unimodal rain season (Table 19.2) need to consider the growing period of selected crops and the timing/volume of supplementary rainfall when making strategic decisions on irrigation. Maximum evapotranspiration rate occurs if the soil is at field capacity (Al-Kaisi, 2000). Drip irrigation is likely to be adopted over longer growing seasons (Negri & Brooks, 1990) as a water-saving and cost-effective measure since it reduces energy costs. It also ensures that water is available throughout the entire growing season to meet the crop water requirements.

An emerging trend is the adoption of drip irrigation systems in annual and perennial cash crops such as citrus, pawpaw, cocoa and rubber irrespective of farm size. Tillage of such cash crops is typically limited, which can influence the adoption of particular irrigation technologies. Heavily mechanized farms tend to adopt systems that can be easily taken off or relocated from the field and installed again when the need arises. These include centre pivot, lateral movement and sprinkler irrigation systems.

TABLE 19.2 Length of Growing Seasons in Agro-Ecological Zones of Ghana

Agro-Ecological Zone	Growing Period (Days)	
	Major Season	Minor Season
Rain forest (evergreen)	150–160	100
Deciduous forest	150–160	90
Transitional	200–220	60
Coastal	100–110	50
Guinea Savanna	180–200	-
Sudan Savanna	150–160	-

Source: Adapted from Boubacar et al. (2005).

19.5.2 External Factors that Influence the Adoption of Irrigation Technology in Ghana

19.5.2.1 Land Tenure

The greatest challenge to resource management and land tenure system is high population growth. Ghana's population growth of 2.7% per annum (Ghana Statistical Service (GSS), 2010) has put pressure on existing land for farming and residential needs. Land is mainly owned communally along ethno-tribal and family lines, with designated traditional authorities responsible for its management in their capacity as trustees (Kasanga & Kotey, 2013). Large-scale irrigation development is constrained by the land tenure system due to weak legal frameworks, institutional structure and the policy environment within which the land administration takes place. Additionally, different land-owning groups may have conflicts of interest, due to inadequate security of tenure making it difficult to use land as an economic asset.

The legal status of a household relative to the land has affected the adoption of agricultural innovations including irrigation technologies. Farms that have lease rights to the land are more inclined to invest in long-term technologies such as irrigation. Farmers who do not own their land may not be willing to invest in the developing permanent irrigation infrastructures, due to the risk of the land being taken from them after incurring development costs. Land ownership in Ghana was correlated with the ownership of wells, which positively affected the adoption of micro-irrigation (Namara et al., 2007). This is because well owners have a high degree of control on the water source and are motivated to effectively use the available water.

19.5.2.2 Initial Investment Cost or Capital

Irrigators often cite financial constraints as major barriers to investing in improved irrigation efficiency (Frisvold & Deva, 2012). The ability of farmers/producers to bear the initial investment cost of irrigation technology depends on the availability of credit or financial resources, as new technologies often come with relatively high investment costs. The income of producers is also an important variable, with off-farm income, especially from the non-agricultural activities, stimulating the adoption of irrigation when an additional source of capital becomes available. However, the likelihood of adoption decreases when off-farm income becomes the main source of revenue (Namara et al., 2007).

Access to credit remains a constraint to adopting improved technologies in developing countries (Adeoti, 2009). The higher the credit availability, the greater the probability that farmers will adopt irrigation (He et al., 2012) more rapidly (Alcon et al., 2011). The public and private sectors provide funds for irrigation development. The public sector in Ghana enjoys support from foreign countries, including China, Japan and Republic of South Korea. Funding is also obtained from international organizations, such as the Food and Agriculture Organization of UN, German Technical Cooperation (GiZ), United States Agency for International Development (USAID), African Development Bank and the World Bank, among others (Miyoshi & Nagayo, 2006). Since irrigation development is an expensive investment, the sector is not attractive to the private sector.

19.5.2.3 Water Resources and Governance

Irrigation is a key input to agricultural growth that relies heavily on surface water in Ghana, with a rapid increase in the demand for irrigation water. For more than a decade, the main objective of farm management especially in countries with limited land and water resources is centred on increasing food production using less irrigation (FAO, 2002). Agricultural water is a key constraint to irrigation development in some (especially northern) parts of the country. Moreover, about 60% of water bodies in Ghana are polluted with household and industrial (mostly from surface mining popularly called *galamsey*) waste with most of them in critical conditions (Ampomah, 2017). "Galamsey" activities release chemicals such as mercury, a heavy metal, from the mercury-gold amalgamation process into the water bodies, thus making them unproductive for the consumptive and even some non-consumptive water use. Several studies (Chaoua et al, 2019; Darko et al, 2019; Kheirabadi et al, 2016; Azevedo and Rodriguez, 2012) have outlined the potential risk to human health of the accumulation of heavy metals in plants. These authors have associated the consumption of heavy-metal-accumulated vegetables with carcinogenic and non-carcinogenic diseases. Despite surface water pollution, rainwater harvesting is becoming more commonplace, as a means to increase the water availability in some parts of the country (WRC Ghana, 2015).

The proximity of a farm to a water body or reservoir influences the decision to the adopt irrigation and the type of irrigation technology, especially with the informal irrigators. Development of an irrigation infrastructure is quite expensive in Ghana and hence, the existence of water source (natural or constructed) will dramatically reduce the cost of development and influence decisions to adopt the irrigation technologies. In Ghana, most private irrigation systems usually emerge along river bodies or areas where shallow wells can easily be developed due to a high-water table. Before 1980, irrigation water was offered free as a social service to farmers; however, reforms were made where farmers had to pay for water charges at rates fixed from time to time (Kyei-Baffour & Ofori, 2007). These rates vary considerably between irrigation schemes. Even within existing irrigation watersheds, farm position within the scheme or basin affects farmer adoption of technologies. Nearness to a water source and upstream location within a watershed were positively correlated with adoption of irrigation in Ghana (Akuffo & Egir, 2013).

One policy option to encourage irrigation technology adoption may be incentive pricing for surface water (Frisvold & Deva, 2015). The GIDA and the ICOUR, managers of public irrigation schemes in Ghana do not charge farmers the marginal cost of water (i.e. the increase or decrease in cost of water when there is an increase or decrease by a unit), but apply quantity allocation, irrigation infrastructure maintenance and farm acreage in determining irrigation service charge. The adoption of tiered pricing schemes by GIDA may encourage more efficient irrigation technologies to enable farmers to produce more with the limited water available. Tiered water pricing, also called increasing block rate pricing, is where the water user/farmer pays different prices per unit of water delivered depending on the amount used, with a higher price charged for larger quantities (Hanemann, 2006).

Some smallholder irrigation schemes, particularly those funded by NGOs and international donors have WUAs comprising the farmer who undertakes crop selection and management tasks. These WUAs elect a small number of officials to carry out fee collection and make day-to-day management and operation decisions such as the distribution of water to farmers' fields. All farmers cultivating a plot within the irrigation system are asked to pay a set fee, per plot, to the WUA. The fees collected are saved to be used for canal repair and maintenance. Such repair and maintenance works include protection of dam embankment and bunds against erosion and grazing animals and repair of canals, spillways and laterals.

19.5.2.4 Technical Expertise

Adoption of irrigation systems is also affected by the availability of technical expertise in the irrigation technologies and infrastructural development. Farmers who obtain knowledge from the irrigation specialists are more likely to adopt irrigation (Alcon et al., 2011; Rossi et al., 2015). Increased collaboration of private initiatives with local institutions such as extension services could also improve the reach of the technology (Adeoti, 2009). In Ghana, most farmers seek the technical expertise from suppliers of irrigation equipment, researchers and extension agents. Some major irrigation equipment suppliers such as Dizengoff and Agrimat are also involved in the installation, capacity building and after-sales services (Mendes et al., 2014).

19.5.2.5 Labour Availability and Requirement

Labour availability is another variable affecting farmers' decisions regarding adoption of new agricultural irrigation technologies. In Ghana, labour availability depends on the cropping season, with peak-season coinciding with the rainy season from April to July for southern Ghana and April to October for northern Ghana. During this period, a labour shortage promotes the adoption of labour-saving practices, but hinders the implementation of technologies that require more labour input. Some irrigation technologies such as the centre pivot and drip systems may be relatively labour-saving, whereas others such as the furrow and basin are deemed labour-intensive. Increased labour availability stimulated adoption of the treadle pump technology in Ghana, which requires labour for operation (Adeoti, 2009).

19.5.2.6 Climate

Climate's influence on irrigation adoption depends on the geographical location (Kabo-Bah et al., 2021). Ghana is characterized by ecological zones defined by their rainfall, temperature and relative humidity. However, increasing climate volatility and more frequent occurrence of extreme climate events affect farmers' decisions on adopting irrigation and other agricultural technologies (Xu et al., 2015). USAID (2017) predicts a 4% decrease in rainfall for the country by 2040 with associated shortening of the growing seasons, increased crop loss or failure and loss of arable land. Already, the number of rainfall events for Ghana has reduced over the years (Buxton, 2018), thus affecting water supply for rainfed agriculture. Rainfall conditions in the Volta delta of Ghana are marked by increased variability, declining rainfall amount and a shift in rainfall regime towards long dry spells, late onset of seasonal rainfall and early cessation of rainfall (Gbangou et al., 2019; Sarku et al., 2020), thereby increasing the vulnerability of farmers who are irrigating. However, specific decisions will depend on whether climatic elements in each specific ecological zone warrant the need to adopt a specific technology. Farmers also consider the indigenous knowledge on changing weather trends for various decision-making such as type of crop to grow and whether to adopt a full or supplementary irrigation system depending on the season-long available moisture.

19.5.2.7 Energy (Fuel/Electricity) Requirement and Ease of Operation of Irrigation System

Irrigation technologies that improve the irrigation efficiency reduce energy use because less water is needed for a comparable area of irrigation, which in turn requires less energy for pumping this water. Thus, farmers consider energy requirements when selecting or adopting an irrigation technology. Public and formal schemes usually rely on grid or solar electricity. Farmers can easily and safely irrigate crops during power disruptions, especially during the night if the irrigation system is automated. A grid electric pump is not an ideal irrigation method in Ghana because many farmers struggle to pay their electricity fees regularly. Diesel or petrol-powered pumps are more suitable options, allowing farmers to pay on demand. An emerging trend is solar power for the commercial and smallholder irrigation. Although cost-intensive in its initial development, it eliminates the reliance on the national grid power and electricity or fuel bills.

19.5.2.8 Social and Personal Incentives

In modern agriculture, farmers are informed about the existence and effective use of any new farming technology mainly through extension personnel as well as from their social interactions with other farmers. Participation in farming organizations constitutes an important social network in which the producer can obtain information on new technologies.

Even without the intervention of extension agents, farmers learn from interacting with the other farmers. Rogers (1995) reported that farmers learn from their "homophilic neighbours": individuals with whom farmers have close social ties and share the common professional or/and personal characteristics (education, age, religious beliefs, farming activities, etc.). Moreover, farmers may also follow or trust the opinion of those that they perceive as being successful in their farming operation, even though they occasionally share quite different characteristics. Farmers also find it prestigious to belong to WUAs. Some of these WUAs have welfare packages that include financial/credit and social benefits (e.g. representation at and donation to social gatherings such as funerals, marriage and naming ceremonies). The social prestige associated with owning irrigation equipment is usually an incentive for smallholder farmers to adopt irrigation technologies. The size of irrigated land cultivated depends on availability and the financial capacity of the farm household for cultivation. It is therefore used as a proxy of the family's wealth status.

19.6 Water-Saving Agriculture to Diminish Irrigation's Environmental Impact

Water-saving agriculture implies the adoption of good agronomic, physiological, biotechnological/ genetic and engineering approaches (Adeyemi et al., 2017) under crop production. Agriculture is responsible for the largest water withdrawals globally and contributes to conditions of local absolute water scarcity and future water conflicts. Managing the environmental impact of agricultural water use must emphasize improving agricultural water use efficiency (crop yield per unit of water use) and on-farm productivity. This includes the improved farm water management and better irrigation system performance within changing national water and irrigation policies. Improved irrigation scheduling and technologies can also ameliorate the negative impacts of irrigation on the environment.

Traditionally, much of the irrigation practiced in Ghana has comprised the gravity-fed systems, where water is transported from surface sources through small channels and used to flood or furrow-feed agricultural land (furrow irrigation). However, sprinkler irrigation using water pumps, often drawing water from subterranean aquifers, is becoming increasingly common. Impacts of water abstraction on the environment in these areas can be severe. Farmer adoption of water-saving agriculture techniques can increase the efficiency of irrigation systems, thereby mitigating the negative impact of irrigated farming (Velasco-Muñoz et al., 2019). Options of water-saving agriculture for increased crop and soil productivity while reducing crop water footprints are discussed below.

19.6.1 Deficit Irrigation

This irrigation practice supplies less water than crop water requirement, or withholds irrigation during certain growth stages of a fully irrigated crop (Al-Omran et al., 2010), and can improve water use efficiency (Koech & Langat, 2018; Evans & Saddler, 2008; Fereres & Soriano, 2007). Rather than maximize yields, deficit irrigation aims to stabilize yields and to obtain the maximum crop water productivity (Zhang & Oweis, 1999). Optimized deficit irrigation techniques maximize water productivity, ideally with improved or unchanged harvest quality (Spreer et al., 2007). Applying less water reduces the leaching of nutrients and agrochemicals from the root-zone, thus preserving groundwater quality (Pandey et al., 2000). By applying less water and limiting ponding on the soil surface or overland flow, deficit irrigation can prevent salinization and reduce soil erosion. A study in Ghana on the effects of deficit irrigation on tomatoes found no significant reduction in tomato yield when 10%–15% less irrigation water was supplied throughout the cropping season (Owusu-Sekyere et al., 2012). Other types of deficit irrigation vary the percentage of the root-zone that receives the irrigation water (partial root-zone drying, PRD) or periodically allows the irrigation water to fall below an identified water level (especially in rice) before flooding the field again, termed alternate wetting and drying (AWD).

19.6.2 Partial Root-Zone Drying

PRD is an irrigation technique that irrigates only half of a crop's root system at a time, while the other half is left to dry. Periodically, the wet and dry parts are switched at subsequent irrigation events (alternate PRD) or half the root-zone is kept dry for the entire growing season (fixed PRD) (Dodd et al., 2011). This irrigation technique has the potential to increase water use efficiency without significantly reducing yields (Giuliani et al., 2017; Adu et al., 2018). In areas with the limited water resources, PRD may be a viable irrigation option to increase water productivity while maintaining crop yield, and increasing the economic yield even without considering the value of water in water-limited environments (Sepaskhah & Ahmadi, 2010). While PRD may be applied empirically by farmers (e.g. only watering every second row in furrow-irrigated crops when water is scarce), a better understanding of plant physiological responses to water deficit may improve its application. Alternating the wet and dry parts of the root system enhance crop yield compared to the fixed PRD (Dodd et al., 2015), but optimal timing of this alternation is unknown, and farmers may typically apply a rigid schedule (e.g. alternate every 1 or 2 weeks) irrespective of crop water demands. Measuring crop physiological responses to PRD may help to determine when the alternation should occur, but such measurements may be beyond the capacity of most farmers.

Within the context of water-saving agriculture and enhancing farmer capacity in irrigation scheduling, Lancaster Environment Centre, Lancaster University and Council for Scientific and Industrial Research (CSIR), Ghana, are implementing a project that hinges on co-design of irrigation scheduling with communities. This project aims to identify the approaches to water-efficient irrigation, including investigation of techniques to optimize the irrigation and food production, and understand socio-economic barriers to adopting the water-saving irrigation techniques such as PRD. Although tomato is widely grown in Ghana, domestic production is currently unable to keep pace with demand, especially in the dry season when irrigation is required for production. Greater use of irrigation may boost production, but clearly needs to be done in a sustainable manner to avoid depleting water resources. Irrigating every second furrow can easily be demonstrated to farmers to encourage adoption of PRD, but adoption requires evidence of beneficial agronomic impacts.

To investigate the optimal frequency of alternation in PRD in tomato, a controlled environment experiment in Lancaster (United Kingdom) on a miniaturized PRD system using the dwarf cultivar, Micro-Tom aimed to understand plants physiological responses, while an analogous field trial was carried out in Kumasi (Ghana) on an improved bush-type cultivar (Petomech) to demonstrate field application (Puértolas et al., 2022). Both studies sought to find out if PRD provides any agronomic or physiological benefits either above or beyond withholding water and if alternation enhanced water productivity compared to fixing the irrigated side. Both field and controlled environment trials assessed the effect of three different frequencies of alternation (none, short cycle and long cycle). The initial results show no effect of alternation and alternation did not affect the plant stress, which simplifies application by farmers. Thus, PRD was suitable to increase the water productivity in tomato.

19.6.3 Alternate Wetting and Drying

AWD is an irrigation technology usually employed in rice cultivation. AWD maintains grain yield and increases water use efficiency compared to the continuous flooding (CF) (Carrijo, 2018), but the physiological changes during drying and rewetting cycles are still poorly understood. AWD has been applied at the different phenological stages, but its physiological effects have been mainly studied in the reproductive stage, especially during grain filling (Zhang et al., 2010, 2012). Although AWD application throughout the entire crop life cycle decreases yield (Carrijo, 2018), the timing and intensity of AWD application can be engineered to enhance final yield. The effects of AWD might also be moderated by nutrient application, in particular nitrogen. Excessive vegetative growth and delayed senescence triggered by high soil N limits stem remobilization of carbohydrates to the grain, which can be overcome by AWD (Wang et al., 2016). Different rice genotypes show variable responses to the AWD application (Sandhu et al., 2017), with some showing increased yield in comparison to CF (Norton et al., 2017). This offers an opportunity to use the modern breeding techniques to select varieties specifically suited for AWD, but requires physiological insights on beneficial plant traits to select.

Rice is a major staple food worldwide and constitutes a major economic activity for smallholder communities. In Ghana, increasing rice production to meet the domestic demand is an important step to achieving food self-sufficiency, but this must occur with more efficient water and nutrient use. With increasing African human population there is unmet demand for rice. Enhancing rice production is a national priority in Ghana. However, appropriate water and nutrient management is necessary to maximize yields while limiting the wasteful losses of scarce nitrogen as nitrous oxide (a potent greenhouse gas) emissions.

Lancaster Environment Centre, Lancaster University and CSIR, Ghana, are implementing a project aimed at managing water and nutrients for increasing crop yields without compromising on environmental quality. One of the key objectives of the RECIRCULATE project is to sustainably increase crop yields despite limiting supplies of water and nutrients. This research investigates the opportunities to develop and optimize water-efficient irrigation techniques for African rice varieties. The research evaluates irrigation type (AWD and CF) and different nutrients (high nitrogen and low nitrogen) and genotype (improved variety cv. CRI-AgraRice and local variety cv. Viwornor). In this research, the AWD treatment was applied throughout the entire cropping cycle and irrigation was carried out when water level has dropped to about 15 cm below the surface of the soil. The improved variety was developed by CSIR-Crops Research Institute. The initial results of this research revealed that although AWD decreased yield of the local variety compared to CF, yield of the improved variety was independent of irrigation treatment. Thus, combining an improved genotype with a water-saving irrigation technique maintained grain yield while supplying *circa* 30% less water. The greenhouse experiment conducted at Lancaster Environment Centre in Lancaster showed that AWD did not affect root dry biomass and leaf area of rice. Whether this technology package is attractive to smallholder rice-farming communities is yet to be evaluated.

19.6.4 Irrigation Scheduling and Management for Schemes and Private Irrigators

Several irrigation scheduling tools have been developed to help the irrigators make informed decisions on crop water needs and irrigation timings. Some of these methods are thermal imaging, irrigation scheduling software such as FAO's CropWat, soil moisture balance methods (moisture meters, water balance/soil water content method, etc.) and tensiometers. However, farmers and/or irrigation staff often lack the relevant technical expertise in these methods and appropriate management skills to regulate irrigation systems properly. Consequences include the reductions in crop yields and a waste of water resources. To improve irrigation performance, it is necessary not only to promote the implementation of irrigation scheduling methods, but concurrently to improve system design and performance and to enhance farmers' skills to control and manage their irrigation system more efficiently. There is the need to develop a systematic training approach for managers and all of the levels of staff, which will lead to improved performance of their irrigation systems and therefore enhance water use efficiency.

19.6.5 Cropping Systems and Cultural Practices

In addition to water scarcity, the environmental impacts of irrigation farming comprise damage to aquatic habitats and displacement of natural ecosystems, increased erosion of cultivated soils on slopes, soil degradation through salinization or sodicity, water pollution from chemical fertilizers and pesticides and contamination of water by minerals from groundwater sources (Environmetal Protection Agency, 1996). Among these, the most significant problems in Ghana are salinization and severe pollution by chemical fertilizers, pesticides and the other farm inputs in significant areas of intensive irrigated agriculture. Continuous application of inorganic fertilizers over long periods of time without proper drainage or leaching of the soils caused salinization of irrigation schemes in Ghana, with recurrent waterlogging also identified (Adongo et al., 2015). Organic manure addition, though not very effective, was widely employed to manage saline soils in Ghana, leading to further recommendations of using halophytes (Asamoah et al., 2013). A better understanding of regional hydrology and better nutrient management advice should prevent the development of more saline soils.

19.7 Prospects for Irrigation Development in Ghana

There are extensive opportunities available in the agricultural and more specifically, irrigation sector in the country. Ghana aims to become a food-secure nation and a net exporter of horticultural crops to the international market. Within this vision, the provision and improvement of irrigation are undeniably a stepping stone. Irrigation development can alleviate poverty among smallholder farmers, improve income for the commercial farmers and ensure national food security in Ghana. This sub-section discusses some policy incentives and allocations made to promote investment in the irrigation subsector in Ghana.

19.7.1 Market Incentives, Subsidies and Access to Credit

There are direct opportunities for agricultural production on a commercial scale when irrigation technologies are adopted or employed. Production can be for the domestic market or can benefit from the trade agreement within the Economic Community of West African States with a market of about 250 million people (World Bank, 2017). Increasing production through the adoption of irrigation technologies would also ensure that those in the food system benefit from increased supply and availability of raw materials to meet demand for agro-processing plants and produce buying and marketing firms.

The government has put in place attractive incentives for irrigation development in the country, including exemption from customs duties and value-added tax for irrigation equipment and machinery as well as fertilizer subsidies (Namara et al., 2011). In 2003, aid agencies contributed 80% of the total GIDA expenditure of US\$ 3,369,295 (Miyoshi & Nagayo, 2006). Concessional funding for agricultural development (including irrigation) can be accessed from the following institutions in the form of loans, grants or equity:

- Export Development and Agricultural Development Investment Fund (EDAIF)
- Export Development and Investment Fund (EDIF)
- Micro and Small Loans Centre (MASLOC)
- Outgrower and Value Chain Fund (OVCF)
- The Venture Capital Trust Fund (VCTF).

19.7.2 Research and Extension

Research on the irrigation technology development is carried out in some universities and mandated institutions of the CSIR such as the Institute of Industrial Research, Crops Research Institute, Water Research Institute and the Savanna Agriculture Research Institute. These institutions continuously work on developing irrigation infrastructure and water-saving agricultural technologies to improve crop production and farmer incomes. Although there is no streamlined system to coordinate public and private research on irrigation technologies, irrigation developers and farmers can access technologies from the research institutions and universities. Field/open days are organized periodically by these organizations, inviting the public and private sector to visit research sites where technologies are showcased for public awareness and subsequent interest in adoption and commercialization. The general public and other stakeholders can also walk in to these research centres for information and direction. Researchers can partner with the local manufacturers or irrigation equipment to commercialize the research outputs for enhanced adoption.

19.7.3 Irrigation-Based Innovation Platforms

The eventual adoption of the Innovation Platform (IP) concept developed by the Forum for Agricultural Research in Africa) under the Integrated Agricultural Research for Development avails a great opportunity for irrigation development in Ghana. IP is an operational instrument which brings together a group of relevant actors within the value chain of a specific commodity or system of production (https://faraafrica.org/iar4d/). Establishing Irrigation-Based Innovation Platforms (IBIP) holds extensive opportunities to promote and eventually adopt irrigation technologies. Platform actors are expected to interact to identify problems and propose or investigate solutions, generating innovations and accompanying socio-economic benefits (https://faraafrica.org/iar4d/). These IPs also bring together all actors on the irrigation value chain to identify common grounds to interact and share their services and expertise. Through the interactions and linkages on the IBIP, farmers can mobilize and develop smallholder irrigation schemes by jointly accessing credit from the government-funded interventions or financial institutions. The platform could also build the capacity of farmers and other relevant stakeholders in irrigation technologies, good agronomic practices and strategic marketing.

19.7.4 Public–Private Partnership

The private sector is the driving force for importing irrigation equipment into the country. The National Policy on Public–Private Partnerships (Government of Ghana Ministry of Finance and Economic Planning, 2011) encourages collaborative investments in agriculture and other areas of the Ghanaian economy. The policy defines the objectives and guiding principles for a public–private partnership (PPP). The policy also outlines the legal, risk sharing and management framework for PPP. The policy environment is very firm to promote the development of irrigation infrastructure in the country. In line with this, GIDA has reviewed its structure and also established an irrigation PPP Unit to leverage private sector resources under PPP arrangements to boost irrigation development in the country (MoFA-GIDA, n.d.). The Government of Ghana has partnered with several private companies in a joint venture to establish commercial irrigation facilities. Notable among them are Kpong Farms Limited, the Volta River Estates and Jei River Farms. These companies produce high-value crops such as Banana, sugarcane and rice for the domestic and export market.

19.8 Conclusions

Intensification is a key to achieving growth in the agricultural economy, with irrigation central to the intensification strategy of Ghana. Fortunately, irrigated agriculture in the country has good prospects to facilitate achieving these goals. Despite the irrigation potential, the adoption of smallholder irrigation technology by smallholder farming households seems low. Policy makers and development partners promoting irrigation technologies to enhance the resilience and increase productivity would benefit from understanding how technologies are used, by whom, and for what purpose. To promote the irrigated farming in Ghana, farmers must be supported with micro-credit facilities to acquire hardware, especially low-cost irrigation pumps. Adoption of equipment for irrigation should be encouraged by the government, and NGOs. Ministry of Food and Agriculture through its agricultural extension agents (AEAs) should empower the farmers by training them in choosing appropriate crops for irrigated agriculture, and appropriate scheduling to avoid the local water scarcity and other environmental problems.

The development, promotion and adoption of irrigation technologies are Ghana's best-bet to achieve the sustainable agriculture growth and improve the livelihood of smallholders.

Acknowledgements

The authors would like to acknowledge the support of the Global Challenges Research Fund (Project RECIRCULATE, ES/P010857/1).

References

Adeoti, A. I. (2009). Factors influencing irrigation technology adoption and its impact on household poverty in Ghana. *Journal of Agriculture and Rural Development in the Tropics and Subtropics, 109*(1), 51–63.

Adeyemi, O., Grove, I., Peets, S., & Norton, T. (2017) Advanced monitoring and management systems for improving sustainability in precision irrigation. *Sustainability, 9,* 353.

Adongo, T. A., Abagale, F., & Kranjac-Berisavljevic, G. (2015). Soil Quality of Irrigable Lands of Irrigation Schemes in Northern Ghana. 2. Accessed online at https://www.researchgate.net/publication/284174123_Soil_Quality_of_Irrigable_Lands_of_Irrigation_Schemes_in_Northern_Ghana

Adu M. O., Yawson, D. O, Armah, F. A, Asare, P. A., & Frimpong K. A. (2018) *Meta Analysis of crop yields of full, deficit and partial rootzone drying.* Accessed from https://www.sciencedirect.com/science/article/abs/pii/S0378377417303785

Agodzo, S. K. & Bobobee, E. Y. H. (1994). Policy issues of irrigation in Ghana: 1960-1990. *Proceedings of the XIIth World Congress on Agricultural Engineering*, Milano, 28 August–1 September 1994. CIGR Vol. 1, pp 335–343.

Akuffo, A., & Egir, S. I. (2013). Modelling the choice of irrigation technologies of urban vegetable farmers in Accra, Ghana. In *AAEA & CAES Joint Annual Meeting*, Washington, DC, USA, August, 4–6.

Alcon, F., Miguel, M. D., & Burton, M. (2011). Duration analysis of adoption of drip irrigation technology in southeastern Spain. *Technological Forecasting & Social Change, 78*(6), 991–1001.

Al-Kaisi, M. (2000). *Crop Water Use or Evapotranspiration. Extension and Outreach*, Iowa State University, Ames, IA.

Al-Omran, A. M., Al-Harbi, A. R., Wahb-Allah, M. A., Mahmoud, N., & Al-Eter, A. (2010). Impact of irrigation water quality, irrigation systems, irrigation rates and soil amendments on tomato production in sandy calcareous soil. *Turkish Journal of Agriculture and Forestry, 34*(1), 59–73. https://doi.org/10.3906/tar-0902-22

Amisigo, B. A., Mccluskey, A., & Swanson, R. (2015). Modeling impact of climate change on water resources and agriculture demand in the Volta basin and other basin systems in Ghana. *Sustainability, 7,* 6957–6975. https://doi.org/10.3390/su7066957

Ampomah, B. (2017). 60% of Ghana's water bodies polluted. Water Resource Commission. Executive Secretary of the Commission at a Workshop in Ho, Ghana, Source: GNA, May 13, 2017.

Asamoah, A., Antwi-Boasiako, C., Frimpong-Mensah, K., & Soma, D. (2013). Adoptable technique(s) for managing Ghanaian saline soils. *Octa Journal of Environmental Research, 1.* Accessed online at https://www.researchgate.net/publication/245022370_ADOPTABLE_TECHNIQUES_FOR_MANAGING_GHANAIAN_SALINE_SOILS

Asante, A. V. (2013). *Smallholder irrigation technology in Ghana: adoption and profitability analysis.* Kwame Nkrumah University of Science and Technology. Retrieved from http://ir.knust.edu.gh/handle/123456789/6674

Asante, F. A., & Amuakwa-Mensah, F. (2015). Climate change and variability in Ghana: Stocktaking. *Climate, 3,* 78–99. https://doi.org/10.3390/cli3010078

Azevedo, R. & Rodriguez E. (2012). Phytotoxicity of mercury in plants: A review. *Journal of Botany*. https://doi.org/10.1155/2012/84861

Baah-Kumi, B. & Ward, F. A. (2020). Poverty mitigation through optimized water development and use: Insights from the Volta Basin, *Journal of Hydrology, 582*, 124548.

Bhatt, N. J., & Kanzariya, B. R. (2017). Experimental investigations on pitcher irrigation: Yield optimization and wetting front advancement. *International Journal of Latest Technology in Engineering, Management and Applied Science (IJLTEMAS), 6*(6), 103–108. http://ijltemas.in/DigitalLibrary/Vol.6Issue6/103-108.pdf

Boubacar, B., Obuobie E., Andreini, M., Andah, W. & Plugueth M. (2005). *Comparative study of river basin development and management: The Volta River Basin.* Accessed online at http://www.iwmi.cgiar.org/assessment/files_new/research_projects/river_basin_development_and_management/VoltaRiverBasin_Boubacar.pdf

Buxton D. N. B (2018). Vulnerability of cocoa production to climate change: a case of the Western and Central Regions in Ghana. Accessed from https://ir.ucc.edu.gh/xmlui/bitstream/handle/123456789/3426/BUXTON%202018.pdf?sequence=1&isAllowed=y

Calatrava-Leyva, J., Franco, J. A., & Gonzalez-Roa, M. C. (2005). Adoption of soil conservation practices in olive groves: The case of Spanish mountainous areas. Paper Prepared for Presentation at the XI International Congress of the European Association of Agricultural Economists, Denmark, August 24–27.

Carrijo, D. R. (2018). *Rice Yields, Water Use and Grain Arsenic Concentration under Alternate Wetting and Drying Irrigation.* University of California, Davis, CA.

Chaoua, S., Boussaa, S., El Gharmali, A. & Boumezzough, A. (2019) Impact of irrigation with wastewater on accumulation of heavy metal in soil and crops in the region of Marrakech in Morocco. Accessed from https://reader.elsevier.com/reader/sd/pii/S1658077X17303521?token=7CFF4401AE40081EB10A62CCDC51FE15C6083E1D18CDAC4BF09AD9B4ED6D4D9FEFB180ECE444355BDC68131B705A01FE

Chuchird, R., Sasaki, N., & Abe, I. (2017) Influencing Factors of the Adoption of Agricultural Irrigation Technologies and the Economic Returns: A Case Study in Chaiyaphum Province, Thailand. *Sustainability, 9*, 1524. https://doi.org/10.3390/su9091524

Critchley, W., Siegert, K., & Chapman, C. (1991). *Water Harvesting: A Manual for the Design and Construction of Water Harvesting Schemes for Plant Production.* Food and Agricultural Organization, Rome, Italy.

Darko, G., Boakye, K. O., Nkansah, M. A., Gyamfi, O., Ansah, E., Yevugah, L. L., Acheampong, A., & Dodd, M. (2019). Human health risk and bioaccessibility of toxic metals in top soils from Gbani mining community in Ghana. *Journal of Health and Pollution, 9*(22), 190602. https://doi.org/10.5696/2156-9614-9.22.190602

Dodd, I. C., Puértolas, J., Huber, K., Pérez-Pérez, J. G., Wright, H. R., & Blackwell, M. S. A. (2015). The importance of soil drying and re-wetting in crop phytohormonal and nutritional responses to deficit irrigation. *Journal of Experimental Botany, 66*, 2239–2252. doi:10.1093/jxb/eru532

Dodd, I. C., Egea, G., Martín-Vertedor, A. I., Romero, P., & Pérez, J. G. P. (2011). Partial rootzone drying: Chemical signalling theory and irrigation practice. *Acta Horticulturae, 922*, 67–74. https://doi.org/10.17660/ActaHortic.2011.922.8

Doss, C. R. (2006). Analyzing technology adoption using micro studies: Limitations, challenges, and opportunities for improvement. *Agricultural Economics, 34*, 207–219.

Dowgert, M. (2010). The impact of irrigated agriculture on a stable food supply. In *Proceedings of the 22nd Annual Central Plains Irrigation Conference*, Kearney, NE., February 24–25, 2010 Available from CPIA, 760 N. Thompson, Colby, Kansas, USA.

Environment Protection Agency (EPA) (1996). Prospect of Spray Irrigation. Making every drop Counts.

Evans, R. G. & Saddler, J. E. (2008) Methods and technologies to improve efficiency of water use. Accessed online at https://agupubs.onlinelibrary.wiley.com/doi/epdf/10.1029/2007WR006200

FAO. (2002). Smallholder Irrigation: Prospects for sub-Saharan Africa. http://www.fao.org/docrep/004/y0969e/y0969e02.htm

Fereres, E. & Soriano, M. A. (2007) Deficit irrigation for reducing agricultural water use. *Journal of Experimental Botany*, 58(2), 147–159. https://doi.org/10.1093/jxb/erl165

Frisvold, G. B., & Deva, S. (2012). Farm size, irrigation practices, and conservation program participation in the US Southwest. *Irrigation and Drainage*, 61, 569–582.

Frisvold, G. B., & Deva, S. (2015). Climate and choice of irrigation technology: implications for climate adaptation. *Journal of Natural Resources Policy Research*, 5(2–3), 107–127. https://doi.org/10.1080/19390459.2013.811854

Gbangou, T., Ludwig, F., van Slobbe, E., Hoang, L., & Kranjac-Berisavljevic, G. (2019) Seasonal variability and predictability of agro-meteorological indices: Tailoring onset of rainy season estimation to meet farmers' needs in Ghana. *Climate Services*, 14, 19–30.

Ghana Maritime (n.d.) Profile of major rivers in Ghana Access online at https://www.ghanamaritime.org/uploads/39536-profile-of-major-rivers-in-ghana.pdf

Ghana Statistical Service (GSS). (2010). *Annual Report*, Ghana Statistical Service, Accra, Ghana.

Giuliani, M. M., Nardella, E., Gagliardi, A., & Gatta, G. (2017). Deficit irrigation and partial root-zone drying techniques in processing tomato cultivated under Mediterranean climate conditions. *Sustainability*, 9, 1–15. https://doi.org/10.3390/su9122197

Gonzalez-Nunez, L. M., Toth, T. & Garcia, D. (2004) Integrated management for the sustainable use of salt-affected soils in Cuba. Vol. 20. Numero 040- Universidad Juarez Autonoma de Tabasco, Villahermosa, Mexico, USA, pp. 85–102.

Government of Ghana Ministry of Finance and Economic Planning. (2011). National Policy on Public Private Partnership (PPP): Private Participation in infrastructure and services for better public service delivery, Ghana.

Government of Ghana Ministry of Food and Agriculture and the Food and Agriculture Organisation. (2010). *National Irrigation Policy, Strategies and Regulatory Measures*. Accra, Ghana. https://doi.org/10.1017/s204047001700108x

Hanemann, M. (2006). *Tiered Pricing of Water*. Bennett Brooks. Work Group April 11, 2006.

He, L., Horbulyk, T. M., Ali, M. K., Le-Roy, D. G., & Klein, K. K. (2012). Proportional water sharing vs. seniority-based allocation in the Bow River Basin of Southern Alberta. *Agricultural Water Management*, 104, 21–31.

Jayne T. S, Chamberlin J., Traub L., Sitko N., Muyanga M., Yeboah F. K., Anseeuwe W., Chapoto A., Wineman A., Nkonde C., Kachule R. (2016). Africa's changing farm size distribution patterns: the rise of medium-scale farms. *Agricultural Economics*, 47, 197–214.

Kabo-Bah, A. T., Diji, C. J., Nokoe, K., Mulugetta, Y., Obeng-Ofori, D., & Akpoti, K. (2016). Multiyear rainfall and temperature trends in the Volta river basin and their potential impact on hydropower generation in Ghana. *Climate*, 4(4), 49–66. https://doi.org/10.3390/cli4040049

Kabo-Bah, A. T., Sedegah, D. D., Antwi, M., Gumindoga, W., & Eslamian, S., (2021). How to increase water harvesting in Africa. In Eslamian, S. & Eslamian, F. (eds.), *Handbook of Water Harvesting and Conservation, Vol. 2: Case Studies and Application Examples* (pp. 141–152), John Wiley & Sons, Inc., New Jersey.

Kasanga, K., & Kotey, N. A. (2013). *Land Management in Ghana: Building on Tradition and Modernity*, IIED, London, UK.

Kheirabadi, H, Afyuni, M, Ayoubi, S, Soffianian, A. (2016). Risk assessment of heavy metals in soils and major food crops in the province of Hamadan. *Journal of Water and Soil Science*, 19(74), 27–38. https://doi.org/10.18869/acadpub.jstnar.19.74.3

Knox, J., Rodriguez-Diaz, J. A., Weatherhead, E. K., & Kay, M. G. (2010). Development of a water strategy for horticulture in England and Wales. *Journal of Horticultural Science and Biotechnology*, 85(2), 89–93. https://doi.org/10.1558/jsrnc.v4i1.24

Koech, R. & Langat, P. (2018) Improving irrigation water use efficiency: A review of advances, challenges and opportunities in the Australian context. *Water*, 10 (12), 1771.

Kyei-Baffour, N., & Ofori, E. (2007). Irrigation development and management in Ghana: Prospects and challenges. *Journal of Science and Technology*, 26(2), 1–11. https://doi.org/10.4314/just.v26i2.32996

Lichtenberg, E. (1989). Land quality, irrigation development, and cropping patterns in the northern high plains. *American Journal of Agricultural Economics*, 71(1), 187–194.

Mendes, D. M., Paglietti, L., Jackson, D., & Altozano, A. G. (2014). *Ghana: Irrigation Market Brief*. FAO Investment Centre. Food and Agriculture Organization of the United Nations, Rome, Italy.

Mendola, M. (2005). *Agricultural technology and poverty reduction. A micro-analysis of causal effect*, Working Paper Number 2005-14.

Ministry of Food and Agriculture (MoFA). (2007). *The Food and Agriculture Sector Development Strategy*. http://mofa.gov.gh/site/wp-content/uploads/2011/06/FASDEP-II-FINAL1.pdf

Ministry of Food and Agriculture (MoFA). (2009). *Brief on Medium Term Agriculture Sector Investment Plan (METASIP)*. Ministry of Food and Agriculture, Accra, Ghana.

Ministry of Food and Agriculture (MoFA). (2010). Public Irrigation Schemes in Ghana (MoFA).

Ministry of Food and Agriculture (MoFA). (2018). *Investing for Food and Jobs (IFJ): An Agenda for Transforming Ghana's Agriculture (2018–2021)*. http://mofa.gov.gh/site/images/pdf/National Agriculture Investment Plan_IFJ.pdf

Ministry of Food and Agriculture (MoFA). (2019). *Ministry of Food and Agriculture: Operational Performance, Planting for Food and Jobs (2017–2018). Planting for Food and Jobs Policy Document*. Ministry of Food and Agriculture, Accra, Ghana.

Miyoshi, T., & Nagayo, N. (2006). *A Study of the Effectiveness and Problems of JICA's Technical Cooperation from A Capacity Development Perspective: Case Study of Support for the Advancement of Ghana's Irrigated Agriculture*. Case Study Report on Capacity Development. Tokyo, Japan Institute for International Cooperation, Japan International Cooperation Agency. Japan.

Mugisa-Mutetikka, M., Opio, A.F., Ugen, M.A. Tukamuhabwa, P. Kayiwa, B.S. Niringiye, C. & Kikoba, E. (2000). *Logistic Regression Analysis of Adoption of New Bean Varieties in Uganda*. MSc Thesis, Makerere University.

Namara, R. & Hope, L., Owusu, E., Fraiture, C. & Owusu, D. (2014). Adoption patterns and constraints pertaining to small-scale water lifting technologies in Ghana. *Agricultural Water Management*, 131, 194–203. https://doi.org/10.1016/j.agwat.2013.08.023

Namara, R. E., Horowitz, L., Kolavalli, S., Kranjac-Berisavljevic, G., Dawuni, B. N., Barry, B., & Giordano, M. (2010). *Typology of irrigation systems in Ghana*. Colombo, Sri Lanka: International Water Management Institute. 35p, IWMI Working Paper 142, https://doi.org/10.5337/2011.200

Namara, R. E., Horowitz, L., Nyamadi, B., & Barry, B. (2011). *Irrigation Development in Ghana: Past Experiences, Emerging Opportunities, and Future Directions*, Ghana Strategy Support Program Working Paper No. 0027, Ghana.

Namara, R. E., Nagar, R. K., & Upadhyay, B. (2007). Economics, adoption determinants, and impacts of micro-irrigation technologies: Empirical results from India. *Irrigation Science*, 25, 283–297.

Negri, D. H., & Brooks, D. H. (1990). Determinants of irrigation technology choice. *Western Journal of Agricultural Economics*, 213–223.

Norton, G. J., Shafaei, M., Travis, A. J., Deacon, C. M., Danku, J., Pond, D., & Dodd, I. C. (2017). Impact of alternate wetting and drying on rice physiology, grain production, and grain quality. *Field Crops Research*, 205, 1–13.

Opoku-Agyeman, M. (2001). Shifting paradigms: Towards the integration of customary practices into the environmental law and policy in Ghana. Paper Presented by the Water Resources Commission at the Conference on Securing the Future Organized by the Swedish Mining Association, 25 May–1 June 2001, Skelloste, Sweden.

Oteng-Darko, P., Frimpong, F., Sarpong, F., & Amenorfe, L. P. (2018). Promoting smallholder irrigation for food Security - A review. *African Journal of Food and Integrated Agriculture, 2*, 1–7. https://doi.org/10.25218/ajfia.2018.01.001.01

Owusu, A. P., Asumadu-Sarkodie S., & Ameyo, P. (2016) A review of Ghana's water resource management and the future prospect. *Cogent Engineering, 3*, 1164275.

wusu-Sekyere, J. D., & Arthur, J. K. (2012). *Effect of Deficit Irrigation on Yield of Tomato Plant.* Unpublished thesis. Department of Agricultural Engineering, University of Cape Coast, Ghana.

Owusu-Sekyere, J. D., Sam-Amoah, L. K., Teye, E., & Osei, B. P. (2012). Crop coefficient (Kc), water requirement and the effect of deficit irrigation on tomato in the coastal savannah zone of Ghana. *International Journal of Science and Nature, 3*(1), 83–87.

Pandey, R. K., Maranville, J. W., & Chetima, M. M. (2000). Deficit irrigation and nitrogen effects on maize in a Sahelian environment II. Shoot growth, nitrogen uptake and water extraction. *Agricultural Water Management, 46*, 15–27. https://doi.org/10.1016/S0378-3774(00)00074-3

Puértolas, J., Oteng-Darko, P., Yeboah, S., Annor, B., Ennin, S.A. & Dodd, I.C. (2022). Does alternation increase water productivity when applying partial root-zone drying to tomato?. Acta Hortic. 1335, 673-680 DOI:10.17660/ActaHortic.2022.1335.85 https://doi.org/10.17660/ActaHortic.2022.1335.85

Rogers, E. M. (1995). *Diffusion of Innovations.* 4th edition, Free Press, New York.

Rossi, F. R., Filho, M. M. d. S. & Carrer J. M. (2015). Determinants of the Adoption of Irrigation Technologies by Citrus Growers of the State Of São Paulo-Brazil. Accessed from https://www.ifama.org/resources/files/2015-Conference/1237_paper_Rossi_Citrus.pdf

Sandhu, N., Subedi, S. R., Yadaw, R. B., Chaudhary, B., Prasai, H., Iftekharuddaula, K., ... & Venkateshwarlu, C. (2017). Root traits enhancing rice grain yield under alternate wetting and drying condition. *Frontiers in Plant Science, 8*, 1879.

Sarku, R., Dewulf, A., van Slobbe, E., Termeer, K & Kranjac-Berisavljevic, G. (2020) Adaptive decision-making under conditions of uncertainty: The case of farming in the Volta delta, Ghana. *Journal of Integrative Environmental Sciences, 17*, 1–13

Sepaskhah, A. R., & Ahmadi, S. H. (2010). A review on partial root-zone drying irrigation. *International Journal of Plant Production, 4*(4), 241–258. https://doi.org/10.22069/ijpp.2012.708

Spreer, W., Nagle, M., Neidhart, S., Carle, R., Ongprasert, S., & Muller, J. (2007). Effect of regulated deficit irrigation and partial rootzone drying on the quality of mango fruits (*Mangifera indica* L., cv. ' Chok Anan'). *Agricultural Water Management, 88*, 173–180. https://doi.org/10.1016/j.agwat.2006.10.012

Tamene, L., Le, Q. B., Brunner, A., & Vlek, P. L. (2008). Estimating soil erosion and sediment yield in the White Volta Basin using GIS. In *Acts of Glowa Volta International Conference in Ouagadougou Burkina Faso, August 25–28.*

USAID (2017) Climate change risks in Ghana: Country fact sheet. Accessed online at https://www.climatelinks.org/sites/default/files/asset/document/2017_USAID_Climate%20Change%20Risk%20Profile%20-%20Ghana.pdf

Velasco-Muñoz, J. F., Aznar-Sánchez, J. A., Batlles-delaFuente, A., & Fidelibus, M. D. (2019). Sustainable irrigation in agriculture: An analysis of global research, *Water, 11*(9), 1758. https://doi.org/10.3390/w11091758

Wang, Y., Jensen, C. R., & Liu, F. (2017). Nutritional responses to soil drying and rewetting cycles under partial root-zone drying irrigation. *Agricultural Water Management, 179*, 254–259.

Wang, Z., Zhang, W., Beebout, S. S., Zhang, H., Liu, L., Yang, J., & Zhang, J. (2016). Grain yield, water and nitrogen use efficiencies of rice as influenced by irrigation regimes and their interaction with nitrogen rates. *Field Crops Research, 193*, 54–69.

Water Resources Commission (WRC, Ghana) (2015) Water resource management and governance. http://www.wrc-gh.org/water-resources-management-and-governance/ground-water-management/

World Bank. (2007). *World Bank Assistance to Agriculture in Sub-Saharan Africa: An IEG Review.* https://openknowledge.worldbank.org/handle/10986/6907 License: CC BY 3.0 IGO.

World Bank. (2017). *Ghana - Agriculture Sector Policy Note: Transforming Agriculture for Economic Growth, Job Creation and Food Security (English)*, World Bank Group, Washington, DC.

Xu, Y., Huang, Q., & West, G. (2015). Adoption of Irrigation Technology and Best Management Practices under Climate Risks: Evidence from Arkansas, United States. *Southern Agricultural Economics Association's 2015 Annual Meeting*, 1–24, USA.

Zhang, H., & Oweis, T. (1999). Water-yield relations and optimal irrigation scheduling of wheat in the Mediterranean region. *Agricultural Water Management*, *38*, 195–211. https://doi.org/10.1016/S0378-3774(98)00069-9

Zhang, H., Chen, T., Wang, Z., Yang, J., & Zhang, J. (2010). Involvement of cytokinins in the grain filling of rice under alternate wetting and drying irrigation. *Journal of Experimental Botany*, 61(13), 3719–3733.

Zhang, Y., Tang, Q., Peng, S., Xing, D., Qin, J., Laza, R. C., & Punzalan, B. R. (2012). Water use efficiency and physiological response of rice cultivars under alternate wetting and drying conditions. *The Scientific World Journal*. https://doi.org/10.1100/2012/287907

20

Over-Irrigation and Adverse Effects

Yuri Nikolskii
Colegio de Postgraduados

Ivan Aidarov
*Russian Academy
of Sciences*

20.1 Introduction

Over-irrigation has many disadvantages. Therefore, it can lead to the following negative adverse effects:

- Excessive water withdrawal from natural sources of surface and subsurface waters, huge water losses because of seepage from irrigation canals (especially from small ones), and deep water percolation through the soil profile in irrigated lands and runoff from them.
- Exhaustion and pollution of natural surface waters (rivers, streams, lakes, estuaries) and deterioration of their aquatic ecosystems and possibly of the health of people due to pesticide and fertilizers residues runoff from irrigated agricultural lands.
- Rise of groundwater table, waterlogging and/or salinization of soils, and accumulation of toxic substances in soils and groundwaters. Lowering subsurface water level, aquifer depletion, and even subsidence of the land surface when subsurface waters are excessively used for irrigation.
- Gradual (over decades) degradation of soils that are long-term irrigated even with fresh water, the loss of their fertility, and reduction of crop productivity.
- Deterioration of the environmental conditions in the watersheds and population welfare.

Therefore, the irrigation techniques and technologies should be developed, taking into account not only the water needs of plants, but also the need to preserve soil fertility and protect environment.

The aim of this book chapter is to determine and discuss on the adverse effects of over-irrigation.

DOI: 10.1201/9781003353928-26

20.2 What Does the Over-Irrigation Mean?

In order to feed the growing world population, it is necessary to increase productivity of agricultural lands by means of the use of highly productive varieties of crops, plant protection from diseases and pests, as well as the improvement of irrigation technology, where it is needed. It is necessary to take into account the limited possibility to increase the area of irrigated lands in the world.

The main purpose of irrigation is to compensate the possible lack of water in the root layer of the soil to be able to completely utilize the biological potential of highly productive crops.

For this, the soil moisture content in the active root layer (approximately 0.3–0.5 m deep) must be maintained in the range from the permissible minimum to the certain maximum level. The basic consumption of water goes to the evaporation from plant leaves or so-called transpiration, which is necessary for the control of plant body temperature, respiration, photosynthesis, and transport of nutrients from soil. A part of soil water evaporates from the soil surface between the plants. The maximum soil moisture content corresponds to the so-called field capacity, which, under the influence of capillary forces, can be retained in the soil after each watering, although a small part of it can then seepage down along the soil profile. During irrigation applications, a part of water can also run off from the surface of irrigated plots. The minimum limit of permissible soil moisture content depending on the crop and soil texture is approximately 0.6–0.7 of field capacity. The intensity of evapotranspiration, that is, of transpiration together with evaporation is less with decreasing soil moisture content. A decrease in the evapotranspiration is accompanied by a certain decrease in the intensity of plant growth and its productivity. The inability to prevent a decrease in soil moisture content is due to the imperfection of the existing irrigation technique. At present, the surface gravity irrigation with furrows, border strips, and flooding is applied in more than 85% of the irrigated lands of the world, and the rest is irrigated with sprinkling, drip, and micro-sprinkler systems. In developing countries, the gravity irrigation is used in 85%–100% of the total irrigated area. The efficiency of water use in irrigation, estimated by the volume of water used to produce a unit mass of agricultural product in developing countries, is 3,500 m³/t on average for cereals, that is, 3.5 thousand times more than the mass of the product, meanwhile, in the developed countries it is 380–1,000 m³/t (Aidarov, 2012).

The agricultural productivity at traditional irrigation during the 20th century was usually lower than potential. An attempt to significantly increase the crop productivity was made in 1960s, and it was called the Green Revolution. The Green Revolution included three main components:

- Breeding and use of highly productive crop varieties.
- The use of high doses of mineral fertilizers and artificially created chemical plant protectors as different pesticides (herbicides, fungicides, insecticides, etc.).
- Irrigation. In some regions without significant limitation of the available water resource, it was possible to increase irrigation water supply, and maintain relatively high soil moisture and intensity of evapotranspiration closed to potential. Such irrigation can conditionally be called Over-Irrigation.

The Green Revolution has been widely applied in developing countries and its positive effects are undeniable. From the 1960s to the beginning of the 21st century, the world production of cereals has increased by about three times. The population growth and increasing food demand have caused the need to expand the area of irrigated lands. Due to the limited possibility to expand the area suitable for irrigation, the lands containing water-soluble toxic salts in soils and groundwaters were also used for irrigation. In order to prevent the accumulation of salts in soils and crop productivity loss, it was necessary to increase water supply for irrigation of these lands in comparison with the lands without salinity problems.

Along with positive effects, the Green Revolution has led in some regions to a number of negative environmental and socio-economic impacts, including the following:

- Reduction of crop diversity on irrigated lands. Rice, wheat, and corn were the main irrigated crops. Other crops were not effective enough for intensive technologies.

- The spread of monocultures has led to the intensive development of pests and plant diseases and to a decrease in the productivity of irrigated lands.

An increase in water supply to irrigated agricultural lands, high doses of mineral fertilizers, and chemical plant protection substances led in some cases to the following negative impacts:

- Increase in water intake from rivers or lakes, increase in water losses due to the growth of seepage from the network of irrigation canals and in irrigated plots along the soil profile into groundwaters, as well as due to the growth of runoff from the surface of irrigated plots.
- Rise of groundwater table, waterlogging and/or salinization of soils, and accumulation of toxic substances in soils and groundwaters.
- Lowering subsurface water level, aquifer depletion, and even subsidence of the land surface when subsurface waters are excessively used for irrigation.
- Increased pesticide and fertilizer residues run off from agricultural plots, pollution of natural, mainly surface waters (rivers, streams, lakes, estuaries), deterioration of their aquatic ecosystems and of the health of people.
- Exhaustion of available water resource for human use in large areas.
- Soils degradation, loss of their natural fertility, and gradual (over decades) and reduction of crop productivity.
- Deterioration of the environmental conditions in the watersheds and population welfare.

Such adverse negative effects of over-irrigation refer mainly to areas irrigated by means of the surface gravity method with furrows, border strips, and flooding.

Let us analyze in more detail these effects and possible measures to prevent them.

20.3 Environmental and Socio-Economic Possible Impacts of Over-Irrigation

The irrigation worldwide consumes more than 80% of the total human use of water. It is believed that the water is a renewable natural resource. Indeed, the annual precipitation replenishes the reserves of surface and groundwater. A part of these reserves is annually lost into the atmosphere in the process of evapotranspiration by natural vegetation, and another part flows down on the land surface to the seas or lakes. Thus, the balance of fresh water is maintained in large territories.

However, irrigation greatly enhances the water loss into the atmosphere during evapotranspiration from irrigated lands and evaporation from constructed reservoirs. This additional loss of water, as a rule, does not return to the region where it was used for irrigation and it was lost to the atmosphere. It does not return to the area even of a large watershed where the irrigated land is located, as water in the atmosphere is transported by air mass very far from the watershed and, possibly, perhaps even from the continent. Some of the fresh water withdrawn from the nature and used for irrigation, as well as for domestic, industrial, and other purposes, is returned to the environment, but in the form of wastewater contaminated with residues of fertilizers, pesticides, and possibly also toxic soil salts and other toxic chemical and biological substances. Therefore, the natural fresh water used for agricultural irrigation is only a partially renewable and exhaustible natural resource.

The greatest concern is the state of land, water resources, and ecosystems of river basins in Central Asia, the Middle East, and North Africa, where progressive salinization and degradation of irrigated lands are observed, and where water withdrawal for irrigation from rivers and lakes can cause a complete exhaustion of water resource. This leads to a decrease in the productivity of the main crops used in the technology of the Green Revolution. The annual increase in wheat production decreased from 5% in 1980 to 2% in 2005, and in rice and corn from 3% to 1% (FAOSTAT, 2019).

The environmental and socio-economic impacts of irrigation depend on:

- The irrigation technology. Is it gravity surface, sprinkling, or drop irrigation, etc.?
- The type of water distribution network. Is it with pipes or canals? Are the canals with or without lining?

- The quality of water. Is it fresh or saline water or wastewater?
- The source of water. Is it surface or subsurface water?

More water loss occurs at surface gravity irrigation, less at sprinkling, and much less at drop and micro-sprinkling techniques. The water loss in the distribution network is less when the canals have lining and much less when it is constructed with pipes. However, the cost of reducing water losses is greatly increased with the improvement of technology and quality of water distribution network.

Excessive pumping of groundwater may cause a significant increase in the cost of pumping and even their depletion.

A decrease in the available water resource without significant investments in increasing efficiency of water distribution and improving the irrigation technique and technology leads to a decrease in the productivity of irrigated lands, and in their profitability. A deterioration of water quality can also be accompanied by pollution of agricultural products with toxic substances contained in irrigation water.

There are water quality standards for irrigation. Fresh water is preferred. Subsurface waters move and replenish very slowly compared to surface waters. Therefore, available subsurface water reserves can be depleted faster than surface water reserves. However, surface waters are more accessible for withdrawal, but less protected from pollution compared to subsurface water. Therefore, pollution of surface waters is widespread over the world. For example, in Mexico 74% of surface water is polluted to different degrees (CONAGUA, 2017). Due to the lack of fresh water, some countries use slightly saline water or treated and even untreated municipal wastewater. Wastewater is used for irrigation in China, Latin America in the Middle East, Central Asia, etc.

The problem of excessive use and depletion of the water resource, as well as the deterioration of its quality in large areas within the basins of rivers or lakes, takes place in different countries.

For example, the Colorado River, flowing through the United States and partially in its lower reaches through Mexico, in the early 20th century has been dumping 22 km³/year into its delta and the Gulf of California (Sea of Cortes). In addition, the river has been carried about 1.23 km³/year of alluvial material, which was important for the river's delta and its ecosystem. At present, Mexico receives about 2.07 km³/year of water contaminated with chemical and organic substances, in particular, selenium, cadmium, lead, mercury, lithium, free-living amebae, *Escherichia coli*, and pesticides, as a result of its mixing with municipal, industrial, and agricultural wastewaters. This water is mainly used in Mexico for irrigation of agricultural lands, which leads to contamination of fodder and some food products. Because of complete use, in most years, the water of the Colorado River does not reach the Gulf. This has implications for the delta and coastal ecosystems (Cohen et al., 2001; García-Hernández et al., 2001; Varady et al., 2001; Angulo, 2004; Samaniego-Lopez, 2008; Stockle, 2012).

The Yellow River in China, one of the largest rivers in Asia and the first in the world to transport suspended sediments, is contaminated with industrial and domestic sewage, as well as agrochemicals, etc. Since the 60s of the 20th century, it is periodically drying up as a result of excessive use of water for irrigation and other economic needs. The degree of drying is gradually increasing. In 2000, the area of irrigated agricultural lands affected by the lack of water increased more than three times compared to 1976. Reducing the runoff of the Yellow River has a negative impact on the socio-economic development of regions located mainly in the middle and lower parts of the river basin.

The construction of dams on the Yangtze River in China and the Nile River in Egypt led not only to a decrease in their runoff due to evaporation from the surface of reservoirs, but also to termination of suspended sediments transport into the lower reaches, to destruction of river deltas, and to reduction in the supply of irrigated lands with nutrients together with sediments. This forced farmers to apply high doses of mineral fertilizers.

Because of excessive use of subsurface waters for irrigation, these can be depleted. For example, because of this, Texas in the USA has lost 14% of its irrigated land (Stockle, 2012). In Punjab, India, aquifer levels decrease with an intensity of at least 1 m/year. In Mexico, 20% of subsurface waters are depleted; in some places the depth of their water table exceeds 300 m (CONAGUA, 2017).

FIGURE 20.1 Location of the Aral Sea in Central Asia.

One of the most impressive and studied examples of the environmental and socio-economic impacts of over-irrigation is the Aral Sea Basin in Central Asia (Figure 20.1).

20.4 Environmental and Socio-Economic Impacts of Over-Irrigation in the Aral Sea Basin

20.4.1 Development of Irrigation

Until the middle of the 20th century, the Aral Sea was the fourth largest lake in the world. Its length was 426 km and width was 284 km. Two large rivers, Amu-Darya and Syr-Darya, with a total annual runoff of about 100 km³, filled the Aral Sea with water and were the main source of irrigation water in its basin. Until 1950, these rivers had been bringing annually to the Sea about 60 km³ of high-quality fresh water obtained mainly from mountain glaciers.

At present, the Sea practically disappeared. The runoff of both rivers is completely used for irrigation and practically no water comes into the Aral Sea.

For many years, the Aral Sea was used for intensive fishing. There are several small settlements around this sea, where fishing and fish processing were a traditional occupation of the population. The flood plains of both rivers and their extensive deltas with numerous small lakes had a specific flora and fauna. Why did the Aral Sea disappear?

There are mainly five republics in the Aral Sea Basin: Uzbekistan and parts of Kazakhstan, Tajikistan, Kyrgyzstan, and Turkmenistan with a total present population (inside the basin borders) of about

25 million people. The average annual air temperature in the main irrigated part of the Aral Sea Basin is about 14°C, annual precipitation is 100–250 mm, and potential evapotranspiration is 1,200–1,400 mm. The soils are mainly Gypsisols, Calcisols, Phaeozems, and Anthrosols in accordance with a modern soil classification (FAO, 2015).

Irrigated agriculture has been used in this basin for centuries. Formerly, irrigated agriculture was practiced mainly in the areas with rich soils, deep water table, or fresh shallow groundwater. The peasants practically did not construct drainage in irrigated lands, generally using an effect of natural drainage in lands with deep water table or effect of subsurface irrigation in areas with shallow fresh groundwaters. The irrigated lands were located principally in river valleys, deltas, and foothills close to the mountains.

At the beginning of the 20th century, the net water volume delivered to the farmed area was rather small, about 3,000–5,000 m³/ha per year. The irrigated plots were small and well leveled, gravity irrigation was being applied. The irrigation systems at that time had a high level of organization and provided not only the efficient use of water, but also the preservation of the ecological equilibrium of the territories. Irrigation canals in river valleys were 1–1.5 m deep and therefore at the same time served as irrigation and drainage network. Irrigation water was being supplied to the plots using primitive water-lifting devices. The water use efficiency was almost 100%, that is, 100% of the water taken from natural sources was consumed for evapotranspiration. Salinity of irrigated lands was excluded. The principle-irrigated crops were cereals and alfalfa. They occupied more than a half of total irrigated area in the Aral Sea Basin. Cotton occupied less than 20%–30% and rice no more than 5%–15% of irrigated area. In those times, the mineral concentration of river water was about 0.1–0.2 g/L upstream and about 0.4 g/L in the downstream parts close to the Aral Sea.

During the 20th century, the population of the Aral Sea Basin has doubled. At the beginning of the century, about 12 million people lived there, and now there are 25 million people. The irrigation was intensively developed, and the irrigated area almost tripled from 2 to 3 million hectares at the beginning of the century to 7 million hectares at its end. In the middle of the 20th century, due to the land ownership transfer from private to state or collective, the following changes in irrigated agriculture have occurred (Yakubova, 1977; Volynov et al., 1980; Samoilenko et al., 1987; Micklin, 2007; Zonn and Glantz, 2008; Aidarov, 2010; Aidarov, 2012):

- The sense of private land property and the desire to make a personal profit from irrigated agriculture have been lost.
- Large irrigation systems with areas of hundred thousands of hectares were built mainly in the steppe and desert parts of the Aral Sea Basin, far from the river beds, where the water table initially was deep (deeper than 30–50 m from a soil surface). During the 1960s–1980s many dams were constructed in the Amu-Darya and Syr-Darya river basins to increase the ability to take water for irrigation.
- The total irrigated area by the end of the 20th century has increased by more than 2.5 times compared to the beginning of the century and reached 7.9 million ha. Irrigation has been developed mainly by means of construction of new irrigation systems and less by means of reconstruction of old ones.
- The size of irrigated plots was enlarged from fractions of a hectare to tens of hectares.
- Cotton became a main crop in order to increase raw cotton production, to satisfy internal needs, to stop its import, to reach the cotton independency, and to sell cotton abroad to get hard currency. The area of cotton in the crop rotation cycles before the 1940s did not exceed 10%–30%, and then it was increased to 70%–80%. Since the 1950s, cotton occupied 70%–80% of the total irrigated area of the Aral Sea Basin. Alfalfa area percentage in the crop rotation cycles decreased accordingly. The area occupied with rice increased considerably since 1970s mainly in both rivers' deltas. At the beginning of the century the rice occupied 0.2–0.4 million hectares and after the 1980s it occupied about 0.7 million ha.
- The volume of water delivered to the irrigated lands increased more than 1.5 times. The application of mineral fertilizers and different pesticides increased significantly.

About 50% of irrigation waters taken from the rivers was being lost annually because of seepage from a network of irrigation canals and a deep water percolation through the soil profile in irrigated plots. The water lost caused the water table rise, soil salinity, and even waterlogging.

- To prevent soil salinity and waterlogging, agricultural drainage was being constructed on a large scale from the 1940s but more slowly than construction of new irrigation systems. Meanwhile, the irrigation depth was being increased gradually to remove soil salts. At the end of the 20th century about 60% of the irrigated lands had drainage. In some areas, it was a pipe drainage and in others in the form of ditches. The drainage accelerated groundwater outflow and lowered the water table. However, this drainage removed not only toxic soil salts, but also part of the fertilizers and pesticides used in agriculture.

At the beginning, when a critical rise of the water table in irrigated plots was observed in the steppe and desert areas, attempts were made to prevent the water table rising by means of less irrigation depths, reconstruction of irrigation canals and decrease in seepage losses. As the irrigated areas were being extended gradually, those approaches turned out to be insufficient and a drainage construction was started on a massive scale. At the beginning, rather shallow drainage (with the depth of 1.5–2 m) was being constructed. However, practice showed that deep drainage (with the depth of 2.5–3 m) which could maintain the water table at the depth of not less than 2 m would be preferable. The shallower the water table, the more irrigation water should be applied to prevent a soil salinity. Drainage with a depth of 2.5–3 m was being constructed mainly from 1960s.

- At the end of 20th century, the estimated total drainage runoff reached about 45% of the water uptake from the rivers. Half of this amount was being returned to the rivers and it was being used again by irrigation systems located downstream of rivers. About 20% was being discharged to the Aral Sea, and about 30% into the dry lands (some part of it was accumulated in artificial lakes in land depressions). Later analysis of experimental data in the Aral Sea Basin showed that practically the only way to save water and at the same time to prevent soil salinity with the same kind of a use of agricultural lands is to keep the water table at the maximum possible depth (not less than 2.0–2.5 m) and to apply fresh water for irrigation (Averianov, 2015; Aidarov et al., 1990). This approach also helps to prevent groundwater contamination. According to FAO (1980) recommendations, in order to prevent soil salinity it is necessary to keep the water table at the depth not less than 1.5–1.7 m and to construct drains at the depth of about 2–2.5 m.

20.4.2 Environmental and Socio-Economic Consequences of Irrigation

As a result of growth in irrigated area and in water consumption, the both rivers' runoff was used completely by the end of the 20th century. The total water consumption in irrigation increased from 10 to 15 km^3/year in 1900–1930s to 40 km^3/year in the 1960s and more than 85 km^3/year since 1985 (Reshetkina, 1991). About 6–7 km^3/year of rivers' runoff were being lost since 1985 because of evaporation from reservoirs in the rivers and additional losses of about 10–20 km^3/year of water because of seepage in irrigation canals and discharge of unused waters (mainly drainage waters) into the desert. Within the basin, drainage waters have been removing from the irrigated lands mainly to the rivers. Downstream, the river waters mixed with drainage waters were being reused for irrigation and for domestic purposes.

The total river runoff into the Aral Sea began to decrease considerably in 1950s and especially since the 1960s–1980s. It dropped from 60 km^3/year before 1940 to 10–15 km^3/year at the end of the 1970s. By 1985, the rivers had ceased their runoff to the Aral Sea. Due to the reduction and, later, cessation of river flow downstream, the level of the Aral Sea began to sink with accelerating intensity since the 1950s. At the end of the 20th century, the sea practically disappeared.

A dried salty desert with an area of about 68,000 km^2 and with a salt content of about 100–300 tons/ha in the upper layer 1 m deep has been formed on the former sea bottom. The salty dust is carried with the wind from this surface every year to a distance more than 300 km from the sea. About 1 ton of salts falls annually on each hectare of rivers' deltas. Salty dusty clouds rise very high and even reach the mountain glaciers, accelerating their melting. Rains with a salt concentration up to 160 mg/L have appeared in the coastal area of the former Aral Sea (Razakov, 1990).

Since the 1980s, fish has disappeared from the sea. All fishing villages and fishing boats are now far away from the sea. The former coastal population has lost jobs. The annual flooding of the flood plains in the lower reaches of both rivers has ceased. Because of this, there is a degradation of rich alluvial soils. The water table in the river deltas has sunk and the soil formation processes have changed from hydromorphic and semihydromorphic to automorphic. There is desertification of the river deltas. Numerous lakes with fresh water (some of them used earlier for fishing) have dried up. Lakes with toxic water have appeared. Natural pastures for domestic and wild animals have disappeared. The endemic flora and fauna in the littoral zone of the former Aral Sea have disappeared. The amplitude of annual air temperature oscillation has risen 2°C–3°C up to a distance of 100 km from the former Sea.

The salt concentration of river waters has been growing through the years because of the enormous discharge of drainage water and its reuse. The quality of river water has deteriorated, especially in the lower reaches. The sharpest increase in salt concentration of river water took place in the 1970s and 1980s. Together with soil salts, the river water transported toxic agricultural substances (different pesticides, nitrates, etc.). The river runoff annually received about 0.3–0.5 kg of pesticides (mainly in clororganic forms) and about 50 kg of nitrogen from each hectare of cotton fields (Yakubova, 1977; Samoilenko et al., 1987). Since the quantity of river water has been reduced and the quality of water has deteriorated, the problem of water distribution for irrigation between the states in the basin has increased. The republics located in the upstream parts of the basin used the water with better quality than those located downstream.

In spite of drainage construction and growth of water volume delivered to the farmed plots, soil salinity has been increasing gradually. It has become noticeable especially in the downstream areas. For example, at the end of 1980s moderately and highly saline soils occupied 30%–40% and 60%–70% of irrigated lands located, respectively, in the upstream (Tadzhikistan and Kirgizstan) and downstream areas (Uzbekistan, Turkmenistan, and Kazakhstan). In the areas close to the Aral Sea, the problem of soil salinity appeared in 70%–90% of irrigated lands (Reshetkina, 1991; Aidarov, 2010). As a result of that, cotton and other crops harvest in the entire Aral Sea Basin began to decline.

To increase the cotton productivity farmers used more and more water, fertilizers, and pesticides. In addition, new drainage was being constructed and irrigation systems were being reconstructed, however, with not sufficient intensity. During period from 1940s to 1980s the net water volume delivered to the farmed plots increased from 5,000–6,000 m^3/ha to 12,000–15,000 m^3/ha and the amount of mineral fertilizers (NPK) increased from 50–100 to 300–500 kg/ha. In 1980s, the amount of nitrogen applied to cotton fields grew up to 250 kg/ha (in comparison with those for cereals of 50 kg/ha) and the amount of pesticides increased up to 20–40 kg/ha annually (Samoilenko et al., 1987). Nevertheless, since the 1970s, cotton yield began to decrease gradually. The problem of the irrigated soils contamination with pesticides has appeared. Since 1980s more than 75% of the total irrigated area of the basin had high levels of soils contamination with pesticides, and 40%–50% of the irrigated lands had very high pesticide concentration in the soils, which exceeded permissible levels by 3–17 times (Yakubova, 1977). The contamination of surface and groundwaters increased in the middle part and even more severe in the lower part of the Aral Sea Basin. It was noticed that in the lower part of the basin, the concentration of pesticides in the meat of domestic animals exceeded a permissible level by eight times, and in vegetables by 16 times (Yakubova, 1977; Bezugli et al., 1987; Nesterov, 1990).

The number of serious diseases, such as viral hepatitis, typhoid fever, and cancer of the esophagus, has increased 5–30 times during the 1960s–1980s in the downstream area of the basin. Infant mortality has doubled there. Extremely severe medical problems appeared in Karakalpak Autonomous Republic of Uzbekistan, which is located in the littoral zone of the Aral Sea. Since the 1980s, the incidence of gall stones and cancer of the esophagus grew 8–10 times, and heart and kidney diseases grew about two times. The salinity of mothers' milk has exceeded the permissible level by 3–4 times (Bezugli et al., 1987; Nesterov, 1990). In the areas close to the former Aral Sea a number of respiratory diseases occurred because of blowing salt and dust from dried sea bottom.

The main causes of the ecological crisis, it seems, are the following:

a. Excessive growth of the irrigated area without modernization of existing irrigation systems;
b. Inefficient water use;
c. Rivers' water pollution and its use for irrigation and domestic needs;
d. Poor sanitary and environmental conditions of human life.

For example, in Uzbekistan (which has more than 60% of total population of the Aral Sea Basin and occupies the principal part of it) 70% of the hospitals did not have piped sewage systems, and 80% of them did not have hot water (Bezugli et al., 1987).

The irrigation area growth and inefficient water use led to the necessity for drainage construction and the management of huge amounts of agricultural drainage waters. Formerly, it was supposed that drainage would prevent not only soil salinity and waterlogging but it would lead to a desalinization of groundwater as well. Therefore, fresh groundwater could be used partly for subirrigation as it was done in ancient times. It was supposed also that the repeated use for irrigation of drainage waters mixed with river waters would increase the effective use of limited water resources. However, it turned out that the salt concentration of drainage water increased. Contaminated river waters and aquifers have been used not only for irrigation but also for a human consumption. This led to growth of a number of severe diseases among the population. The rise of saline groundwater in irrigated lands led to the necessity of increasing the irrigation depths to prevent soil salinity in spite of the drainage being constructed. Together with low efficiency of irrigation systems, poor control of water use and growth of irrigated area have led to depletion of water resources, disappearance of Aral Sea, growth of salinity of agricultural lands, and some other impacts.

20.5 The Basic Ways to Reduce Adverse Effects of Over-Irrigation

Many suggestions have been made by a number of specialists from different countries to solve the problem of the Aral Sea Basin. Some of them are the following:

a. It is necessary to purify the drinking water in order to reduce the incidence of human diseases in the lower reaches of the Amu-Darya and Syr-Darya rivers.
b. It is necessary to reduce seepage losses in the irrigation canals and to control strictly water consumption in agriculture.
c. It is necessary to reduce consumption of water resource by means of the following:

- Reduce the irrigated areas (first of all in the lands with low productivity and with certain difficulties of reclamation);
- Create other enterprises with a smaller water consumption to employ people and to save water;
- Reduce the area dedicated to cotton;
- Grow less-water-consuming crops;
- Improve, step-by-step, irrigation techniques, substituting the furrow and border irrigation systems with better methods (sprinkling, drop irrigation, micro-sprinkling, etc.) which use water with better efficiency, and reduce seepage loss through the soil profile in irrigated plots. Reconstruction of irrigation canal networks, improvement in their management and in irrigation technology, as well as drainage reconstruction, should reduce drainage runoff.

    d. Drainage waters should not be mixed directly with river water. They must be channeled into the Aral Sea or into the natural depressions.

    e. The Aral Sea cannot be restored in its former view. However, by means of saving water it could be possible to supply annually the Aral Sea with 20–30 km³ of water including drainage waters as a main component. It is necessary to use special methods (like a cascade of bordered dams and areas between them periodically flooded by influent waters) to prevent sand, dust, and salt removal from the dried sea bottom. The experience of Kazakhstan has shown that by means of construction of bordered dams and flooding of fenced areas, it is possible to gradually improve the condition of certain parts of the sea.

20.6 Soil Degradation Due to Over-Irrigation

Irrigation increases the productivity of agricultural lands and inevitably affects soil properties. Irrigation can increase soil fertility, but can also reduce it. The process of soil properties modification, as a rule, is rather slow, measured over a number of years or even tens of years (Arnold et al., 1990; Schlesinger and Bernhardt, 2013). It is known that agricultural activities modify the natural soil formation processes and soil fertility. Irrigation not only enhances soil formation processes but also causes soil leaching. The impact of irrigation on soils is related to type of land use, practices of crop and soil management, as well as irrigation technology and the quality of irrigation water. When irrigation uses water of a good quality and does not cause soil salinity or sodicity or contamination, and when there is no soil erosion, a gradual increase, or at least conservation of soil fertility is, as a rule, expected (Hagan et al., 1987; Molden, 2007). However, there are publications indicating that long-term freshwater over-irrigation under different climatic conditions can lead to a gradual deterioration of the basic properties of irrigated soils and a decrease in their fertility, even if neither erosion, nor salinization or sodicity is observed. This occurs, because of leaching of organic matter (OM) and a number of useful exchange cations into the deep layers of the soil profile below the root zone.

Traditionally, irrigation scheduling and irrigation depths are determined based on the water needs of agricultural crops, particularity for plant growth, applied agricultural technology, hydrophysical soil properties, and on the availability and quality of water resource.

The positive impact of irrigation on soil fertility is associated with an increase in soil moisture content, intensification of microbiological and macrobiological activities, accumulation of OM, increase in cation exchange capacity (CEC), and better efficiency in the use of organic and mineral fertilizers (Aidarov, 1985; Baldock et al., 2000; Oriola, 2003; Schjonning et al., 2004; Aidarov and Nikolskii-Gavrilov, 2016). It should be noted that OM is the main component of natural soil fertility.

The negative impact of irrigation on soil is usually associated with water erosion of soil, the use of water of poor quality (high salinity or poor chemical composition, as well as treated or untreated wastewater), soil salinity, or sodicity. There are various recommendations for reducing the risk of soil and plant contamination, preventing soil salinity and sodicity and the possible decline in the productivity of irrigated lands (Aidarov, 1985; Alguacil del Mar et al., 2012; Artigao et al., 2002; Skaggs and van Shilfgaarde, 1999).

Studies, unfortunately few and heterogeneous in terms of goals and methods, on impact of long-term freshwater over-irrigation on soil fertility were carried out in North and South America, Europe, Russia, Southeast Asia (mainly in China and India), Israel, North Africa, Australia, and New Zealand, but did not have a systemic character and were not generalized.

For example, the studies performed in California and Arizona, USA, in arid and semi-arid climates, showed that conventional freshwater gravity irrigation of sandy loam and clay loam calcareous soils during 90 years led to deterioration of the soil fertility. In particular, it was noted a 56%–62% reduction of the organic carbon (OC) content, as a main component of OM, in the 0–30 cm soil layer in comparison with non-irrigated soil as a result of leaching into deep layers of the soil profile (Eshel et al., 2007; Artiola and Walworth, 2009). Soil was slightly alkaline with pH=7–8. The irrigation water was of good quality with 0.1–0.6 g/L total dissolved solids (TDS) and electrical conductivity (EC) of 0.2–1 ds/m. It was also noted that the greatest decrease in OC and OM corresponded to irrigation with water with TDS less than 0.2 g/L or EC less than 0.3 ds/m in comparison with relatively saline waters. Presence of sodium ions (Na) significantly increased leaching of OC and OM. The degree of OC and OM leaching strongly correlated with the solubility of soil carbonates. A slight increase in silt content (up to 5%) was also observed under long-term freshwater over-irrigation.

Getahun et al. (2011) studied the impact of freshwater over-irrigation for 20–30 years on sandy loam, clay loam, and clay soils (Alisols, Fulvisols, Cambisols, Vertisols) in Ethiopia. These soils were formed under conditions of semi-humid and humid climates with average annual precipitation of 850–2,000 mm and intensive leaching. Soil pH was 5–7 in the 0–20 cm layer, and CEC=30–50 cmol(+)/kg. TDS of irrigation water varied at different sites from 0.1 to 0.7 g/L and EC from 0.2 to 0.9 ds/m. Irrigation and increased leaching had different effects on soils, depending on their texture and salinity of the irrigation water. In sandy loam and clay loam soils, long-term freshwater irrigation caused deterioration of the soil fertility, decreasing OM content by a factor of 1.2–1.5 times. In clay soils, irrigation led to an increase in OM content by a factor of 1.5–2.0. Plant-available phosphorus (P), exchangeable magnesium (Mg), and soil bulk density increased at all the sites. Exchangeable calcium (Ca) and CEC noticeably decreased in sites where EC of irrigation water was less than 0.3 ds/m and slightly increased in sites where EC=0.9 ds/m. Soil pH did not change significantly at any of the sites.

In New Zealand, it was found that in regions with a semi-humid and humid subtropical climate and annual precipitation of 600–1,600 mm, even drip irrigation of orchards with rainwater for 18 years led to a decrease in the soil fertility. In particular, OC content reduced from 5.4 to 5.1 g/kg and CEC from 3.8 to 2.9 cmol(+)/kg in the 0–15 cm layer of sandy soils (Siggins et al., 2016).

In India, in the region of a semi-humid tropical climate with annual precipitation of 1,100 mm, gravity freshwater irrigation for 15 years resulted in less soil fertility due to a decrease in OC content from 6.6 to 5.0 g/kg and CEC from 21.2 to 16.0 cmol(+)/kg in the 0–22.5 cm layer of sandy loam soil (Ghosh et al., 2011).

The impact of long-term freshwater gravity and sprinkling irrigation on Chernozem and Kastanozem soils was also studied in the semi-arid zone of Russia with annual precipitation of 300–650 mm. Irrigation water had TDS=0.4–0.7 g/L and EC=0.3–0.4 ds/m (Aidarov, 1985; Aidarov, 2012). The results showed that freshwater irrigation of Chernozem and Kastanozem soils for 20 years deteriorated the soil fertility. The changes in the properties of Chernozems and Kastanozems in the 0–20 cm layer were as follows:

- Decrease in *OM* content from approximately 6.7% and 2.7% to 5.9% and 2.3% in Chernozems and Kastanozems, respectively;
- Increase in the content of fulvic acids (FA), as a easily soluble and leachable component of *OM*; the ratio between slightly soluble humic acids and *FA* decreased from 1.0 and 1.2 to 0.6 and 0.8, respectively;
- Decrease in *CEC*, accumulation of *Na* and *Mg*, and reduction in Ca;
- Reduction in the content of plant-available mineral nutrients;
- Increase in soil bulk density.

Increasing annual irrigation depth caused more intensive deep water percolation through the soil profile from 0.14–0.17 to 0.30–0.40 of the mean annual water supply (irrigation depth Ir plus precipitation Pr) and led to the following effects after 20 years:

- *OM* leaching into the deep soil layers increased from 5%–10% to 30%–40% of its initial content;
- *CEC* decreased by 2%–5% and 15%–20%, respectively, in comparison with its value before irrigation;
- The content of *Na* and *Mg* in the *CEC* increased up to two-fold; *Ca* decreased by a factor of 1.5.

Special field experiments were realized in Brazil in order to determine the effect of irrigation depths on leaching of OC in sandy loam soil, classified as Quartzarenic Neosol (Diogenes et al., 2017). The basic soil properties in the 0–20 cm layer were as follows: OM=0.37%, CEC=3.19 cmol(+)/kg, pH=5; absorbed bases (as a percentage of their total amount): Ca=18%, Mg =12%, Na =1%, K=4%. Climate is semi-arid with mean annual precipitation (Pr) of 900 mm. Irrigation was carried out in the dry period of the year with irrigation depths (Ir) of 108, 215, 288, and 426 mm in different plots, which corresponded proportionally to 0.3, 0.6, 0.9, and 1.2, respectively, of the potential evapotranspiration. The agricultural plant was cowpea. The results showed that annual application of each 1 mm of irrigation water in those conditions resulted in the accumulation of about 8 kg/ha/year of OC from plant residues. At the same time, as the irrigation depth increased, the annual decomposition and leaching of OM were enhanced, despite the growth of plant residue biomass. As the irrigation depth increased, the intensity of OC leaching progressively increased to about 23% of its initial content. Overall, the OC balance was positive: the final OC content was more than at the beginning for all irrigation depths except where the irrigation depths were equal or more than potential evapotranspiration.

The impact of long-term freshwater irrigation on soil fertility at a regional level in different climatic zones was assessed also in Mexico by means of statistical comparison of the properties of soils irrigated during more than 50 years and non-irrigated virgin soils within large territories in different climatic zones. The selected irrigated and non-irrigated soils are located in flat, geomorphologically homogeneous lands with slopes less than 1%. The water table is located rather deep, at the depth of more than 5 m. The basic technique of irrigation is conventional gravity irrigation. It is applied in more than 85% of the total area of irrigated lands. The principle source of irrigation is surface water from reservoirs located in mountainous regions, supplying more than 80% of the total irrigated area. The typical quality of surface water is good.

The mean annual climatic conditions in these zones are characterized by the ratio between the values of potential evapotranspiration (ET) and Pr. In the tropical humid zone, the ET/Pr ratio is approximately less than 0.7; in the semi-humid and semi-arid zones, the ET/Pr ratio varies from 0.7 to 1.0 and from 1.0 to 3.0, respectively; in the arid zone ET/Pr is more than 3.0.

The compared irrigated and non-irrigated lands do not have problems neither with water or wind erosion, nor with salinity and sodicity of soils. The fertility of irrigated agricultural soils and non-irrigated virgin soils in the humid tropical, semi-humid, semi-arid, and arid zones of Mexico was compared (Nikolskii et al., 2019).

The soils are mainly Phaeozems in the semi-arid and semi-humid regions of the central part of the country; Xerosols (or Calcisols), Rendzinas (Leptosols), and Kastanozems in the arid and semi-arid zones of the central and northern regions; and Luvisols and Vertisols (FAO, 2015) in the temperate and tropical humid regions in the central and southeastern part of the country. Soil depth usually exceeds 1.5 m, and all soils have good agricultural potential, high saturation of exchangeable bases, and the presence of stable aggregates. The Rendzinas have a predominance of calcium cations in an adsorbed complex. Vertisols are clay-based and characterized by expansion–contraction and cracking as a result of annual wetting and drying of the soil. The problems of soil salinity or sodicity are not observed; there is also no problem of soil erosion in analyzed territories.

The Mexican study showed the following outcomes (SEMARNAT-CP, 2002; Ortiz-Solorio, 2011; Nikolskii et al., 2019):

1. The long-term freshwater over-irrigation can cause a decline in soil fertility in different climatic conditions because of leaching of OM into deep layers of the soil profile. Such an effect is typical for soils with relatively high permeability, and mainly with sandy, sandy loamy, and loamy textures. The loss of soil fertility in Mexico occurs in about 1.6×10^6 ha or 20% of the irrigated lands.
2. The observed effect of reducing soil fertility under long-term freshwater over-irrigation is typical for soils with relatively high permeability, and mainly with sandy, sandy loamy, and loamy textures.

   It can be assumed that in order to preserve soil fertility, it is necessary to reduce the intensity of deep water percolation through the soil profile in irrigated lands. This can be achieved by improving techniques and technology of irrigation. In humid tropical conditions, where there is also a negative effect of irrigation on soil fertility, it is advisable to increase the surface runoff of rainwater by applying surface drainage, leveling the surface of agricultural lands, and other measures.

3. The OM content in irrigated soils after more than 50 years of irrigation is significantly less than that in non-irrigated soils in the semi-arid and semi-humid climatic zones of Mexico, where ET/Pr ratio is between 0.7 and 2.0, approximately. The most significant loss of OM is observed in the semi-humid zone, where ET/Pr ratio is between 0.7 and 1.0. The OM loss in humid conditions is also observed but in less extent than that in the semi-arid and semi-humid climatic zones. The main reason for the loss of OM and soil fertility during long-term freshwater over-irrigation is a change in the historical conditions of soil formation. An increase in soil moisture content generates a number of interrelated soil processes, including intensification of microbiological activity, increase in solubility and mobility of organic compounds, increase in the intensity of deep percolation of irrigation water and rainwater through the soil profile below the root zone, leaching of mineral and organic substances, etc. The most intensive annual water percolation through the soil profile occurs in irrigated soils in semi-humid and humid conditions, where percolation of irrigation water is added to the natural annual percolation of rainwater. In the humid tropical zone, irrigation is applied in the dry seasons. Therefore, the contribution of irrigation to the change in the mean annual intensity of natural seepage of rainwater is much less. In the semi-arid and arid zones, the probability of coincidence of irrigation and precipitation is less than that in semi-humid conditions. Therefore, the mean annual intensity of deep seepage of water through the soil profile is less than that in semi-humid conditions. The amount of OM leaching from the top soil to deep layers is proportional to the intensity of soil water percolation and the content of OM in the top soil. The application of increased doses of mineral fertilizers in irrigated soils reduces the pH of soils and often enhances the OM leaching (Black, 1968; Schlesinger and Bernhardt, 2013; Strawn et al., 2015).

Considering that OM is the main component of soil fertility, on which the productivity of irrigated lands depends, it can be concluded that over-irrigation most noticeably affects soil fertility in semi-arid and semi-humid conditions.

In the arid zone, the content of OM in irrigated soils is about twice more than that in non-irrigated soils. This can be explained by the fact that in the annual balance of the OM of irrigated soil in these zones compared with others, the amount of newly formed organic mass exceeds its decrease because of mineralization and leaching into deep layers of the soil profile.

It is important to note that the process of soil degradation under long-term freshwater over-irrigation occurs in both developing and developed countries. The term "over-irrigation" is associated mainly with imperfect techniques and technology of irrigation and therefore with the excessive use of water. The direct cause of irrigated soil degradation is a violation of natural biochemical processes. It is known that the biological productivity of agricultural plants and their consumption of soil nutrients are significantly higher than that by natural plants. However, up to 80% of the agricultural plant biomass is removed from the field through harvest. Therefore, a negative balance of OM is formed even in non-irrigated soil. Irrigation enhances the leaching of OM and soil nutrients. Natural plant communities are characterized by an equilibrium or a slight excess of crop residues introduced into the soil compared to their decomposition and removal. Underestimation of the problem of possible degradation of irrigated soils seems to be because the degradation processes proceed slowly over decades. Therefore, little attention is given to this problem. Meanwhile, it is necessary to take into account that the soil is an exhaustible natural resource, which on a lifetime scale of one generation is practically non-renewable.

20.7 Conclusions

1. The irrigation worldwide consumes more than 80% of the total human use of water. It is believed that water is a renewable natural resource. Indeed, the annual precipitation replenishes the reserves of surface and groundwaters. However, irrigation greatly enhances the water loss into the

atmosphere during evaporation from irrigated lands and constructed reservoirs. Moreover, some of the fresh water withdrawn from the nature and used for irrigation is returned to the environment in the form of wastewater contaminated with residues of fertilizers, pesticides, and possibly also toxic soil salts. Therefore, the natural fresh water used for agricultural irrigation is only a partially renewable and exhaustible natural resource.

2. Over-irrigation refers mainly to the surface gravity irrigation with furrows, border stripes, and flooding, which is most widespread in the world, especially in regions without significant limitation of the available water resource. This irrigation is characterized by imperfection of techniques and technology of irrigation and is often accompanied by an imperfect state of the distribution canal network when it does not have lining to prevent water seepage losses.

3. Over-irrigation can lead to the following negative adverse effects (Albaji et al., , 2020)

   - Excessive water withdrawal from natural sources of surface and subsurface waters, huge water losses because of seepage from irrigation canals (especially from small ones), and deep water percolation through the soil profile in irrigated lands and runoff from them.
   - Exhaustion and pollution of natural surface waters (rivers, streams, lakes, estuaries) and deterioration of their aquatic ecosystems and possibly of the health of people due to pesticide and fertilizers residues runoff from irrigated agricultural lands.
   - Rise of groundwater table, waterlogging and/or salinization of soils, and accumulation of toxic substances in soils and groundwaters. Lowering subsurface water level, aquifer depletion, and even subsidence of the land surface when subsurface waters are excessively used for irrigation.
   - Gradual (over decades) degradation of soils that are long-term irrigated even with fresh water, the loss of their fertility, and reduction of crop productivity.
   - Deterioration of the environmental conditions in the watersheds and population welfare.

4. Irrigation techniques and technology should be developed, taking into account not only the water needs of plants, but also the need to preserve soil fertility and protect environment.

References

Aidarov I.P. 1985. *Patterns in Water, Salt and Fertility Regimes of Irrigated Soils.* Publ. Agropromizdat, Moscow, USSR (in Russian).

Aidarov, I.P. 2010. Ways to solve regional water problems, using as an example, the Aral Sea basin. *Meliotratcia y vodnoye khozyaistvo*, 5: 43–48 (in Russian).

Aidarov, I.P. 2012. *Ecological Fundamentals of Land Reclamation.* MGUP Publ., Moscow, Russia, p. 177 (in Russian).

Aidarov, I.P., and Nikolskii-Gavrilov, I. 2016. Modeling and its application for interpretation of soil quality. In: Lucke, B., Bäumler, R., and Schmidt, M. (Eds.). *Soils and Sediments as Archives of Landscape Change, Geoarchaeology and Landscape Change in the Subtropics and Tropics.* Publ. Selbstverlag der Fränkischen Geographischen Gesellschaft in Kommission bel Palm&Enke, Germany, pp. 329–348.

Aidarov, I.P., Golovanov, A.L., and Nikolski, Y.N. 1990. *Optimization of Water, Heat and Fertilizer Regimes in Irrigated and Drained Agricultural Lands.* Ed. Agropromizdat, Moscow, p. 60 (in Russian).

Albaji, M., Eslamian, S., Naseri, A. and Eslamian, F. 2020. *Handbook of Irrigation System Selection for Semi-Arid Regions.* Taylor and Francis, CRC Group, Boca Raton, USA, 317 Pages.

Alguacil del Mar, M., Torrecillas E., Torres P., Garcia-Orenes F., and Roldan A. 2012. Long-term effects of irrigation with waste water on soil fungi diversity and microbial activities: the implications for agroecosystem resilience. *PLoS One* 7(10): 1–7. DOI: 10.1371/journal.pone.0047680

Angulo, C. 2004. El delta del río Colorado, situación crítica. Available in: http://dignidadysupervivencia003. blogspot.com/

Arnold, R.W., Szabolcs, I., and Targulian, V.O. (Eds.). 1990. *Global Soil Change.* IIASA Publ., Laxenburg, Austria.

Artigao, A., Ortega, J.F., Tarjuelo, J.M., and Juan, J.A. 2002. The impact of irrigation application upon soil physical degradation in Castilla-La Mancha (Spain). *Advances in Geoecology* 35: 83–90.

Artiola, J.F., and Walworth, J.L. 2009. Irrigation water quality effects on soil carbon fractionation and organic carbon dissolution and leaching in a semiarid calcareous soil. *Soil Science* 174(7): 356–371. DOI: 10.1097/SS.0b013e3181aea7b4

Averianov, S.F. 2015. *Water Management of Reclaimed Agricultural Lands*. RGAU Publ., Moscow, Russia, p. 523. (in Russian).

Baldock, D., Caraveli, H., Dwyer, J., Einschutz, S., Petersen, J.E., Sumpsi-Vinas, J., and Varela-Ortega, C. 2000. *The Environmental Impacts of Irrigation in the European Union*. Publ. Institute for European Environmental Policy, London.

Bezugli, V.P., Barda, L.K., and Gorskaya, N.Z. 1987. Role of pesticides in human diseases in a hot climate. In: *Hygienic and Biologic Aspects of Pesticide Application in Central Asia and Kazahstan*. Dushanbe, Kazahstan, pp. 254–286.

Black, C.A. 1968. *Soil-Plant Relationships*. Second edition. Publ. Wiley Publ., New York.

Cohen, M.J., Henges-Jeck, C., and Castillo-Moreno, G. 2001. A preliminary water balance for the Colorado River delta, 1992–1998. *Journal of Arid Environments*, 49(1): 35–48. DOI: 10.1006/jare.2001.0834

CONAGUA. 2017. *Estadísticas del Agua en Mexico*. Publ. Comision Nacional del Agua. Available in: https://agua.org.mx/biblioteca/estadisticas-del-agua-en-mexico-edicion-2016/

Diogenes, L.C., Filho, J.F.L., Da Silva A.F.T., Nobrega J.C.A., Nobrega R.S.A., Filho J.I., and De Andrade Junior A.S. 2017. Microbial activities, carbon, and nitrogen in an irrigated Quartzarenic Neosol cultivated with cowpea in southwest Piauí. *Revista Brasileira de Engenharia Agrícola e Ambiental*, 4(3): 348–354. DOI: 10.5433/1679-0359.2017v38n4p1765

Eshel, G., Fine, P., and Singer, M.J. 2007. Total soil carbon and water quality: an implication for carbon sequestration. *Soil Science Society of America Journal*, 71(2): 397–405. DOI:10.2136/sssaj2006.0061

FAOSTAT. 2019. *World Food and Agriculture. Statistical Pocketbook*. Publ. FAO, Italy, p. 248.

Food and Agriculture Organization of the United Nations (FAO). 1980. *Drainage Design Factors*. Irrigation and Drainage Paper, No. 38, Rome, Italy.

Food and Agriculture Organization of the United Nations (FAO). 2015. *World Reference Base for Soil Resources*. World Soil Res. Rep. No. 103. Publ. IUSS Working Group WRB, Rome, Italy.

García-Hernández, J., King, K.A., Velasco, A.L., Shumilin, E., Mora, M.A., Edward, P., and Glenn, E.P. 2001. Selenium, selected inorganic elements, and organochlorine pesticides in bottom material and biota from the Colorado River delta. *Journal of Arid Environments*, 49(1): 55–89. DOI: 10.1006/jare.2001.0836

Getahun, M., Adgo, E., and Atalay, A. 2011. Impacts of irrigation on soil characteristics of selected irrigation schemes in the Upper Blue Nile. In: A.M. Melesse (Ed.), *Nile River Basin. Hydrology, Climate and Water Use*, Springer Science+Business Media Publ., London; New York, pp. 383–389. DOI: 10.1007/978-94-007-0689-7_19

Ghosh, A. K., Bhatt, M.A., and Agrawal, H.P. 2011. Effect of long-term application of treated sewage water on heavy metal accumulation in vegetables grown in Northern India. *Environmental Monitoring and Assessment*, 184: 1025–1036. DOI: 10.1007/s10661-011-2018-6

Hagan, R.M., Haise, H.R., and Edminster, T.W. (Eds.). 1987. *Irrigation of Agricultural Lands*. Am. Soc. Agron. Publ, No. 11, Madison, USA.

Micklin, P. 2007. The Aral Sea disaster. *Annual Review of Earth and Planetary Sciences*, 35: 47–72.

Molden, D. (Ed.) 2007. *Water for food, Water for life: A Comprehensive Assessment of Water Management in Agriculture*. Earthscan/IWMI Publ., London; Colombo.

Nesterov, E.A. 1990. In the zone of ecological crisis. *Melioratcia y Vodnoe Hoziaistvo*, 2: 6–8 (in Russian).

Nikolskii, Y.N., Aidarov, I.P., Landeros-Sanchez, C., and Pchyolkin, V.V. 2019. Impact of long-term freshwater irrigation on soil fertility. *Journal of Irrigation and Drainage*, 68(5): 993–1001. DOI: 10.1002/ird.2381

Oriola E.O. 2003. Effects of irrigation on soils of a sub-humid part of Kwara state, Nigeria. *Centrepoint (Science Edition)* 12: 52–62.

Ortiz-Solorio, C.A. 2011. Cartografía de la degradación de suelos en la República Mexicana; evolución y perspectivas. In: P. Krasilnikov, F.J. Jiménez-Nava, T. Reyna-Trujillo, N.E. García-Calderón (Eds.), *Geografía de suelos de México,* UNAM, Ciudad de México, México, Capitulo 10.

Razakov, R.M. 1990. Investigations and programs to improve an ecological situation in the littoral zone of the Aral Sea. *Melioratcia y Vodnoe Hoziaistvo*, 1: 6–8 (in Russian).

Reshetkina, N.M. 1991. The Aral Sea basin-self managed system. *Melioratcia y Vodnoe Hoziaistvo*, 10: 13–18 (in Russian).

Samaniego-Lopez, M.A. 2008. El control del río Colorado como factor histórico. *La necesidad de estudiar la relación tierra/agua. Frontera Norte*, 20(40): 49–78. Available in: http://www.scielo.org.mx/scielo.php?script=sci_arttext&pid=S0187-73722008000200002

Samoilenko, V., Yacubova, R., and Kaharov, A. 1987. *Protection of Subsurface Waters from Contamination with Pesticides.* Ed. Mehnat, Tashkent, p. 189 (in Russian).

Schjonning, P., Elmholt, S., and Christensen, B.T. (Eds.). 2004. *Managing Soil Quality: Challenges in Modern Agriculture.* CABI Publ., Wallingford, UK; Cambridge, USA.

Schlesinger, W.H., and Bernhardt, E.S. 2013. *Biogeochemistry. An Analysis of Global Change.* Academic Press Elsevier Publ., New York, London, USA-UK.

SEMARNAT-CP. 2002. *Evaluacion de la Degradacion del Suelo Causada por el Hombre en la República Mexicana.* Memoria Nacional. Publ. Diamante, Edo. de Mexico, Mexico.

Siggins, A., Burton, V., Ross, C., Lowe, H., and Horswell, J. 2016. Effects of long-term greywater disposal on soil: A case study. *Science of the Total Environment*, 557–558: 627–635. DOI: 10.1016/j.scitotenv.2016.03.084

Skaggs, R.W., and van Shilfgaarde, J. (Eds). 1999. *Agricultural Drainage.* Agronomy Publ., 38, Madison, Wisconsin, USA.

Stockle C.O. 2012. *Environmental Impact of Irrigation: Review.* Washington State Univ. Publ., USA, 15 p. Available in: https://www.researchgate.net/publication/252698502_ENVIRONMENTAL_IMPACT_OF_IRRIGATION_A_REVIEW

Strawn, D.G., Bohn, H.L., and O'Connor, G.A. 2015. *Soil Chemistry.* John Wiley & Sons Publ., London, UK, p. 357.

Varady, R.G., Hankins, K.B., Kaus, A., Young, E., and Merideth, R. 2001. ...to the Sea of Cortés: nature, water, culture, and livelihood in the Lower Colorado River basin and delta—an overview of issues, policies, and approaches to environmental restoration. *Journal of Arid Environments*, 49(1): 195–209. DOI: 10.1006/jare.2001.0842

Volynov, A.V., Zabelin, V.A., Kiyatkin, A.K., and Lunezhkova, M.S. 1980. *Irrigation in the Central Asia.* Ed. Kolos, Moscow, Russia (in Russian).

Yakubova, R.A. 1977. *Natural Waters of Uzbekistan and their Protection from Contamination with Pesticides.* Ed. FAN, Tashkent, Uzbekistan (in Russian).

Zonn, I.S., and Glantz, M.G. 2008. *Aral Encyclopedia.* Mezhdunarodnye Otnosheniya Publ., Moscow; Russia, p. 256 (in Russian).

Index

Note: **Bold** page numbers refer to tables; *Italic* page numbers refer to figures and page numbers followed by "n" denote endnotes.

For Product Safety Concerns and Information please contact our EU
representative GPSR@taylorandfrancis.com
Taylor & Francis Verlag GmbH, Kaufingerstraße 24, 80331 München, Germany

www.ingramcontent.com/pod-product-compliance
Lightning Source LLC
Chambersburg PA
CBHW080137220326
41598CB00032B/5098

* 9 7 8 1 0 3 2 4 2 9 1 0 6 *